Tauberian Theorems for Generalized Functions

V. S. Vladimirov, Yu. N. Drozzinov, and B. I. Zavialov

*Steklov Institute of Mathematics, Moscow, U.S.S.R.*

# Tauberian Theorems for Generalized Functions

Kluwer Academic Publishers

DORDRECHT / BOSTON / LONDON

**Library of Congress Cataloging-in-Publication Data** CIP

Vladimirov, V. S. (Vasiliĭ Sergeevich)
Tauberian theorems for generalized functions.

(Mathematics and its applications (Soviet series))
Translation of: Mnogomernye tauberovy teoremy dlia
obobshchennykh funktsii.
Bibliography: p.
1. Tauberian theorems. I. Drozhzhinov, IU. N.
(IUriĭ Nikolaevich) II. Zav'ialov, B. I. (Boris
Ivanovich) III. Title. IV. Series: Mathematics and
its applications (D. Reidel Publishing Company).
Soviet series.
QA295.V5713 1988 515'.243 88-6814

ISBN-13: 978-94-010-7774-3 e-ISBN-13: 978-94-009-2831-2
DOI: 10.1007/978-94-009-2831-2

---

Published by Kluwer Academic Publishers,
P.O. Box 17, 3300 AA Dordrecht, The Netherlands.

Kluwer Academic Publishers incorporates
the publishing programmes of
D. Reidel, Martinus Nijhoff, Dr W. Junk and MTP Press.

Sold and distributed in the U.S.A. and Canada
by Kluwer Academic Publishers,
101 Philip Drive, Norwell, MA 02061, U.S.A.

In all other countries, sold and distributed
by Kluwer Academic Publishers Group,
P.O. Box 322, 3300 AH Dordrecht, The Netherlands.

This is a translation of the original work: *Mnvgomernye Taubberovy dlja Vbobščennykh Funckcy.*
Published by Nauka, Moscow.

CONTENTS

## EDITOR'S PREFACE

Approach your problems from the right end and begin with the answers. Then one day, perhaps you will find the final question.

'The Hermit Clad in Crane Feathers' in R. van Gulik's *The Chinese Maze Murders*.

It isn't that they can't see the solution. It is that they can't see the problem.

G.K. Chesterton. *The Scandal of Father Brown* 'The point of a Pin'.

Growing specialization and diversification have brought a host of monographs and textbooks on increasingly specialized topics. However, the "tree" of knowledge of mathematics and related fields does not grow only by putting forth new branches. It also happens, quite often in fact, that branches which were thought to be completely disparate are suddenly seen to be related.

Further, the kind and level of sophistication of mathematics applied in various sciences has changed drastically in recent years: measure theory is used (non-trivially) in regional and theoretical economics; algebraic geometry interacts with physics; the Minkowsky lemma, coding theory and the structure of water meet one another in packing and covering theory; quantum fields, crystal defects and mathematical programming profit from homotopy theory; Lie algebras are relevant to filtering; and prediction and electrical engineering can use Stein spaces. And in addition to this there are such new emerging subdisciplines as "experimental mathematics", "CFD", "completely integrable systems", "chaos, synergetics and large-scale order", which are almost impossible to fit into the existing classification schemes. They draw upon widely different sections of mathematics. This programme, Mathematics and Its Applications, is devoted to new emerging (sub)disciplines and to such (new) interrelations as exempla gratia:

- a central concept which plays an important role in several different mathematical and/or scientific specialized areas;
- new applications of the results and ideas from one area of scientific endeavour into another;
- influences which the results, problems and concepts of one field of enquiry have and have had on the development of another.

The Mathematics and Its Applications programme tries to make available a careful selection of books which fit the philosophy outlined above. With such books, which are stimulating rather than definitive, intriguing rather than encyclopaedic, we hope to contribute something towards better communication among the practitioners in diversified fields.

Because of the wealth of scholarly research being undertaken in the Soviet Union, Eastern Europe, and Japan, it was decided to devote special attention to the work emanating from these particular regions. Thus it was decided to start three regional series under the umbrella of the main MIA programme.

Tauber theorems, the topic of this volume in the MIA (USSR) series, are rather difficult to place in the general scheme of things mathematical I have always found. They are theorems which connect the asymptotic behaviour of a (generalized) function in the neighborhood of zero with the behaviour of its Fourier- or Laplace- (or other integral-) transform near infinity. That should put them somewhere in the section on integral transforms, possibly Fourier transforms. But then the applications: these range from statistics to number theory from harmonic analysis, filtering and signal processing to differential equations and mathematical physics. All this is for one variable Tauberian theory.

For more variable Tauberian theory the applications have, until fairly recently, not amounted to very much. Quite possibly because the available theory did not go beyond rather trivial straightforward generalizations. On the other hand, many of the applications areas above have natural more dimensional extensions; think of filtering and prediction, for example, or statistics, or ... Other demands came from mathematical physics (both for Tauberian theory for several variables and for distributions in one and more variables) and, as the authors report in their preface, the subject took off: both in theory and applications. This is the first and only book on the subject which is no wonder given the origin of the main theorems. It seems more than likely to spawn many more applications, e.g. to the more variable or distributional cases of the areas listed for the one-variable case.

The unreasonable effectiveness of mathematics in science ...

Eugene Wigner

Well, if you know of a better 'ole, go to it.

Bruce Bairnsfather

What is now proved was once only imagined.

William Blake

As long as algebra and geometry proceeded along separate paths, their advance was slow and their applications limited.

But when these sciences joined company they drew from each other fresh vitality and thenceforward marched on at a rapid pace towards perfection.

Joseph Louis Lagrange.

Bussum, October 1986

Michiel Hazewinkel

## PREFACE

Tauberian theorems are usually assumed to connect the asymptotic behaviour of a generalized function (or a distribution) in the neighbourhood of zero with that of its Fourier transform, Laplace transform, or some other integral transform at infinity. The inverse theorems are usually called 'Abelian'.

The term 'Tauberian theorems' was introduced to mathematics by A. Tauber in 1897. He proved the first theorem of this type and all such theorems are now known by the general name 'Tauberian theorems'. Tauberian theory was intensively developed in the first half of this century and the results were gathered by N. Wiener in his extensive work 'Tauberian theorems' (1932) and in **Divergent series** by G.H. Hardy (1949).

For the case of only one variable, and for measures, Tauberian theory is rather advanced. It has many applications in number theory, in harmonic analysis, in probability theory, in differential equations, and in mathematical physics. One can find a survey of the main results and literature in A.G. Postnikov's 'Tauberian Theory and its Applications' (1979), in M.A. Subhankulov's **Tauberian Theorems with Remainder** (1979), and in T.H. Ganelius' 'Tauberian remainder theorems' (1971).

To date, in contrast with the one-dimensional case, many-dimensional Tauberian theory was no essential application and its development is characterized by straightforward and more or less trivial generalizations of one-dimensional results to a many-dimensional situation. In this direction only a few results related to multiple series or measures, with support in a positive octant, have been obtained.

The requirements of modern mathematics, and especially of mathematical physics, make it imperative that a Tauberian theory for the distributions (generalized functions) of many variables should be developed. In other words, it is necessary to extend the classical Tauberian theory to the case of more general objects - namely, for distributions defined on a many-dimensional space. This problem was initiated in the work of N.N. Bogolyubov, V.S. Vladimirov and A.N. Tavkhelidze (1972), in connection with a theoretical explanation of experimentally observed phenomena - the so-called automodel behaviour in quantum field theory, typical, say, for deep inelastic lepton-hadron scattering (Fig. 1).

This paper was followed by a systematic study of the many-dimensional Tauberian theory for distributions, and of its applications, in the Department of Mathematical Physics at the Steklov Mathematical Institute. In 1973, B.I. Zavialov introduced the important concept of the quasi-asymptotics of distributions and applied it to a study of the asymptotic properties of form factors and their Jost-Lehmann-Dyson spectral functions. In 1976, V.S. Vladimirov gave a many-dimensional generalization of the well-known Hardy-Littlewood Tauberian theorem for measures located on a nonnegative half axis. In a series of papers (1977-1985) by Yu.N. Drozzinov and B.I. Zavialov, the many-dimensional theorem

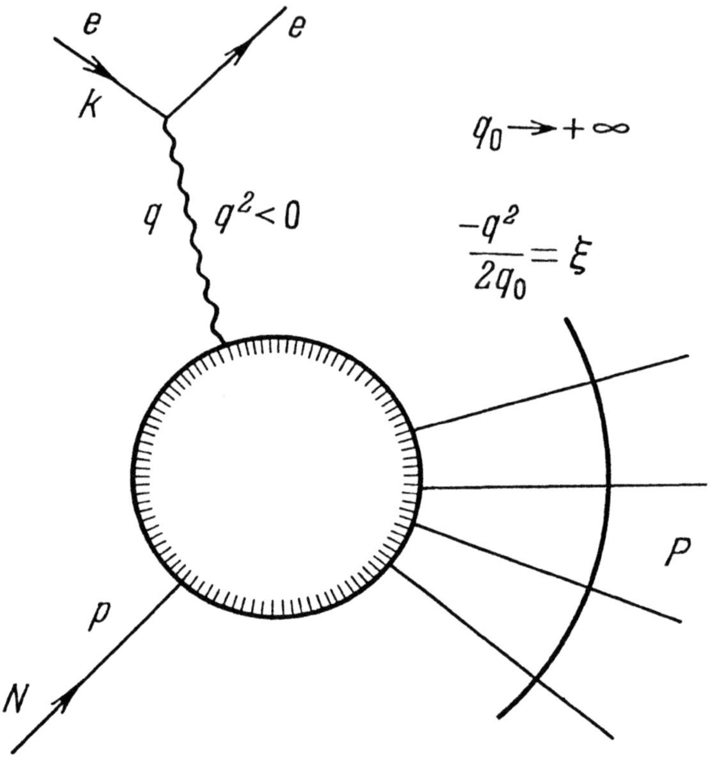


Figure 1.

of Vladimirov was extended to some classes of distribution, and the notion of quasi-asymptotics was shown to be the most suitable tool for the formulation of Abelian and Tauberian distribution theorems.

These results will be discussed in Chapter 2 (the many-dimensional case) and in Chapter 3 (the one-dimensional case). In Chapter 2 we also formulate the so-called 'comparison' Tauberian theorems (for example, by Karamata and Keldysh) on the basis of a many-dimensional generalization of regularly varying functions. In Chapter 3 some of the many-dimensional results of Chapter 2 are reformulated for the one-dimensional case in a more precise and detailed fashion. We also prove a theorem on the decomposition of a tempered distribution, defined on the whole axis, into a sum of distributions from $\mathcal{S}'_+$ and $\mathcal{S}'_-$, which preserves the quasi-asymptotic properties. Some applications of this theorem are also included.

In a series of papers (1979-1981), V.S. Vladimirov and B.I. Zavialov extended the results on the automodel behaviour of form factors to arbitrary causal (distributions) functions and expressed these results in the form of Tauberian and Abelian theorems. Vladimirov and Zavialov connected the automodel behaviour of a causal function with the asymptotic behaviour of its Fourier transform in the neighbourhood of a light cone. It is shown that in a description of the automodel behaviour of a causal function the fundamental role is played by the Radon transform of distributions, with support in a three-dimensional ball. This, in turn, implies the so-called 'sum rules' for limit distributions. All these results are discussed in Chapter 5, where we also give a more precise version and new proof of the well-known Jost-Leymann-Dyson representation, study the questions of uniqueness, and give the inversion formulae of this representation (Section 13).

Among other applications of the many-dimensional Tauberian theory for distributions we mention only the following: the Abel and Česaro summation of divergent series with respect to an automodel weight (Section 9); the asymptotic behaviour of the Cauchy problem solution for parabolic and hyperbolic equations and that of the fundamental solutions of linear passive systems (Chapter 4); the many-dimensional generalizations of the Lindelöf theorem in the complex analysis of several variables (see Yu.N. Drozzinov and B.I. Zavialov (1977)); the study of the Bellman-Harris branch processes (see A.L. Yakymiv (1981, 1983)), and (Section 5); the study of many-dimensional spectral asymptotics for elliptic operators (see S.M. Kozlov (1983)); the asymptotic properties of the two-point Wightman function (Section 14); and in statistical physics, the existence of spontaneous magnetization in the Ising model, which is a typical Tauberian condition, implies some additional restrictions on the Lee-Yan measure (see V.S. Vladimirov, and I.V. Volovich (1982)).

In this book we do not touch the so-called Tauberian theorems with the remainder term, but refer the reader to the works of M.A. Subhankulov (1976), T. Ganelius (1971), L. Frennemo (1965, 1966) and K.A. Bukin (1981).

All the necessary tools for the theory of distributions and related topics in analsysis, that are used in this book are given in Chapter 1, where, in particular, we give the theory of the space $\mathcal{S}'(F)$, where $F$ is a closed regular set.

As well as the more traditional aspects of distribution theory, we discuss in this book in more detail some integral transformations of distributions (such as Fourier transform, Laplace transform, Radon transform, $B$-transform, fractional derivation, integration with respect to a regular cone, and so on). We also discuss questions related to the quasi-asymptotic properties of distributions with respect to a given automodel function and to give family of automorphisms of a cone.

The Authors

## Chapter 0

# NOTATION AND DEFINITIONS

0.1 We denote the *points* of an $n$-dimensional real space $\mathbf{R}^n$ by $x, y, \xi, \eta, p, q, \ldots$; $x = (x_1, x_2, \ldots, x_n)$. The points of an $n$-dimensional complex space $\mathbf{C}^n$ are given by $z, \zeta, \ldots$; $z = (z_1, z_2, \ldots, z_n) = x + iy$; $x = \operatorname{Re} z$ is the real part of $z$ and $y = \operatorname{Im} z$ is the imaginary part of $z$, $\bar{z} = x - iy$ is the *complex conjugate* of $z$; $\zeta = p + iq$. We introduce in $\mathbf{R}^n$ and $\mathbf{C}^n$ the *scalar products*

$$(x, \xi) = x_1\xi_1 + \cdots + x_n\xi_n, \quad <z, \zeta> = z_1\bar{\zeta}_1 + \cdots + z_n\bar{\zeta}_n$$

and the *norms*

$$|x| = \sqrt{(x, x)} = (x_1^2 + \cdots + x_n^2)^{1/2},$$

$$|z| = \sqrt{<z, z>} = (|z_1|^2 + \cdots + |z_n|^2)^{1/2}.$$

0.2. *Sets in* $\mathbf{R}^n$ are denoted by $A, B, \mathcal{O}, G, \ldots$; $\phi$ is the *empty* set. We denote $\bar{A}$ the *closure* of a set $A$ in $\mathbf{R}^n$ (in $\mathcal{O}$), by $\operatorname{int} A$ the set of *interior points* of $A$, by $\partial A = \bar{A} \setminus \operatorname{int} A$ the *boundary* of $A$. We shall say that a set $A$ is *compact* in a set $B$ (or is *strictly contained* in $B$) if $A$ is bounded and $\bar{A} \subset B$; then we write $A \subset\subset B$.

Let $A$ and $B$ be sets in $\mathbf{R}^n$. We shall denote by $A + B$ the set of all $\xi \in \mathbf{R}^n$ such that $\xi = x + y$, $x \in A$, $y \in B$; that is,

$$A + B = [\xi \in \mathbf{R}^n : \xi = x + y, x \in A, y \in B].$$

The following notation is used:

$$B(x_0; R) = [x : |x - x_0| < R]$$

is an *open ball* of radius $R$ with centre at the point $x_0$; $S(x_0; R) = \partial B(x_0; R)$ is a *sphere* of radius $R$ with centre at the point $x_0$; $B_R = B(O; R)$, $S_R = S(O, R)$, $S_1 = S^{n-1}$.

We use $\Delta(A, B)$ to denote the *distance* between the sets $A$ and $B$ in $\mathbf{R}^n$; that is,

$$\Delta(A, B) = \inf_{x \in A, y \in B} |x - y|.$$

In the case $B = \partial A$ and $x \in A$ we write $\Delta(x, \partial A) = \Delta_A(x)$.

We use $A^\epsilon$ to denote the $\epsilon$-*neighbourhood* of a set $A$; that is,

$$A^\epsilon = A + B_\epsilon = [x : x = y + \xi, y \in A, |\xi| < \epsilon];$$

$A_\epsilon$ denotes the set of those points of $A$ which are separated from $\partial A$ by a distance greater than $\epsilon$; that is,

$$A_\epsilon = [x : x \in A, \Delta_A(x) > \epsilon].$$

The *characteristic function* of a set $A$ is the function $\theta_A(x)$ which is equal to 1 when $x \in A$ and is equal to 0 when $x \notin A$. The characteristic function $\theta_{[0,\infty)}(x)$ of the semi-axis $x \geqslant 0$ is called the *Heaviside function* and is denoted by $\theta(x)$:

$$\theta(x) = 0,\ x < 0;\ \theta(x) = 1,\ x \geqslant 0.$$

We write $\theta_n(x) = \theta(x_1)\cdots\theta(x_n)$; $\epsilon(x) = \operatorname{sign} x = \theta(x) - \theta(-x)$. A set $A$ is called *solid* if $\operatorname{int} A \neq \varnothing$.

The set $A$ is said to be *convex* if for any points $x'$ and $x''$ in $A$ the line segment joining them, $tx' + (1-t)x''$, $0 \leqslant t \leqslant 1$, lies entirely in $A$. We shall use $\operatorname{ch} A$ to denote the *convex hull* of a set $A$.

A *cone* in $\mathbf{R}^n$ (with the vertex at 0) is a set $\Gamma$ with the property that if $x \in \Gamma$, then $\lambda x$ also belongs to $\Gamma$ for all $\lambda > 0$. Denote by $\operatorname{pr}\Gamma$ the intersection of $\Gamma$ with the unit sphere $S^{n-1}$ (Fig. 2). A cone $\Gamma'$ is said to be *compact* in the cone $\Gamma$ if $\operatorname{pr}\Gamma' \subset\subset \operatorname{pr}\Gamma$ (fig. 2); then we write $\Gamma' \subset\subset \Gamma$.

The cone

$$\Gamma^* = [\xi : (\xi, x) \geqslant 0,\ x \in \Gamma]$$

is said to be *dual* to the cone $\Gamma$ (Fig. 2). Clearly, $\Gamma^*$ is a closed convex cone with its vertex at 0 and $(\Gamma^*)^* = \overline{\operatorname{ch}\Gamma}$. A cone $\Gamma$ is said to be *acute* if there exists a supporting plane for $\overline{\operatorname{ch}\Gamma}$ that has a unique common point 0 with $\overline{\operatorname{ch}\Gamma}$ (Fig. 2). This condition is equivalent to the condition of the cone $\Gamma^*$ being solid (see Section 1.9). We write $\operatorname{int}\Gamma^* = C$.

Let $\Gamma$ be an acute cone in $\mathbf{R}^n$ and $C = \operatorname{int}\Gamma^*$. A smooth $(n-1)$-dimensional surface $S$ without a border is said to be *C-like* if for any $x \in S$ the cone $\overline{\operatorname{ch}\Gamma} + x$ intersects $S$ at the single point $x$. We denote by $S_+$ and $S_-$ those (open) parts of $\mathbf{R}^n$ which contain the cones $\overline{\operatorname{ch}\Gamma} + x$, $x \in S$, and $\overline{\operatorname{ch}\Gamma} - x$, $x \in S$, respectively, so that $S_+ \cup S \cup S_- = \mathbf{R}^n$ (Fig. 3).

**Examples of convex acute cones.** (1) The *positive octant* in $\mathbf{R}^n$:

$$\mathbf{R}_+^n = [x : x_1 > 0, \ldots, x_n > 0],\ \left(\mathbf{R}_+^n\right)^* = \overline{\mathbf{R}}_+^n.$$

(2) The *future light cone* in $\mathbf{R}^{n+1}$:

$$V_n^+ = [x = (x_0, \mathbf{x}) : x_0 > |\mathbf{x}|],\ \left(V_n^+\right)^* = \overline{V}_n^+.$$

Here $\mathbf{x} = (x_1, \ldots, x_n)$.

(3) The *origin of coordinates* $\{0\}$; $\{0\}^* = \mathbf{R}^n$.

Let $A$ be a set in $\mathbf{R}_y^n$. The *tube set* $\mathbf{R}^n + iA = [z = x + iy : x \in \mathbf{R}^n, y \in A]$ in $C^n$ is denoted by $T^A$.

0.3. The *Lebesgue integral* of a (complex-valued, measurable) function $f(x)$ over $\mathbf{R}^n$ is denoted by

$$\int_{\mathbf{R}^n} f(x)\,dx_1\cdots dx_n = \int f(x)\,dx;$$

and over a measurable set $A$ by

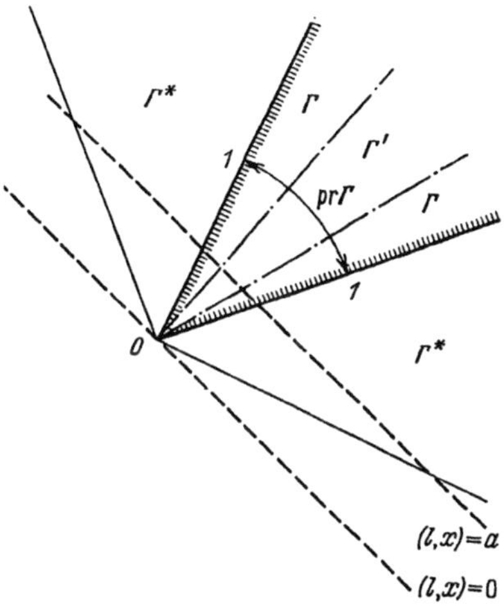


Figure 2.

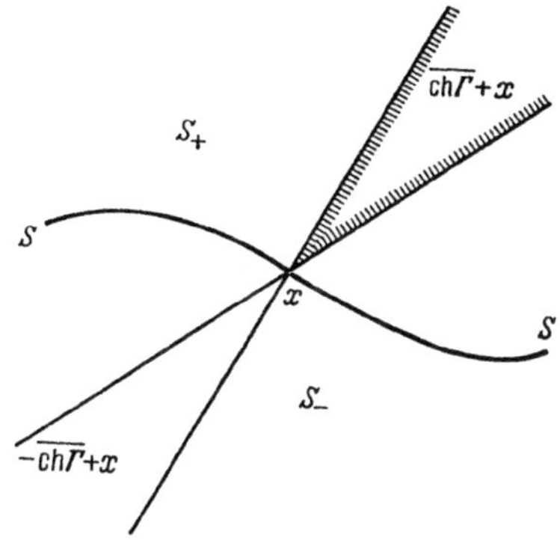


Figure 3.

$$\int_A f(x)\,dx = \int \theta_A(x) f(x)\,dx.$$

The set of all functions $f$ for which the norm

$$\|f\|_{\mathcal{L}^p(A)} = \begin{cases} \left[\int_A |f(x)|^p\,dx\right]^{1/p}, & 1 \leqslant p < \infty; \\ \operatorname*{ess\,sup}_{x \in A} |f(x)|, & p = \infty. \end{cases}$$

is finite, will be denoted by $\mathcal{L}^p(A)$; we write $\|f\| = \|f\|_{\mathcal{L}^2(\mathbf{R}^n)}$, $\mathcal{L}^p(\mathbf{R}^n) = \mathcal{L}^p$.

If $f \in \mathcal{L}^p(A')$ for every $A' \subset\subset A$, then $f$ is said to be *p-locally summable* in $A$ (for $p = 1$ we say that it is *locally summable* in $A$). The set of all functions $p$-locally summable in $A$ is denoted by $\mathcal{L}^p_{loc}(A)$; $\mathcal{L}^p_{loc}(\mathbf{R}^n) = \mathcal{L}^p_{loc}$.

A measure $d\mu(x)$ defined on a Borel set $A \subset \mathbf{R}^n$ is said to be *locally finite* in $A$ if for every compact $K \subset\subset A$, $\int_K |d\mu(x)| < \infty$; and is said to be *finite* on $A$ if

$$\int_A |d\mu(x)| < \infty.$$

A measure $d\mu(x)$ (or a function $f(x)$), defined on an unbounded Borel set $A \subset \mathbf{R}^n$, is said to be *tempered* on $A$ if for some $m \geqslant 0$

$$\int_A (1+|x|)^{-m}\,|d\mu(x)| < \infty \quad (\text{resp.} \int_A (1+|x|)^{-m}\,|f(x)|\,dx < \infty)$$

0.4. Let $\alpha = (\alpha_1, \alpha_2, \ldots, \alpha_n)$ be a *multi-index;* that is, the components $\alpha_j$ are nonnegative integers, $\alpha \in \mathbf{Z}_+^n$. We use the following notation:

$$\alpha! = \alpha_1!\alpha_2!\cdots\alpha_n!,\quad x^\alpha = x_1^{\alpha_1}x_2^{\alpha_2}\cdots x_n^{\alpha_n},$$

$$\begin{pmatrix}\beta\\ \alpha\end{pmatrix} = \begin{pmatrix}\beta_1\\ \alpha_1\end{pmatrix}\begin{pmatrix}\beta_2\\ \alpha_2\end{pmatrix}\cdots\begin{pmatrix}\beta_n\\ \alpha_n\end{pmatrix} = \frac{\alpha!}{\beta!(\alpha-\beta)!},$$

$$|\alpha| = \alpha_1 + \alpha_2 + \cdots + \alpha_n .$$

Let $\partial = (\partial_1, \partial_2, \ldots, \partial_n)$, $\partial_j = \partial/\partial x_j$, $j = 1, 2, \ldots, n$. Then

$$\partial^\alpha f(x) = \frac{\partial^{|\alpha|} f(x_1, x_2, \ldots, x_n)}{\partial x_1^{\alpha_1}\partial x_2^{\alpha_2}\cdots\partial x_n^{\alpha_n}} = \partial_1^{\alpha_1}\partial_2^{\alpha_2}\cdots\partial_n^{\alpha_n} f(x).$$

We write $\alpha \geqslant \beta$ iff $\alpha_j \geqslant \beta_j$ $j = 1, \ldots, n$. Sometimes we shall use the notation $\alpha = (\alpha_1, \ldots, \alpha_n)$ for a multi-index with components $\alpha_j$ of any sign, $\alpha \in \mathbf{Z}^n$.

Let $\mathcal{O}$ be an open set in $\mathbf{R}^n$. We use $C^k(\mathcal{O})$ to denote the set of all functions $f(x)$ which are continuous in $\mathcal{O}$ together with all derivatives $\partial^\alpha f(x)$, $|\alpha| \leqslant k$; $C^\infty(\mathcal{O})$ is the set of all infinitely differentiable functions in $\mathcal{O}$. The set of all functions $f(x)$ from $C^k(\mathcal{O})$, $0 \leqslant k \leqslant \infty$, for which all derivatives $\partial^\alpha f(x)$, $|\alpha| \leqslant k$, admit continuous extension on $\bar{\mathcal{O}}$ will be denoted by $C^k(\bar{\mathcal{O}})$. We introduce the norm $C^k(\bar{\mathcal{O}})$, $0 \leqslant k \leqslant \infty$, by the formula

$$\|f\|_{C^k(\bar{\mathcal{O}})} = \sup_{x \in \bar{\mathcal{O}},\, |\alpha| \leqslant k} |\partial^\alpha f(x)|.$$

We also denote by $C_0^k(\bar{\mathcal{O}})$, $0 \leqslant k \leqslant \infty$, the set of all functions from $C^k(\bar{\mathcal{O}})$ which vanish on $\partial\mathcal{O}$ together with all their derivatives up to the order $k$.

The *support* of a function $f(x)$ continuous in $\mathcal{O}$ is the closure in $\mathcal{O}$ of the set of those points for which $f(x) \neq 0$; the support of $f$ is denoted by $\operatorname{supp} f$; that is

$$\operatorname{supp} f = \overline{[x : f(x) \neq 0, x \in \mathcal{O}]}.$$

If $\operatorname{supp} f \subset\subset \mathcal{O}$ then $f$ is called *finite* in $\mathcal{O}$.

We denote by $C_0^k(\mathcal{O})$ the set of all functions of the class $C^k(\mathcal{O})$, which are finite in $\mathcal{O}$, $0 \leqslant k \leqslant \infty$. We write $C(\mathcal{O}) = C^0(\mathcal{O})$, $C(\bar{\mathcal{O}}) = C^0(\bar{\mathcal{O}})$, $C_0(\mathcal{O}) = C_0^0(\mathcal{O})$, $C_0(\bar{\mathcal{O}}) = C_0^0(\bar{\mathcal{O}})$, $C^k = C^k(\mathbf{R}^n)$.

We shall say that a vector function (or a matrix function) possesses some property if each of its components possesses this property.

A map $x \to y(x) = (y_1(x), \ldots, y_n(x))$ from an open set $\mathcal{O} \subset \mathbf{R}_x^n$ onto an open set $\mathcal{O}_1 \subset \mathbf{R}_y^n$ is called a $C^k$*-diffeomorphism* (or $C^k$*-smooth diffeomorphism),* $0 \leqslant k \leqslant \infty$, if $y \in C^k(\mathcal{O})$, it is one-to-one, and the inverse map $y \to x(y)$ belongs to $C^k(\mathcal{O}_1)$; an $C^0$-diffeomorphism is called a *homeomorphism.*

For $C^k$-diffeomorphisms, $k \geqslant 1$, the Jacobian is nonzero:

$$J\left(\frac{y}{x}\right) = \det\left(\frac{\partial y_i(x)}{\partial x_j}\right) \neq 0, \; x \in \mathcal{O}$$

0.5. *Miscellaneous notation.*

$$\sigma_n = \int_{S^{n-1}} d\sigma = \frac{2\pi^{n/2}}{\Gamma\left(\frac{n}{2}\right)}$$

is the surface area of the unit sphere $S^{n-1}$; $d\sigma$ is the Lebesgue measure on the sphere $S^{n-1}$.

The transpose of a matrix $A = (a_{kj})$ is denoted by $A^T = (a_{jk})$. We denote by $I$ the unit matrix, and by $g$ a diagonal with elements $[1, -1, \ldots, -1]$ on the diagonal. For every $n \times n$ real matrix $A$ we associate the linear transformation $x \to Ax$ of the space $\mathbf{R}^n$.

We denote by $O(n)$ the *rotation group* of $\mathbf{R}^n$ (the set of all real linear transformations $x \to Ax$ such that $A^T = A^{-1}$; $SO(n)$ is the *proper rotation group* of $\mathbf{R}^n$ ($A \in O(n)$, $\det A = 1$); $L_+^\uparrow(n)$ is the *proper orthochronous Lorentz group* of $\mathbf{R}^{n+1}$ ($A^T gA = g$, $\det A = 1$, $a_{00} > 0$); and $T(n)$ denotes the *translation group* of $\mathbf{R}^n$ ($x \to x + b, b \in \mathbf{R}^n$).

We denote the *uniform convergence* of a sequence of functions $\{\phi_k(x), k \to \infty\}$ to a function $\phi(x)$ on a set $A$ by:

$$\phi_k(x) \overset{x \in A}{\Longrightarrow} \phi(x), \; k \to \infty;$$

if $A = \mathbf{R}^n$, then instead of $\overset{x \in \mathbf{R}^n}{\Longrightarrow}$ we shall write $\overset{x}{\Longrightarrow}$.

We denote the convergence of a sequence of elements $\{f_k, k \to \infty\}$ to an element $f$ in a topological vector space $B$ by: $f_k \to f$, $k \to \infty$ in $B$.

We denote by $\omega_\epsilon(x)$ the *normed bell-shaped function* (Fig. 4)

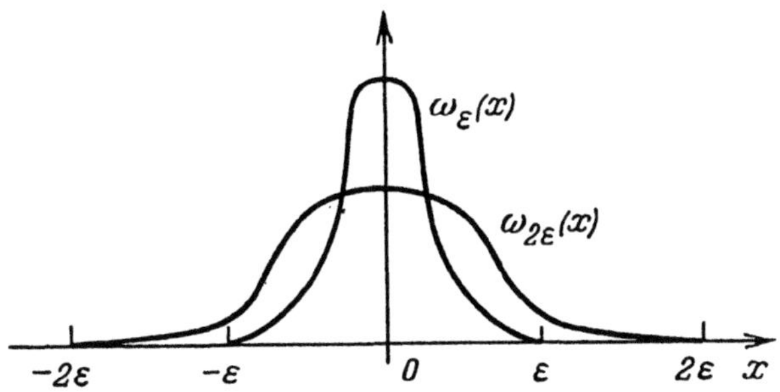


Figure 4.

$$\omega_\epsilon(x) = \begin{cases} C_\epsilon \exp\left(-\dfrac{\epsilon^2}{\epsilon^2 - |x|^2}\right), & |x| \leqslant \epsilon, \\ 0 & , |x| > \epsilon, \end{cases} \quad \int \omega_\epsilon(x)\,dx = 1.$$

Let $\alpha$ be a real number; we denote by $[\alpha]$ the *greatest integer* which is not larger than $\alpha$.

The *wave operator (d'Alembertian)* is denoted by

$$\Box_n = \partial_0^2 - \partial_1^2 - \cdots - \partial_n^2.$$

Let $A$ and $B$ be two arbitrary sets. We shall denote by $A \times B$ the set of all pairs $(a, b)$, $a \in A$, $b \in B$.

The sections are numbered in a single sequence. Each section consists of subsections; references to a subsection contain the number of the corresponding section. Formulae are numbered separately in each subsection, and they also contain the number of the subsection. When a reference is made to a formula in another section, the number of that section is also given.

Chapter 1

# SOME FACTS ON THE THEORY OF DISTRIBUTIONS

The general results from the theory of distributions presented in this chapter are contained in the books by L. Schwartz (1950–1951), by I.M. Gel'fand and G.E. Shilov (1964, 1968), and by V.S. Vladimirov (1979). References on special results will be given in the text.

## 1. Distributions and their properties

In this section we shall define the spaces of distributions $\mathcal{D}'(\mathcal{O})$ and $\mathcal{S}'(F)$, and their simplest properties will also be given.

**1. Spaces of test functions.** Let $K$ be a compact set in $\mathbf{R}^n$ such that $\overline{\operatorname{int} K} = K$. We define the space of test functions $\mathcal{D}(K)$ as follows: $\mathcal{D}(K)$ is a set of $C_0^\infty(K)$ functions equipped with the topology of a projective limit (intersection) on a decreasing sequence of Banach spaces

$$C_0(K) \supset C_0^1(K) \supset \cdots \supset C_0^p(K) \supset \ldots,$$

where $C_0^p(K)$ is defined in Section 0.4, with $\mathcal{O} = \operatorname{int} K$. Every embedding $C_0^{p+1}(K) \subset C_0^p(K)$ is compact, $p = 0, 1, \cdots$. The space $\mathcal{D}(K)$ is defined by the countable family of norms (see Section 0.4)

$$\|\phi\|_{C(K)} \leqslant \|\phi\|_{C^1(K)} \leqslant \ldots, \quad \phi \in \mathcal{D}(K).$$

The space $\mathcal{D}(K)$ is also the Fréchet-Schwartz space ($FS$- space).

Let $\mathcal{O}$ be an open set in $\mathbf{R}^n$. We define the space $\mathcal{D}(\mathcal{O})$ as follows: $\mathcal{D}(\mathcal{O})$ is a set of $C_0^\infty(\mathcal{O})$-functions equipped with the topology of an inductive limit (union) on an increasing sequence of countable normed spaces

$$\mathcal{D}(\overline{\mathcal{O}}_1) \subset \mathcal{D}(\overline{\mathcal{O}}_2) \subset \cdots \subset \mathcal{D}(\overline{\mathcal{O}}_k) \subset \cdots$$

where $\{\mathcal{O}_k, k = 1, 2, \cdots\}$ is a sequence of open sets *exhausting* $\mathcal{O}$: $\mathcal{O}_1 \subset\subset \mathcal{O}_2 \subset\subset \cdots$, $\bigcup_{1 \leqslant k < \infty} \mathcal{O}_k = \mathcal{O}$. Each embedding $\mathcal{D}(\overline{\mathcal{O}}_k) \subset \mathcal{D}(\overline{\mathcal{O}}_{k+1})$ is compact, $k = 1, 2, \cdots$. The introduced topology does not depend on $\{\mathcal{D}_k, k = 1, 2, \cdots\}$ being an exhausting sequence of a special type.

We write $\mathcal{D}(\mathbf{R}^n) = \mathcal{D}$. The bell-shaped function $\omega_\epsilon(x)$ is an example of functions from $\mathcal{D}$ (see Section 0.5).

The space $\mathcal{D}(\mathcal{O})$ is a $DFS$-space (strongly conjugate to a $FS$- space). *Convergence* in $\mathcal{D}(\mathcal{O})$: $\phi_k \to 0$, $k \to \infty$ in $\mathcal{D}(\mathcal{O})$ iff there exists a set $\mathcal{O}' \subset\subset \mathcal{O}$ such that $supp\ \phi_k \subset \mathcal{O}'$ and for every $\alpha$

$$\partial^\alpha \phi_k(x) \overset{x}{\rightrightarrows} 0, \ k \overset{x}{\rightrightarrows} \infty.$$

We define the space $\mathcal{E}(\mathcal{O})$ as follows: $\mathcal{E}(\mathcal{O})$ is a set of $C^\infty(\mathcal{O})$-functions equipped with the topology of a projective limit on a decreasing sequence of countable normed spaces

$$\mathcal{E}(\bar{\mathcal{O}}_1) \supset \mathcal{E}(\bar{\mathcal{O}}_2) \supset \cdots \supset \mathcal{E}(\bar{\mathcal{O}}_p) \supset \cdots$$

where $\{\mathcal{O}_k, k = 1, 2, \cdots\}$ is a sequence of open sets $\mathcal{O}_k$ exhausting $\mathcal{O}$ and $\mathcal{E}(\bar{\mathcal{O}}_p)$ is the projective limit of a decreasing sequence of Banach spaces

$$C(\bar{\mathcal{O}}_p) \supset C^1(\bar{\mathcal{O}}_p) \supset \cdots \supset C^k(\bar{\mathcal{O}}_p) \supset \ldots, \ p = 1, 2, \cdots$$

Convergence in $\mathcal{E}(\mathcal{O})$: $\phi_k \to 0$, $k \to \infty$ in $\mathcal{E}(\mathcal{O})$ iff for every $\mathcal{O}' \subset\subset \mathcal{O}$ and $\alpha$

$$\partial^\alpha \phi_k(x) \overset{x \in \mathcal{O}'}{\rightrightarrows} 0, \ k \to \infty.$$

$\mathcal{D}(\mathcal{O}) \subset \mathcal{E}(\mathcal{O})$ and $\mathcal{D}(\mathcal{O})$ is dense in $\mathcal{E}(\mathcal{O})$.

We write $\mathcal{E} = \mathcal{E}(\mathbf{R}^n)$.

We say that a system of open sets $\{\mathcal{O}_k, k = 1, 2, \cdots\}$ exhausting an open set $\mathcal{O}$ forms a *locally finite covering* of $\mathcal{O}$ if every compact set $K \subset\subset \mathcal{O}$ intersects only a finite number of sets $\{\mathcal{O}_k, k = 1, 2, \cdots\}$.

For any locally finite covering $\{\mathcal{O}_k, k = 1, 2, \cdots\}$ of $\mathcal{O}$ there exists a system of functions $\{e_k(x), k = 1, 2, \cdots\}$ such that

$$e_k \in \mathcal{D}(\mathcal{O}_k), 0 \leqslant e_k \leqslant 1, \ \sum_{1 \leqslant k < \infty} e_k(x) \equiv 1, x \in \mathcal{O}.$$

The system $\{e_k, k = 1, 2, \cdots\}$ is called the *decomposition of unity corresponding to a given locally finite covering* $\{\mathcal{O}_k, k = 1, 2, \cdots\}$ of the open set $\mathcal{O}$.

The following statements hold: $\mathcal{D}(\mathcal{O})$ *is dense in* $\mathcal{L}^p(\mathcal{O})$, $1 \leqslant p < \infty$; $\mathcal{D}(\mathcal{O})$ *is dense in* $C_0^k(\bar{\mathcal{O}})$ (in the norm $C^k(\bar{\mathcal{O}})$).

A set $F \subset \mathbf{R}^n$ is said to be *regular* if there exist positive numbers $d$, $\omega$, and $q$ ($q \leqslant 1$), such that any two points $x'$ and $x''$ from $F$, for which $|x' - x''| \leqslant d$, can be joined in $F$ by a rectifiable curve of length $l \leqslant \omega |x' - x''|^q$. It is clear that the union of finite numbers of points is a regular set; every convex set is regular.

Let $F$ be the closure of a regular domain $G$, $F = \bar{G}$. (In this case $F$ is a regular closed set, see H. Whitney (1934b)). We define the space of test functions $\mathcal{S}(F)$ (fast decreasing on $F$) as the set of all $C^\infty(F)$-functions for which all norms

$$\|\phi\|_{p,F} = \sup_{x \in F, |\alpha| \leqslant p} (1 + |x|^2)^{p/2} |\partial^\alpha \phi(x)|, \ p = 0, 1, \cdots \tag{1.1}$$

are finite. We write $\mathcal{S}(\mathbf{R}^n) = \mathcal{S}$, $\|\cdot\|_{p,\mathbf{R}^n} = \|\cdot\|_p$.

In $\mathcal{S}(F)$ we introduce the topology of a projective limit on a decreasing sequence of Banach spaces

$$\mathcal{S}_0(F) \supset \mathcal{S}_1(F) \supset \cdots \supset \mathcal{S}_p(F) \supset \ldots,$$

where $\mathcal{S}_p(F)$ is the completion of $\mathcal{S}(F)$ in the $p$-th norm. Each embedding $\mathcal{S}_{p+1}(F) \subset \mathcal{S}_p(F)$ is continuous, $p = 0, 1, \cdots$. Furthermore, for every $p$ there exists a number $p' = p'(p) > p$ such that the embedding $\mathcal{S}_{p'}(F) \subset \mathcal{S}_p(F)$ is compact.

The space $\mathcal{S}(F)$ is a $FS$-space; its topology is defined by the countable system of norms (1.1):

$$\|\phi\|_{0,F} \leqslant \|\phi\|_{1,F} \leqslant \ldots, \quad \phi \in \mathcal{S}(F).$$

Convergence in $\mathcal{S}(F)$ is a convergence in very norm (1.1); $\mathcal{D} \subset \mathcal{S}$ and $\mathcal{D}$ is dense in $\mathcal{S}$; $\mathcal{S} \subset \mathcal{S}(F)$.

The following theorems on the extension of differentiable functions hold:

(i) *Let $F$ be the closure of a regular domain and let $p \geqslant 0$ be an integer. Then there exists an integer $p' \geqslant p$ such that for any function $\phi$ from $\mathcal{S}_{p'}(F)$ there exists a function $\psi$ from $C^p$ such that*

$$\partial^\alpha\phi(x) = \partial^\alpha\psi(x), \quad x \in F, \quad |\alpha| \leqslant P$$

*and the correspondence $\phi \to \psi$ is linear* (H. Whithney, 1934a, b); see also B. Malgrange (1966)).

(ii) *Let $F$ be a closed regular set and let $p \geqslant 0$ be an integer. Then there exists numbers $C$ and $p' \geqslant p$ such that for every function $\phi$ from $C^p \bigcap \mathcal{S}_{p'}(F)$ there exists a function $\psi$ from $C^{p'} \bigcap \mathcal{S}_p$ with the properties:*

$$\partial^\alpha\phi(x) = \partial^\alpha\psi(x), \quad x \in F, \quad |\alpha| \leqslant p', \quad \|\psi\|_p \leqslant C\|\phi\|_{p',F} \tag{1.2}$$

*and the correspondence $\phi \to \psi$ is linear* (V.S. Vladimirov, 1981).

Let $G$ be a domain in $\mathbf{R}^n$ and $a \in \partial G$. We say that $\mathcal{O}$ is a *Lip $\gamma$-domain at the point $a$* if there exists a neighbourhood $U$ of the point $a$, a system of coordinates in $U$ $(\xi_1, \xi_2, \ldots, \xi_n) = (\xi_1, \tilde{\xi})$, $\tilde{\xi} = (\xi_2, \ldots, \xi_n)$, and a function $f(\tilde{\xi})$ satisfying the Lipschitz (Hölder for $0 < \gamma < 1$) condition

$$\left|f(\tilde{\xi}') - f(\tilde{\xi}'')\right| \leqslant M \left|\tilde{\xi}' - \tilde{\xi}''\right|^\gamma$$

for some $M > 0$ such that

$$G \bigcap U = \left[\xi \in U : \xi_1 > f(\tilde{\xi})\right].$$

If $G$ is a Lip $\gamma$-domain at every point of $\partial G$ then we say that $G$ is a *Lip $\gamma$-domain in $\mathbf{R}^n$*.

A domain $G$ is called a *Lip-domain in $\mathbf{R}^n$* if it is a Lip $\gamma$-domain in $\mathbf{R}^n$ for some $\gamma \in (0, 1]$.

(iii) *Let $G$ be a bounded Lip-domain in $\mathbf{R}^n$. Then there exists a linear continuous map*

$$E : \mathcal{E}(\overline{G}) \to \mathcal{E}(\mathbf{R}^n)$$

*such that for every $\phi \in \mathcal{E}(\overline{G})$*

$$E[\phi](x) = \phi(x), \quad x \in \overline{G}.$$

*In addition, if a domain $G$ is a Lip $1/k$-domain in $\mathbf{R}^n$ for some integer $k > 0$ then for every compact $K \subset \mathbf{R}^n$ and integer $m \geqslant 0$ there exists a constant $C > 0$ such that*

$$\|E[\phi]\|_{C^m(K)} \leqslant C\|\phi\|_{C^{km}(G)}, \quad \phi \in \mathcal{E}(\overline{G})$$

(E. Bearstoun (1978))*.

If a domain $G$ is unbounded then a point at infinity belongs to the boundary of $G$, and then for any point $x_0 \in \mathbf{R}^n$ the inversion

$$i_{x_0}: x \to \frac{x - x_0}{|x - x_0|^2} + x_0 = x'$$

maps the infinite point at $x_0$. We shall say that $G$ is a *Lip $\gamma$-domain at the infinite point* if the domain $i_{x_0}[G]$ is a Lip $\gamma$-domain at the point $x_0$ $(0 < \gamma \leqslant 1)$. This is an accurate definition because it does not depend on a choice of the point $x_0$. If a domain $G$ is a Lip $\gamma$-domain at every boundary point including the infinite point, then we shall say that $G$ is a *Lip $\gamma$-domain* (or *Lip-domain* ) in $\mathbf{R}^n$ .

**Examples.** The half-space $\mathbf{R}^1_+ \times \mathbf{R}^{n-1}$, the future light cone $V^+_{n-1}$, and the semi-cylinder $\mathbf{B}(x_0; R) \times \mathbf{R}^1_+$ are Lip-domains in $\mathbf{R}^n$. We note also that the closure of a bounded Lip-domain in $\mathbf{R}^n$ is a regular set.

(iv) **Theorem.** *Let $F$ be the closure of a Lip-domain in $\mathbf{R}^n$. Then for every $\phi \in \mathcal{S}(F)$ there exists a function $\psi \in \mathcal{S}$ such that*

$$\psi(x) = \phi(x), \; x \in \Gamma,$$

*and for any integer $p \geqslant 0$ there exist numbers $p' \geqslant p$ and $C > 0$ such that*

$$\|\psi\|_p \leqslant C \|\phi\|_{p', F}. \tag{1.3}$$

*and the correspondence $\phi \to \psi$ is linear.*

**Proof.** If $F = \mathbf{R}^n$, then the theorem is trivial. Suppose that $F \neq \mathbf{R}^n$ and let $x_0 \in \mathbf{R}^n \setminus F$. Choose a number $r > 0$ such that $\overline{B(x_0; r)} \cap F = \emptyset$. The set $i_{x_0}(F)$ is compact in $\mathbf{R}^n$ and is the closure of some bounded Lip-domain. We denote by $\mathcal{E}_{x_0}(i_{x_0}[F])$ the subspace of $\mathcal{E}(i_{x_0}[F])$ which consists of all functions $g \in \mathcal{E}(i_{x_0}[F])$ vanishing at the point $x_0$ together with all their derivatives. It is clear that $\mathcal{E}_{x_0}(i_{x_0}[F])$ is a closed subspace of the Fréchet space $(i_{x_0}[F])$ and therefore it is also a Fréchet space. The mapping

$$h: \phi \to g(x) = \phi(i_{x_0} x) \tag{1.4}$$

is linear and continuous from the Fréchet space $\mathcal{S}(F)$ onto the Fréchet space $\mathcal{E}_{x_0}(i_{x_0}[F])$. Therefore, by the open mapping theorem it is a topological isomorphism between the spaces $\mathcal{S}(F)$ and $\mathcal{E}_{x_0}(i_{x_0}[F])$. On the other hand, according to statement (iii), there exists a linear and continuous mapping

$$E: \mathcal{E}_{x_0}(i_{x_0}[F]) \to \mathcal{E}_{x_0}(\mathbf{R}^n) \subset \mathcal{E}_{x_0}(\overline{B(x_0; \tfrac{2}{r})})$$

where $\mathcal{E}_{x_0}(\mathbf{R}^n)$ (resp. $\mathcal{E}_{x_0}(\overline{B(x_0; 2/r)})$) is a closed subspace of $\mathcal{E}(\mathbf{R}^n)$ (resp. $\mathcal{E}(\overline{B(x_0; 2/r)})$) which consists of all functions vanishing at the point $x_0$ together with all their derivatives. Similarly, using the open mapping theorem we prove

* For other conditions which ensure the possibility of a $C^\infty$- extension of functions with an estimate, see M. Tidten (1979).

that the inversion mapping (1.4) is a topological isomorphism from the space $(\mathbf{R}^n \backslash B(x_0; r/2))$ onto the space $_{x_0}(\overline{B(x_0; 2/r)})$. Now let $\eta(x)$ be a $C^\infty$- function such that $0 \leqslant \eta(x) \leqslant 1$, $\eta(x) \equiv 1$ if $x \in \mathbf{R}^n \backslash B(x_0; r)$ and $\eta(x) \equiv 0$ if $x \in B(x_0; r/2)$. If $\phi$ belongs to $(F)$ then as $\psi(x)$ we can take the function

$$\psi(x) = \eta(x) E\left[\phi\left(i_{x_0}^{-1} x^1\right)\right](i_{x_0} x).$$

From the previous arguments it follows that the mapping $\phi \to \psi$ is linear and continuous from the space $\mathcal{S}(F)$ onto the space $(\mathbf{R}^n)$. From this, in particular, follows inequality (1.3). □

We note the following corollary from the (iv) theorem.

**Theorem.** *If $F$ is the closure of a Lip-domain in $\mathbf{R}^n$ then the space $\mathcal{S}(F)$ is topologically isomorphic to the factor-space $\mathcal{S}(\mathbf{R}^n)/\mathcal{S}_0(\overline{\mathbf{R}^n \backslash F})$, where $\mathcal{S}_0(\overline{\mathbf{R}^n \backslash F})$ is a subspace of $\mathcal{S}(\mathbf{R}^n)$ consisting of those functions $\phi$ for which* $\operatorname{supp} \phi \subset \overline{\mathbf{R}^n \backslash F}$.

We denote: $\mathcal{S}_+(n) = (\overline{\mathbf{R}}^+ \times \mathbf{R}^n)$, $\mathcal{S}_+ = \mathcal{S}_+(0)$, $\mathcal{S}_- = \mathcal{S}(-\overline{\mathbf{R}}_+^1)$; $x = (x_0, \mathbf{x}) \in \mathbf{R}^{n+1}$, $\mathbf{x} = (x_1, \ldots, x_n)$, see Section 0.2.

2. **The space of distributions $\mathcal{D}'(\mathcal{O})$.** A *distribution (generalized function)* defined on an open set $\mathcal{O}$ is a *continuous linear functional* on the space of tests functions $\mathcal{D}(\mathcal{O})$. The set of all distributions in $\mathcal{O}$ is denoted by $\mathcal{D}'(\mathcal{O})$; $\mathcal{D}'(\mathbf{R}^n) = \mathcal{D}'$.

Let us equip $\mathcal{D}'(\mathcal{O})$ with the strong topology of the space adjoint to $\mathcal{D}(\mathcal{O})$.

We shall denote the value of a functional $f$ from $\mathcal{D}'(\mathcal{O})$ (distribution) on a test function $\phi$ from $\mathcal{D}(\mathcal{O})$ by $(f, \phi)$. By analogy with ordinary functions, sometimes we formally write $f(x)$ instead of $f$, and we consider $x$ as the argument of the test functions on which the functional $f$ operates. A distribution $f$ from $\mathcal{D}'(\mathcal{O})$ is called *null in* $\mathcal{O}$ iff $(f, \phi) = 0$, $\phi \in \mathcal{D}(\mathcal{O})$. We shall then write $f = 0$ in $\mathcal{O}$ or $f(x) = 0$, $x \in \mathcal{O}$. Distributions $f$ and $g$ from $\mathcal{D}'(\mathcal{O})$ are called *equal in* $\mathcal{O}$ if $f - g = 0$ in $\mathcal{O}$; we then write $f = g$ in $\mathcal{O}$, or $f(x) = g(x)$, $x \in \mathcal{O}$.

**Example 1.** The Dirac $\delta$-function,

$$(\delta, \phi) = \phi(0), \quad \phi \in \mathcal{D}.$$

We define the *convergence in* $\mathcal{D}'(\mathcal{O})$ as the *weak convergence* of functionals: $f_k \to f$, $k \to \infty$, in $\mathcal{D}'(\mathcal{O})$ iff

$$(f_k, \phi) \to (f, \phi), \quad k \to \infty, \quad \phi \in \mathcal{D}(\mathcal{O}).$$

**Example 2.** $\omega_\epsilon(x) \to \delta(x)$, $\epsilon \to +0$ in $\mathcal{D}'$ where $\omega_\epsilon$ is the normed bell-shaped function (see Section 0.5).

If $\mathcal{O}_1 \subset \mathcal{O}_2$ then $\mathcal{D}'(\mathcal{O}_1) \subset \mathcal{D}'(\mathcal{O}_2)$. Hence, it follows that for any distribution $f \in \mathcal{D}'(\mathcal{O})$ and an open set $\mathcal{O}_1 \subset \mathcal{O}$ there exists a (unique) restriction $f_{\mathcal{O}_2}$ from $\mathcal{D}'(\mathcal{O}_1)$, that is

$$(f, \phi) = (f_{\mathcal{O}_1}, \phi), \quad \phi \in \mathcal{D}(\mathcal{O}_1).$$

Conversely, the following *piecewise sewing theorem* holds.

**Theorem.** *Let $\{\mathcal{O}_k, k = 1, 2, \cdots\}$ be a locally finite covering of an open set $\mathcal{O}$. Let distributions $f_k$ from $\mathcal{D}'(\mathcal{O}_k)$, $k = 1, 2, \cdots$, be compatible in $\mathcal{O}$ that is, $f_k(x) = f_j(x)$ if $x \in \mathcal{O}_k \bigcap \mathcal{O}_j \neq \varnothing$ then there exists a (unique) distribution $f$ from $\mathcal{D}'(\mathcal{O})$ such that $f_{\mathcal{O}_k}(x) = f(x)$, $x \in \mathcal{O}_k$, $k = 1, 2, \cdots$.*

This theorem follows from the existence of a decomposition of unity corresponding to a given locally finite covering $\{\mathcal{O}_k, k = 1, 2, \cdots\}$ of an open set $\mathcal{O}$ (see Section 1.1).

The *null set* of a distribution $f$ from $\mathcal{D}'(\mathcal{O})$ is the largest open set $\mathcal{O}_f \subset \mathcal{O}$ such that $f_{\mathcal{O}_f} = \mathcal{O}$ in $\mathcal{O}_f$. By the 'piecewise sewing theorem' the set $\mathcal{O}_f$ always exists and is unique. The *support* of $f \in \mathcal{D}'(\mathcal{O})$ is defined as the complement of the set $\mathcal{O}_f$ in $\mathcal{O}$. The support of $f$ is deonted by supp $f$ so that supp $f = \mathcal{O} \setminus \mathcal{O}_f$. If supp $f \subset\subset \mathcal{O}$ then $f$ is said to be *finite in* $\mathcal{O}$ (with compact support). Distributions which are finite in $\mathbf{R}^n$ are called *finite*.

The space $\mathcal{D}'(\mathcal{O})$ is complete with respect to a weak convergence. This means that if a sequence $\{f_k, k \to \infty\}$ from $\mathcal{D}'(\mathcal{O})$; such that for every $\phi$ from $\mathcal{D}(\mathcal{O})$ the numerical sequence $(f_k, \phi)$ converges as $k \to \infty$, then the functional $\lim_{k\to\infty}(f_k, \phi)$ is a distribution from $\mathcal{D}'(\mathcal{O})$.

A set M of distributions from $\mathcal{D}'(\mathcal{O})$ is called *weakly bounded* in $\mathcal{D}'(\mathcal{O})$ if for every $\phi \in \mathcal{D}(\mathcal{O})$ there exists a number $C_\phi$ such that

$$|(f, \phi)| \leqslant C_\phi, \quad f \in \mathrm{M}.$$

*If a set* M *is weakly bounded in $\mathcal{D}'(\mathcal{O})$ then for every open set $\mathcal{O}' \subset\subset \mathcal{O}$ there exists numbers* C *and* $m$ *such that*

$$|(f, \phi)| \leqslant C \|\phi\|_{C^m(\bar{\mathcal{O}}')}, \quad f \in \mathrm{M}, \quad \phi \in \mathcal{D}(\bar{\mathcal{O}}') \tag{2.1}$$

*In inequality (2.1), numbers* C *and* $m$ *depend in general on* $f$ *and* $\mathcal{O}'$. *If the integer* $m$ *can be chosen independently of* $\mathcal{O}' \subset\subset \mathcal{O}$ *then we say that the distribution* $f$ *has a finite order in* and the smallest integer $m_f$ is called the *order of* $f$ *in* $\mathcal{O}$.

We denote by $\mathcal{E}'(\mathcal{O})$ the space that is strongly conjugate to the space of test functions $\mathcal{E}(\mathcal{O})$ (see Section 1.1); $\mathcal{E}' = \mathcal{E}'(\mathbf{R}^n)$.

If $f \in \mathcal{D}'(\mathcal{O})$ and supp $f \subset\subset \mathcal{O}$ then $f$ has a finite order in $\mathcal{O}$ so that in inequality (2.1) the constant $C$ does not depend on $\mathcal{O}$, and therefore the functional $f$ admits an extension $\tilde{f}$ as an element of the space $\mathcal{E}'(\mathcal{O})$. Conversely, if $f \in \mathcal{E}(\mathcal{O})$, then $f \in \mathcal{D}'(\mathcal{O})$ and supp $f \subset\subset \mathcal{O}$. Hence, $\mathcal{E}'(\mathcal{O})$ is the space of distributions from $\mathcal{D}'(\mathcal{O})$ with compact support in $\mathcal{O}$. Convergence in $\mathcal{E}'(\mathcal{O})$ is defined as follows: $f_k \to f$, $k \to \infty$ in $\mathcal{E}'(\mathcal{O})$ iff there exists an open set $\mathcal{O}' \subset\subset \mathcal{O}$ such that supp $f_k \subset \mathcal{O}'$ and $f_k \to f$, $k \to \infty$ in $\mathcal{D}'(\mathcal{O})$.

If $f \in \mathcal{L}^1_{loc}(\mathcal{O})$ (see Section 0.3) then the functional

$$(f, \phi) = \int f(x)\phi(x)\,dx, \quad \phi \in \mathcal{D}(\mathcal{O}) \tag{2.2}$$

belongs to $\mathcal{D}'(\mathcal{O})$. Distributions from $\mathcal{D}'(\mathcal{O})$, defined by $\mathcal{L}^1_{loc}(\mathcal{O})$-functions $f$ according to formula (2.2), are called as *regular in* $\mathcal{O}$; all other distributions are said to be *singular in* $\mathcal{O}$.

There is a one-to-one correspondence between the $\mathcal{L}^1_{loc}(\mathcal{O})$-functions and the regular distributions in $\mathcal{O}$. For this reason we shall henceforth indentify a $\mathcal{L}^1_{loc}(\mathcal{O})$-function $f(x)$ with the regular distribution from $\mathcal{D}'(\mathcal{O})$ which is generated by it by (2.2).

An example of a singular distribution from $\mathcal{D}'$ is the $\delta$-function, $\delta(x) = 0$, $x \neq 0$; its order is equal to 0.

More generally, all distributions from $\mathcal{D}'(\mathcal{O})$ of order 0 are generated by the measures $d\mu$, which are locally finite in $\mathcal{O}$ (see Section 0.3), according to the formula

$$([d\mu], \phi) = \int \phi(x) d\mu(x), \quad \phi \in \mathcal{D}(\mathcal{O}). \tag{2.3}$$

Here $[d\mu]$ is the distribution generated by the measure $d\mu$. We note that the formula (2.2) is a particular case of formula (2.3) when $d\mu(x) = f(x)dx$.

Let $f$ be a distribution from $\mathcal{D}'(\mathcal{O})$. We define a distribution $\overline{f}$ complex conjugate to $f$ by the formula

$$(\overline{f}, \phi) = \overline{(f, \overline{\phi})}, \quad \phi \in \mathcal{D}(\mathcal{O}).$$

The distributions

$$\operatorname{Re} f = \frac{f + \overline{f}}{2}, \quad \operatorname{Im} f = \frac{f - \overline{f}}{2i}$$

are called the *real* and the *imaginary* parts of $f$, respectively, so that

$$f = \operatorname{Re} f + i \operatorname{Im} f, \quad \overline{f} = \operatorname{Re} f - i \operatorname{Im} f.$$

If $\operatorname{Im} f = 0$ then we call $f$ a *real* distribution.

Let $f \in \mathcal{D}'(\mathcal{O})$ and let $\mathcal{O}_1$ be an open subset of $\mathcal{O}$. We shall say that the distribution $f$ *belongs to* the class $C^k(\mathcal{O}_1)$ (resp. $C^k(\overline{\mathcal{O}}_1)$) iff $f_{\mathcal{O}_1} \in C^1(\mathcal{O}_1)$ (resp. $C^k(\overline{\mathcal{O}}_1)$).

We consider open sets $\mathcal{O}_1 \subset \mathbf{R}^n_x$, $\mathcal{O}_2 \subset \mathbf{R}^m_y$, a distribution $f(x, y)$ from $\mathcal{D}'(\mathcal{O}_1 \times \mathcal{O}_2)$ and a test function $\phi(x)$ from $\mathcal{D}(\mathcal{O}_1)$. We define a distribution $f_\phi(y)$ from $\mathcal{D}'(\mathcal{O})$ by the formula

$$(f_\phi, \psi) = (f(x, y), \phi(x)\psi(y)), \quad \psi \in \mathcal{D}(\mathcal{O}_2).$$

We shall say that a distribution from $\mathcal{D}'(\mathcal{O}_1 \times \mathcal{O}_2)$ *belongs to* the class $C^k(\mathcal{O}_2)$ (resp. $C^k(\overline{\mathcal{O}}_2)$) with respect to $y$ iff for every $\phi \in \mathcal{D}(\mathcal{O}_1)$ the distribution $f_\phi$ belongs to the class $C^k(\mathcal{O}_2)$ (resp. $C^k(\overline{\mathcal{O}}_2)$).

**3. The space of distribution $\mathcal{S}'(F)$.** Let $F$ be the closure of a regular domain (see Section 1.1). The space of distributions $\mathcal{S}'(F)$ is the space that is strongly conjugated to the space of test functions $\mathcal{S}(F)$. The space $\mathcal{S}'(F)$ is called the space of *tempered distributions on $F$*.

The space $\mathcal{S}'(F)$ is a *DFS*-space. It is the inductive limit (union) on an increasing sequence of Banach spaces

$$\mathcal{S}'_0(F) \subset \mathcal{S}'_1(F) \subset \cdots \subset \mathcal{S}'_p(F) \subset \cdots$$

where $\mathcal{S}'_p(F)$ is the conjugate space to the Banach space $\mathcal{S}_p(F)$, $p = 0, 1, \cdots$.

Thus every distribution $f$ from $\mathcal{S}'(F)$ has a finite order $m_f$ (see Section 1.2); that is $f \in \mathcal{S}'_m(F)$ for some (smallest) $m = m_f$ and

$$|(f, \phi)| \leqslant ||f||_{-m} ||\phi||_{m,F}, \quad \phi \in \mathcal{S}(F)$$

where $||f||_{-m}$ is the norm of the functional $f$ in $\mathcal{S}'_m(F)$.

Similarly, as in the case of the space $\mathcal{D}'(\mathcal{O})$, in the space $\mathcal{S}'(F)$ we define weak convergence and weakly bounded sets. The following statements are valid: *every weakly bounded set of distributions from $\mathcal{S}'(F)$ is contained in some $\mathcal{S}'_p(F)$ and is bounded by the norm in it, so $(f_k, \phi_k) \to (f, \phi)$, $k \to \infty$ if $f_k \to f$, $k \to \infty$ in $\mathcal{S}'(F)$ and $\phi_k \to \phi$, $k \to \infty$, in $\mathcal{S}(F)$;* every weak convergent sequence of distributions from $\mathcal{S}'(F)$ converges in the norm of some $\mathcal{S}'_p(F)$; the space $\mathcal{S}'(F)$ is complete with respect to weak (and strong) convergence.

Every measure $d\mu$ tempered on $F$ (see Section 0.3) defines by formula (2.3) the distribution $[d\mu]$ from $\mathcal{S}'(F)$ of order 0.

The space $\mathcal{S}' = \mathcal{S}'(\mathbf{R}^n)$ of tempered distributions in $\mathbf{R}^n$ is important; $\mathcal{S}' \subset \mathcal{D}'$ and $\mathcal{S}'$ is dense in $\mathcal{D}'$; $\mathcal{S}'(F) \subset \mathcal{S}'$.

All notions for the space $\mathcal{D}'$ are also valid for the space $\mathcal{S}'$; in particular, the notions of null set and the support of a distribution.

Let $F$ be a closed set. We define by $\mathcal{D}'_F$ (resp. $\mathcal{S}'_F$) the set of distributions from $\mathcal{D}'$ (resp. from $\mathcal{S}'$) with supports in $F$ and with the topology induced by that of $\mathcal{D}'$ (resp. $\mathcal{S}'$).

From the results of Section 1.1 the following statement follows. *If $F$ is the closure of a regular domain (or Lip-domain in $\mathbf{R}^n$) then the spaces $\mathcal{S}'(F)$ and $\mathcal{S}'_F$ are isomorphic, $\mathcal{S}'(F) \sim \mathcal{S}'_F$.* We denote by $\dot{J}$ the mapping which realizes this isomorphism, $\dot{J}: \mathcal{S}'(F) \to \mathcal{S}'_F$, and we denote its inverse mapping by $\dot{J}^{-1}: \mathcal{S}'_F \to \mathcal{S}'(F)$.

We denote:

$$\mathcal{S}'_+(n) = \mathcal{S}'(\mathbf{R}^1_+ \times \mathbf{R}^n) \sim \mathcal{S}'_{\mathbf{R}^1_+ \times \mathbf{R}^n},$$

$$\mathcal{S}'_+(0) = \mathcal{S}'_+,$$

$$\mathcal{S}'_- = \mathcal{S}'(-\mathbf{R}^1_+),$$

$$\mathcal{D}'_+(n) = \mathcal{D}'_{\mathbf{R}^1_+ \times \mathbf{R}^n},$$

$$\mathcal{D}'_+(0) = \mathcal{D}'_+ = \mathcal{D}'_{\mathbf{R}^1_+},$$

$$\mathcal{D}'_- = \mathcal{D}'_{-\mathbf{R}^1_+}.$$

4. **Linear operations on distributions** are defined as operations that are conjugate to the corresponding operations on test functions. Let $\phi \to A\phi$ be a linear and continuous operation from $\mathcal{D}(\mathcal{O}_1)$ into $\mathcal{D}(\mathcal{O}_2)$ and let $f$ be a distribution from $\mathcal{D}'(\mathcal{O}_2)$. We define a distribution $A^* f$ from $\mathcal{D}'(\mathcal{O}_1)$ by the formula

$$(A^* f, \phi) = (f, A\phi), \quad \phi \in \mathcal{D}(\mathcal{O}_1). \tag{4.1}$$

The operation $f \to A^* f$ is *linear and continuous from $\mathcal{D}'(\mathcal{O}_2)$ into $\mathcal{D}'(\mathcal{O}_1)$*; in other words, if $f_k \to 0$, $k \to \infty$ in $\mathcal{D}'(\mathcal{O}_2)$ then $A^* f_k \to 0$, $k \to \infty$, in $\mathcal{D}'(\mathcal{O}_1)$.

Linear and continuous operations from $\mathcal{S}'(F_2)$ into $\mathcal{S}'(F_1)$ are defined in a similar way.

**Multiplication.** Let $a \in C^\infty(\mathcal{O})$ and $f \in \mathcal{D}'(\mathcal{O})$. As the operation $\phi \to a\phi$ is linear and continuous from $\mathcal{D}(\mathcal{O})$ into $\mathcal{D}(\mathcal{O})$ then the product $af$ from $\mathcal{D}'(\mathcal{O})$ is defined according to (4.1) by the formula

$$(af, \phi) = (f, a\phi), \quad \phi \in \mathcal{D}(\mathcal{O}). \tag{4.2}$$

The operation $f \to af$ is *linear and continuous from* $\mathcal{D}'(\mathcal{O})$ *into* $\mathcal{D}'(\mathcal{O})$; therefore the function $a \in C^\infty(\mathcal{O})$ is called a *multiplier in* $\mathcal{D}'(\mathcal{O})$ (and *in* $\mathcal{D}(\mathcal{O})$).

From definition (4.2) it follows that *if* $f \in \mathcal{D}'(\mathcal{O})$ *then*

$$f = \eta f \tag{4.3}$$

*where* $\eta$ *is any* $C^\infty(\mathcal{O})$*-function which is equal to* 1 *in some neighbourhood of* supp $f$.

If the order $f$ in $\mathcal{O}$ is finite, $m_f = m < \infty$, then the product $af$, where $a \in C^m(\mathcal{O})$, is well defined, and also according to (4.1) and inequality (2.1) of Section 1.2, the order of $af$ is equal to $m_f$. In addition, equality (4.3) is valid for $\eta \in C^m(\mathcal{O})$, $\eta(x) = 1$, $x \in \operatorname{supp} f$.

**Example 1.** $a(x)\delta(x) = a(0)\delta(x)$, $a \in C$.

If $f \in \mathcal{D}'(\mathcal{O})$ and supp $f \subset\subset \mathcal{O}$ then $f$ admits an extension $\tilde{f}$ on $\mathcal{E}(\mathcal{O})$ as an element of $\mathcal{E}'(\mathcal{O})$ (see Section 1.2) and $\tilde{f}$ is defined by the formula

$$(\tilde{f}, \phi) = (f, \eta\phi), \quad \phi \in \mathcal{E}(\mathcal{O}) \tag{4.4}$$

and also, owing to (4.3), the right-hand side does not depend on an auxiliary function $\eta$ with prescribed properties; the extension $\tilde{f}$ is unique.

Let $F$ be a closed set such that $\overline{\int F} = F$. We denote by $\theta_M(F)$ the set of $C^\infty(F)$-tempered functions on $F$ (see Section 0.3 and Section 0.4) together with all their derivatives. We denote $\theta_M(\mathbf{R}^n) = \theta_M$.

The operation $f \to af$, where $a \in \theta_M(F)$, is linear and continuous from $\mathcal{S}'(F)$ into $\mathcal{S}'(F)$ (and from $\mathcal{S}(F)$ into $\mathcal{S}(F)$) so that $\theta_M(F)$ is exactly the *set of the multipliers from* $\mathcal{S}'(F)$ *into* $\mathcal{S}'(F)$.

**Differentiation.** Let $f \in \mathcal{D}'(\mathcal{O})$ and $\alpha$ be a multi-index. A generalized (weak) derivative $\partial^\alpha f$ is defined according to (4.1) by the equality

$$(\partial^\alpha f, \phi) = (-1)^{|\alpha|}(f, \partial^\alpha \phi), \quad \phi \in \mathcal{D}(\mathcal{O}). \tag{4.5}$$

Since the operation $\phi \to (-1)^{|\alpha|}\partial^\alpha\phi$ is linear and continuous from $\mathcal{D}(\mathcal{O})$ into $\mathcal{D}(\mathcal{O})$, the functional $\partial^\alpha f$, which is defined by the right-hand side of equality (4.5), is a distribution from $\mathcal{D}'(\mathcal{O})$ and the operation $f \to \partial^\alpha f$ is *linear and continous from* $\mathcal{D}'(\mathcal{O})$ *into* $\mathcal{D}'(\mathcal{O})$.

From this definition it follows that *every distribution is infinitely differentiable* (in the generalized sense); *the differentiation does not depend on its order,*

$$\partial^\alpha(\partial^\beta f) = \partial^\beta(\partial^\alpha f) = \partial^{\alpha+\beta} f\,;$$

*the Leibniz formula is valid for the differentiation of the product* $af$, *where* $a \in C^\infty(\mathcal{O})$; *any distribution from* $\mathcal{D}'(\mathcal{O})$ *in any open set* $\mathcal{O}' \subset\subset \mathcal{O}$ *is a derivative of* *some continuous function in $\mathcal{O}$; any tempered distribution from $\mathcal{S}'$ is a derivative of a tempered continuous function in $\mathbf{R}^n$.*

Any distribution $f$ from $\mathcal{S}'_F$, where $F$ is a regular closed set, can be represented in the form

$$f(x) = \sum_{|\alpha| \leqslant m} \partial^\alpha \left[d\mu_\alpha\right](x) \tag{4.6}$$

*where $d\mu_\alpha$ are tempered measures in $F$ and* $\operatorname{supp} d\mu_\alpha \subset F$.

In particular, *any distribution with a support at the point* 0 *can be represented (uniquely) in the form*

$$f(x) = \sum_{|\alpha| \leqslant m} a_\alpha \partial^\alpha \delta(x). \tag{4.7}$$

**Example 2.** $\theta'(x) = \delta(x)$, where $\theta$ is the Heaviside function (Section 0.2).

Let $n = 1$. We shall denote the derivative $\partial^\alpha f$ by $f^{(\alpha)}$; $f^{(1)} = f'$, $f^{(2)} = f''$,... Let $f \in \mathcal{D}'(a,b)$. We denote by $f^{(-1)}$ the *primitive* of the distribution $f$ in $(a,b)$ so that $f^{(-1)} \in \mathcal{D}'(a,b)$ and $f^{(-1)\prime} = f$ in $(a,b)$,

$$(f^{(-1)}, \phi') = -(f, \phi), \quad \phi \in \mathcal{D}(a,b). \tag{4.8}$$

Equality (4.8) shows that the functional $f^{(-1)}$ is defined only on the first derivatives of test functions from $\mathcal{D}(a,b)$. To find the primitive it is necessary to extend this functional on the whole space $\mathcal{D}(a,b)$ in such a way that the extended functional should be linear and continuous on $\mathcal{D}(a,b)$. It is necessary to determine the degree of arbitrariness of this extension.

*The primitive $f^{(-1)}$ of any $f$ from $\mathcal{D}'(a,b)$ exists in $(a,b)$, is defined up to an arbitrary additive constant $C$, and is given by the formula*

$$(f^{(-1)}, \phi) = -(f, \psi) + C \int \phi(\xi)\, d\xi, \quad \phi \in \mathcal{D}(a,b) \tag{4.9}$$

*where*

$$\psi(x) = \int_{-\infty}^{x} \left[\phi(x') - \omega_\epsilon(x' - x_0) \int \phi(\xi)\, d\xi\right] dx',$$

$x_0 \in (a,b)$ and $\omega_\epsilon$ is the normed bell-shaped function with $\epsilon < \min(x_0 - a, b - x_0)$ (see Section 0.5).

Let $\mathcal{O} = (a,b) \times \tilde{\mathcal{O}}$ where $\tilde{\mathcal{O}}$ is an open set in the space $\mathbf{R}^{n-1}$ of variables $(x_2, \ldots, x_n)$ and let $f \in \mathcal{D}'(\mathcal{O})$. The primitive $f^{(-1,0,\ldots,0)}$ on the variable $x_1$ of the distribution $f$ is defined as a solution of the differential equation

$$\partial_1 f^{(-1,0,\ldots,0)} = f, \quad f^{(-1,0,\ldots,0)} \in \mathcal{D}'(\mathcal{O}).$$

The primitive $f^{(-1,0,\ldots,0)}$ always exists; it can be constructed by a formula similar to (4.9); its arbitrariness will be discussed in Section 1.7.

Let $F$ be the closure of a regular domain. For a derivative $\partial^\alpha f$ of a distribution $f$ from $\mathcal{S}'(F)$ the following formula is valid (see Section 1.3):

$$\partial^\alpha f = \dot{J}^{-1} \partial^\alpha (\dot{J} f). \tag{4.10}$$

**5. Change of variables.** Let a mapping $x \to y(x)$ be a $C^\infty$-diffeomorphism of an open set $\mathcal{O} \subset \mathbf{R}_x^n$ onto an open set $\mathcal{O}_1 \subset \mathbf{R}_y^n$; let $x = x(y)$ be the inverse mapping and let $J(x/y)$ be its Jacobian (see Section 0.4). Let $f(y)$ be from $\mathcal{D}'(\mathcal{O}_2)$. A distribution $f(y(x))$ from $\mathcal{D}'(\mathcal{O})$ is defined according to (4.1) by the formula

$$(f(y(x)), \phi(x)) = \left( f(y), \phi(x(y)) \left| J\left(\frac{x(y)}{y}\right) \right| \right), \quad \phi \in \mathcal{D}(\mathcal{O}). \tag{5.1}$$

Since the operation $\phi \to \phi(x(y)) | J((x(y))/y) |$ is an isomorphism from $\mathcal{D}(\mathcal{O})$ onto $\mathcal{D}(\mathcal{O}_1)$, then the operation $f(y) \to f(y(x))$ is an *isomorphism* from $\mathcal{D}'(\mathcal{O}_1)$ onto $\mathcal{D}'(\mathcal{O})$.

If a mapping $x \to y(x)$ is a local $C^{in}$-diffeomorphism in $\mathcal{O}$ then by means of a suitable decomposition of unity (see Section 1.1) we can construct by formula (5.1) the local elements $\{f_k(y(x)), x \in \mathcal{O}_k\}$. If they are compatible in $\mathcal{O}$, then with the help of the 'piecewise sewing theorem' (see Section 1.2) we can construct a unique distribution from $\mathcal{D}'(\mathcal{O})$ which is denoted by $f(y(x))$.

**Example 1.** $\delta(x^2 - a^2) = 1/(2|a|)[\delta(x - a) + \delta(x + a)]$, $a \neq 0$.

**Example 2.** $\delta(\sin x) = \sum_{k=-\infty}^{\infty} \delta(x - k\pi)$.

In particular, if $y = Ax + b$, where $A$ is a non-singular matrix, then the distribution $f(Ax + b)$ is defined according to (5.1) by the formula

$$(f(Ax + b), \phi(x)) = \left( f(y), \frac{\phi[A^{-1}(y - b)]}{|\det A|} \right), \quad \phi \in \mathcal{D}(\mathcal{O}). \tag{5.2}$$

**6. $L$-invariant distributions.** Let $L$ be a group which consists of non-singular linear transformations $x \to Ax$ of the space $\mathbf{R}^n$; $A \in L$. A distribution $f$ from $\mathcal{D}'(\mathcal{O})$ (resp. from $\mathcal{S}'$, $\mathcal{D}(\mathcal{O})$, $\mathcal{S}$, ...) is called *$L$-invariant* if $\mathcal{O}$ is $L$-invariant and $f(Ax) = f(x)$, $A \in L$. We denote by $\mathcal{D}'_L(\mathcal{O})$, $\mathcal{S}'_L$,... the sets of $L$-invariant distributions from $\mathcal{D}'(\mathcal{O})$, $\mathcal{S}'$,..., $\mathcal{D}'_L(\mathcal{O})$, $\mathcal{S}'_L$,... are closed subspaces of the spaces $\mathcal{D}'(\mathcal{O})$, $\mathcal{S}'$,... resp.

**Spherically-symmetric distributions:** $\mathcal{D}'_{SO(n)}(\mathcal{O})$, $\mathcal{S}'_{SO(n)}$ where $SO(n)$ is the proper rotation group of $\mathbf{R}^n$ (see Section 0.5). It is intuitively clear that $SO(n)$-invariant distributions depend only on the radius $r = |x|$. For their definition we consider the space $\mathcal{S}'_{SO(n)}$. The mapping

$$\phi(x) \to \frac{1}{2\pi} \int_{S^{n+1}} \phi(|x| s)\, ds = \int_{SO(n)} \phi(Ax)\, d\mu(A) \tag{6.1}$$

where $d\mu(A)$ is the normed Haar measure on the group $SO(n)$ and the number $\sigma_n$, defined in Section 0.5, is linear and continuous from $\mathcal{S}(\mathbf{R}^n)$ onto $\mathcal{S}_{SO(n)}$. The mapping $\phi(x) \to \tilde{\phi}(t)$, where

$$\phi(x) \equiv \phi(\pm x_1, 0, \ldots, 0) = \tilde{\phi}(x_1^2), \tag{6.2}$$

is linear and continuous from $\mathcal{S}_{SO(n)}$ in $\mathcal{S}_+$; its inverse mapping

$$\phi(t) \to \phi(|x|^2) \tag{6.3}$$

is linear and continuous from $\mathcal{S}_+$ in $\mathcal{S}_{SO(n)}$. Thus mappings (6.2) and (6.3) define an isomorphism $\mathcal{S}_{SO(n)} \sim \mathcal{S}_+$. Therefore, the spaces $\mathcal{S}'_+$ and $\mathcal{S}'_{SO(n)}$ are isomorphic and the mappings

$$f \to f_+,\ (f_+,\phi) = \frac{2}{\sigma_n}(f(x), \phi(|x|^2)),\ \phi \in \mathcal{S}_+, \tag{6.4}$$

$$f_+ \to f,\ (f, \phi) = (f_+, \int_{SO(n)} \phi(Ax)d\mu(A)),\ \phi \in \mathcal{S} \tag{6.5}$$

define their isomorphism.

**Example 1.** $\delta \in \mathcal{S}'_{SO(n)}$.

**Lorentz-invariant distributions** $\mathcal{D}'_{L_+^\uparrow(n)}(\mathcal{O}), \mathcal{S}'_{L_+^\uparrow(n)}$ where $L_+^\uparrow(n)$ is the proper orthochrone Lorentz group of $\mathbf{R}^{n+1}$.

**Example 2.** $\delta \in \mathcal{S}'_{L_+^\uparrow(n)}$.

**Example 3.** $\delta_\pm(x^2 - m^2) \in \mathcal{S}'_{L_+^\uparrow(n)}$ where $\delta_\pm(x^2 - m^2)$ are defined by the formula

$$(\delta_\pm(x^2 - m^2), \phi) = \frac{1}{2}\int \frac{\phi(\pm\sqrt{|\mathbf{x}|^2 + m^2}, \mathbf{x})}{\sqrt{|\mathbf{x}|^2 + m^2}}d\mathbf{x},\ \phi \in \mathcal{S}(\mathbf{R}^{n+1})$$

(for $m = 0$ we assume $n \geqslant 2$).

**Example 4.**

$$\delta(x^2 - m^2) = \delta_+(x^2 - m^2) + \delta_-(x^2 - m^2) \in \mathcal{S}'_{L_+^\uparrow(n)},$$

$$\epsilon(x_0)\delta(x^2 - m^2) = \delta_+(x^2 - m^2) - \delta_-(x^2 - m^2) \in \mathcal{S}'_{L_+^\uparrow(n)}.$$

A description of $L_+^\uparrow(n)$-invariant distributions was given by P.D. Methée (1954). We reproduce his results for the case $n = 3$ and for the distributions from $\mathcal{S}'_{L_+^\uparrow}$, $L_+^\uparrow = L_+^\uparrow(3)$.

Every $L_+^\uparrow$-invariant distribution can be represented uniquely as a sum of an even distribution $f_+$ (invariant with respect to the proper Lorentz group $L_+$) and an odd $f_-$ distribution (changing the sign under $L_+$-transformations which change the direction of time $x_0$). The set of even distributions from $\mathcal{S}'_{L_+^\uparrow}$ is denoted by $\mathcal{L}^+$, and the set of odd distributions by $\mathcal{L}^-$.

It is clear that $\operatorname{supp} f \subset \bar{V}_3$ is $f \in \mathcal{L}^-$, where $V_3 = V_3^+ \cup V_3^-$.

Let $\phi$ be from $\mathcal{S}(\mathbf{R}^4)$. We introduce the mapping $\mathbf{M}: \phi \to \mathbf{M}\phi$ by the formula ('averaging on hyperboloids')

$$(\mathbf{M}\phi)(\tau) = (\delta(x^2 - \tau), \phi) = \int_{x^2=\tau} \phi\, dS_x .$$

The function $(\mathbf{M}\phi)(\tau)$ belongs to the class $\mathbf{H}$ which consists of functions $\psi(\tau)$ of the form

$$\psi(\tau) = \psi_1(\tau) + \psi_2(\tau)\ln|\tau|,$$

where $\psi_1$ and $\psi_2$ belong to $\mathcal{S}(\mathbf{R}^1)$ and $\psi_2(0) = 0$. This representation is not unique. The topology in $\mathbf{H}$ is introduced with the help of the countable family of norms

$$\|\psi\|'_p = \left\| \psi(\tau) - \eta(\tau)\ln|\tau| \sum_{k=1}^{p} \frac{\psi_2^{(k)}(0)}{k!}\tau^k \right\|_p + \sum_{k=1}^{p} |\psi_2^{(k)}(0)|,\ p = 0, 1, \cdots$$

where the norm $\|\cdot\|_p$ was defined in (1.1); $\eta$ is any function from $\mathcal{D}$ which is equal to 1 in a neighbourhood of 0. The introduced topology depends on neither the representation of $\psi$ nor the choice of $\eta$. The space **H** is complete.

Let $\phi$ be from $\mathcal{S}(\bar{V}_3)$. We introduce the mapping $M_1 : \phi \to M_1\phi$ by the formula

$$(M_1\phi)(\tau) = (\epsilon(x_0)\delta(x^2 - \tau), \phi) = \frac{1}{2}\int \frac{\phi(\sqrt{|\mathbf{x}|^2 + \tau}, \mathbf{x}) - \phi(-\sqrt{|\mathbf{x}|^2 + \tau}, \mathbf{x})}{\sqrt{|\mathbf{x}|^2 + \tau}}\, d\mathbf{x}$$

if $\tau \geqslant 0$.

The mappings $M$ and $M_1$ are *linear and continuous* from $\mathcal{S}(\mathbf{R}^4)$ onto **H** and from $\mathcal{S}(\bar{V}_3)$ onto $\mathcal{S}_+$, respectively.

We denote by **H'** the space adjoint to **H**. We define the conjuate mappings $M^*$ from $H^1$ in $\mathcal{S}'(\mathbf{R}^4)$, and $M_1^*$ from $\mathcal{S}_+'$ into $\mathcal{S}'(\bar{V}_3^+)$, according to formula (4.1).

The following statement is valid: $M^*$ *is an isomorphism from* $H'$ *onto* $\mathcal{L}^+$, *and* $M_1^*$ *is an isomorphism from* $\mathcal{S}_+'$ *onto* $\mathcal{L}^+$.

The last statement means that every $L_+^\uparrow$-invariant odd-tempered distribution $f$ is represented uniquely in the form

$$f(x) = \quad \varepsilon(x_0) f_0(x^2)$$

where $f_0 \in \mathcal{S}_+'$, and in addition

$$(f, \phi) = (f_0(\tau), \int \check{\phi}(\tau - |\mathbf{x}|^2, \mathbf{x})\, d\mathbf{x}), \quad \phi \in \mathcal{S}(\mathbf{R}^4)$$

where

$$\check{\phi}(x_0^2, \mathbf{x}) = \frac{1}{2x_0}\left[\phi(x_0, \mathbf{x}) - \phi(-x_0, \mathbf{x})\right],$$

$$\hat{\phi}(x_0^2, \mathbf{x}) = \frac{1}{2}\left[\phi(x_0, \mathbf{x}) + \phi(-x_0, \mathbf{x})\right],$$

$$\phi(x) = x_0\check{\phi}(x_0^2, \mathbf{x}) + \hat{\phi}(x_0^2, \mathbf{x}).$$

**Example 5.** $\epsilon(x_0)\delta(x^2 - m^2) \in \mathcal{L}^-$,

$$(\epsilon(x_0)\delta(x^2 - m^2), \phi) = \int \check{\phi}(m^2 - |\mathbf{x}|^2, \mathbf{x}). d\mathbf{x}, \quad \phi \in \mathcal{S}(\mathbf{R}^4).$$

We note that the general form of an $L_+^\uparrow$-invariant nonnegative tempered measure $d\mu$ with a support in $\bar{V}_3^+$ is

$$d\mu(x) = \int_0^\infty \delta_+(x^2 - \tau)\, d\rho(\tau) + C\,\delta(x)$$

where $C \geqslant 0$ and $d\rho$ is a nonnegative tempered measure with a support in $[0, \infty)$, $[d\mu] \in \mathcal{S}'(\bar{V}_3^+)$, $[d\rho] \in \mathcal{S}_+'$.

**Translationally-invariant distributions** $\mathcal{D}'_{T(n)}$ where $T(n)$ is the group of translations in $\mathbf{R}^n$ (see Section 0.5):

$$f(x + h) = f(x), \quad h \in \mathbf{R}^n.$$

**Symmetric and antisymmetric distributions:** $f(x) = f(-x)$, $f(x) = -f(-x)$ respectively.

We denote by $\mathcal{S}'_s$ and $\mathcal{S}'_a$ (resp. $\mathcal{S}_s$ and $\mathcal{S}_a$) the closed subspaces of symmetric and antisymmetric distributions from $\mathcal{S}'(\mathbf{R}^1)$ (resp. test functions from $\mathcal{S}(\mathbf{R}^1)$).

Let $f_+ \in \mathcal{S}'_+$. We introduce the distribution $f^a$ from $\mathcal{S}'_a$ by the formula

$$(f^a, \phi) = \left(f_+(t), \frac{\phi(\sqrt{t}) - \phi(-\sqrt{t})}{2\sqrt{t}}\right), \quad \phi \in \mathcal{S}(\mathbf{R}^1). \tag{6.6}$$

*The mapping* $f_+ \to f^a$ *is an isomorphism from* $\mathcal{S}'_+$ *onto* $\mathcal{S}'_a$.

Thus, *the mapping* $f \to f_+ \to f^a$ *is an isomorphism from* $\mathcal{S}'_{SO(n)}$ *onto* $\mathcal{S}'_a$; the corresponding *inverse mappings* are determined from (6.4) and (6.6) by the formulae

$$(f, \phi) = \frac{\sigma_n}{2}(f_+, \tilde{\phi}) = \frac{\sigma_n}{2}(f^a(t), t\tilde{\phi}(t^2)), \quad \phi \in \mathcal{S}_{SO(n)} \tag{6.7}$$

where $\phi(x) = \tilde{\phi}(|x|^2)$.

Example 6. $\delta_+ = \delta$, $\delta^a = -\dfrac{\delta'}{2\pi}$ $(n = 1)$.

**Homogeneous distributions** of degree $\alpha$, $-\infty < \alpha < \infty$,

$$f(kx) = k^\alpha f(x), \quad k > 0.$$

For example,

$$f_\alpha(t) = \begin{cases} \dfrac{\theta(t)t^{\alpha-1}}{\Gamma(\alpha)}, & \alpha > 0; \\ f_{\alpha+N}^{(N)}(t), & \alpha \leqslant 0, \ \alpha + N > 0, \quad N \text{ integer}, \end{cases} \tag{6.8}$$

is a homogeneous distribution from $\mathcal{S}'_+$ of degree $\alpha - 1$.

Conversely, every homogeneous distribution $f$ from $\mathcal{S}'_+$ of degree $\alpha$ is proportional to $f_{\alpha+1}$:

$$f(t) = \frac{(f, e^{-t})}{(f_{\alpha+1}, e^{-t})} f_{\alpha+1}(t). \tag{6.9}$$

This statement follows from the fact that the set of functions $\{e^{-kt}, k > 0\}$ is dense in $\mathcal{S}_+$.

**A translationally-invariant linear operator** is an operator $L$ from $\mathcal{D}'$ into $\mathcal{D}'$ which commutes with a translation operator; that is

$$Lf(x+h) = (Lf)(x+h), \quad f \in \mathcal{D}', \ h \in \mathbf{R}^n. \tag{6.10}$$

For example, the differentiation operator $\partial^\alpha$ is translationally-invariant: $\partial^\alpha f(x+h) = (\partial^\alpha f)(x+h)$.

Let $F$ be the closure of a regular domain, let $\mathcal{O}$ be a neighbourhood of $F$ and let a mapping $x \to y(x)$ be the diffeomorphism of $\mathcal{O}$ onto $\mathcal{O}_1$, such that $y \in \theta_M(F)$ (see Section 1.4). Further, let $F_1 \subset \mathcal{O}_1$ be the image of $F$ under this mapping, and let $y \to x(y)$ be the inverse mapping such that $x \in \theta_M(F_1)$. Then the mapping

$$\phi(x) \to \phi(x(y)) \left| J\left(\frac{x(y)}{y}\right) \right|$$

is an *isomorphism of* $\mathcal{S}(F)$ *onto* $\mathcal{S}(F_1)$. Therefore, the change of variables $x \to y(x)$ in distributions $f$ from $\mathcal{S}'(F_1)$, $f(y) \to f(y(x))$, defined by formula (5.1), is an *isomorphism of* $\mathcal{S}'(F_1)$ *onto* $\mathcal{S}'(F)$ and by virtue of the isomorphism $\dot{J}$ (see Section 1.3) $\mathcal{S}'_{F_1} \sim \mathcal{S}'_F$.

**7. Direct product of distributions.** Let $\mathcal{O} \subset \mathbf{R}_x^n$ and $\mathcal{O}_2 \subset \mathbf{R}_y^n$ be open sets, $f \in \mathcal{D}'(\mathcal{O}_1)$, and $g \in \mathcal{D}'(\mathcal{O}_2)$. We define distributions $f(x) \times g(y)$ and $g(y) \times f(x)$ by the formulae

$$(f(x) \times g(y), \phi) = (f(x), (g(y), \phi(x, y))), \quad \phi \in \mathcal{D}(\mathcal{O}_1 \times \mathcal{O}_2). \tag{7.1}$$

$$(g(y) \times f(x), \phi) = (g(y), (f(x), \phi(x, y))), \quad \phi \in \mathcal{D}(\mathcal{O}_1 \times \mathcal{O}_2). \tag{7.1'}$$

The functionals in the right-hand side of (7.1) and (7.1') are linear. Their continuity follows from the following

**Lemma.** *Let* $\mathcal{O}' \subset\subset \mathcal{O}_1 \times \mathcal{O}_2$ *be an open set and* $g \in \mathcal{D}'(\mathcal{O}_2)$. *Then there exists an open set* $\tilde{\mathcal{O}}_1 \subset\subset \mathcal{O}_1$, *numbers* $C$, *and an integer* $m \geqslant 0$ *such that for any* $\phi \in (\mathcal{O}')$

$$\psi(x) = (g(y), \phi(x, y)) \in \mathcal{D}(\tilde{\mathcal{O}}_1),$$

$$\partial^\alpha \psi(x) = (g(y), \partial_x^\alpha \phi(x, y)), \quad |\partial^\alpha \psi(x)| \leqslant C \|\phi\|_{C^m(\bar{\mathcal{O}}')}, \quad x \in \tilde{\mathcal{O}}_1.$$

Thus, the functionals $f(x) \times g(y)$ and $g(y) \times f(x)$ belong to $\mathcal{D}'(\mathcal{O}_1 \times \mathcal{O}_2)$; that is, they are *distributions* in $\mathcal{O}_1 \times \mathcal{O}_2$. They are called the *direct products* of distributions $f(x)$ and $g(y)$.

The direct product has the following properties:

$f(x) \times g(y) = g(y) \times f(x)$ — commutativity,
$[f(x) \times g(y)] \times h(x) = f(x) \times [g(y) \times h(z)]$ — associativity.

Henceforth, taking into account the commutativity and associativity of the direct product, we shall write $f \times g \times h$ instead of $(f \times g) \times h$, $f \times (g \times h)$, ...

**Example 1.** $\delta(x) = \delta(x_1) \times \delta(x_2) \times \cdots \times \delta(x_n)$.

From the above lemma it also follows that the operation $(f, g) \to f \times g$ is *linear and continuous from* $\mathcal{D}'(\mathcal{O}_1) \times \mathcal{D}'(\mathcal{O}_2)$ *in* $\mathcal{D}'(\mathcal{O}_1 \times \mathcal{O}_2)$.

The following formulae hold:

$$\operatorname{supp}(f \times g) = \operatorname{supp} f \times \operatorname{supp} g, \tag{7.2}$$

$$\partial_x^\alpha \partial_y^\beta [f(x) \times g(y)] = \partial^\alpha f(x) \times \partial^\beta g(y), \tag{7.3}$$

$$a(x) b(y)[f(x) \times g(y)] = a(x) f(x) \times b(y) g(y), \quad a, b \in C^\infty, \tag{7.4}$$

$$(f \times g)(x + x_0, y + y_0) = f(x + x_0) \times g(y + y_0), \quad x_0 \in \mathbf{R}^n, \ y_0 \in \mathbf{R}^m \tag{7.5}$$

(translational-invariance, see Section 1.6).

We shall say that a distribution $f(x, y)$ from $\mathcal{D}'(\mathcal{O}_1 \times \mathcal{O}_2)$ *does not depend on the variable* $y$ if it can be represented in the form $f(x, y) = f_0(x) \times 1(y)$, where $f_0 \in \mathcal{D}'(\mathcal{O}_1)$ (and then $f \in \mathcal{D}'(\mathcal{O}_1 \times \mathbf{R}^m)$). The distribution $f_0(x) \times 1(y) = 1(y) \times f_0(x)$ acts on the test functions $\phi$ from $\mathcal{D}(\mathcal{O}_1 \times \mathbf{R}^m)$ by the rule

$$(f_0(x) \times 1(y), \phi) = (f_0(x), \int \phi(x, y)\, dy) =$$

$$= (1(y) \times f_0(x), \phi) = \int (f_0(x), \phi(x, y))\, dy.$$

Thus we have obtained the equality

$$(f(x), \int \phi(x, y)\, dy) = \int (f(x), \phi(x, y))\, dy, \tag{7.6}$$

$$f \in \mathcal{D}'(\mathcal{O}_1), \ \phi \in \mathcal{D}(\mathcal{O}_1 \times \mathbf{R}^m).$$

Let $f(x, y) \in \mathcal{D}'(\mathcal{O} \times (a, b))$. The following statements are equivalent:

(i) *$f(x, y)$ does not depend on $y$;*

(ii) *$f(x, y)$ is invariant in $\mathcal{O} \times (a, b)$ with respect to the translations along $y$;* that is, $f(x, y + h) = f(x, y)$, $a < y, y + h < b$;

(iii) *$f(x, y)$ satisfies the equation $\partial_y f(x, y) = 0$ in $\mathcal{O} \times (a, b)$.*

In particular, if $f \in \mathcal{D}'(\mathcal{O})$ and $\partial_{x_j} f(x) = 0$, $j = 1, \ldots, n$ in $\mathcal{O}$, then $f$ = constant in $\mathcal{O}$; if $f$ is invariant with respect to the translation group (see Section 1.6) then $f$ = constant in $\mathcal{O}$; the primitive $f^{(0,0,\ldots,-1)}$ of a distribution $f$ from $\mathcal{D}'(\mathcal{O} \times (a, b))$ (see Section 1.4) is defined up to an arbitrary additive distribution of the form $f_0(x) \times 1(y)$, $f_0 \in \mathcal{D}'(\mathcal{O})$.

Let a distribution $f(x, y)$ from $\mathcal{D}'(\mathcal{O}_1 \times \mathcal{O}_2)$ belong to the class $C^k(\mathcal{O}_2)$ on $y$ (see Section 1.2). Then for every $y \in \mathcal{O}_2$ there exists a restriction $f_y(x)$ from $\mathcal{D}'(\mathcal{O}_1)$ of the distribution $f(x, y)$, and

$$f_\phi(y) = (f_y, \phi), \ \phi \in \mathcal{D}(\mathcal{O}_1), \ y \in \mathcal{O}_2, \tag{7.7}$$

and the following equalities are valid

$$\partial^\alpha (f_y, \phi) = \left( \left( \partial_y^\alpha f \right)_y, \phi \right), \ \phi \in \mathcal{D}(\mathcal{O}_1), \ y \in \mathcal{O}_2, \ |\alpha| \leqslant k, \tag{7.8}$$

$$(f, \phi) = \int (f_y, \phi(x, y))\, dy, \ \phi \in \mathcal{D}(\mathcal{O}_1 \times \mathcal{O}_2), \tag{7.9}$$

so that $f(x, y) = f_y(x)$.

**Example 2.** Let the distribution $f(x, y) = f_0(x) \times 1(y)$ be such that it does not depend on $y$. Then $f \in C^\infty(\mathbf{R}^m)$ on $y$ and $f_y(x) = f_0(x)$.

The collected results, with suitable modifications, are also valid for the space $\mathcal{S}'(F)$ of tempered distributions on $F$, where $F$ is the closure of a regular domain.

**8. Convolution of distributions.** Let $f$ and $g$ be from $\mathcal{L}^1_{loc}$. If the integral $\int f(y) g(x - y)\, dy$ exists for almost all $x \in \mathbf{R}^n$ and defines a function from $\mathcal{L}^1_{loc}$, then it is called the *covolution* of the functions $f$ and $g$ and is denoted by $f * g$ so that

$$(f * g)(x) = \int f(y) g(x - y)\, dy = \int f(y) f(x - y)\, dy = (g * f)(x). \tag{8.1}$$

We note two cases when the convolution $f * g$ definitely exists.

(a) Let $f \in \mathcal{L}^1_{loc}$, $g \in \mathcal{L}^1_{loc}$, $\operatorname{supp} f \subset A$, $\operatorname{supp} g \subset B$, and let the sets $A$ and $B$ be such that for any $\mathbf{R} > 0$ the set

$$T_{\mathbf{R}} = [(x, y) : x \in A, y \in B, |x + y| \leqslant \mathbf{R}]$$

is bounded in $\mathbf{R}^{2n}$ (Fig. 5). Then $f * g \in \mathcal{L}^1_{loc}$. In particular, if $f$ or $g$ is finite then

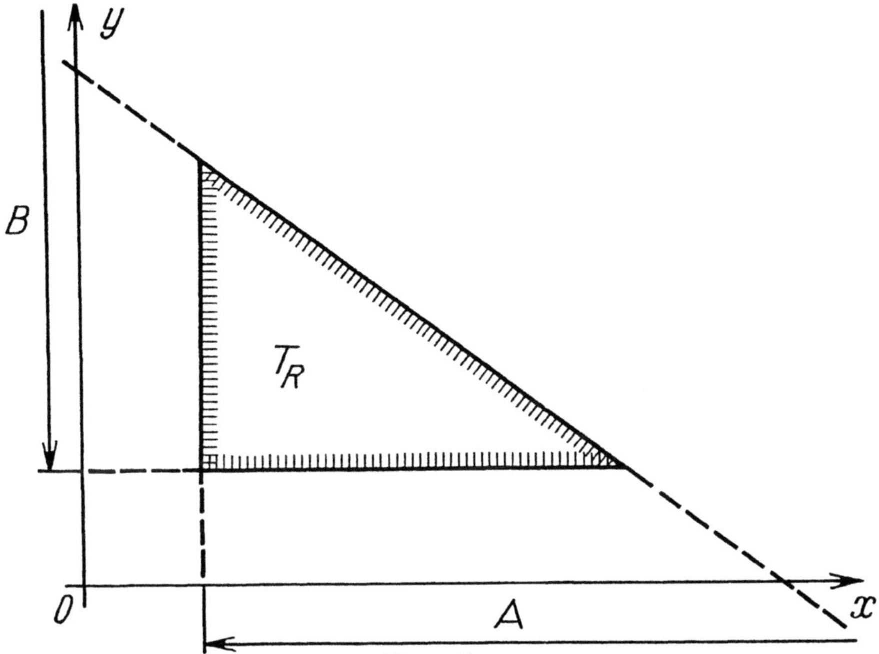


Figure 5.

$T_R$ is bounded.

(b) Let $f \in \mathcal{L}^p$ and $g \in \mathcal{L}^q$ if $1/p + 1/q \geqslant 1$. Then $f * g \in \mathcal{L}^r$ where $1/r = 1/p + 1/q - 1$, and the Jung inequality is valid

$$\|f * g\|_{\mathcal{L}^r} \leqslant \|f\|_{\mathcal{L}^p} \|g\|_{\mathcal{L}^q}. \tag{8.2}$$

From (8.1) it follows that the convolution $f * g$ defines a regular functional on by the formula

$$(f * g, \phi) = \int f(x) g(y) \phi(x + y)\, dx dy, \quad \phi \in \quad . \tag{8.3}$$

We shall say that a sequence $\{\eta_k(x), k \to \infty\}$ of functions from $\mathcal{D}(\mathbf{R}^n)$ *converges to 1 in* $\mathbf{R}^n$ if (i) for any compact $K$ there is a number $\mathbf{N}$ such that $\eta_k(x) = 1$, $x \in K$, $k \geqslant \mathbf{N}$, and (ii) the functions $\{\eta_k\}$ are uniformly bounded together with all their derivatives, $|\partial^\alpha \eta_k(x)| \leqslant C_\alpha$, $x \in \mathbf{R}^n$, $k \to \infty$. Such sequences always exist; for example, $\eta_k(x) = \eta(x/k)$, $k = 1, 2, \cdots$ where $\eta \in \quad$, $\eta(x) = 1$, $|x| < 1$.

Now the functional (8.3) can be rewritten as

$$(f * g, \phi) = \lim_{k \to \infty} (f(x) g(y), \eta_k(x; y) \phi(x + y)), \quad \phi \in \mathcal{D} \tag{8.4}$$

where $\{\eta_k, k \to \infty\}$ is any sequence of functions from $\mathcal{D}(\mathbf{R}^{2n})$ which converges to 1 in $\mathbf{R}^{2n}$.

Starting from equality (8.4) we shall introduce the following definition of the convolution of distributions from $\mathcal{D}'$. Let $f$ and $g$ be from $\mathcal{D}'(\mathbf{R}^n)$ such that their direct product $f(x) \times g(y)$ admits an extension to functions of the form $\phi(x + y)$, where $\phi$ is any functions from $\mathcal{D}(\mathbf{R}^n)$ in the following sense: for any sequence

$\{\eta_k, k \to \infty\}$ of functions from $\mathcal{D}(\mathbf{R}^{2n})$ which converge to 1 in $\mathbf{R}^{2n}$ there exists a limit to the numerical sequence

$$(f(x) \times g(y), \eta_k(x;y)\phi(x+y)), \quad k \to \infty$$

and this limit does not depend on the sequence $\{\eta_k, k \to \infty\}$.

*The convolution* $f * g$ is the functional

$$(f * g, \phi) = \lim_{k \to \infty}(f(x) \times g(y), \eta_k(x;y)\phi(x+y)), \quad \phi \in \mathcal{D}. \tag{8.5}$$

It follows from the completeness of the space of distribution $\mathcal{D}'$ (see Section 1.2) that $f * g \in \mathcal{D}'$.

Note that since $\phi(x+y)$ does not belong to $\mathcal{D}(\mathbf{R}^{2n})$ (it is not finite in $\mathbf{R}^{2n}$!), the right-hand side of (8.5) does not exist in general for all pairs of distributions $f$ and $g$ and, thus, the convolution does not always exist.

The convolution of any number of distributions is similarly defined; for example, $f * g * h$.

**Properties of a convolution.** If the convolution $f * g$ exists then does the convolution $g * f$ and they are equal ( *commutativity* of a convolution):

$$f * g = g * f. \tag{8.6}$$

Convolution with the δ-function:

$$f * \delta = f. \tag{8.7}$$

**Translational-invariance** of a convolution (see Section 1.6):

$$f(x+h) * g = (f * g)(x+h), \quad h \in \mathbf{R}^n. \tag{8.8}$$

The **reflection** of a convolution:

$$f \hat{*} g = \hat{f} * \hat{g} \quad \text{where } \hat{f}(x) = f(-x). \tag{8.9}$$

**Differentiation** of a convolution

$$\partial^\alpha f * g = \partial^\alpha (f * g) = f * \partial^\alpha g. \tag{8.10}$$

**Support** of a convolution

$$\operatorname{supp}(f * g) \subset \overline{\operatorname{supp} f + \operatorname{supp} g}. \tag{8.11}$$

**Associativity** of a convolution. Generally, the operation of convolution is not associative; for example,

$$(1 * \delta') * \theta = 1' * \theta = 0 \quad \theta = \theta$$

but

$$1 * (\delta' * \theta) = 1 * \delta = 1.$$

However, this unpleasantness does not arise if the convolutions $f * g * h$, $f * g$, $f * h$, $g * h$ exist. In this case other convolutions exist, and they are equal:

$$f * g * h = (f * g) * h = f * (g * h) = (f * h) * g.$$

**Some sufficient conditions of the existence of a convolution.**

(c) *Let* $f \in \mathcal{D}'_A$ *and* $g \in \mathcal{D}'_B$ *where* $A$ *and* $B$ *are closed sets such that the set* $T_R$ *is bounded in* $\mathbf{R}^{2n}$ *for any* $R > 0$. *Then the convolution* $f * g$ *exists in* $\mathcal{D}'_{\overline{A+B}}$ *and is represented by the form*

$$(f * g, \phi) = (f(x) \times g(y), \xi(x)\eta(y)\phi(x+y)), \quad \phi \in \mathcal{D} \tag{8.12}$$

*where* $\xi$ *and* $\eta$ *are any* $C^\infty$*-functions which are equal to* 1 *in* $A^\epsilon$ *and* $B^\epsilon$ *and are equal to* 0 *outside* $A^\delta$ *and* $B^\delta$, *respectively* ($\epsilon$ *and* $\delta$ *are any numbers,* $\delta > \epsilon > 0$). *In addition, the operation* $(f, g) \to f * g$ *is linear and continuous from* $\mathcal{D}'_A \times \mathcal{D}'_B$ *into* $\mathcal{D}'_{\overline{A+B}}$.

In particular, for any $f \in \mathcal{E}'$ and $g \in \mathcal{D}'$ the convolution $f * g$ exists in $\mathcal{D}'$ and the operation $f \to f * g$ is *linear* and continuous from $\mathcal{E}'$ into $\mathcal{D}'$.

(d) *If* $f \in \mathcal{D}'$ *and* $g \in \mathcal{D}$ *then* $f * g \in C^\infty$ *and is represented by the form*

$$(f * g)(x) = (f(y), g(x-y)) \tag{8.13}$$

(e) *Let a distribution* $g \in$ ' *be such that the operation*

$$\phi \to \hat{g} * \phi = (g(y), \phi(x+y))$$

*is linear and continuous from* $\mathcal{D}$ *into* $\mathcal{D}$ (for instance, if $g \in \mathcal{E}'$). Such a distribution $g$ is called a *convolutor.*

*If* $g$ *is a convolutor then for any* $f$ *from* $\mathcal{D}'$ *a convolution* $f * g$ *exists in* $\mathcal{D}'$ *and is represented by the form*

$$(f * g, \phi) = (f, \hat{g} * \phi), \quad \phi \in \mathcal{D} \tag{8.14}$$

*and the operation* $f \to f * g$ *is continuous (and linear) from* $\mathcal{D}'$ *into* $\mathcal{D}'$.

**Remark.** Formulae (8.14) and (4.1) show that the convolution opeation $f \to f * g$ is a conjugate operation to the linear continuous convolution operation $\phi \to \hat{g} * \phi$ in the test space $\mathcal{D}$. The definition of convolution given by formula (8.5) is an extension of the convolution operation. But by such an extension the convolution operation loses its continuity property in $\mathcal{D}'$ as the following example shows: $\delta(x-k) \to 0$, $k \to \infty$ in $\mathcal{D}'$ but $\delta(x-k) * 1 = 1$.

**Convolution with respect to some of the variables.** Let distributions $f(x, y)$ from $\mathcal{D}'(\mathbf{R}^{n+m})$ and $g \in \mathcal{D}'(\mathbf{R}^n)$ be given. Their convolution is defined by the equality

$$f(x, y) * g(x) = f(x, y) * [g(x) \times \delta(y)]. \tag{8.15}$$

According to definition (8.5), equality (8.15) takes the following form: for any $\phi \in \mathcal{D}(\mathbf{R}^{n+m})$

$$(f(x, y) * g(x), \phi) = \lim_{k \to \infty} (f(x, y) \times g(x'), \eta_k(x; x')\phi(x + x', y)) \tag{8.16}$$

if the limit in the right-hand side of (8.16) exists for any sequence $\{\eta_k, k \to \infty\}$ of functions from $\mathcal{D}(\mathbf{R}^{2n})$ which converges to 1 in $\mathbf{R}^{2n}$, and if this limit does not depend on the sequence $\{\eta_k, k \to \infty\}$. Then $f(x, y) * g(x) \in \mathcal{D}'(\mathbf{R}^{n+m})$.

The variables $y \in \mathbf{R}^m$ have the meaning of parameters in definition (8.16). For this reason all properties of the convolution given by formulae (8.6)–(8.11) and criteria of its existence remain valid.

The convolution operator $f \to f * g$ is linear and continuous from $\mathcal{E}'$ into $\mathcal{D}'$ for any $g \in \mathcal{D}'$ (see (c)) and translationally-invariant (see (8.8)). In fact, these properties are characteristic of the convolution operator. More precisely, the following theorem is valid.

**Theorem** *An operator $L$ is linear, continuous, and translationally-invariant from $\mathcal{E}'$ to $\mathcal{D}'$ if and only if it is a convolution operator; that is to say, it can be represented in the form $L = f_0*$, where $f_0$ is some distribution from $\mathcal{D}'$; then $f_0$, the kernel of the operator $L$, is unique and is expressed by the formula $f_0 = L\delta$.*

The proof of this theorem is based on the following lemma which is interesting for its own sake.

**Lemma.** *An operator $L$ is linear, continuous, and translationally-invariant from $\mathcal{D}$ into $\mathcal{E}$ if and only if it is the convolution operator $L = f_0*$, where $f_0 \in \mathcal{D}'$; here the kernel $f_0$ is unique.*

The convolution of distributions $f$ from $\mathcal{D}'(\mathcal{O})$ and $g \in \mathcal{E}'$, $\operatorname{supp} g \subset \bar{B}_\epsilon$ is defined in accordance with (8.14) by the formula

$$(f * g, \phi) = (f, \hat{g} * \phi), \quad \phi \in \mathcal{D}(\mathcal{O}_\epsilon). \tag{8.17}$$

Since the operation $\phi \to \hat{g} * \phi$ is linear and continuous from $\mathcal{D}'(\mathcal{O}_\epsilon)$ into $\mathcal{D}'(\mathcal{O})$, the operation $f \to f * g$ is *linear and continuous from $\mathcal{D}'(\mathcal{O})$ into $\mathcal{D}'(\mathcal{O}_\epsilon)$*.

Let $\omega_\epsilon(x)$ be the normed bell-shaped function (see Section 0.5) and $f \in \mathcal{D}'(\mathcal{O})$. The convolution $f_\epsilon = f * \omega_\epsilon$ is called a *regularization* of the distribution $f$.

The regularization $f_\epsilon \in C^\infty(\mathcal{O}_\epsilon)$, and

$$f_\epsilon \to f, \quad \epsilon \to +0 \quad \text{in } \mathcal{D}'(\mathcal{O}). \tag{8.18}$$

From (8.18) it follows that $\mathcal{D}(\mathcal{O})$ is *dense in* $\mathcal{D}'(\mathcal{O})$.

All results of this subsection are valid for the spaces $\mathcal{S}'$ and $\mathcal{S}$ without any essential changes; for example, in statement (d) we have $f * g \in \theta_M$ (about $\theta_M$ see Section 1.4).

The convolution equation has the form

$$a * u = f \tag{8.19}$$

where $a$ and $f$ are given distributions and $u$ is an unknown distribution. The *fundamental solution* of the convolution operator $a*$ is a distribution $\mathcal{E}$ that satisfies equation (8.19) for $f = \delta$,

$$a * \mathcal{E} = \delta. \tag{8.20}$$

*If a distribution $f$ is such that the convolutions $\mathcal{E} * f$ and $a * \mathcal{E} * f$ exist, then a solution of equation (8.19) can be evaluated by the formula $u = \mathcal{E} * f$.*

**Example.** For

$$a(x) = \sum_{|\alpha| \leq m} a_\alpha \partial^\alpha \delta(x)$$

equation (8.19) becomes a linear differential equation with constant coefficients.

9. **Convolution algebras of distributions.** At first we note the following

lemmas.

**Lemma 1.** *The following statements are equivalent:*

(i) *a cone* $\Gamma$ *is acute;*

(ii) *the cone* $\overline{\mathrm{ch}\,\Gamma}$ *does not contain a whole straight line;*

(iii) $\int \Gamma^* = C \neq \emptyset$. *In addition*

(iv) *for any* $C' \subset\subset C$ *there exists a number* $\sigma > 0$ *such that*

$$(\xi, x) \geqslant \sigma |\xi| \, |x|, \quad \xi \in C', \; x \in \overline{\mathrm{ch}\,\Gamma}; \tag{9.1}$$

(v) *for any* $e \in C$ *and* $a \geqslant 0$ *the set*

$$B_e = [x : 0 \leqslant (e, x) \leqslant a, \; x \in \overline{\mathrm{ch}\,\Gamma}] \tag{9.2}$$

*is convex and bounded in* $\mathbf{R}^n$ (Fig. 2).

**Lemma 2.** *Let* $\Gamma$ *be a convex cone. Then* $\Gamma = \Gamma + \Gamma$.

**Lemma 3.** *If* $S$ *is a* $C$*-like surface then* $\bar{S}_+ = S + \overline{\mathrm{ch}\,\Gamma}$ (Fig. 3).

**Lemma 4.** *Let* $S$ *be a* $C$*-like surface. Then for any* $R > 0$ *there is a number* $R' = R'(R) > 0$ *such that the set*

$$T_R = [(x, y) : x \in S, y \in \Gamma, |x + y| \leqslant R]$$

*is contained in the ball* $B_{R'} \subset \mathbf{R}^{2n}$.

We shall say that a $C$-like surface $S$ is a *strictly $C$-like surface* iff, under the conditions of lemma 4,

$$R'(R) \leqslant a(1 + R)^{\nu}, \quad \nu \geqslant 1, \; a > 0.$$

Let $\Gamma$ be a closed convex acute cone in $\mathbf{R}^n$ (with its vertex at 0). We shall say that a set $A$ is *bounded on the side of the cone* $\Gamma$ if there exists a compact set $K$ such that $A \subset \Gamma + K$ (Fig. 6).

We denote by $\mathcal{D}'_{\Gamma+}$ (resp. $\mathcal{S}'_{\Gamma+}$) the set of distributions from $\mathcal{D}'$ (resp. from $\mathcal{S}'$) whose supports are bounded on the side of the cone $\Gamma$. It is clear that $\mathcal{D}'_{\Gamma} \subset \mathcal{D}'_{\Gamma+}$, $\mathcal{S}'_{\Gamma} \subset \mathcal{S}'_{\Gamma+}$, $\mathcal{D}'_{\{0\}+} = \mathcal{E}'$.

The convergence in $\mathcal{D}'_{\Gamma+}$ (resp. in $\mathcal{S}'_{\Gamma+}$) means the following: $f_k \to 0$, $k \to \infty$ in $\mathcal{D}'_{\Gamma+}$ (resp. in $\mathcal{S}'_{\Gamma+}$) if there exists a compact set $K$ such that $\mathrm{supp}\, f_k \subset \Gamma + K$ and $f_k \to 0$, $k \to \infty$ in $\mathcal{D}'$ (resp. in $\mathcal{S}'$).

From lemma 4 and the criterion (c), Section 1.8, we have

**Theorem.** *If* $S$ *is a* $C$*-like surface then for any* $f \in \mathcal{D}'_{\Gamma+}$ *and* $g \in \mathcal{D}'_{S+}$ *the convolution* $f * g$ *exists in* $\mathcal{D}'$ *and can be represented in the form* (8.12); *in addition, if* $\mathrm{supp}\, f \subset \Gamma + K$, *where* $K$ *is the compact, then* $\mathrm{supp}(f * g) \subset \bar{S}_+ + K$ *and the operation* $(f, g) \to f * g$ *is continuous (and linear) from* $\mathcal{D}'_{\Gamma+K} \times \mathcal{D}'_{S_+}$ *into* ${}_{S_+ + K}$.

From this theorem in particular it follows that *if* $\Gamma$ *is a closed convex acute cone then for any* $f$ *and* $g$ *from* $\mathcal{D}'_{\Gamma+}$ *the convolution* $f * g$ *exists in* $\mathcal{D}'_{\Gamma+}$ *and can be represented in the form* (8.12); *in addition, the operation* $(f, g) \to f * g$ *is continuous (and linear) from* $\mathcal{D}'_{\Gamma+} \times \mathcal{D}'_{\Gamma+}$ *into* $\mathcal{D}'_{\Gamma+}$.

In a similar way we can prove that the convolution of any number of distributions from $\mathcal{D}'_{\Gamma+}$ exists in $\mathcal{D}'_{\Gamma+}$ and this can be represented by a formula similar to (8.12). From this it follows that the convolution operation of distributions from

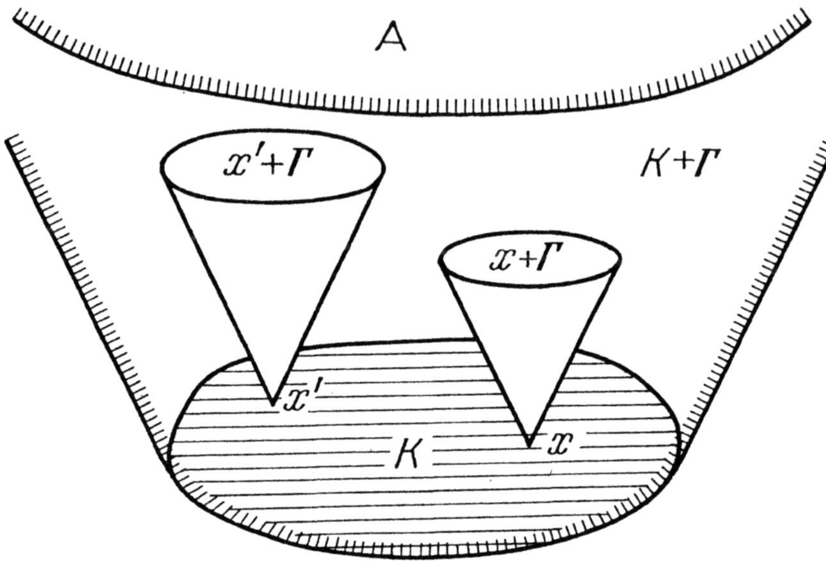


Figure 6.

$\mathcal{D}'_{\Gamma+}$ is *associative* (and commutative, see (8.6), Section 1.8).

Thus, the set of distributions $\mathcal{D}'_{\Gamma+}$ forms a topological *algebra*, associative and commutative, with the unit element if for the operation of multiplication we take the convolution operation $*$. Such algebras are called *convolution algebras;* the unit element here is the $\delta$-function (see (8.7), Section 1.8).

Note that the set of distributions $\mathcal{D}_\Gamma$ also forms a convolution algebra, a subalgebra of the algebra $\mathcal{D}'_{\Gamma+}$.

The results listed above can be extended without essential changes on the spaces of tempered distributions $\mathcal{S}'_{\Gamma+}$ and $\mathcal{S}'_\Gamma$; the only difference is that, in the conditions of the theorem, it is necessary to assume that $S$ is a strictly C-like surface. In particular, $\mathcal{S}'_{\Gamma+}$ and $\mathcal{S}'_\Gamma$ are convolution algebras and $\mathcal{S}'_\Gamma$ is a subalgebra of the algebras $\mathcal{S}'_{\Gamma+}$ and $\mathcal{D}'_\Gamma$, and $\mathcal{S}'_{\Gamma+}$ is a subalgebra of the algebra $\mathcal{D}'_{\Gamma+}$. We note also that $\mathcal{E}'$ and $\mathcal{L}^1$ form convolution algebras (see Section 1.8 and Section 1.9).

Let $\Gamma$ be a closed acute solid cone so that $\Gamma = \overline{\text{int}\,\Gamma}$, int $\Gamma$ is a convex (and hence also regular) domain. We shall define the convolution operation of distributions $f$ and $g$ from $\mathcal{S}'(\Gamma)$ by the formula

$$(f * g, \phi) = (f(x) \times g(y), \phi(x+y)), \quad \phi \in \mathcal{S}(\Gamma). \tag{9.3}$$

Then in accordance with formula (8.12), and the fact that the spaces $\mathcal{S}'_\Gamma$ and $\mathcal{S}'(\Gamma)$ are isomorphic as stated in Section 1.3, the space $\mathcal{S}'(\Gamma)$ becomes a convolution algebra isomorphic to the convolution algebra $\mathcal{S}'_\Gamma$.

**Equations in convolution algebras.** Let $A$ be a convolution algebra and let $a \in A$, $f \in A$. Let us consider equation (8.19), $a * u = f$, in algebra $A$. The fundamental solution (see Section 1.8) of the operator $a^*$ will be conveniently denoted as $a^{-1}$, so that

$$a^{-1} * a = \delta. \tag{9.4}$$

The following useful proposition is valid:

**Proposition.** If $a_1^{-1}$ and $a_2^{-1}$ exists in $A$ then $(a_1 * a_2)^{-1}$ exists in $A$ and the equality

$$(a_1 * a_2)^{-1} = a_1^{-1} * a_2^{-1}. \tag{9.5}$$

holds.

**Fractional differentiation and integration.** Let $f_\alpha$, $-\infty < \alpha < \infty$, be the family of distributions from $\mathcal{D}'_+$ which is defined by formulae (6.8). It forms an abelian group

$$f_\alpha * f_\beta = f_{\alpha+\beta}, \quad -\infty < \alpha, \ \beta < \infty. \tag{9.6}$$

Let us consider the convolution operator $f \to f_\alpha * f$ in the algebra $\mathcal{D}'_+$ (resp. in $\mathcal{S}'_+$). Since $f_0 = \theta' = \delta$, it follows from (9.6) that the fundamental solution $f_\alpha^{-1}$ of the operator $f_\alpha *$ exists and is equal to $f_{-\alpha}$: $f_\alpha^{-1} = f_{-\alpha}$. Furthermore, for integers $n < 0$, $f_n = f^{(-n)}$, and for integers $n > 0$, $f_n^{(n)} = \delta$. For this reason the operator $f_\alpha *$ is called the operator of *fractional differentiation* of order $|\alpha|$ for $\alpha < 0$, and the operator of *fractional integration* of order $\alpha$ for $\alpha > 0$ (it is also called the *Riemann-Liouville operator*) and, respectively, the distribution $f_\alpha$ is called the *canonical kernel of the operator of franctional differentiation and integration*. We shall write $f^{(-\alpha)} = f_\alpha * f$.

Similarly, for distributions from $\mathcal{D}'_+(n)$ (or $\mathcal{S}'_+(n)$ resp.) (see Section 1.3) we introduce the operations of franctional differentiation or integration over the variable $x_0$ as the convolution $f_\alpha(x_0)$ with $f(x_0, x)$ on $x_0$ (see Section 1.8); that is,

$$f^{(-\alpha)} = f_\alpha(x_0) * f(x_0, \mathbf{x}) \in \mathcal{D}'_+(n) \ \ (\text{or } \mathcal{S}'_+(n))$$

if $f \in \mathcal{D}'_+(n)$ (or $\mathcal{S}'_+(n)$ resp.).

## 2. Integral transformations of distributions

In this section we shall present the theories of Fourier transform, $B$-transform, Radon transform, Laplace transform, and other, for tempered distributions.

**1. The Fourier transform of tempered distributions.** Let $F$ be the operation of the Fourier transform of test functions from $\mathcal{S}$:

$$\phi \to F[\phi] = \int \phi(x) e^{i(x,\xi)} dx, \quad \phi \in \mathcal{S}.$$

*This operation is a linear and continuous isomorphism $\mathcal{S}$ onto $\mathcal{S}$; its inverse is given by the classical inversion formula for Fourier transform*

$$\psi \to F^{-1}[\psi] = \frac{1}{(2\pi)^n} \int \psi(\xi) e^{-i(x,\xi)} d\xi, \quad \psi \in \mathcal{S}.$$

In accordance to formula (4.1) of Section 1.4, the Fourier transform of a distribution $f$ from $\mathcal{S}'$ is defined by the equality

$$(F[f],\phi) = (f, F[\phi]),\ \phi \in \mathcal{S}. \tag{1.1}$$

The operation $f \to F[f]$ is a *linear and continuous isomorphism from* $\mathcal{S}'$ *onto* $\mathcal{S}'$; its inverse operation $F^{-1}[g]$ is expressed by the formula

$$F^{-1}[g] = \frac{1}{(2\pi)^n} F[\hat{g}],\ g \in \mathcal{S}'. \tag{1.2}$$

**Example 1.** $F[\delta] = 1$, $F[1] = (2\pi)^n \delta$;
**Example 2.**

$$F[e^{i\alpha x^2}] = \sqrt{\frac{\pi}{|\alpha|}} \exp\frac{i}{4}\left\{\pi\epsilon(\alpha) - \frac{\xi^2}{4\alpha}\right\},\ \alpha \neq 0; \tag{1.3}$$

**Example 3.** If $f \in$ ' then $F[f] \in \theta_M$ and

$$F[f](\xi) = (f(x), e^{i(x,\xi)}).$$

**Example 4.**

$$F[f_\alpha] = [-i(x+i0)]^{-\alpha} = \begin{cases} \Gamma(1-\alpha)\left[e^{i(\pi/2)\alpha} f_{-\alpha+1}(x) + e^{-i(\pi/2)\alpha} f_{-\alpha+1}(-x)\right], & \alpha \neq 1, 2, \ldots; \\ \dfrac{e^{-i(\pi/2)\alpha}}{\Gamma(\alpha)}\left[i\pi\delta^{(\alpha-1)}(x) - \left(\mathfrak{P}\dfrac{1}{x}\right)^{(\alpha-1)}\right], & \alpha = 1, 2, \ldots. \end{cases} \tag{1.4}$$

where $f_\alpha$ is defined by (6.10) of Section 1.6, and the distribution $(x+i0)^{-\alpha}$ is defined by the formula

$$(x+i0)^{-\alpha} = \lim_{\epsilon\to+0} \frac{1}{(x+i\epsilon)^\alpha} \quad \text{in } \mathcal{S}'.$$

The existence of this limit, and formula (1.4), follow from the limit representation of $f_\alpha$,

$$f_\alpha(t)e^{-\epsilon t} \to f_\alpha(t),\ \epsilon \to +0 \quad \text{in } \mathcal{S}'_+,$$

from the equality

$$F\left[f_\alpha(t)e^{-\epsilon t}\right] = [-i(x+i\epsilon)]^{-\alpha},\ \epsilon > 0$$

and from the Sochozki formula

$$\frac{1}{x+i0} = -i\pi\delta(x) + \mathfrak{P}\frac{1}{x}. \tag{1.5}$$

Here the distribution $\mathfrak{P}1/x$ is defined by the formula

$$(\mathfrak{P}\frac{1}{x}, \phi) = Vp\int_{-\infty}^{\infty} \frac{\phi(x)}{x}\,dx,\ \phi \in \mathcal{S}.$$

**The following properties of the Fourier transform are valid:**

$$\partial^\alpha F[f] = F[(ix)^\alpha f], \tag{1.6}$$

$$F[\partial^\alpha f] = (-i\xi)^\alpha F[f]. \tag{1.7}$$

$$F[f(x-x_0)] = e^{i(\xi,x_0)}F[f]. \tag{1.8}$$

$$F\left[e^{i(\xi_0,x)}f\right](\xi) = F[f](\xi+\xi_0). \tag{1.9}$$

$$F[f(Ax)](\xi) = \frac{1}{|\det A|}F[f](A^{-1T}\xi), \ \det A \neq 0, \tag{1.10}$$

$$F[f(x)\times g(y)] = F[f](\xi)\times F[g](\eta). \tag{1.11}$$

$$F[f * g] = F[f]F[g], \ g \in \mathcal{E}'. \tag{1.12}$$

**Fourier transform with respect to some variables.** Let $f(x,y)$ be a tempered distribution from $\mathcal{S}'(\mathbf{R}^{n+m})$ where $x \in \mathbf{R}^n$ and $y \in \mathbf{R}^m$. We introduce the Fourier transform $F_x[f]$ with respect to $x$ according to formula (1.1) by the equality

$$(F_x[f],\phi) = (f, F_\xi[\phi]), \ \phi \in \mathcal{S}(\mathbf{R}^{n+m}). \tag{1.13}$$

where

$$F_\xi[\phi](x,y) = \int \phi(\xi,y)e^{i(x,\xi)}d\xi.$$

The inverse transform $F_\xi^{-1}[g]$ is given by the equality

$$F_\xi^{-1}[g] = \frac{1}{(2\pi)^n}F_\xi[\hat{g}_\xi], \ \hat{g}_\xi = g(-\xi,y). \tag{1.14}$$

Formulae similar to (1.6)-(1.12) are also valid for the Fourier transform $F_x$.

**Positive definite distributions.** Suppose $f \in \mathcal{D}'(\mathbf{R}^n)$; the distribution $f^*(x) = \bar{f}(-x)$ in $\mathcal{D}'(\mathbf{R}^n)$ is said to be the *$*$-Hermite conjugate* of $f$; if $f = f^*$ then $f$ is said to be the *$*$-Hermite distribution*.

A distribution $f \in \mathcal{D}'(\mathbf{R}^n)$ is said to be *positive definite* if it satisfies the condition

$$(f, \phi * \phi^*) \geqslant 0, \ \phi \in \mathcal{D}. \tag{1.15}$$

The property of positive definiteness of a distribution $f$ will be denoted by $f >> 0$.

**The Bochner-Schwartz theorem.** *A distribution from $\mathcal{D}'$ is positive definite if and only if it is the Fourier transform of a nonnegative tempered measure.*

From this follows the **Bochner theorem.** *If a function $f$ is continuous in $\mathbf{R}^n$ then the following conditions are equivalent:*

(i) $f >> 0$;

(ii) $\sum_{1\leqslant j,k\leqslant l} f(x_j - x_k)z_j\bar{z}_k \geqslant 0$ *for any* $x_1, \ldots, x_l$ *from* $\mathbf{R}^n$ *and* $z_1, \ldots, z_n$ *from* $\mathbf{C}^1$;

(iii) $f(x) = \int e^{i(x,\xi)}d\mu(\xi), \ d\mu \geqslant 0, \int d\mu(x) = f(0)$.

**Example 5.** $\delta >> 0$.

**2. Fourier series of periodic distributions.** A distribution $f$ from $\mathcal{D}'(\mathbf{R}^n)$ is called *periodic with an $n$-period* $T = (T_1, T_2, \ldots, T_n)$, $T_j > 0$, $j = 1, \ldots, n$, if it is periodic with respect to each argument $x_j$ with the period $T_j$; that is,

$$f(x+kT) = f(x), \ x \in \mathbf{R}^n, \ k \in \mathbf{Z}^n \tag{2.1}$$

where $kT = (k_1T_1, \ldots, k_nT_n)$. We use $\mathcal{D}'_T$ to denote the set of all periodic disributions with the $n$-period $T$.

For every $n$-period $T$ there exists a nonnegative function $e_T$ from $\mathcal{D}$ such that

$$e_T(x) = 1, \quad |x_j| < \frac{1}{4}T_j, \quad j = 1, \ldots, n;$$

$$e_T(x) = 0, \quad |x_j| > \frac{3}{4}T_j, \quad \text{for some } j = 1, \ldots, n$$

and the following decomposition of the unity in $\mathbf{R}^n$ holds

$$\sum_{k \in \mathbf{Z}^n} e_T(x + kT) = 1 \tag{2.2}$$

such that every $f \in \mathcal{D}_T$ can be represented in the form

$$f = (e_T f) * \delta_T \tag{2.3}$$

where

$$\delta_T(x) = \sum_{k \in \mathbf{Z}^n} \delta(x - kT). \tag{2.4}$$

From (2.3) it follows in particular that $\mathcal{D}'_T \subset \mathcal{S}'$.

Let $f \in \mathcal{D}'_T$ and $\phi \in C^\infty \bigcap \mathcal{D}'_T$. We define their scalar product $(f, \phi)_T$ by the formula

$$(f, \phi)_T = (f, e_T \phi) \tag{2.5}$$

and the right-hand side does not depend on the auxiliary function $e_T(x)$ with the properties stated above.

Denoting $\omega = (2\pi/T_1, \ldots, 2\pi/T_n)$ we obtain from (2.5) the equality

$$\left(e^{i(k\omega, x)}, e^{-i(k'\omega, x)}\right)_T = \frac{(2\pi)^n}{\omega_1 \cdots \omega_n}\delta_{kk'} = T_1 \cdots T_n \delta_{kk'}. \tag{2.6}$$

**Theorem 1.** *For any distribution $f$ in $\mathcal{D}'_T$ the following Fourier-series expansions are valid:*

$$\partial^\alpha f(x) = \sum_{k \in \mathbf{Z}^n} (ik\omega)^\alpha C_k(f) e^{i(k\omega, x)},$$

$$C_k(f) = \frac{\omega_1 \cdots \omega_n}{(2\pi)^n}\left(f, e^{-i(k\omega, x)}\right)_T \tag{2.7}$$

*where the Fourier series converge in $\mathcal{S}'$ and for any $\phi \in C^\infty \bigcap \mathcal{D}'_T$ the following generalized Parseval-Steklov equality holds true:*

$$(f, \phi)_T = \sum_{k \in \mathbf{Z}^n} C_k(f) C_k(\phi). \tag{2.8}$$

**Example 1.**

$$\delta_T(x) = \frac{\omega_1 \cdots \omega_n}{(2\pi)^n} \sum_{k \in \mathbf{Z}} e^{-i(k\omega, x)}. \tag{2.9}$$

We introduce, on the set $\mathcal{D}'_T$, the convolution operation $\circledast$ by the formula

$$f \circledast g = (e_T f) \circledast g = (e_T g) \circledast f = g \circledast f, \quad f, g \in \mathcal{D}'_T \tag{2.10}$$

since the right-hand side of (2.10) does not depend on the auxiliary function $e_T$. The set $\mathcal{D}'_T$ forms the convolution algebra; it sunity is $\delta_T$ (cf. Section 1.9). In addition,

$$C_k(f \circledast g) = \frac{(2\pi)^n}{\omega_1 \cdots \omega_n} C_k(f) C_k(g), \quad f, g \in \mathcal{D}'_T. \tag{2.11}$$

**Example 2.**

$$f \circledast e^{i(k\omega, x)} = \frac{(2\pi)^n}{\omega_1 \cdots \omega_n} C_k(f) e^{i(k\omega, x)}. \tag{2.12}$$

**Theorem 2.** *The convolution operator $a*$, $a \in \mathcal{D}'_T$, has a inverse $a^{-1\circledast}$ in the algebra $\mathcal{D}'_T$ if and only if for certain $L > 0$ and $m$ the inequality*

$$|C_k(a)| \geq L(1 + |k|)^{-m}, \quad k \in \mathbf{Z}^n. \tag{2.13}$$

*holds. Here, the fundamental solution $a^{-1}$ is unique and its Fourier expansion is*

$$a^{-1}(x) = \frac{1}{T_1^2 \cdots T_n^2} \sum_{k \in \mathbf{Z}} \frac{1}{C_k(a)} e^{i(k\omega, x)}. \tag{2.14}$$

**Example 3.**

$$(\delta_T^1 - \lambda\delta_T)^{-1} = \frac{\omega}{2\pi} \sum_{k=-\infty}^{\infty} \frac{1}{ik\omega - \lambda} e^{ik\omega x}, \quad \lambda \neq ik\omega, \quad k = 0, \pm 1, \cdots \tag{2.15}$$

3. **The $B$-transform of distributions.** For functions $\phi$ from $\mathcal{S}_+$ (see Section 1.1) we introduce the modified Fourier-Bessel transform $b[\phi]$, *$b$-transform*, by the formula

$$b[\phi](t) = -\int_0^\infty \phi'(\tau) J_0(\sqrt{t\tau})\, d\tau = \phi(0) + \frac{1}{2}\int_0^\infty \sqrt{\frac{t}{\tau}} J_0'(\sqrt{t\tau}) \phi(\tau)\, d\tau, \tag{3.1}$$

where $J_0$ is the Bessel function.

We recall that the Fourier-Bessel transform of order 0 and its inverse transform are expressed by the formulae

$$H[f](\rho) = \int_0^\infty f(r) J_0(r\rho) r\, dr = \frac{1}{2\pi} F[f(|x|)], \quad x \in \mathbf{R}^2, \tag{3.2}$$

$$f = H[H[f]]. \tag{3.3}$$

Owing to (3.1) and (3.3) the $b$-transform is the composition of the following transformations

$$\phi(\tau) \to \phi'(\tau) \to \phi'(|x|^2) \to \frac{1}{\pi} F[\phi'(|x|^2)](\xi)\Big|_{|\xi|^2 = t} \to b[\phi](t)$$

which are the linear and continuous transformations $\mathcal{S}_+ \to \mathcal{S}_+ \to \mathcal{S}_{SO(2)} \to \mathcal{S}_{SO(2)} \to \mathcal{S}_+$ (see Section 1.1 and Section 1.6). Hence, the *$b$-transform is linear and continuous from $\mathcal{S}_+$ into $\mathcal{S}_+$.*

**Example 1.**

$$b\left[e^{i\tau z}\right] = e^{-(it)/(4z)}, \quad \mathrm{Im}\, z > 0. \tag{3.4}$$

In fact, this formula follows from (3.1) and (3.2)

$$b\left[e^{i\tau z}\right](t) = -iz\int_0^\infty e^{i\tau z} J_0(\sqrt{t\tau})\, d\tau = -2iz\int_0^\infty e^{i\rho^2 z} J_0(\sqrt{t}\,\rho)\rho\, d\rho =$$

$$= -\frac{iz}{\pi} F\left[e^{i|x|^2 z}\right](\sqrt{|}\xi|) = e^{-(it)/(4z)}, \quad t = |\xi|.$$

Here we have used formula (6.6) from V.S. Vladimirov (1979, p. 11).

Now we are going to establish an inversion formula for the $b$-transform:

$$b[b[\phi]] = \phi, \quad \phi \in \mathcal{S}_+. \tag{3.5}$$

**Proof.** By differentiating the equality (3.1) using the equation for the Bessel function $J_0(x)$,

$$J_0''(x) + \frac{1}{x}J_0'(x) + J_0(x) = 0,$$

and formula (3.2) (for $t = \rho^2$, $\tau = r^2$) we obtain

$$b[\phi](t) = -\frac{1}{4}\int_0^\infty \phi(\tau)J_0(\sqrt{t\tau})\, d\tau = -\frac{1}{2}H[\phi(\tau^2)](\rho).$$

From this, owing to (3.3) formula (3.5) follows:

$$b[b[\phi]] = \frac{1}{2}\int_0^\infty H[\phi(r^2)](\sqrt{t}\,)J_0(\sqrt{t\tau})\, dt =$$

$$= \frac{1}{2}\int_0^\infty H[\phi(r^2)](\rho)J_0(r\rho)\rho\, d\rho = H[H[\phi(r^2)]](\rho) = \phi(\rho^2) = \phi(t).$$

The inversion formula (3.5) shows that the *B-transform is an isomorphism from $\mathcal{S}_+$ onto $\mathcal{S}_+$.*

This result enables us to introduce, on distributions $f$ from $\mathcal{S}_+'$, the conjugate operation $b^* = B$ in accordiance with formula (4.1) of Section 1.4:

$$(B[f], \phi) = (f, b[\phi]), \quad \phi \in \mathcal{S}_+. \tag{3.6}$$

The *$B$-transform provides the isomorphism from $\mathcal{S}_+'$ onto $\mathcal{S}_+'$ and the inversion formula valid* (cf. (3.5)),

$$B[B[f]] = f, \ f \in \mathcal{S}'_+. \tag{3.7}$$

**Remark.** On functions $f \in \mathcal{L}^1(0, \infty)$ the $B$-transform takes the form

$$B[f](t) = \frac{d}{dt}\left[\theta(t)\int_0^\infty J_0(\sqrt{t\tau})f(\tau)\,d\tau\right] =$$

$$= \delta(t)\int_0^\infty f(\tau)\,d\tau + \frac{1}{2}\theta(t)\int_0^\infty -\sqrt{\frac{\tau}{t}}J_0(\sqrt{t\tau})f(\tau)\,d\tau.$$

**Example 2.** $B[\delta] = \delta$.

**Example 3.**

$$B\left[\theta(\tau)e^{i\tau z}\right] = \frac{i}{z}\delta(t) + \frac{\theta(t)}{4z}e^{-(it)/(4z)}, \ \operatorname{Im} z > 0. \tag{3.8}$$

**Proof.** Formula (3.8) follows from fromula (3.4):

$$\left(B_\tau\left[\theta(\tau)e^{i\tau z}\right], \phi\right) = \left(\theta(\tau)e^{i\tau z}, b[\phi]\right) =$$

$$= -\int_0^\infty e^{i\tau z}\int_0^\infty \phi'(t)J_0(\sqrt{t\tau})\,dt\,d\tau =$$

$$= \frac{1}{iz}\int_0^\infty \phi'(t)b[e^{i\tau z}](t)\,dt =$$

$$= \frac{1}{iz}\int_0^\infty \phi'(t)e^{-(it)/(4z)}\,dt =$$

$$= \frac{i}{z}\phi(0) + \frac{1}{4z^2}\int_0^\infty \phi(t)e^{-(it)/(4z)}\,dt, \ \phi \in \mathcal{S}_+.$$

**Properties of the $B$-transform.**

$$(a)\ B[f(k\tau)](t) = \frac{1}{k^2}B[f]\left(\frac{t}{k}\right), \ f \in \mathcal{S}'_+, \ k > 0. \tag{3.9}$$

**Proof.** It follows from the definitions and from the equality

$$b[\phi(kt)] = b[\phi]\left(\frac{\tau}{k}\right), \ k > 0 \tag{3.10}$$

$$(b)\ B[f] * B[g] = B[f * g], \ f, g \in \mathcal{S}'_+. \tag{3.11}$$

**Proof.** In fact, using formula (9.3) of Section 1.9 for all $\phi \in \mathcal{S}_+$ we have

$$(B[f * g], \phi) = (f * g, b[\phi]) = (f(t) \times g(\tau), b[\phi](t + \tau)). \tag{3.12}$$

We verify the equality

$$b[\phi](t + \tau) = b_\xi\left[b_\eta[\phi(\xi + \eta)]\right], \ \phi \in \mathcal{S}_+. \tag{3.13}$$

as

$$b_\xi\left[b_\eta[\phi(\xi+\eta)]\right] = \int_0^\infty\int_0^\infty \phi''(\xi+\eta)J_0(\sqrt{t\xi})J_0(\sqrt{\tau\eta})\,d\xi\,d\eta =$$

$$= 2\int_0^\infty z\phi''(z)\int_0^{\pi/2} J_0(\sqrt{tz}\,\cos\theta)J_0(\sqrt{\tau z}\,\sin\theta)\sin\theta\cos\theta\,d\theta\,dz =$$

$$= 2\int_0^\infty z\phi''(z)\frac{J_1(\sqrt{z(t+\tau)})}{\sqrt{z(t+\tau)}}\,dz =$$

$$= -2\int_0^\infty \phi'(z)\frac{\partial}{\partial z}\left[\sqrt{\frac{z}{t+\tau}}J_1(\sqrt{z(t+\tau)})\right]dz =$$

$$= -\int_0^\infty \phi'(z)J_0(\sqrt{z(t+\tau)})\,dz = b[\phi](t+\tau).$$

Here we have changed the order of integration, made a change in the variables of integration

$$\xi = 2z\cos^2\theta,\ \eta = 2z\sin^2\theta,\ \frac{\partial(\xi,\eta)}{\partial(z,\theta)} = 2z\sin\theta\cos\theta,$$

and we have used formula 6.683.2 of I.S. Gradshtein and I.M. Ryzhik (1965) and the equality $J_1 = -J_0'$.

Using formula (3.13), we continue the equalities (3.12):

$$(B[f*g],\phi) = \left(f(t)\times g(\tau), b_\xi\left[b_\eta[\phi(\xi+\eta)]\right]\right) =$$

$$= (B[f](\xi)\times B[g](\eta), \phi(\xi+\eta)) = (B[f]*B[g],\phi). \qquad \square$$

$$\text{(c) } B[f_\alpha] = 4^\alpha f_{-\alpha},\ -\infty<\alpha<\infty, \tag{3.14}$$

where the distribution $f_\alpha$ has been defined in formula (6.10) of section 1.6.

**Proof.** By virtue of (3.9) $B[f_\alpha]$ is a homogeneous distribution from $\mathcal{S}'_+$ of degree $-\alpha-1$ and, therefore (see Section 1.6),

$$B[f_\alpha] = C_\alpha f_{-\alpha},\ C_\alpha \neq 0.$$

Further, using formulae (3.11), (3.14), and (9.6) of Section 1.9 we have

$$B\left[f_\alpha * f_\beta\right] = B\left[f_{\alpha+\beta}\right] = C_{\alpha+\beta}f_{-\alpha-\beta} = B\left[f_\alpha\right]*B\left[f_\beta\right] =$$

$$= \left(C_\alpha C_\beta\left(f_{-\alpha}*f_{-\beta}\right)\right) = C_\alpha C_\beta f_{-\alpha-\beta}$$

from where it follows that

$$C_{\alpha+\beta} = C_\alpha C_\beta.$$

This means that $C_\alpha = a^\alpha$, $a\neq 0$. To determine $a$ we put $\alpha = 1$. As the operator $B$ is continuous in $\mathcal{S}'_+$, from (3.8) we have

$$B[\theta] = \lim_{\epsilon \to +0} \left[ \frac{1}{\epsilon}\delta(t) - \frac{1}{4\epsilon^2} e^{-(t)/(4\epsilon)} \right] = 4\delta'(t)$$

in $\mathcal{S}'_+$ so that $B[f_1] = 4f_{-1}$. Therefore, $C_1 = a = 4$, and formula (3.14) has been proved.

**The Fourier transform of odd $L_+^\uparrow$-invariant tempered distributions.** Every such a distribution $f$ has the form (see Section 1.6)

$$f(x) = \epsilon(x_0) f_0(x^2), \quad f_0 \in \mathcal{S}'_+.$$

Its Fourier transform $F[f]$ is also an odd $L_+^\uparrow$-invariant distribution from $\mathcal{S}'(\mathbf{R}^4)$ and thus it has the form

$$F[f](\xi) = \epsilon(\xi_0) \ (\xi^2), \quad \in \mathcal{S}'_+ \tag{3.15}$$

where

$$(t) = 4\pi^2 i B[f_0](t). \tag{3.16}$$

Formula (3.16) follows from the more general formulae (1.11) and (1.12) of Section 13.1 if we put

$$\tilde{F}(x) = \tilde{F}_a(x) = \epsilon(x_0) f_0(x^2), \quad \phi_a(k, \mathbf{x}) = f_0(k)$$

and then, owing to (1.11) and (1.12) of Section 13.1,

$$\psi_a(\lambda, \mathbf{u}) = \frac{1}{4\pi^2 i} B[f_0](\lambda) \times \delta(\mathbf{u}),$$

$$F(q) = F_a(q) = \frac{1}{4\pi^2 i} B[f_0](q^2).$$

**Example 4.**

$$F\left[\epsilon(x_0) f_\alpha(x^2)\right] = 4^{\alpha+1}\pi^2 i \ \epsilon(x_0) f_{-\alpha}(\xi^2). \tag{3.17}$$

This follows from formulae (3.14)-(3.16).

**Example 5.**

$$F\left[\epsilon(x_0)\delta^{(n)}(x^2)\right] = \begin{cases} 4\pi^2 i \, \epsilon(\xi_0)\delta(\xi^2), & n = 0, \\ \dfrac{4^{1-n}}{(n-1)!}\pi^2 i \, \epsilon(\xi_0) \left[\xi_+^2\right]^{n-1}, & n = 1, 2, \cdots \end{cases}$$

This follows from formula (3.17) if we note that $\delta^{(n)}(\tau) = f_{-n}(\tau)$.

**4. Fractional derivatives (primitives).** In Section 1.9 we introduced notion of fractional derivatives (primitive) for distributions from the algebra $\mathcal{D}'_+$. Here we extend this notion onto other classes of distributions.

Let $f_\alpha(t)$ be the canonical kernel of the operator of fractional differentiation for $\alpha < 0$ and fractional integration for $\alpha > 0$ which is defined by formula (6.8) of Section 1.6. The right $f^{(\alpha)}$ and left ${}^{(\alpha)}f$ (fractional) derivatives (primitives) of order $\alpha$ are defined by the following formulae (cf. (8.9), Section 1.8):

$$f^{(\alpha)} = f * f_\alpha, \quad {}^{(\alpha)}f = f * \check{f}_{-\alpha} = (\check{f} * f_{-\alpha}) = (\check{f}^{(\alpha)}). \tag{4.1}$$

where $\check{f}(t)=f(-t)$.

**Example 1.** If $f \in \mathcal{L}^1(\mathbf{R}^1)$ then

$$f^{(-1)}(t)=\int_{-\infty}^{t} f(\tau)\,d\tau, \quad {}^{(-1)}f(t)=\int_{t}^{\infty} f(\tau)\,d\tau.$$

**Example 2.** If $f \in C^n(\mathbf{R}^1)$ then

$${}^{(n)}f(t)=(-1)^n f^{(n)}(t).$$

We shall use the following criterion of existence and some properties of the operations introduced.

**Lemma.** *Let $\phi \in C^\infty(\mathbf{R}^1)$ satisfy the conditions*

$$\int_{-\infty}^{\infty} \left|\phi^{(j)}(t)\right| |t|^k\,dt < \infty, \quad k, j = 0, 1, \ldots, \tag{4.2}$$

$$\int_{-\infty}^{\infty} \phi(t) t^k\,dt = 0, \quad k = 0, 1, \ldots N, \quad N \geqslant 1. \tag{4.3}$$

*Further, let $\alpha$ and $\beta$ be real numbers which satisfy the conditions $M = N-2+[\alpha]+[\beta] \geqslant 0$, $N+[\alpha] \geqslant 1$ and $N+[\beta] \geqslant 1$. Then the following statements are valid*

$${}^{(\beta)}\left(\phi^{(\alpha)}\right) = \left({}^{(\beta)}\phi\right)^{(\alpha)} = {}^{(\beta)}\phi^{(\alpha)}; \tag{4.4}$$

$$\int_{-\infty}^{\infty} \left|{}^{(\beta)}\phi^{(\alpha+j)}(t)\right| |t|^k\,dt < \infty, \quad k = 0, 1, \ldots M, \quad j = 0, 1, \ldots; \tag{4.5}$$

$$\int_{-\infty}^{\infty} {}^{(\beta)}\phi^{(\alpha)}(t) t^k\,dt = 0, \quad k = 0, 1, \ldots M. \tag{4.6}$$

**Proof.** To prove these statements it is sufficient to apply a Fourier transform and to use formula (1.4) of Section 2.1

$$F\left[{}^{(\beta)}\phi^{(\alpha)}\right](x) = F[\phi](x)(x+i0)^\alpha(-i)^\alpha i^\beta (x-i0)^\beta,$$

$$F[\phi]^{(k)}(0) = 0, \quad k = 1, 2, \ldots N.$$

**5. The Laplace transform of tempered distributions.** Here we expose the theory of the Laplace transform of distributions from the algebras $\mathcal{S}'_{\Gamma+}$ and $\mathcal{S}'_{\Gamma}$, where $\Gamma$ is a closed convex acute cone with its vertex at 0 (see Section 1.9); we put $C = \operatorname{int}\Gamma^*$ and $T^C = \mathbf{R}^n + iC$ (see Section 0.2).

Let $g \in \mathcal{S}'_{\Gamma+}$. We denote the *Laplace transform* of the distribution $g$ by the expression

$$L[g] = F\left[g(\xi)e^{-(y,\xi)}\right](x) \tag{5.1}$$

where $F$ is the Fourier transform (see Section 2.1). The right-hand side of (5.1) can be rewritten in the form

$$L[g] = \left(g(\xi), e^{i(z,\xi)}\right), \quad z = x+iy. \tag{5.2}$$

Since the function $e^{i(z,\xi)}$ belongs to $\mathcal{S}(\Gamma + K)$ for every $z \in T^C$ and each compact set $K \subset \mathbf{R}^n$, then the right-hand side of (5.2) makes sense (see Section 1.3) and defines the functions $f(z) = L[g]$ which is holomorphic in the tube domain $T^C$ and is tempered in $x$ for every $y \in C$. In addition, the following inversion formula for the Laplace transform is valid

$$g(\xi) = L^{-1}[f] = e^{(y,\xi)}F_x^{-1}[f(x+iy)](\xi) \tag{5.3}$$

and the right-hand side in (5.3) does not depend on $y \in C$.

The distribution $g(\xi)$ is called the *spectral function* of $f(z)$, and the function $f(z)$ is called the *image* of $g$.

**Example 1.** $L[\delta(\xi - \xi_0)] = e^{i(z,\xi_0)}$.

**Example 2.** $L[f_\alpha] = (-iz)^{-\alpha}$, $y > 0$, where $f_\alpha$ is defined in (6.8) of Section 1.6 (cf. (1.5), Section 2.1).

**Properties of the Laplace transform.**

$$\partial^\alpha L[g] = L\Big[(i\xi)^\alpha g\Big]. \tag{5.4}$$

$$L\Big[\partial^\alpha g\Big] = (-iz)^\alpha L[g]. \tag{5.5}$$

$$L\Big[g(\xi)e^{i(a,\xi)}\Big] = L[g](z+a). \tag{5.6}$$

$$L\Big[g(\xi-\xi_0)\Big] = e^{i(z,\xi_0)}L[g]. \tag{5.7}$$

$$L[g(A\xi)] = \frac{1}{|\det A|}L[g]\Big(A^{-1T}z\Big),\quad z \in T^{A^T C}. \tag{5.8}$$

$$L\Big[g * g_1\Big] = L[g]L[g_1],\quad g, g_1 \in \mathcal{S}'_{\Gamma+}. \tag{5.9}$$

If $g_1 \in \mathcal{S}'_{\Gamma_1+}$ and $g_2 \in \mathcal{S}'_{\Gamma_2+}$ then (denoting $C_i = \operatorname{int}\Gamma_2^*$)

$$L[g_1 \times g_2](z,\zeta) = L[g_1](z)L[g_2](\zeta),\quad z \in TC_1,\ \zeta \in TC_2. \tag{5.10}$$

We denote by $H_a^{(\alpha,\beta)}(C)$, $\alpha \geqslant 0$, $\beta \geqslant 0$, $a \geqslant 0$, the set of all functions $f(z)$ which are holomorphic in $T^C$ and which satisfy the following growth condition:

$$|f(z)| \leqslant M_f e^{a|y|}(1+|z|^2)^{\alpha/2}\Big[1+\Delta_C^{-\beta}(y)\Big],\quad z \in T^C \tag{5.11}$$

where $\Delta_C(y)$ is the distance from the point $y \in C$ to $\partial C$ (see Section 0.2)

$$\Delta_C(y) = \Delta(y, \partial C) = \min_{\sigma \in \operatorname{pr}\Gamma}(y, \sigma). \tag{5.12}$$

We introduce the topology in $H_a^{(\alpha,\beta)}(C)$ in accordance with estimate (5.11) by means of the norm

$$\|f\|_a^{(\alpha,\beta)} = \sup_{z \in T^C} \frac{|f(z)|e^{-a|y|}}{(1+|z|^2)^{\alpha/2}[1+\Delta_C^{-\beta}(y)]},\quad f \in \mathbf{H}_a^{(\alpha,\beta)}.$$

The spaces $H_a^{(\alpha,\beta)}(C)$ are Banach spaces and Montel spaces and

$$H_a^{(\alpha,\beta)}(C) \subset H_{a'}^{(\alpha',\beta')}(C),\quad \alpha' \geqslant \alpha,\ \beta' \geqslant \beta,\ a' \geqslant a$$

by virtue of the obvious inequality

$$\| f \|_a^{(\alpha',\beta')} \leqslant 2 \| f \|_a^{(\alpha,\beta)}.$$

We denote by $H_+(C)$ the inductive limit (union) on an incresing sequence of Banach spaces $H_a^{(\alpha,\beta)}(C)$ when $\alpha \to \infty$, $\beta \to \infty$, $a \to \infty$; we denote by $\mathbf{H}(C)$ the inductive limit of Banach spaces $H_0^{(\alpha,\beta)}(C)$ when $\alpha \to \infty$, $\beta \to \infty$. Convergence in $H_+(C)$ (resp. in $H(C)$) means convergence with respect to one of the norms $\| f \|_a^{(\alpha,\beta)}$ (resp. $\| \ \|_0^{(\alpha,\beta)}$). The spaces $H_+(C)$ and $\mathbf{H}(C)$ are *DFS*-spaces. They form algebras with respect to the operation of ordinary multiplication which are associative, commutative, with the unit element, and without zero divisors; $H(C)$ is a subalgebra of the algebra $H_+(C)$.

If $\Gamma$ is a closed convex acute and solid cone, then the following isomorphisms exist

$$\mathcal{S}'_{\Gamma+} \sim H_+(C), \quad \mathcal{S}_\Gamma \sim H(C),$$

and these isomorphisms are given by the Laplace transform $L$ and by its inverse $L^{-1}$.

As a **corollary** we have the result: *every $f(z)$ from $H_+(C)$ has a boundary value $f_+(x)$ from $\mathcal{S}'$,*

$$f(x+iy) \to f_+(x), \ y \to 0, \ y \in C' \subset\subset C \ \text{in} \ \mathcal{S}'$$

*and $f_+ = F[g]$, where $g$ is the spectral function of $f$, and the operation $f \to f_+$ is continuous from $H_+(C)$ into $\mathcal{S}'$ (and, resp., from $H(C)$ into $\mathcal{S}'$).*

In the case when $\Gamma = \{0\}$, $C = \mathbf{R}^n$ the set of distributions $\mathcal{S}'_{\{0\}+} = \mathcal{E}'$ forms a convolution algebra (see Section 1.9) which is isomorphic to the algebra $H_+(\mathbf{R}^n)$. More precisely, the following **Paley-Wiener-Schwartz theorem** is valid: *a function $f(z)$ is entire and satisfies the growth condition* (5.11) *for $\Delta_C(y) = \infty$,*

$$|f(z)| \leqslant M_f \, e^{a|y|}(1+|z|^2)^{\alpha/2}, \ z \in C^n \tag{5.13}$$

*if and only if its spectral function $g$ belongs to $\mathcal{E}'_{\overline{B}_a}$.*

**Corollary.** *A function $f(z)$ is entire and satisfies the growth condition*

$$|f(z)| \leqslant K_N e^{a|y|}(1+|z|^2)^{-N}, \ z \in C^n, \ N = 0, 1, \cdots \tag{5.14}$$

*if and only if its spectral function $g$ belongs to $\mathcal{D}(\overline{B}_a)$; in addition,*

$$K_N = \int_{|\xi|<a} \left|(1-\Delta)^N g(\xi)\right| d\xi. \tag{5.15}$$

6. **The Cauchy kernel of the tube domain $T^C$.** Let $C$ be an open (convex) acute cone in $\mathbf{R}^n$ (with vertex at 0) and $\Gamma = C^*$ its conjugate cone; $\Gamma$ is a closed convex acute solid cone in $\mathbf{R}^n$ (see Section 1.9). The function

$$\mathcal{K}_C(z) = L\left[\theta_\Gamma\right] = \int_\Gamma e^{i(z,\xi)} d\xi \tag{6.1}$$

is called the *Cauchy kernel* of the tube domain $T^C = \mathbf{R}^n + iC$; here $\theta_\Gamma(\xi)$ is the characteristic function of the cone $\Gamma$ (see section 0.2).

The Cauchy kernel $\,_C(z)$ belongs to the algebra $H(C)$ and is represented by the integral

$$\,_C = i^n \Gamma(n) \int\limits_{\mathrm{pr}\,\Gamma} \frac{d\sigma}{(z,\sigma)^n}, \quad z \in T^C \tag{6.2}$$

where $d\sigma$ is the Lebesgue measure on the sphere $S^{n-1}$ (see Section 0.5). From the representation (6.2) it follows that $\mathcal{K}_C(z)$ is holomorphic in the domain

$$D = \mathbf{C}^n \setminus \bigcup_{\sigma \in \mathrm{pr}\,\Gamma} [z : (z,\sigma) = 0].$$

and satisfies the equations

$$\begin{aligned} \mathcal{K}_{-C}(z) &= (-1)^n \mathcal{K}_C(z) = \\ &= \overline{\mathcal{K}_C(\bar{z})} = \\ &= \mathcal{K}_C(-z), \quad z \in T^C \bigcup T^{-C}. \end{aligned} \tag{6.3}$$

$$\mathcal{K}_C(\lambda z) = \lambda^{-n} \mathcal{K}_C(z), \quad \lambda \in \mathbf{C}^1 \setminus \{0\}$$

and the estimates

$$\left|\partial^\alpha \mathcal{K}_C(z)\right| \leqslant \mathrm{M}_\alpha \left[|x|^2 + \Delta_C^2(y)\right]^{-1/2} \Delta_C^{1-n-|\alpha|}(y),$$
$$z \in T^C \bigcup T^{-C}, \quad \alpha \in \mathbf{Z}_+^n. \tag{6.4}$$

**Proof.** It is sufficient to prove estimate (6.4) for $\alpha = 0$, $n \geqslant 3$ and $z \in T^C$. Using (6.1) and (5.12) we have, for $z \in T^C$,

$$\begin{aligned} \left|\mathcal{K}_C(z)\right| &\leqslant \Gamma(n) \int\limits_{\mathrm{pr}\,\Gamma} \frac{d\sigma}{[(x,\sigma)^2 + (y,\sigma)^2]^{n/2}} \leqslant \\ &\leqslant \Gamma(n) \Delta_C^{3-n}(y) \int\limits_{\mathrm{pr}\,\Gamma} \frac{d\sigma}{[(x,\sigma)^2 + \Delta_C^2(y)]^{3/2}} \leqslant \\ &\leqslant M_0 \Delta_C^{3-n}(y) \int\limits_0^{\pi/2} \frac{\sin\theta\, d\theta}{[|x|^2 \cos^2\theta + \Delta_C^2(y)]^{3/2}} = \\ &= \frac{M_0 \Delta_C^{1-n}(y)}{\sqrt{|x|^2 + \Delta_C^2(y)}}. \end{aligned}$$

Let $U$ be a linear nonsingular transformation of the space $\mathbf{R}^n$ which leaves the cone $\Gamma$ invariant (automorphism of the cone $\Gamma$),

$$U\Gamma = \Gamma. \tag{6.5}$$

Denote

$$J = \det U, \quad V = (U^T)^{-1}. \tag{6.6}$$

The following formula holds

$$\mathcal{K}_C(Vz) = J\mathcal{K}_C(z). \tag{6.7}$$

**Proof.** In fact,

$$\mathcal{K}_C(Vz) = \int\limits_\Gamma e^{i(Vz,\xi')} d\xi' = \int\limits_\Gamma e^{i(z,U^{-1}\xi')} d\xi' =$$

$$= J\int_{\Gamma} e^{i(z,\xi)}\,d\xi = J\mathcal{K}_C(z).$$

A convex cone $\Gamma$ is called *homogeneous* if for any two points $\xi'$ and $\xi''$ from $\operatorname{int}\Gamma$ there exists a nonsingular, linear transformation $U$ of the space $\mathbf{R}^n$ (depending on the points $\xi'$ and $\xi''$) which leaves the cone $\Gamma$ invariant and such that $U\xi' = \xi''$.

**Example 1.** The positive octant $\mathbf{R}_+^n$ is a homogeneous cone and

$$\mathcal{K}_{\mathbf{R}_+^n}(z) = \frac{i^n}{z_1 \cdots z_n}.$$

**Example 2.** The future light cone $V_n^+$ is a homogeneous cone and

$$\mathcal{K}_{V_n^+}(z) = 2^n \pi^{(n-1)/2}\Gamma\left(\frac{n+1}{2}\right)(-z^2)^{-(n+1)/2}$$

where $z^2 = z_0^2 - z_1^2 - \cdots - z_n^2$ (see Section 0.2).

7. **Regular cones.** An open (convex) acute cone $C$ is called *regular* if its Cauchy kernel $\mathcal{K}_C(z)$ is a divisor of the unity in the algebra $H(C)$; that is $1/\mathcal{K}_C \in H(C)$; in this case the closed convex solid cone $C^*$ will be also called *regular*. For a cone $C$ to be regular it is necessary that $\mathcal{K}_C(z) \neq 0$ in $T^C$. As the following example shows,

$$e^{-(i/z)} \in H(0,\infty) \quad \text{but} \quad e^{i/z} \notin H(0,\infty),$$

from the conditions $f \in H(C)$ and $f(z) \neq 0$ in $T^C$ it does not follow that $1/f \in H(C)$. From the results of Section 2.5 we have the following

**Theorem.** *For a function $f \in H(C)$ to be a divisor of the unity in the algebra $H(C)$, it is necessary and sufficient that there exist numbers $m > 0$, $\alpha \geqslant 0$ and $\beta \geqslant 0$ such that*

$$|f(z)| \geqslant m(1+|z|^2)^{-(\alpha/z)}\Delta_C^{\beta}(y),\quad z \in T^C. \tag{7.1}$$

From this theorem it follows that if $\mathcal{K}_C(z)$ is a rational function which does not vanish in $T^C$ then $C$ is a regular cone (N.N. Boglyubov and V.S. Vladimirov, 1973). In particular, the cones $\mathbf{R}_+^n$ and $V_n^+$ are regular; for $n = 1$ and $n = 2$ all acute cones are regular.

If a cone $C$ is self-conjugate, $\bar{C} = C^*$, and homogeneous then $\mathcal{K}_C^2(z)$ is a rational function which does not vanish in $T^C$ (O.S. Rothaus, 1960) and thus $C$ is a regular cone.

*For $n = 3$ all open (convex) acute cones are regular; for $n \geqslant 4$ there exist non-regular open (convex) acute cones* (L.I. Danilov, 1985).

**Proof.** Let be $n \geqslant 4$. We construct a nonregular cone:

$$C = \left[y : y_1 > \pm a y_2 \pm \frac{1}{2}y_3 \pm \cdots \pm \frac{1}{2}y_n\right],\quad a > 0,$$

$$\Gamma = C^* = \left[\xi : |\xi_2| \leqslant a\xi_1,\ |\xi_j| \leqslant \frac{1}{2}\xi_1,\ j = 3,4,\ldots,n\right].$$

Putting $z_0 = (i, 1, 0, \ldots, 0)$ we have

$$\mathcal{K}_C(z_0) = \int_\Gamma e^{i(z_0,\xi)}\,d\xi = \int_0^\infty e^{-\xi_1}\int_{-a\xi_1}^{a\xi_1} e^{i\xi_2}\int_{-(\xi_1/2)}^{(\xi_1/2)}\cdots\int_{-(\xi_1/2)}^{\xi_1/2} d\xi_1\,d\xi_2\cdots d\xi_n =$$

$$= 2\int_0^\infty \xi_1^{n-2}e^{-\xi_1}\,\mathrm{Im}\,e^{ia\xi_1}\,d\xi_1 = 2\,\mathrm{Im}\int_0^\infty \xi_1^{n-2}e^{-\xi_1(1-ia)}\,d\xi_1 =$$

$$= \frac{2\Gamma(n-1)}{(1+a^2)^{n-1}}\mathrm{Im}(1+ia)^{n-1} = \frac{2\Gamma(n-2)}{(1+a^2)^{(n-1)/2}}\sin\left[(n-1)\,\mathrm{arctg}\,a\right].$$

The last expression vanishes if $a = \mathrm{tg}(\pi/n-1)$.

Let $n = 3$, and let $C$ be an open (convex) acute cone in $\mathbf{R}^3$; let $z = x + iy$ be in $T^C$. We suppose at first that the vectors $0x$ and $0y$ are not parallel. Without loss of generality we can choose a system of coordinates $\xi = (\xi_1, \xi_2, \xi_3)$ such that $y = (0, 0, |y|)$, $|y| > 0$, and $x = (x_1, 0, x_3)$, $x_1 > 0$. In this case

$$\mathcal{K}_C(z) = \int_\Gamma e^{ix_1\xi_1 + i(x_3+i|y|)\xi_3}\,d\xi_1\,d\xi_2\,d\xi_3.$$

Since $\xi_3 \geqslant 0$ for $\xi \in \Gamma$ and $(0,0,1) \in \int\Gamma$ then the intersection $\mathbf{D}$ of the cone $\Gamma$ with the plane $\xi_3 = 1$ is a closed, bounded, convex and solid domain in $\mathbf{R}^2$ (see Lemma 1, Section 1.9). Making the change of variables of integration,

$$\xi_1 = \alpha_1\alpha_2,\ \ \xi_2 = \alpha_2\alpha_3,\ \ \xi_3 = \alpha_3,\ \ d\xi = \alpha_3^2\,d\alpha$$

we get

$$\mathcal{K}_C(z) = \int_0^\infty\int_D e^{i\alpha_3(x_1\alpha_1+x_3+i|y|)}\alpha_3^2\,d\alpha = \tag{7.2}$$

$$= \int_D\int_0^\infty e^{i\alpha_3(x_1\alpha_1+x_3+i|y|)}\alpha_3^2\,d\alpha_3\,d\alpha_1\,d\alpha_2 =$$

$$= -2i\int_D \frac{d\alpha_1 d\alpha_2}{(x_1\alpha_1+x_3+i|y|)^3}.$$

We denote by $f(\alpha_1)$ the length of the intersection of the domain $D$ with the line $\alpha_1 = \text{const}$, and by $[a, b]$ the projection of $D$ onto the axis $\alpha_1$. Then the function $f(\alpha_1)$ is concave and nonnegative on the segment $[a, b]$, and is positive and continuous in the interval $(a, b)$. Therefore the integral (7.2) takes the form

$$\mathcal{K}_C(z) = -2i\int_a^b \frac{f(\alpha_1)\,d\alpha_1}{(x_1\alpha_1+x_3+i|y|)^3}.$$

To estimate this integral we use estimate (7.5) from the lemma cited below

$$|\mathcal{K}_C(z)| = \frac{2}{x_1^3}\left|\int_a^b \frac{f(\alpha_1)\,d\alpha_1}{\left[\alpha_1 + \dfrac{x_3+i|y|}{x_1}\right]^3}\right| \geqslant \frac{|y|\,m(D)}{2\left[x_3^2+|y|^2+C^2x_1^2\right]} \tag{7.3}$$

where $C = \max[|a|, |b|]$, $m(\mathbf{D}) = \int_a^b f(\alpha_1)\,d\alpha_1$ is the area of $\mathbf{D}$. Taking into account the estimate $(\xi, y) \geqslant |\xi| \Delta_C(y)$, $\xi \in \Gamma$, $y \in C$, for $\xi = \left|\xi_1, \xi_2, 1\right|$ and $y = (0, 0, |y|)$ we obtain

$$|\xi| \leqslant \frac{|y|}{\Delta_C(y)} \geqslant 1$$

so that

$$C \leqslant \frac{|y|}{\Delta_C(y)}.$$

With the help of this estimate we proceed with inequality (7.3)

$$\left|K_C(z)\right| \geqslant \frac{\Delta_C^4(y) m(\mathrm{D})}{2|y|^3 |z|^4} \geqslant \frac{\Delta_C^4(y) m(\mathrm{D})}{2|z|^7}. \tag{7.4}$$

Estimate (7.4) can be extended by a limiting process to the case when the vectors $0x$ and $0y$ are parallel. Further, we notice that the areas of intersections of the cone $\Gamma$ (convex solid with vertex at 0) with various planes which are on distance 1 from the origin are bounded from below by a positive number $\sigma(C)$; that is,

$$\inf_{z \in T^C} m(\mathrm{D}) = \sigma(C) > 0.$$

Then it follows from (7.4) that

$$\left|K_C(z)\right| \geqslant \frac{\sigma(C)\Delta_C^4(y)}{2|z|^7}, \quad z \in T^C,$$

which shows that inequality (7.1) is fulfilled for $\alpha = 7$ and $\beta = 4$, and thus, by virtue of the theorem, $K_C(z)$ is a divisor of the unity in the algebra $H(C)$, so that $\mathbf{C}$ is a regular cone.

**Lemma.** *Let $f(x)$ be a concave nonnegative function on $[a, b]$. Then for* $\mathrm{Im}\, z > 0$ *the following estimate is valid*

$$\left|\int_a^b \frac{f(x)\,dx}{(x+z)^3}\right| \geqslant \frac{\mathrm{Im}\, z}{4(|z|^2 + C^2)} \int_a^b f(x)\,dx, \tag{7.5}$$

*where* $C = \max(|a|, |b|)$.

**Proof.** Since any concave nonnegative function $f(x)$ on $[a, b]$ can be approximated by a sequence $\{f_k(x), k \to 0\}$ of functions $f_k$ which are polygonal, concave, nonnegative, vanish at the points $a$ and $b$, and for any $\phi \in C([a, b])$

$$\int_a^b f_k(x)\phi(x)\,dx \to \int_a^b f(x)\phi(x)\,dx, \quad k \to \infty$$

it is sufficient to prove the lemma for such polygonal functions $f$.

Let $x_i$, $i = 1, 2, \ldots, N$, be the breaking points of a function $f$. We construct functions

$$f_i(x) = \begin{cases} A_i \dfrac{x-a}{x_i - a}, & 0 \leqslant x \leqslant x_i, \\ A_i \dfrac{b-x}{b-x_i}, & x_i \leqslant x \leqslant b, \quad i = i, 2, \ldots, \mathrm{N} \end{cases} \tag{7.6}$$

where the $A_i$ are chose so that

$$A_i\left[\frac{1}{b-x_i}+\frac{1}{x_i-a}\right]=f'(x_i-0)-f'(x_i+0). \tag{7.7}$$

Thus

$$f(x)=\sum_{i=1}^{N}f_i(x). \tag{7.8}$$

Since $f$ is a concave function, and $-f''(x)$ is a nonnegative measure on $(a.b)$, then $f'(x_i-0)-f'(x_i+0)>0$, and hence, by virtue of (7.7), $A_i>0$ and $f_i(x)\geqslant 0$.

From (7.6) and (7.8) we can represent the integral (7.5) in the form

$$\int_a^b\frac{f(x)\,dx}{(x+z)^3}=\sum_{i=1}^{N}\int_a^b\frac{f_i(x)\,dx}{(x+z)^3}=\frac{1}{(a+z)(z+b)}\sum_{i=1}^{N}\frac{1}{x_i+z}\int_a^b f_i(x)\,dx. \tag{7.9}$$

Taking into account the inequalities

$$|(a+z)(z+b)|\leqslant(|z|+C)^2\leqslant 2(|z|^2+C^2),\quad |x_i+z|^2\leqslant 2(|z|^2+C^2),$$

$$-\mathrm{Im}\frac{1}{x_i+z}=\frac{\mathrm{Im}\,z}{|x_i+z|^2}\geqslant\frac{\mathrm{Im}\,z}{2(|z|^2+C^2)}$$

we estimate (7.9) from

$$\left|\int_a^b\frac{f(x)\,dx}{(x+z)^3}\right|\geqslant\frac{-1}{|(z+z)(z+b)|}\sum_{i=1}^{N}\mathrm{Im}\frac{1}{x_i+z}\int_a^b f_i(x)\,dx\geqslant$$

$$\geqslant\frac{\mathrm{Im}\,z}{4(|z|^2+C^2)^2}\sum_{i=1}^{N}\int_a^b f_i(x)\,dx=\frac{\mathrm{Im}\,z}{4(|z|^2+C^2)^2}\int_a^b f(x)\,dx.$$

**8. Fractional derivatives (primitives) with respect to a cone.** Let $C$ be a regular cone. Since $\mathcal{K}_C(z)$ is a divisor of the unity in the algebra $H(C)$, then for all real $\alpha$ the function $\mathcal{K}_C^\alpha(z)$ also belongs to the algebra $H(C)$. We denote by $\theta_\Gamma^\alpha(\xi)$ its spectral function (see Section 2.5), $\theta_\Gamma^\alpha\in\ {}'_\Gamma$ where $\Gamma=C^*$ so that, by definition,

$$L\left[\theta_\Gamma^\alpha\right]=\mathcal{K}_C^\alpha(z),\ z\in T^C,\ -\infty<\alpha<\infty.$$

Since

$$\mathcal{K}_C^\alpha(z)\mathcal{K}_C^\beta(z)=\mathcal{K}_C^{\alpha+\beta}(z)$$

then the family of distributions $\{\theta_\Gamma^\alpha, -\infty<\alpha<\infty\}$ forms the abelian group

$$\theta_\Gamma^\alpha*\theta_\Gamma^\beta=\theta_\Gamma^{\alpha+\beta}. \tag{8.1}$$

From (8.1), in particular, it follows that $\theta_\Gamma^0=\delta$, $\theta_\Gamma^1=\theta_\Gamma$, the fundamental solution of the operator $\theta_\Gamma^\alpha*$ in the algebra ${}'_\Gamma$ is equal to $\theta_\Gamma^{-\alpha}$, so that $\theta_\Gamma^\alpha*\theta_\Gamma^{-\alpha}=\delta$.

**Example 1.**

$$\theta_{\mathbb{R}_+^n}^\alpha(\xi)=f_\alpha(\xi_1)\times\cdots\times f_\alpha(\xi_n)$$

where $f_\alpha$ has been introduced in (6.8), Section 1.6.

**Example 2.**

$$\theta^{\alpha}_{V_n^+}(\xi) = 2^{n\alpha}\pi^{(n-1)\alpha/2}\Gamma^{\alpha}\left(\frac{n+1}{2}\right)\Box_n^{-(n+1)\alpha/2}\delta(\xi).$$

Fractional powers of the operator $\Box_n$ by other means have been introduced by M. Riesz (1949).

Let $f \in \mathcal{S}'_\Gamma$. The distribution

$$f^{(-\alpha)} = \theta^{\alpha}_{\Gamma} * f$$

is called the *primitive* of $f$ of order $\alpha > 0$ (the *derivative* of order $|\alpha|$ if $\alpha < 0$) with respect to the cone $\Gamma$ (cf. Section 1.9). It is clear that $f^{(-\alpha)} \in \mathcal{S}'_\Gamma$ and $(f^{(-\alpha)})^{(-\beta)} = f^{(-\alpha-\beta)}$.

*If $f \in \mathcal{S}'_\Gamma$, then there exists a real number $\alpha_0$, depending only on the order of $f$, such that $f^{(-\alpha)}(\xi)$ is a continuous tempered function with support in $\Gamma$, together with all its derivatives up to the order $k$ for all $\alpha \geqslant \alpha_0 + k$.*

**Proof.** This statement follows from the equality (cf. (5.3))

$$\partial^{\beta} f^{(-\alpha)}(\xi) = e^{(y,\xi)}F^{-1}\left[i^{\beta}(x+iy)^{\beta}L[f](x+iy)\mathcal{K}^{\alpha}_{C}(x+iy)\right],$$

$$y \in C, \quad |\beta| \leqslant k$$

and from the estimates (5.11) and (6.4), according to which, for a sufficiently large $\alpha$, the function

$$(x+iy)^{\beta}L[f](x+iy)\mathcal{K}^{\alpha}_{C}(x+iy)$$

belongs to $\mathcal{L}^1 \bigcap \mathcal{L}^2$ and satisfies the estimate

$$\left\|(x+iy)^{\beta}L[f](x+iy)\mathcal{K}^{\alpha}_{C}(x+iy)\right\| \leqslant M_f\left[1 + \Delta_C^{-\gamma}(y)\right], \quad y \in C$$

for some $\gamma > 0$.

**Corollary.** *For any $p \geqslant 0$ there exists $\alpha_0$ such that for all $\alpha \geqslant \alpha_0$ the following statements are valid:*

(i) *For any $\xi_0 \in \mathbf{R}^n$, $\theta^{\alpha}_{\Gamma}(\xi_0 - \xi) \in \mathcal{S}_p(\Gamma)$ and $\theta^{\alpha}_{\Gamma}(\xi_0 - \xi)$ depends continuously on $\xi_0$ in the topology of $\mathcal{S}_p(\Gamma)$;*

(ii) *For any compact $K \subset \mathbf{R}^n$ the set $\{\theta^{\alpha}_{\Gamma}(\xi_0 - \xi), \xi_0 \in K\}$ is bounded in $\mathcal{S}_p(\Gamma)$.*

**Proof.** To prove these statements it is sufficient to note that

$$\theta^{\alpha}_{\Gamma}(\xi_0 - \xi) = \delta^{(-\alpha)}(\xi_0 - \xi)$$

and to use the previous statement.

*If $U$ is a linear nonsingular transformation which leaves the cone $\Gamma$ invariant* (see (6.5) and (6.6)), *then*

$$\theta^{\alpha}_{\Gamma}(U\xi) = J^{\alpha-1}\theta^{\alpha}_{\Gamma}(\xi). \tag{8.2}$$

**Proof.** In fact, using (6.7) we have

$$\left(\theta^{\alpha}_{\Gamma}(U\xi), e^{i(z,\xi)}\right) = \frac{1}{J}\left(\theta^{\alpha}_{\Gamma}(\xi), e^{i(Vz,\xi)}\right) =$$

$$= \frac{1}{J}\mathcal{K}_C^\alpha(Vz) =$$

$$= J^{\alpha-1}\mathcal{K}_C^\alpha(z) = J^{\alpha-1}\left(\theta_\Gamma^\alpha(\xi), e^{i(z,\xi)}\right).$$

and formula (8.2) follows from the fact that the set of functions $\{e^{i(z,\xi)}, z \in T^C\}$ is dense in the space $\mathcal{S}(\Gamma)$.

**9. The Radon transform of distributions with compact support in an odd-dimensional space.** The theory of the Radon transform of tempered distributions has been presented in I.M. Gel'fand, M.I. Graev and N.Ya. Vilenkin (1966) and S. Helgason (1980). Here we expose this theory for distributions with compact support in an odd-dimensional space $\mathbf{R}^n$ (see also V.S. Vladimirov and B.I. Zavialov, 1979).

Let $\phi \in \mathcal{S}(\mathbf{R}^n)$. The function

$$\check{\phi}(\lambda, \sigma) = \int_{(\sigma, x)=\lambda} \phi(x)\, dS = \int \phi(x)\delta(\lambda - (\sigma, x))\, dx, \quad \lambda \in \mathbf{R}^1, \quad \sigma \in S^{n-1}, \tag{9.1}$$

is called the *Radon transform* of the function $\phi$. This definition is equivalent to:

$$\check{\phi}(\lambda, \sigma) = F_\nu^{-1}[F[\phi](\nu\sigma)]. \tag{9.2}$$

From this definition it follows that the function $\check{\phi}(\lambda, \sigma)$ is infinitely differentiable on $\mathbf{R}^1 \times S^{n-1}$ and fast decreasing in $\lambda$. The set of such functions $\{\psi\}$ is denoted by $\mathcal{S}(\mathbf{R}^1 \times S^{n-1})$ (cf. Section 1.1). The topology in $\mathcal{S}(\mathbf{R}^1 \times S^{n-1})$ is introduced by means of an increasing sequence of norms

$$\|\psi\|_{p, \mathbf{R}^1 \times S^{n-1}}, \quad p = 0, 1, \cdots$$

(cf. (1.1), Section 1.1).

*The mapping $\phi \to \check{\phi}$ is linear and continuous from $\mathcal{S}(\mathbf{R}^n)$ into $\mathcal{S}(\mathbf{R}^1 \times S^{n-1})$, and the following inversion formula holds*

$$\phi(x) = \frac{2^{n-1}}{2(2\pi)^{n-1}} \int_{S^{n-1}} \check{\phi}_\lambda^{(n-1)}((\sigma, x), \sigma)\, d\sigma, \quad \phi \in \mathcal{S}. \tag{9.3}$$

Therefore the *mapping $\phi \to \check{\phi}$ is one-to-one but not onto.* The image of that mapping can be specified by the following theorem.

**Theorem.** *A function $\psi(\lambda, \sigma)$ from $\mathcal{S}(\mathbf{R}^1 \times S^{n-1})$ is the Radon transform of a function $\phi$ from $\mathcal{S}(\mathbf{R}^n)$, if and only if it satisfies the following conditions:*

(i) $\psi(\lambda, \sigma) = \psi(-\lambda, -\sigma)$,

(ii) $\int_{-\infty}^{\infty} \psi(\lambda, \sigma)\lambda^k\, d\lambda = \sum_{|\alpha|=k} a_\alpha \sigma^\alpha$, $k = 0, 1, \cdots$.

In accordinace with definition (9.1), the *Radon transform* of a distribution $f$ from $\mathcal{E}'(\mathbf{R}^n)$ is a linear continuous functional $\check{f}$ defined on $\phi \in \mathcal{S}(\mathbf{R}^1 \times S^{n-1})$ by the formula

$$(\check{f}, \phi) = \int_{S^{n-1}} (f, \phi((\sigma, x), \sigma))\, d\sigma = \left(f, \int_{S^{n-1}} \phi((\sigma, x), \sigma)\, d\sigma\right). \tag{9.4}$$

From this definition it follows that on functions of the form $\phi_\lambda^{(n-1)}(\lambda,\sigma)$, $\phi \in \mathcal{S}(\mathbf{R}^n)$, by virtue of the inversion formula (9.3), the following formula, similar to the Plancherel formula, holds

$$(f,\phi) = \frac{i^{n-1}}{2(2\pi)^{n-1}}\left(\check{f},\phi_\lambda^{(n-1)}\right), \quad \phi \in \mathcal{S}(\mathbf{R}^n). \tag{9.5}$$

Formula (9.4) selects a unique (finite) Radon transform from the set of various Radon transforms of distribution $f \in \mathcal{E}'(\mathbf{R}^n)$ if, as an initial definition of that transform, we take the Plancherel formula (9.5).

**Properties of the Radon transform.**

(a) $\check{f}$ belongs to $\mathcal{S}'(\mathbf{R}^1 \times S^{n-1})$, and is finite in $\lambda$ and $C^\infty(S^{n-1})$ on $\sigma$ (cf. Section 1.2), and $\operatorname{supp}\check{f} \subset [(\lambda,\sigma)]: |\lambda| \leqslant a, \sigma \in S^{n-1}]$ if $\operatorname{supp} f \subset \bar{B}_a$, and

$$(\check{f}(\lambda,\sigma),\phi(\lambda)) = (f(x),\phi((\sigma,x))), \quad \phi \in \mathcal{S}(\mathbf{R}^1). \tag{9.6}$$

(b) $\check{f}(\lambda,\sigma) = \check{f}(-\lambda,-\sigma)$.

(c) $IF_\lambda[\check{f}(\lambda,\sigma)] = f_1(\xi)$ admits an analytic extension in $\mathbf{C}^n$ as an entire function satisfying the following growth condition

$$|f_1(\xi + i\eta)| \leqslant Me^{a|\eta|}(1 + |\xi|^\alpha),$$

where $I$ is the operation which every even function $\psi(\lambda,\sigma)$ defined on $\mathbf{R}^1 \times S^{n-1}$ transforms into the function $\psi(|\xi|,\xi/|\xi|)$ defined on $\mathbf{R}^n \backslash \{0\}$.

$$\text{(d) } \check{f}(\lambda,\sigma) = F_\nu^{-1}[F[f](\nu\sigma)] \tag{9.7}$$

where, by (1.4),

$$F[f](\nu\sigma) = \left(f(x), e^{i\nu(x,\sigma)}\right). \tag{9.8}$$

$$\text{(e) } \{f(Ax)\}^\vee = \check{f}(\lambda, A\sigma), \quad A \in SO(n). \tag{9.9}$$

$$\text{(f) } \{f(x+b)\}^\vee = \check{f}(\lambda + (b,\sigma),\sigma). \tag{9.10}$$

$$\text{(g) } \{\partial^\alpha f(x)\}^\vee = \sigma^\alpha \check{f}_\lambda^{(|\alpha|)}(\lambda,\sigma). \tag{9.11}$$

$$\text{(h) } \{f_1 * f_2\}^\vee = (\check{f}_1(\cdot,\sigma) * \check{f}_2(\cdot,\sigma))(\lambda,\sigma), \quad f_i \in \mathcal{E}'. \tag{9.12}$$

**Proof.** Properties (a), (b), (e)-(h) easily follow from the definitions. Properties (c) and (d) follow from the formula

$$F[f](\xi) = IF_\lambda[\check{f}] \tag{9.13}$$

and from the Paley-Wiener-Schwartz theorem (see Section 2.5). Formula (9.13) follows from (9.5) and (9.2). In fact, for $\phi \in \mathcal{S}(\mathbf{R}^n)$ we have

$$(f,\phi) + \left(F^{-1}[f], F[\phi]\right) = \frac{i^{n-1}}{2(2\pi)^{n-1}}\left(\check{f},\check{\phi}_\lambda^{(n-1)}\right) =$$

$$= \frac{i^{n-1}}{2(2\pi)^{n-1}}\left(\check{f}, F_\nu^{-1}\left[(-i\nu)^{n-1}F[\phi](\nu\sigma)\right]\right) =$$

$$= \frac{i^{n-1}}{2(2\pi)^{n-1}}\left(F_\lambda^{-1}[\check{f}], (-i\nu)^{n-1}F[\phi](\nu\sigma)\right) =$$

$$= \frac{1}{(2\pi)^{n-1}} \left( IF_{\lambda}^{-1}[\check{f}\,], F[\phi] \right).$$

Hence we obtain the equality

$$F^{-1}[f\,] = \frac{1}{(2\pi)^{n-1}} IF_{\lambda}^{-1}[\check{f}\,]$$

which is equivalent to (9.13).

**Theorem.** *(an analogy of the Paley-Wiener-Schwartz theorem). A distribution* $g(\lambda, \sigma)$ *from* $\mathcal{S}'(\mathbf{R}^1 \times S^{n-1})$ *is the Radon transform of a distribution* $f$ *from* $\mathcal{E}'(\mathbf{R}^n)$, $\operatorname{supp} f \subset \bar{B}_a$, *if and only if it satisfies conditions* (a), (b), (c), *and then*

$$f(x) = F^{-1}[IF_{\lambda}[g\,]]. \tag{9.14}$$

**Proof.** The necessity of conditions (a), (b), (c) has just been proved. To prove their sufficiency we construct the distribution $f$ by formula (9.14). We then use the Paley-Wiener-Schwartz theorem (see Section 2.5) and calculate $\check{f}$ by formula (9.7). As a result we obtain $g = \check{f}$.

Condition (c) has an equivalent formulation in terms of moments of the distribution $g$:

*Let* $g(\lambda, \sigma)$ *satisfy the conditions* (a) *and* (b). *In order to satisfy condition* (c) *it is necessary and sufficient that the following equalities are fulfilled ('sum rules')*

$$(g(\lambda, \sigma), \lambda^k) = \sum_{|\alpha|=k} a_\alpha \sigma^\alpha, \quad k = 0, 1, \cdots \tag{9.15}$$

*where* $a_\alpha$ *are moments of a distribution* $f \in \mathcal{E}'(\mathbf{R}^n)$, $\operatorname{supp} f \subset \bar{B}_a$,

$$a_\alpha = (f(x), x^\alpha). \tag{9.16}$$

*In addition, the following equalities are valid:*

$$g = \check{f}. \tag{9.17}$$

$$a_\alpha = \frac{|\alpha|!}{\alpha!} i^{-|\alpha|} \partial^\alpha f_1(0), \quad f_1 = F[f\,]. \tag{9.18}$$

**Proof.** This statement follows from the previous theorem and from the following chain of equalities:

$$f_1(\nu\sigma) = \sum_{|\alpha| \geq 0} \frac{\partial^\alpha f_1(0)}{\alpha!} \nu^{|\alpha|} \sigma^\alpha = \sum_{k=0}^{\infty} \nu^k \sum_{|\alpha|=k} \frac{\partial^\alpha f_1(0)}{\alpha!} \sigma^\alpha =$$

$$= F_\lambda[g\,] = \left( g(\lambda, \sigma), e^{i\nu\lambda} \right) = \left( g(\lambda, \sigma), \sum_{k=0}^{\infty} \frac{(i\nu\lambda)^k}{k!} \right) =$$

$$= \sum_{k=0}^{\infty} \frac{(i\nu)^k}{k!} \left( g(\lambda, \sigma), \lambda^k \right) = F[f\,](\nu\sigma) = \left( f(x), e^{i\nu(x,\sigma)} \right) =$$

$$= \left( f(x), \sum_{k=0}^{\infty} \frac{(i\nu)^k}{k!} (x, \sigma)^k \right) = \sum_{k=0}^{\infty} \frac{(i\nu)^k}{k!} \sum_{|\alpha|=k} \sigma^\alpha (f(x), x^\alpha) =$$

$$= \sum_{k=0}^{\infty} \frac{(i\nu)^k}{k!} a_\alpha \sigma^\alpha.$$

To characterize the coefficients $\{a_\alpha, |\alpha| \geqslant 0\}$ it is necessary to solve the many-dimensional moment problem: *find the necessary and sufficient conditions on numbers* $\{a_\alpha, |\alpha| \geqslant 0\}$ *under which they are moments of a distribution* $f \in \mathcal{E}'(\mathbf{R}^n)$, $\operatorname{supp} f \subset \overline{B}_a$; *that is*, $a_\alpha = (f, x^\alpha)$.

For $n = 1$, with $f$ a finite measure (or continuous, or bounded, or from $\mathcal{L}^2$), this problem is solved by the Hausdorff theorems (see, for instance, I.P. Natanson, 1950). For $n > 1$, with $f$ a nonnegative measure, this problem has been solved by I.E. Ovcharenko (1983).

The following theorem on 'support' is valid (see S. Helgason, 1980).

**Theorem.** *If the Radon transform* $\check{f}(\lambda, \sigma)$ *of a distribution* $f$ *from* $\mathcal{E}'(\mathbf{R}^n)$ *vanishes for* $\lambda > a \geqslant 0$, *then* $\operatorname{supp} f \subset \overline{B}_a$.

## 3. Quasi-asymptotics of distributions

**1. General definitions and basic properties.** Let $\Gamma$ be a closed convex acute solid cone in $\mathbf{R}^n$ (with vertex 0) and let $\{U_k, k \in I\}$ be a family of linear nonsingular transformations of $\mathbf{R}^n$ which leave the cone $\Gamma$ invariant (automorphisms of $\Gamma$, see Section 2.6):

$$U_k \Gamma = \Gamma, \quad J_k = \det U_k, \quad k \in I.$$

We shall assume that $J_k > 0$, $k \in I$, and that the set $I$ is contained in $\mathbf{R}^1$ and has a single limit point $+\infty = \infty$. Further, let $\rho_k$ be a positive function of the parameter $k \in I$.

**Definition 1.** Let $f \in \mathcal{L}^1_{loc}(\Gamma)$. We say that $f$ *has an asymptotic in the cone* $\Gamma$ *over a family of transformations* $\{U_k, k \in I\}$ *with respect to a function* $\rho_k$, $k \in I$ if there exists a function $g(\xi) \not\equiv 0$ such that

$$\lim_{k \to \infty} \frac{1}{\rho_k} f(U_k \xi) = g(\xi) \quad \text{for a.e. } \xi \in \Gamma \tag{1.1}$$

and the following estimate is fulfilled:

$$\left| \frac{1}{\rho_k} f(U_k \xi) \right| \leqslant \psi(\xi), \quad k \in I \tag{1.2}$$

where $\psi$ is a tempered function on $\Gamma$ (see Section 0.3).

**Definition 2.** Let $f \in \mathcal{S}'(\Gamma)$. we say that $f$ *has a quasi-asymptotic in the cone* $\Gamma$ *over a family of transformations* $\{U_k, k \in I\}$ *with respect to a function* $\rho_k$, $k \in I$ if there exists a distribution $g(\xi) \not\equiv 0$ such that*

$$\frac{1}{\rho_k} f(U_k \xi) \to g(\xi), \quad k \to \infty, \quad k \in I \quad \text{in } \mathcal{S}'(\Gamma). \tag{1.3}$$

If $\rho_k \equiv 1$ then sometimes we shall call $g(\xi)$ the *quasi-limit*.

*Later, if $g \equiv 0$, we shall say that $f$ has a *trivial quasi-asymptotic*.

*If a locally integrable function $f(\xi)$ has an asymptotics $g(\xi)$ in $\Gamma$ over a family $\{U_k, k \in I\}$ with respect to $\rho_k$, $k \in I$ then it has a quasi- asymptotics $g(\xi)$.*

**Proof.** In fact, by virtue of (1.3) it is sufficient to verify that for any $\phi \in \mathcal{S}(\Gamma)$ there exists the limit

$$\lim_{k \to \infty, k \in I} \frac{1}{\rho_k} \left( f(U_k \xi), \phi(\xi) \right) = (g, \phi).$$

But

$$\frac{1}{\rho_k} \left( f(U_k \xi), \phi(\xi) \right) = \int \frac{1}{\rho_k} f(U_k \xi) \phi(\xi)\, d\xi, \quad k \in I,$$

and condition (1.2) allows the limit to be passed under the integral by the Lebesgue theorem.

The converse statement is not valid in general. As an example we put

$$\Gamma = \mathbb{R}^1_+, \quad I = \{k > 0\}, \quad U_k \xi = k\xi, \quad \rho_k = \frac{1}{k}, \quad f(\xi) = \theta(\xi) e^{i\xi}.$$

For any $\phi \in \mathcal{S}_+$ we have

$$k(f(k\xi), \phi(\xi)) = k \int_0^\infty e^{ik\xi} \phi(\xi)\, d\xi = \frac{1}{i}\phi(0) - \frac{1}{i} \int_0^\infty e^{ik\xi} \phi'(x)\, d\xi \to$$

$$\to \frac{1}{i}\phi(0), \quad k \to \infty.$$

Therefore

$$kf(k\xi) \to -i\,\delta(\xi), \quad k \to \infty \quad \text{in } \mathcal{S}'_+$$

and hence $f$ has a quasi-asymptotics in $\mathbb{R}_1^+$ over the family of transformations $\xi \to k\xi$ with respect to $\rho_k = 1/k$. Nevertheless, it is clear that $f(\xi)$ has no asymptotics over a given family of transformations with respect to $1/k$.

In this section we suppose that the cone $\Gamma$ is regular (see Section 2.7), so that for distributions from $\mathcal{S}'(\Gamma)$ the notion of a derivative (primitive) of any real order with respect to the cone $\Gamma$ is determined (see Section 2.8).

**Theorem 1.** *Let $f \in \mathcal{S}'(\Gamma)$ and let $\gamma$ be a real number. Then $f$ has a quasi-asymptotics in $\Gamma$ over a family of transformations $\{U_k, k \in I\}$ with respect to $\rho_k$, $k \in I$ if, and only if, when its primitive, of order $\gamma$, $f^{(-\gamma)}(\xi)$ has a quasi-asymptotics in $\Gamma$ over a family of transformations $\{U_k, k \in I\}$ with respect to $J_k^\gamma \rho_k$, $k \in I$.*

**Proof.** For any $\phi \in \mathcal{S}(\Gamma)$ and $k \in I$, we obtain, using formula (8.2), Section 2.8:

$$\left( f^{(-\gamma)}(U_k \xi), \phi(\xi) \right) = \frac{1}{J_k} \left( f^{(-\gamma)}(\xi), \phi\left(U_k^{-1}\xi\right) \right) =$$

$$= \frac{1}{J_k} \left( f(\xi), \left( \theta_\Gamma^\gamma(\tau), \phi\left(U_k^{-1}(\xi + \tau)\right) \right) \right) =$$

$$= J_k^\gamma \left( f(U_k \xi), \left( \theta_\Gamma^\gamma(\tau), \phi(\xi + \tau) \right) \right).$$

Hence

$$\frac{1}{J_k^{\gamma}\rho_k}\Big(f^{(-\gamma)}(U_k\xi),\phi(\xi)\Big) = \frac{1}{\rho_k}\Big(f(U_k\xi),\Big(\theta_\Gamma^{\gamma}(\tau),\phi(\xi+\tau)\Big)\Big).$$

We notice now that the mapping $\phi(\xi)\to\Big(\theta_\Gamma^{\gamma}(\tau),\phi(\xi+\tau)\Big)$ is an isomorphism of the space $\mathcal{S}(\Gamma)$, as the cone $\Gamma$ is regular and the proof is complete.

**Theorem 2.** *$f\in\mathcal{S}'(\Gamma)$ has a quasi-asymptotics in $\Gamma$ over a family $\{U_k, k\in I\}$ with respect to $\rho_k$, $k\in I$, if and only if there exists an integer $N$ such that the primitive $f^{(-N)}(\xi)$ is continuous and has in $\Gamma$ an asymptotics over the family $\{U_k, k\in I\}$ with respect to $J_k^N\rho_k$, $k\in I$.*

**Proof.** The sufficiency of the conditions follows easily from theorem 1. We now prove the necessity of the conditions. Since the sequence of functionals $1/(\rho_k)f(U_k\xi)$, $k\to\infty$, $k\in I$ converges in the space $\mathcal{S}'(\Gamma)$, then there exists an integer $p\geqslant 0$ such that this sequence converges in the space $\mathcal{S}_p'(\Gamma)$ (see Section 1.3). On the other hand, there exists an integer $N_0$ such that for any $\tau\in\mathbf{R}^n$ and $N\geqslant N_0$ the function $\theta_\Gamma^N(\tau-\xi)\in\mathcal{S}_p(\Gamma)$ (see Section 2.8), the distribution $f^{(-N)}$ is continuous and

$$f^{(-N)}(\xi) = \Big(f(\tau),\theta_\Gamma^N(\xi-\tau)\Big)$$

(cf. (8.13), Section 1.8) from which it follows that

$$\frac{1}{J_k^N\rho_k}f^{(-N)}(U_k\xi) = \frac{1}{J_k^N\rho_k}\Big(f(\tau),\theta_\Gamma^N(U_k\xi-\tau)\Big) =$$

$$= \Big(\frac{1}{\rho_k}f(U_k\tau),\theta_\Gamma^N(\xi-\tau)\Big). \tag{1.4}$$

Here we use again formula (8.2) of Section 2.8. From (1.4) for any $\xi\in\Gamma$ it follows that

$$\lim_{k\to\infty,k\in I}\frac{1}{J_k^N\rho_k}f^{(-N)}(U_k\xi) = \Big(g(\tau),\theta_\Gamma^N(\xi-\tau)\Big).$$

$$\Big|\frac{1}{J_k^N\rho_k}f^{(-N)}(U_k\xi)\Big|\leqslant C\Big\|\theta_\Gamma^N(\xi-\tau)\Big\|_{p,\Gamma},\quad k\in I.$$

To complete the proof it is sufficient to note that the estimate (1.2) immediately follows from the following lemma.

**Lemma.** *Let $\Gamma$ be a regular cone in $\mathbf{R}^n$. Then for any integer $p\geqslant 0$ there exists a number $N_0>0$ such that for any $N>N_0$ the following estimate is valid*

$$\Big\|\theta_\Gamma^N(\xi-\tau)\Big\|_{p,\Gamma}\leqslant A(1+|\xi|)^q,\quad \xi\in\Gamma \tag{1.5}$$

*for some $A$ and $q>0$.*

**Proof.** By virtue of the results of Section 2.8, estimate (1.5) is sufficient to prove for $|\xi|>0$. Inverting the Laplace transform for $\theta_\Gamma^N$ (see Section 2.8) we have for some fixed $y\in C$

$$\theta_\Gamma^N(\xi - \tau) = \int\limits_{\mathbf{R}^n + i(y/|\xi|)} e^{-i(\xi-\tau,z)} \mathcal{K}_C^N(z)\, dx$$

from which, by virtue of estimate (6.4) of Section 2.6 and of the equality

$$\Delta_C \left( \frac{y}{|\xi|} \right) = \frac{1}{|\xi|} \Delta_C(y)$$

we get

$$\left\| \theta_\Gamma^N(\xi - \tau) \right\|_{p,\Gamma} \leqslant \max_{|\alpha| \leqslant p} \sup_{\tau \in \Gamma} (1 + |\tau|)^p \int\limits_{\mathbf{R}^n + i(y/|\xi|)} \left| e^{-i(\xi-\tau,z)} z^\alpha \mathcal{K}_C^N(z) \right| dx \leqslant$$

$$\leqslant A_1 \sup_{\tau \in \Gamma} (1 + |\tau|)^p e^{-(\tau,y)/(|\xi|)} \left( \frac{|\xi|}{|y|} \right)^{N(n-1)} \int\limits_{\mathbf{R}^n} \frac{1 + \left( |x|^2 + \dfrac{|y|^2}{|\xi|^2} \right)^{p/2}}{\left( |x|^2 + \dfrac{|y|^2}{|\xi|^2} \right)^{n/2}} dx$$

where $A_1$ does not depend on $\xi$. Now taking into account the inequality $(\tau, y) > \sigma_y |\tau|$, $\tau \in \Gamma$ (see Section 1.9) we obtain estimate (1.5) for a sufficiently large $N$.

In many cases it is important to know exactly which primitive of a distribution $f$, which has a quasi-asymptotics, already has an asymptotics. There is no simple or universal criterion. For instance, it would be natural to suppose that the condition "$f$ belongs to some space $\mathcal{S}_p'(\Gamma)$" guarantees the existence of such an $N$ (depending perhaps on $p$, on a family $\{U_k, k \in I\}$, and on a function $\rho_k$, $k \in I$) for which the primitive $f^{(-N)}$ has an asymptotics. The following example shows that it is not the case, in general.

Let $\Gamma = \mathbb{R}_+^1$ and let $\{f_n(\xi) = \theta(\xi)\exp(i\xi^{1/n}), n = 1, 2, \cdots\}$ be the family of functions from $\mathcal{S}_+'$. At first we prove that each function $f_n(\xi)$ has a quasi-asymptotics $-i^n n!\delta(\xi)$ over the family of transformations $\{U_k \xi = k\xi, k > 0\}$ with respect to the function $\rho_k = 1/k$, $k > 0$. In fact, for any $\phi \in \mathcal{S}_+$ we have

$$k(f_n(k\xi), \phi(\xi)) = k\int_0^\infty \exp\left[ i(k\xi)^{1/n} \right] \phi(\xi)\, d\xi =$$

$$= kn\int_0^\infty \exp(ik^{1/n}\tau)\phi(\tau^n)\tau^{n-1}\, d\tau =$$

$$= \frac{n}{i^n}\int_0^\infty \phi(\tau^n)\tau^{n-1} \left( \frac{d}{d\tau} \right)^n \exp(ik^{1/n}\tau)\, d\tau =$$

$$= -i^n n!\phi(0) + i^n n\int_0^\infty \exp(ik^{1/n}\tau) \left( \frac{d}{d\tau} \right)^n [\phi(\tau^n)\tau^{n-1}]\, d\tau.$$

The last integral in this chain of equalities obviously tends to zero as $k \to \infty$, which proves our assertion.

Now we find out which primitive of $f_n$ has a normal asymptotics.

For that we use a formula which is valid for every natural $p$ and $n$:

$$\int_0^\tau \xi^{p/n} \exp(i\xi^{1/n})\, d\xi = n\int_0^{\tau^{1/n}} e^{i\xi}\xi^{p+n-1}\, d\xi = a + \sum_{j=0}^{p+n-1} b_j \tau^{j/n} \exp(i\tau^{1/n}) \qquad (1.6)$$

where all coefficients $a$ and $b_j$ are not equal to zero. Applying formula (1.6) $N$ times we obtain

$$f_n^{(-N)}(\xi) = \theta(\xi)\left[\alpha_1 \xi^{N-1} + \cdots + \alpha_n + \sum_{j=0}^{(n-1)N} \beta_j \xi^{j/n} \exp(i\,\xi^{1/n})\right]$$

where $\alpha_1$ and $\beta_{(n-1)N}$ differs from zero. So the function $f_n^{(-N)}$ has an asymptotics over the family of transformations $\{U_k \xi = k\xi, k > 0\}$ with respect to the function $k^{N-1}$, $k > 0$ if, and only if, $N-1 > ((n-1)/n)N$; that is, when $N > n$ and, hence, $N \to \infty$, $n \to \infty$. On the other hand, all the functions $f_n(\xi)$, $n = 1, 2, \cdots$ are infinitely differentiable for $\xi > 0$ and are uniformly bounded.

Nevertheless, in some particular cases one can obtain such criteria. Now we consider the case when $f = [d\mu]$, where $d\mu$ is a nonnegative measure in the cone $\Gamma$. Another examples will be considered later.

If $d\mu(\xi)$ is a nonnegative measure with a support in $\Gamma$ then for its first primitive with respect to the cone $\Gamma$,

$$\mu(\xi) = \theta_\Gamma * [d\mu] = \int \theta_\Gamma(\xi - \xi')\, d\mu(\xi').$$

we can take any function $\mu(\xi)$ such that (Fig. 7)

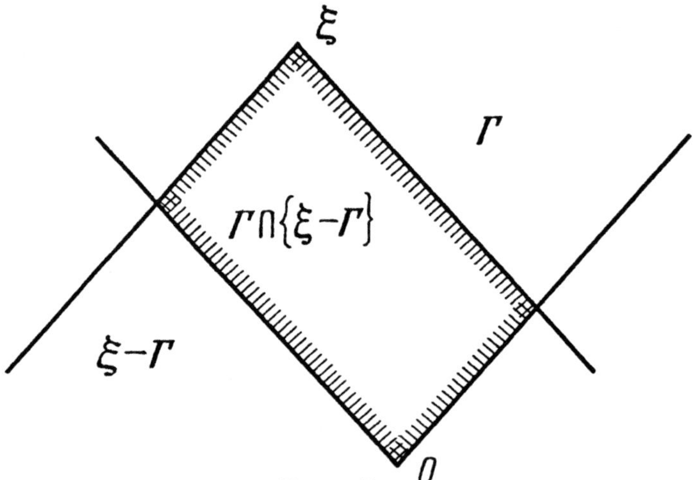


Figure 7.

$$\int_{\Gamma \cap \operatorname{int}\{\xi-\Gamma\}} d\mu(\xi') \leqslant \mu(\xi) \leqslant \int_{\Gamma \cap \{\xi-\Gamma\}} d\mu(\xi'). \tag{1.7}$$

We shall call $\mu(\xi)$ the *reproducing function* of the measure $d\mu(\xi)$. The function $\mu(\xi)$ is uniquely defined almost everywhere and $\mu(\xi \geqslant 0$.

**Theorem 3.** *Let $f = [d\mu]$ where $d\mu$ is a nonnegative tempered measure with a support in the cone $\Gamma$ and has a quasi-asymptotics $g$ in $\Gamma$ over a family of transformations $\{U_k, k \in I\}$ with respect to $\rho_k$, $k \in I$. Then for any compact $K \subset \mathbf{R}^n$, in some neighbourhood of which the first primitive $g^{(-1)}(\xi)$ is continuous, the following limit relation is valid*

$$\frac{\mu(U_k\xi)}{J_k\rho_k}\overset{\xi\in K}{\Longrightarrow} g^{(-1)}(\xi),\ k\to\infty,\ k\in I, \tag{1.8}$$

*and, in addition, for some* $A>0$ *and* $q\geqslant 0$ *the following estimate holds*

$$\left|\frac{\mu(U_k\xi)}{J_k\rho_k}\right|\leqslant A(1+|\xi|)^q,\ k\in I. \tag{1.9}$$

**Proof.** Choose some point $e\in \operatorname{int}\Gamma$ and let $L=\Delta_\Gamma(\Gamma+e)$ be the distance from the cone $\Gamma+e$ to $\partial\Gamma$ (see Section 0.2). It is clear that $\Delta_\Gamma(\Gamma+\epsilon e)=\epsilon L$. We put $\phi_\epsilon=\theta_\Gamma*\omega_\epsilon$ where $\omega_\epsilon(\xi)$ is the normed bell-shaped function (see Section 0.5). Owing to $d\mu\geqslant 0$ and $\phi_\epsilon\geqslant 0$ from (1.7) we have the inequalities

$$\left(\frac{f(U_k\tau)}{\rho_k},\phi_\epsilon(\xi-\tau-\frac{\epsilon}{L}e)\right)\leqslant\frac{\mu(U_k\xi)}{J_k\rho_k}\leqslant$$
$$\leqslant\left(\frac{f(U_k\xi)}{\rho_k},\phi_\epsilon(\xi-\tau+\frac{\epsilon}{L}e\right). \tag{1.10}$$

Under the hypotheses of theorem 3 for any fixed $\epsilon>0$ we have

$$\left(\frac{f(U_k\tau)}{\rho_k},\phi_\epsilon(\xi-\tau\pm\frac{\epsilon}{L}e)\right)\overset{\xi\in K}{\Longrightarrow}(g(\tau),\phi(\xi-\tau\pm\frac{\epsilon}{L}e)),\ k\to\infty,\ k\in I. \tag{1.11}$$

and in addition the following inequalities are valid

$$g^{(-1)}(\xi-\frac{2\epsilon}{L}e)\leqslant(g(\tau),\phi_\epsilon(\xi-\tau-\frac{\epsilon}{L}e))\leqslant$$
$$\leqslant(g(\tau),\phi_\epsilon(\xi-\tau+\frac{\epsilon}{L}e))\leqslant g^{(-1)}(\xi+\frac{2\epsilon}{L}e). \tag{1.12}$$

Let $\epsilon_1$ be an arbitrary positive number. As the function $g^{(-1)}(\xi)$ is continuous in some neighbourhood of $K$ there exists some $\epsilon=\epsilon(\epsilon_1)$ such that

$$\left|g^{(-1)}(\xi)-g^{(-1)}\left(\xi\pm\frac{2\epsilon}{L}\right)\right|\leqslant\frac{\epsilon_1}{2},\ \xi\in K. \tag{1.13}$$

Choose $k_0=k_0(\epsilon_1)$ such that for $k>k_0$, $k\in I$, owing to (1.11), the following estimates are fulfilled

$$\left|\left(\frac{f(U_k\tau)}{\rho_k},\phi_\epsilon(\xi-\tau\pm\frac{\epsilon}{L}e)\right)-(g(\tau),\phi_\epsilon(\xi-\tau\pm\frac{\epsilon}{L}e))\right|<\frac{\epsilon_1}{2}. \tag{1.14}$$

Now from (1.14)-(1.12) and (1.10) it follows that

$$\left|\frac{\mu(U_k\xi)}{J_k\rho_k}-g^{(-1)}(\xi)\right|<\epsilon_1,\ \xi\in K,\ k>k_0(\epsilon),\ k\in I.$$

Thus we have just proved the limit relation (1.8). Inequality (1.9) follows from the right-hand side of inequality (1.10) and from the fact that for any integer $p\geqslant 0$ there are numbers $q\geqslant 0$ and $A_1>0$ such that

$$\left\|\phi_\epsilon(\xi-\tau+\frac{\epsilon}{L}e)\right\|_{p,\Gamma}\leqslant A_1(1+|\xi|)^q$$

**2. Automodel (regularly varying) functions.** Let $I = (0, \infty)$. For a function $\rho_k$, $k > 0$, which can now be conveniently denoted by $\rho(k)$, we shall suppose that $\rho(k)$ is continuous and positive for $k > k_0 \geqslant 0$ ($k_0$ depends on $\rho$, in general). In that case we shall say for brevity that $\rho(k)$ is *continuous and positive as* $k \to \infty$.

Now we discuss the condition that $\rho(k)$ is continuous as $k \to \infty$. If this is not the case, then $\rho(k)$ can be changed by an equivalent continuous function $\rho_1(k)$ as $k \to \infty$. More precisely, the following statement is valid: if a distribution $f \in \mathcal{S}'(\Gamma)$ has a quasi-asymptotics $g$ over a family of transformations $\{U_k, k > 0\}$ which continuously depends on $k$ with respect to a positive (as $k \to \infty$) function $\rho(k)$, then there exists a positive and continuous (as $k \to \infty$) function $\rho_1(k)$, such that $\rho(k)/\rho_1(k) \to 1$ as $k \to \infty$. In fact, as $\rho_1(k)$ we can take the function

$$\rho_1(k) = \left|\mathrm{Re}(g, \phi)\right|^{-1} \left|\mathrm{Re}(f(U_k \xi), \phi(\xi))\right|$$

where $\phi \in \mathcal{S}(\Gamma)$ such that $\mathrm{Re}(g, \phi) \neq 0$.

**Definition.** Let $\rho$ be a positive and continuous function as $k \to \infty$. We say that $\rho$ is an *automodel (or regular varying)* function if for any $a > 0$ there exists the limit

$$\lim_{k \to \infty} \frac{\rho(ak)}{\rho(k)} = C(a)$$

and, in addition, the convergence is uniform in $a$ on each compact subset of semi-axis $(0, \infty)$.

It is easy to see that $C(a)$ satisfies the functional equation $C(a)C(b) = C(ab)$, from which it follows that $C(a) = a^\alpha$ for some real $\alpha$. In this case we call the function $\rho(k)$ an automodel of *order* $\alpha$ or that $\alpha$ is the *order of automodelity* of $\rho$.

**Remark.** The regularly varying functions were introduced by J. Karamata (1930a). Their theory has been presented in detail by E. Seneta (1976).

Note that in the definition of an automodel function we deliberately impose stronger requirements than usual. So, instead of the continuity of functions one requires only its measurability. As for the requirement of the existence of a limit for all $a > 0$, it may be possible to change it on a weaker, but in fact equivalent, requirement that the limit exists on a set of positive measure. Since in what follows we are interested in automodel functions as natural objects for studying the quasi-asymptotics of distributions, we shall not discuss these problems in detail, but shall refer the reader to E. Seneta (1976) mentioned above.

We require some properties of automodel functions which we formulate in the form of the following lemmas.

**Lemma 1.** *Let $\rho$ be an automodel function of order $\alpha$. Then for any $\epsilon > 0$ there exists a number $k_\epsilon > 0$ such that*

$$a^{\alpha-\epsilon} \leqslant \frac{\rho(ak)}{\rho(k)} \leqslant a^{\alpha+\epsilon} \quad \text{for } k > k_\epsilon,\ a > 1+\epsilon,$$

$$a^{\alpha+\epsilon} \leqslant \frac{\rho(ak)}{\rho(k)} \leqslant a^{\alpha-\epsilon} \quad \text{for } \frac{k_\epsilon}{k} < a < 1-\epsilon.$$

*In particular, if $\alpha > 0$ then $\rho(k) \to \infty$ as $k \to \infty$; if $\alpha < 0$, then $\rho(k) \to 0$ as $k \to \infty$;*

*if* $\alpha < -1$ *then* $\rho \in \mathcal{L}^1((k_0, \infty))$.

The following lemma gives a convenient criterion of automodelity.

**Lemma 2.** Let a function $\rho(k)$ be continuous and positive for $k \geqslant k_0 \geqslant 0$. Then

(i) *For* $\rho$ *to be automodel of order* $\alpha > -1$, *it is necessary and sufficient that there exists a limit*

$$\frac{1}{k\,\rho(k)}\int_{k_0}^{k}\rho(t)\,dt \to \beta \neq 0,\ k \to \infty,$$

*and then* $\alpha = -1 + 1/\beta$.

(ii) *For* $\rho$ *to be automodel of order* $\alpha < -1$, *it is necessary and sufficient that there exists a limit*

$$\frac{1}{k\,\rho(k)}\int_{k}^{\infty}\rho(t)\,dt \to \beta \neq 0,\ k \to \infty$$

*and then* $\alpha = -1-1/\beta$.

(iii) *If* $\rho(k)$ *is differentiable for* $k \geqslant k_0$ *and there exists a limit*

$$\frac{k\,\rho'(k)}{\rho(k)} \to \alpha,\ k \to \infty$$

*then* $\rho$ *is automodel of order* $\alpha$.

**Examples.** The functions $k^\alpha$, $k^\alpha \ln k$, $k^\alpha(2 + \sin\sqrt{\ln k}\,)$ are automodel of order $\alpha$. On the other hand, all these functions are mutually asymptotically none-quivalent in the sense that limits of their ratio do not exist or are equal to zero.

**3. Quasi-asymptotics over one-parameter groups of transformations.** In this subsection we consider the case when a family of transformations $\{U_k, k \in I\}$ which leaves the cone $\Gamma$ invariant forms a one-parameter group (see Section 3.1). Namely, let $I = \{0 < k < \infty\}$ and the linear transformations $U_k$ from the family $\{U_k, k > 0\}$ depend continuously on the parameter $k$ and satisfy the condition

$$U_k U_s = U_{ks},\ k, s > 0. \tag{3.1}$$

In this case it is clear that $J_k J_s = J_{ks}$. Therefore, there exists a real number $\beta$ such that

$$J_k = \det U_k = k^\beta. \tag{3.2}$$

The following theorem shows the role of automodel functions when considering quasi-asymptotics over one-parameter groups.

**Theorem 1.** *Let a distribution* $f \in \mathcal{S}'(\Gamma)$ *have a quasi- asymptotics in* $\Gamma$ *over the one-parameter group* $\{U_k, k > 0\}$ *with respect to a positive continuous function* $\rho$; *that is,*

$$\frac{1}{\rho(k)} f(U_k \xi) \to g(\xi) \not\equiv 0,\ k \to \infty \ \text{ in } \mathcal{S}'(\Gamma). \tag{3.3}$$

*Then* $\rho$ *is an automodel function. In addition, if the order* $\rho$ *is equal to* $\alpha$ *then the distribution* $g$ *is homogeneous with respect to the group* $\{U_k, k > 0\}$ *of degree* $\alpha$; *that is,*

$$g(U_k \xi) = k^\alpha g(\xi). \tag{3.4}$$

**Proof.** Let condition (3.3) be fulfilled. If we choose $\phi \in \mathcal{S}(\Gamma)$ such that $(g, \phi) \neq 0$ then for any compact $K \subset (0, \infty)$ we have

$$\frac{1}{\rho(k)}\left(f(U_k \xi), \frac{1}{J_a}\phi(U_{1/a}\xi)\right) \overset{a \in K}{\Longrightarrow} \left(g(\xi), \frac{1}{J_a}\phi(U_{1/a}\xi)\right), \quad k \to \infty.$$

Therefore,

$$\frac{\rho(ak)}{\rho(k)}\left(\frac{1}{\rho(ak)}f(U_k \xi), \phi(\xi)\right) \overset{a \in K}{\Longrightarrow} \left(g(\xi), \frac{1}{J_a}\phi(U_{1/a}\xi)\right), \quad k \to \infty.$$

On the other hand,

$$\left(\frac{1}{\rho(ak)}f(U_{ak}\xi), \phi(\xi)\right) \overset{a \in K}{\Longrightarrow} (g, \phi) \neq 0, \quad k \to \infty.$$

Hence,

$$\frac{\rho(ak)}{\rho(k)} \overset{a \in K}{\Longrightarrow} \frac{\left(g(\xi), \frac{1}{J_a}\phi(U_{1/a}\xi)\right)}{(g, \phi)} = C(a) \tag{3.5}$$

which means that $\rho$ is automodel. If the order of the automodelity of $\rho$ is $\alpha$ then $C(a) = a^\alpha$ and from (3.5) we obtain (3.4).

Thus, it makes sense to consider a quasi-asymptotics over one-parameter groups only with respect to automodel functions.

Any cone $\Gamma$ (see Section 0.2) has at least one one-parameter group of transformations leaving it invariant; namely, the dilatation group $\{k\xi, k > 0\}$. In the cases in which we deal with asymptotics or quasi-asymptotics over that group, we shall drop the words "over the dilatation group". Thus a distribution $f$ has a quasi-asymptotics in $\Gamma$ with respect to an automodel function $\rho$ if, and only if,

$$\frac{1}{\rho(k)}f(k\xi) \to g(\xi) \not\equiv 0, \quad k \to \infty \quad \text{in } \mathcal{S}'(\Gamma). \tag{3.6}$$

In addition, according to formula (3.4) the distribution $g$ is homogeneous of degree $\alpha$ (see Section 1.6)

$$g(k\xi) = k^\alpha g(\xi) \tag{3.7}$$

where $\alpha$ is the order of the automodelity of $\rho$.

**Theorem 2.** *Let $\Gamma$ be a closed convex acute solid cone in $\mathbf{R}^n$, $f \in \mathcal{S}'(\Gamma) \bigcap \mathcal{L}^1_{loc}(\Gamma \bigcap [|\xi| > R])$ for some $R > 0$ and let $\rho$ be an automodel function of order $\alpha > -n$. If, for any $e \in \text{pr}\,\Gamma$, there exists the limit*

$$\lim_{|\xi| \to \infty} \frac{1}{\rho(|\xi|)} f(|\xi| e) = g(e) \not\equiv 0 \tag{3.8}$$

*and the estimate*

$$\frac{|f(\xi)|}{\rho(|\xi|)} \leqslant g_1\left(\frac{\xi}{|\xi|}\right), \quad |\xi| > R, \quad \xi \in \Gamma, \quad g_1 \in \mathcal{L}^1(\text{pr}\,\Gamma) \tag{3.9}$$

*is satisfied, then f has a quasi-asymptotics in* $\Gamma$ *with respect to* $\rho$.

First we prove the following lemma.

**Lemma 1.** *If a distribution* $f \in \mathcal{S}'(\Gamma)$ *has a compact support and* $\rho$ *is an automodel function of order* $\alpha > -n$ *then*

$$\frac{1}{\rho(k)} f(k\xi) \to 0, \ k \to \infty \ \text{ in } \mathcal{S}'(\Gamma).$$

**Proof.** Let $\eta \in \mathcal{S}(\Gamma)$ and $\eta(\xi) = 1$, $\xi \in \operatorname{supp} f$. Then for any $\phi \in \mathcal{S}(\Gamma)$ we have

$$\frac{1}{\rho(k)}(f(k\xi), \rho(\xi)) = \frac{1}{\rho(k)}(\eta(k\xi) f(k\xi), \phi(\xi)) =$$

$$= \frac{1}{\rho(k)k^n}\left(f(\xi), \eta(\xi)\phi\left(\frac{\xi}{k}\right)\right).$$

But the set $\{\eta(\xi)\phi(\xi/k), k \to \infty\}$ is bounded in $\mathcal{S}(\Gamma)$. Thus the function $(f(\xi), \eta(\xi)\phi(\xi/k))$ is bounded as $k \to \infty$. On the other hand, according to lemma 1 of Section 3.2, $\rho(k)k^n \to \infty$ as $k \to \infty$ so that lemma 1 follows.

Now we proof theorem 2.

**Proof.** Take $\epsilon > 0$ such that $\alpha - \epsilon > -n$ and choose $k_0 = k_0(\epsilon) > R$ such that the estimates of lemma 1 of Section 3.2 are satisfied. Let $\eta(\xi)$ be a $C^\infty$-function such that $0 \leqslant \eta(\xi \leqslant 1$; $\eta(\xi) \equiv 1$ for $|\xi| > k_0 + 1$ and $\eta(\xi) \equiv 0$ for $|\xi| < k_0$. According to lemma 1 it is sufficient to prove that there exists a limit of the sequence of functions $\{(1/\rho(k))\eta(k\xi)f(k\xi), \ k \to \infty\}$ in $\mathcal{S}'(\Gamma)$ and that this limit differs from zero. In turn it is sufficient to prove that the function $\eta f$ has an asymptotics in $\Gamma$ with respect to $\rho$ in the sense of definition 1 of Section 3.1. First, let us verify the limit ration (1.1) of Section 3.1. From (3.8) we have

$$\lim_{k \to \infty} \frac{\eta(k\xi)f(k\xi)}{\rho(k)} = \lim_{k \to \infty} \frac{f(k\xi)}{\rho(k)} = \lim_{k \to \infty} \frac{f(k|\xi|e)}{\rho(k|\xi|)} \frac{\rho(k|\xi|)}{\rho(k)} =$$

$$= g(e)|\xi|^\alpha.$$

Now we prove estimate (1.2) of Section 3.1. From (3.9) we have

$$\left|\frac{\eta(k\xi)f(k\xi)}{\rho(k)}\right| = \left|\frac{f(k|\xi|e)}{\rho(k|\xi|)} \cdot \frac{\eta(k\xi)\rho(k|\xi|)}{\rho(k)}\right| \leqslant$$

$$\leqslant g_1(e)\frac{\eta(k\xi)\rho(k|\xi|)}{\rho(k)}. \tag{3.10}$$

But from lemma 1 of Section 3.2 and from the properties of the function $\eta$ we obtain, for $k > k_0$,

$$0 \leqslant \frac{\eta(k\xi)\rho(k|\xi|)}{\rho(k)} \leqslant \begin{cases} |\xi|^{\alpha-\epsilon}, & 0 \leqslant |\xi| \leqslant 1-\epsilon, \\ C, & 1-\epsilon \leqslant |\xi| \leqslant 1+\epsilon, \\ |\xi|^{\alpha+\epsilon}, & |\xi| > 1+\epsilon. \end{cases} \tag{3.11}$$

Since $\alpha - \epsilon > -n$ and $g_1 \in \mathcal{L}^1(\operatorname{pr}\Gamma)$ so that from (3.10) and (3.11) it follows that for the function $\eta f$ the estimate (1.2) of Section 3.1 is satisfied.

The next theorem will be useful later.

**Theorem 3.** *Let* $f \in \mathcal{D}'_\Gamma$ *where* $\Gamma$ *is a regular cone and for some automodel function* $\rho$

$$\frac{1}{\rho(k)} f(k\xi) \to g(\xi) \not\equiv 0, \ k \to \infty \quad \text{in } \mathcal{D}'(\mathbf{R}^n).$$

*Then* $f \in {}'(\Gamma)$ *and has a quasi-asymptotics in* $\Gamma$ *with respect to* $\rho$.

**Proof.** Choose a function $\eta \in$ such that $\eta(\xi) \equiv 1$ for $\xi \in \Gamma$, $|\xi| \leqslant 1$. Then the family of distributions $\{1/(\rho(k))\eta(\xi)f(k\xi), k \to \infty\}$ converges in the space $\mathcal{S}'(\Gamma)$ and hence in the space $\mathcal{S}'_p(\Gamma)$ for some $p \geqslant 0$. Therefore according to the lemma of Section 3.1 there exists a number $N_0 > 0$ such that for any $N \geqslant N_0$ and $e \in \operatorname{pr}\Gamma$

$$\frac{1}{\rho(k)}(\eta(\xi)f(k\xi), \theta_\Gamma^N(e-\xi)) \to g_N(e), \ k \to \infty.$$

As $[1-\eta(\xi)]\theta_\Gamma^N(e-\xi) = 0$ then (see Section 2.8)

$$(\eta(\xi)f(k\xi), \theta_\Gamma^N(e-\xi)) = (f(k\xi), \theta_\Gamma^N(e-\xi)) = \frac{1}{k^{nN}} f^{(-N)}(ke) \tag{3.12}$$

and hence

$$\frac{1}{k^{nN}\rho(k)} f^{(-N)}(ke) \to g_N(e), \ k \to \infty.$$

Similarly, from (3.12) and the lemma of Section 3.1 it follows that $f^{(-N)} \in \mathcal{L}^1_{loc}$ and

$$\left|\frac{1}{k^{nN}\rho(k)} f^{(-N)}(ke)\right| \leqslant C\|\theta_\Gamma^N(e-\xi)\|_{p,\Gamma} \leqslant$$

$$\leqslant C_1, \ k \to \infty.$$

Let a number $N \geqslant N_0$ be such that $\alpha + nN > -n$, where $\alpha$ is the order of the automodelity of $\rho$. By Theorem 2 we conclude that $f^{(-N)}$ has a quasi-asymptotics in $\Gamma$ with respect to $k^{nN}\rho(k)$ and thus by theorem 1 of Section 3.1, $f$ has a quasi-asymptotics in $\Gamma$ with respect to $\rho$.

The following theorem specifies theorem 3 of Section 3.1 for the case of the dilatation group.

**Theorem 4.** *Let* $f = ]d\mu]$, *where* $d\mu$ *is a nonnegative tempered measure with a support in a regular cone* $\Gamma$. *If* $[d\mu]$ *has a quasi-asymptotics* $g$ *in* $\Gamma$ *with respect to an automodel function* $\rho$ *then for any compact* $K \subset\subset \operatorname{int}\Gamma$ *the following limit relation holds*

$$\frac{\mu(k\xi)}{k^n\rho(k)} \overset{\xi \in K}{\rightrightarrows} g^{(-1)}(\xi), \ k \to \infty$$

*where* $\mu = [d\mu] * \theta_\Gamma$ *is the reproducing function of the measure* $d\mu$.

The theorem follows immediately from theorem 3 of Section 3.1 and the following lemma.

**Lemma 2.** *Let $g$ be a nonnegative tempred measure with a support in a regular cone $\Gamma$ which is homogeneous of degree $\alpha$. Then its reproducing function $g^{(-1)}$ is continuous in* $\operatorname{int}\Gamma$.

**Proof.** The function $g^{(-1)}(\xi)$ is homogeneous of degree $\alpha + n$:

$$g^{(-1)}(k\xi) = k^{\alpha+n} g^{(-1)}(\xi), \quad k > 0 \tag{3.13}$$

and moreover it is a monotone non-decreasing function with respect to the cone $\Gamma$. In other words, if $\xi_1 - \xi_2 \in \Gamma$ then $g^{(-1)}(\xi_1) \geqslant g^{(-1)}(\xi_2)$. Let $\xi_k \in \Gamma$, $k = 1, 2, \cdots$ be an arbitrary sequence which converges to some point $\xi_0 \in \operatorname{int}\Gamma$, so that $\Delta_\Gamma(\xi_0) > 0$. Then we have

$$\left(1 + 2\frac{|\xi_k - \xi_0|}{\Delta_\Gamma(\xi_0)}\right)\xi_0 - \xi_k \in \Gamma, \tag{3.14}$$

$$\xi_k - \left(1 - 2\frac{|\xi_k - \xi_0|}{\Delta_\Gamma(\xi_0)}\right)\xi_0 \in \Gamma, \quad k = 1, 2, \cdots$$

because

$$\Delta_\Gamma\left(2\frac{|\xi_k - \xi_0|}{\Delta_\Gamma(\xi_0)}\xi_0\right) = 2|\xi_k - \xi_0|.$$

From (3.14) it follows that

$$g^{(-1)}\left(\left(1 - 2\frac{|\xi_k - \xi_0|}{\Delta_\Gamma(\xi_0)}\right)\xi_0\right) \leqslant g^{(-1)}(\xi_k)$$

$$\leqslant g^{(-1)}\left(\left(1 + 2\frac{|\xi_k - \xi_0|}{\Delta_\Gamma(\xi_0)}\right)\xi_0\right), \quad k = 1, 2, \cdots$$

But, by virtue of (3.13), the left- and right-hand sides of this inequality tend to $g^{(-1)}(\xi_0)$ as $k \to \infty$ and lemma 2 follows.

**4. The one-dimensional case. Quasi-asymptotics at infinity and at zero.** In the space $\mathbf{R}^1$ there are only two acute closed solid cones: $\mathbf{R}^1_+ = [\xi : \xi \geqslant 0]$ and $\mathbf{R}^1_- = [\xi : \xi \leqslant 0]$.

If $f \in \mathcal{S}'_+$ has a quasi-asymptotics in $\mathbf{R}^1_+$ with respect to an automodel function $\rho$, then we say that $f$ *has a quasi-asymptotics at infinity with respect to* $\rho$. Similarly, if $f \in \mathcal{S}'_-$ has a quasi-asymptotics in $\mathbf{R}^1_-$ then we say that $f$ *has a quasi-asymptotics at minus infinity.*

If $f \in \mathcal{S}'_+$ has a quasi-asymptotics at infinity with respect to an automodel function $\rho$ of order $\alpha$, then according to theorem 1 of Section 3.3:

$$\frac{f(k\xi)}{\rho(k)} \to C f_{\alpha+1}(\xi), \quad k \to \infty \quad \text{in} \quad \mathcal{S}'_+, \quad C \neq 0 \tag{4.1}$$

where $f_{\alpha+1}$ is the canonical kernel of the operator of fractional differentiation and integration (see Section 1.6 and Section 1.9).

In the case of a quasi-asymptotics at infinity the statements of many results of the previous subsections are essentially simplier. For instance, theorem 2 of Section 3.3 can be reformulated in the following way.

**Theorem 1.** *Let* $f \in \mathcal{S}'_+$, *and let* $\rho$ *be an automodel function of order* $\alpha > -1$. *Also, let* $f \in \mathcal{L}^1_{loc}(\xi > R)$ *for som* $R$ *and* $f(\xi)/\rho(\xi) \to C \neq 0$, $\xi \to \infty$. *Then* $f$ *has a quasi-asymptotics at infinity with respect to* $\rho$ *and*

$$\frac{f(k\xi)}{\rho(k)} \to C\Gamma(\alpha+1)f_{\alpha+1}(\xi),\ k \to \infty \text{ in } \mathcal{S}'_+.$$

Taking theorem 1 into account we can now formulate theorem 2 of Section 3.1 in the following way.

**Theorem 2.** *A distribution* $f \in \mathcal{S}'_+$ *has a quasi-asymptotics at infinity with respect to an automodel function* $\rho$ *of order* $\alpha$ *if and only if there exists a number* $N > -\alpha - 1$ *such that*

$$\frac{f^{(-N)}(\xi)}{\xi^N \rho(\xi)} \to C \neq 0,\ \xi \to \infty.$$

**Remark 1.** The condition $\alpha > -n$ in theorem 2 of Section 3.3, as well as the conditions $\alpha > -1$ in theorem 1 and $\alpha + N > -1$ in theorem 2, are essential. For instance, the function $f(\xi) = \theta(\xi)(1+\xi^2)^{-1}$ satisfies the property $f(\xi)/\rho_1(\xi) \to 1$, $\xi \to \infty$ for $\rho_1 = \xi^2$, but $f$ has a quasi-asymptotics at infinity with respect to $\rho(k) = 1/k$. Thus for $\alpha \leqslant -1$ the property of distributions from $\mathcal{S}'_+$ of having or of not having a quasi-asymptotics at infinity is not purely asymptotic but depends on the global properties of distribution. In particular, if $f \in \mathcal{S}'_+$ has a compact support, then, as can easily be seen, it has a quasi-asymptotics at infinity with respect to $k^{-m}$ for some integer $m \geqslant 1$.

**Remark 2.** Let $f(t)$ be a local integrable function, $f(t) = 0$ for $t < 0$. Then

$$S^N_\xi(f) = \frac{N}{\xi^N}\int_0^\xi (\xi - t)^{N-1} f(t)\,dt = \frac{N!}{\xi^N} f^{(-N)}(\xi)$$

is the Česaro mean value of order $N \geqslant 0$. According to theorem 2, the function $f$ has a quasi-asumptotics at infinity with respect to an automodel function $\rho$ if and only if there exists a number $N \geqslant 0$ such that its Česaro mean value of order $N$ has an asymptotics at infinity with respect to $\rho$:

$$\frac{1}{\rho(k)} S^N_\xi(f) \to C \neq 0,\ \xi \to \infty.$$

Thus, the notion of quasi-asymptotics adequately reflects the asymptotic behaviour of the Česaro mean values.

Now we give a notion of the quasi-asymptotics of distributions of one variable at zero.*

**Definitions.** Let $f \in \mathcal{S}'$. We shall say that $f$ *has a quasi-asymptotics at zero with respect to an automodel function* $\rho$ if there exists the limit

*In the case $\rho \equiv 1$ this definition is consistent with the definition of a value of $f(x)$ at the point $x = 0$ in the sense of Lojasiewicz (1957).

$$\frac{1}{\rho(k)} f\left(\frac{x}{k}\right) \to g(x) \quad 0,\ k \to \infty \quad \text{in } \mathcal{S}'. \tag{4.2}$$

If for any real $\alpha$

$$k^{\alpha} f\left(\frac{x}{k}\right) \to \infty \quad \text{in } \mathcal{S}'$$

then we shall say that the *order of $f$ at zero is equal to $-\infty$.*

Note that in the definition of a quasi-asymptotics at zero one does not impose any conditions upon the support of $f$. It is clear that if $f \in \mathcal{S}'_{+}$, then the existence of its quasi-asymptotics at zero is equivalent to the existence for $f$ of a quasi-asymptotics at infinity over the transformation group $x \to x/k$. In this case the general results of the previous subsections are valid. However, in what follows, it is convenient that we do not restrict ourselves to such a condition.

The following lemma shows that the quasi-asymptotic properties of a distribution of zero are completely defined by its behaviour in arbitrarily small neighbourhoods of the origin.

**Lemma.** *If $f \in \mathcal{S}'$ vanishes in some neighbourhood of the point $x = 0$ then its order at zero is equal to $-\infty$.*

**Proof.** Let $\phi \in \mathcal{S}$ and let $\eta(x)$ be a $C^{\infty}$- function which vanishes in some neighbourhood of the origin and is equal to 1 for $x \in \operatorname{supp} f$. Then we have

$$\left(k^{\alpha} f\left(\frac{x}{k}\right), \phi(x)\right) = k^{\alpha+1}(f(x), \phi(kx)) =$$

$$= (f(x), k^{\alpha+1}\eta(x)\phi(kx)). \tag{4.3}$$

We shall now show that for any $\alpha$

$$k^{\alpha}\eta(x)\phi(kx) \to 0,\ k \to \infty \quad \text{in } \mathcal{S}. \tag{4.3'}$$

In fact, for any integer $m \geqslant 0$ we have for some $C_m$ and $k \geqslant 1$

$$\| k^{\alpha}\eta(x)\phi(kx) \|_m = \max_{0 \leqslant i \leqslant m} \sup_{x} (1 + x^2)^{m/2} \left| \frac{d^i}{dx^i}[\eta(x)\phi(kx)k^{\alpha}] \right| \leqslant$$

$$\leqslant \max_{0 \leqslant i \leqslant m} \sup_{|x| \geqslant a > 0} (1 + x^2)^{m/2} k^{\alpha} \sum_{j=0}^{i} \binom{j}{i} k^{j} \left| \phi^{(j)}(kx)\eta^{(i-j)}(x) \right| \leqslant$$

$$\leqslant C_m k^{\alpha+m} \max_{0 \leqslant i \leqslant m} \sup_{|y| \geqslant ka} (1 + |y|^2)^{m/2} |\phi^{(i)}(y)|.$$

Since $\phi \in \mathcal{S}$, the last expression tends to 0 as $k \to \infty$. From (4.3') and (4.3) it follows that for any real $\alpha$

$$k^{\alpha} f\left(\frac{x}{k}\right) \to 0,\ k \to \infty \quad \text{in } \mathcal{S}$$

and the lemma follows from the definition.

Now we prove a theorem that is similar to theorem 1 in the case of a quasi-asymptotics at zero.

**Theorem 3.** *Let $f \in \mathcal{S}'$ be a locally integrable function in some neighbourhood of the point $x = 0$, and for some automodel function $\rho$ of order $\alpha < 1$*

$$\frac{f(x)}{\rho\left(\frac{1}{|x|}\right)} \to C_{\pm}, \quad x \to \pm 0, \quad (C_+, C_-) \neq 0. \tag{4.4}$$

*Then $f$ has a quasi-asymptotics at zero with respect to $\rho$ and*

$$\frac{1}{\rho(k)} f\left(\frac{x}{k}\right) \to C_+\theta(x)x^{-\alpha} + C_-\theta(-x)(-x)^{-\alpha}, \quad k \to \infty \quad \text{in } \mathcal{S}'. \tag{4.5}$$

**Proof.** Let a number $\delta \in (0, 1)$ be sufficiently small, so that the segment $[-\delta, \delta]$ is contained in the neighbourhood of zero defined in the conditions of theorem 3. According to the lemma it is sufficient to prove that for any $\phi \in \mathcal{S}$ there exist limits of the following expressions

$$\frac{1}{\rho(k)} \int_0^{\delta} f\left(\frac{x}{k}\right)\phi(x)\,dx, \tag{4.6}$$

$$\frac{1}{\rho(k)} \int_{-\delta}^{0} f\left(\frac{x}{k}\right)\phi(x)\,dx$$

as $k \to \infty$. We consider, for instance, the first one:

$$\frac{1}{\rho(k)} \int_0^{\delta} f\left(\frac{x}{k}\right)\phi(x)\,dx = \int_0^{\delta} \frac{\rho\left(\frac{k}{x}\right)}{\rho(k)} \frac{f\left(\frac{x}{k}\right)}{\rho\left(\frac{k}{x}\right)} \phi(x)\,dx =$$

$$= \int_{1/\delta}^{\infty} \frac{\rho(k\xi)}{\rho(k)} \frac{f\left(\frac{1}{k\xi}\right)}{\rho(k\xi)} \phi\left(\frac{1}{\xi}\right) \frac{d\xi}{\xi^2}. \tag{4.7}$$

The integrand of the last integral, by vritue of (3.4), tends to $C_+\xi^{\alpha}$ as $k \to \infty$. Let $\epsilon > 0$ be such that $\alpha + \epsilon < 1$. Then according to lemma 1 of Section 3.3 there exists a number $k_0$ such that

$$\frac{\rho(k\xi)}{\rho(k)} \leqslant \xi^{\alpha+\epsilon}, \quad k > k_0, \quad \xi \geqslant \frac{1}{\delta}.$$

On the other hand, the expression $(f(1/(k\xi))/\rho(k\xi))$, owing to (4.4), has a limit as $k \to \infty$ and therefore it is bounded for $\xi \geqslant 1/\delta$, $k > 0$. From these arguments we conclude that it is possible to pass to the limit under the last integral in (4.7). Considering in the same way the second expression in (4.6) we conclude that the limit ratio (4.5) is valid where the right-hand side differs from zero if $C_+$ and $C_-$ are not zero simultaneously.

Note, that unlike the quasi-asymptotcs at infinity, the distribution $g$ in (4.2) is not defined (at least up to a constant factor) by the order of automodelity of the function $\rho$; so, for example, in (4.7) two constants occur, not one. This is connected with the fact that the set of homogeneous distributions in $\mathcal{S}'$ is essentially wider than in $\mathcal{S}'_+$. This 'obstacle' can be overcome by narrowing the class of distributions under consideration. For instance, the following two simple theorems hold.

**Theorem 4.** *Let $\tilde{f}$ be the Fourier transform of a distribution $f \in \mathcal{S}'_+$. Then $\tilde{f}$ has a quasi-asymptotics $\tilde{g}$ at zero with respect to an automodel function $\rho$ of order $\alpha$ if, and only if, $f$ has a quasi-asymptotics $g$ at infinity with respect to $\rho(k)/k$. In addition, $\tilde{g} = F[g]$, so that*

$$\tilde{g}(x) = \frac{C}{(x + i0)^{\alpha+1}}, \quad C \neq 0. \tag{4.8}$$

**Theorem 5.** *Let $\tilde{f}$ be the $B$-transform* of a distribution $f \in \mathcal{S}'_+$. Then $\tilde{f}$ has a quasi-asymptotics $\tilde{g}$ at zero with respect to an automodel function $\rho$ of order $\alpha$ if, and only if, $f$ has a quasi-asymptotics $g$ at infinity with respect to $\rho(k)/k^2$. In addition, $\tilde{g} = B[g]$, so that*

$$\tilde{g}(x) = C f_{-\alpha+1}(x), \quad C \neq 0. \tag{4.9}$$

**Proof.** The proof of theorem 4 immediately follows from formulae (1.10) and (1.5) of Section 2.1, and also from (4.1). Similarely, theorem 5 at once follows from formulae (3.9) and (3.14) of Section 2.3, and (4.1).

5. **The one-dimensional case. Asymptotics by translations.** The notion of quasi-asymptotics of distributions at infinity generalizes the notion of the usual automodel asymptotics of functions (at least for $\alpha > -1$). However, there is another generalization, namely, the so-called *asymptotics by translations* (see L. Schwartz, 1950-1951), which is closely connected with the concept of a wave form (see L. Hörmander, 1971).

**Definition.** Let $f \in \mathcal{S}'(\mathbf{R}^1)$ and let $\rho$ be an automodel function. We shall say that $f$ *has an asymptotics by translations at infinity* if

$$\frac{1}{\rho(a)} f(\xi + a) \to g(\xi) \not\equiv 0, \quad a \to \infty \quad \text{in } \mathcal{S}'. \tag{5.1}$$

It is easy to verify that in this case $g(\xi + a) = g(\xi)$ for any real $a$. Therefore, for any $\phi \in \mathcal{S}$ (see Section 1.7)

$$(g, \phi) = C \int \phi(\xi)\, d\xi, \quad C \neq 0, \tag{5.2}$$

so that in (5.1), $g(\xi) = C \neq 0$.

The next theorem shows that for $\alpha > -1$ the class of distributions which possesses quasi-asymptotics at infinity is wider than the class of distributions which possesses an asymptotics by translations at infinity.

---

*For the $B$-transform, see Section 2.3.

**Theorem 1.** *Let $f \in \mathcal{S}'_+$ and let $\rho$ be an automodel funciton of order $\alpha > -1$. If for some $\phi_1 \in \mathcal{S}$, such that $\int \phi_1(\xi)\, d\xi \neq 0$,*

$$\frac{1}{\rho(a)}(f(\xi + a), \phi_1(\xi)) \to C \neq 0, \quad a \to \infty$$

*then $f$ has a quasi-asymptotics at infinity with respect to $\rho$.*

**Proof.** Consider the function $f_1 = f * \hat{\phi}_1$ where $\hat{\phi}_1(\xi) = \phi_1(-\xi)$. By the arguments of theorem 1, $f_1(\xi)/\rho(\xi) \to C \neq 0$ as $\xi \to +\infty$ and $f_1(\xi)$ fast decreases as $\xi \to -\infty$. Therefore, as in the proof of theorem 1 of Section 3.4, one can conclude that for some $g \in \mathcal{S}'_+$

$$\frac{1}{\rho(k)} f_1(k\xi) \to g(\xi) \not\equiv 0, \quad k \to \infty \quad \text{in } \mathcal{S}'.$$

Hence using formula (1.12) of Section 2.1 we conclude that the Fourier transform $\tilde{f}_1 = \tilde{f}\, \tilde{\hat{\phi}}_1$ has a quasi- asymptotics at zero with respect to $k\rho(k)$. Owing to theorem 4 of Section 3.4 it remains to prove that $\tilde{f}$ has a quasi- asymptotics at zero with respect to $k\rho(k)$. In turn, owing to the lemma in Section 3.4 it remains to be proved that for some multiplicator $\eta \in \theta_M$ (see Section 1.4) which is equal to 1 in some neighbourhood of the origin the distribution $\eta\tilde{f}$ has a quasi-asymptotics at zero with respect to $k\rho(k)$. Since by the arguments of theorem 1 $\tilde{\hat{\phi}}_1(0) \neq 0$ then there exist numbers $\delta > 0$ and $\epsilon > 0$ such that $|\tilde{\hat{\phi}}_1(x)| > \epsilon$ for $|x| < \delta$. We suppose in addition that $\eta(x) \equiv 0$ for $|x| > \delta$. Then for any $\phi \in \mathcal{S}$ we have

$$\frac{\eta\left(\dfrac{x}{k}\right)}{\tilde{\hat{\phi}}\left(\dfrac{x}{k}\right)}\phi(x) \to \frac{\phi(x)}{\tilde{\hat{\phi}}_1(0)}, \quad k \to \infty \quad \text{in } \mathcal{S},$$

so our statement follows from the equality

$$\frac{1}{k\rho(k)}\left(\eta\left(\frac{x}{k}\right)\tilde{f}\left(\frac{x}{k}\right), \phi(x)\right) = \left(\frac{1}{k\rho(k)}\tilde{f}_1\left(\frac{x}{k}\right), \frac{\eta\left(\dfrac{x}{k}\right)}{\tilde{\hat{\phi}}_1\left(\dfrac{x}{k}\right)}\phi(x)\right)$$

because the right-hand side has a limit as $k \to \infty$ (see Section 1.3).

The next theorem will prove to be useful later.

**Theorem 2.** *Let $f \in \mathcal{S}'$ and for any $\phi \in \mathcal{D}$ let there exist the limit*

$$\frac{1}{\rho(a)}(f(\xi + a), \phi(\xi)) \to C\int \phi(\xi)\, d\xi, \quad a \to \infty, \quad C \neq 0 \tag{5.3}$$

*where $\rho$ is some automodel function. Then $f$ has an asymptotics by translations at infinity with respect to $\rho$.*

**Proof.** It is necessary to prove that (5.3) takes place for all functions $\phi$ from $\mathcal{S}$. Since $\mathcal{D}$ is dense in $\mathcal{S}$ (see Section 1.1) it is sufficient to prove that the family of distributions $\{1/(\rho(a))f(\xi + a), a \to \infty\}$ is bounded in $\mathcal{S}'$ (see Section 1.3). For that we use the decomposition of unity in $\mathbf{R}^1$ which was described in Section 2.2, namely

$$1 \equiv \sum_{k=-\infty}^{\infty} e(\xi - k)$$

where $e \in ((-R, R))$ for some $R > 0$. For some fixed $\epsilon > 0$ we choose $k_0 > 0$ as in lemma 1 of Section 3.2. Then for any $\phi \in \mathcal{S}$ we have

$$\frac{1}{\rho(a)}(f(\xi + a), \phi(\xi)) = \sum_{k=-\infty}^{\infty} \frac{1}{\rho(a)}(f(\xi + a), \phi(\xi)e(\xi - k)) =$$

$$= \sum_{k \leqslant k_0 - a} \frac{1}{\rho(a)}(f(\xi), \phi(\xi - a)e(\xi - a - k)) +$$

$$+ \sum_{k > k_0 - a} \frac{1}{\rho(a)}(f(\xi + a + k), \phi(\xi + k)e(\xi)). \tag{5.4}$$

Initially, we estimate the first sum in the right-hand side of (5.4). For a sufficiently large $a > 0$ and $k \leqslant k_0 - a$, and for some $p > 0$, and $C_i > 0$ which does not depend on $a$ and $k$, the following inequalities hold

$$|(f(\xi), \phi(\xi - a)e(\xi - a - k))| \leqslant C_1 \|\phi(\xi - a)e(\xi - a - k)\|_p \leqslant$$

$$\leqslant C_2 \max_{0 \leqslant i \leqslant p} \sup_{\xi} (1 + \xi^2)^{p/2} \left| \frac{d^i}{d\xi^i} [\phi(\xi - a)e(\xi - a - k)] \right| \leqslant$$

$$\leqslant C_3 \max_{0 \leqslant i, j \leqslant p} \sup_{\xi} (1 + \xi^2)^{p/2} \left| \phi^{(i)}(\xi - a)e^{(j)}(\xi - a - k) \right| \leqslant$$

$$\leqslant C_4 \max_{0 \leqslant i \leqslant p} \sup_{|\xi - a - k| \leqslant R} (1 + \xi^2)^{p/2} \left| \phi^{(i)}(\xi - a) \right| \leqslant$$

$$\leqslant C_5 \sup_{|\xi - a - k| \leqslant R} \frac{(1 + \xi^2)^{p/2}}{[1 + (\xi - a)^2]^{p/2+1}} \leqslant \frac{C_5}{k^2}.$$

From this it follows that the first sum in the right-hand side of (5.4) is bounded as $a \to \infty$.

Next we estimate the second sum. At first we note that from (5.3) it follows that there exist numbers $q > 0$ and $C > 0$ which do not depend on $a$, such that for every $\phi \in ((-R, R))$ (see Section 1.2)

$$\left| \frac{1}{\rho(a)}(f(\xi + a), \phi(\xi)) \right| \leqslant C \|\phi\|_q, \quad a \to \infty.$$

From this, as $\operatorname{supp} e \subset (-R, R)$, we have inequality

$$\left\| \left( \frac{1}{\rho(a)} f(\xi + a + k), \phi(\xi + k)e(\xi) \right) \right\|$$

$$= \frac{\rho(a + k)}{\rho(a)} \frac{1}{\rho(a + k)} \left| (f(\xi + a + k), \phi(\xi + k)(e(\xi)) \right| \leqslant$$

$$\leqslant \frac{\rho(a + k)}{\rho(a)} C \|\phi(\xi + k)e(\xi)\|_q .$$

From lemma 1 of Section 3.2 it follows that for some $m \geqslant 0$ the estimate

$$\frac{\rho(a+k)}{\rho(a)} \leqslant C_6(1+|k|)^m, \; a > k_0, \; k > k_0 - a$$

holds, from which, and from the previous estimate, it follows that

$$\left|\frac{1}{\rho(a)}(f(\xi+a+k), \phi(\xi+k)e(\xi))\right| \leqslant C_7(1+|k|)^m \|\phi(\xi+k)e(\xi)\|_q \leqslant$$

$$\leqslant C_9(1+|k|)^{-2}, \; a > k_0, \; k > k_0 - a.$$

Hence we conclude that the second sum in the right-hand side of (5.4) is bounded as $a \to \infty$.

**Remark.** In fact it was proved that in the arguments of theorem 2 it is sufficient to require that supports of test functions for which (5.3) is fulfilled should lie in any fixed finite interval.

In conclusion we prove a theorem which we shall use in Chapter 5.

**Theorem 3.** *Let* $f \in \mathcal{S}'_+ \bigcap \mathcal{L}^1_{loc}(\xi \geqslant \xi_0)$ *and* $g \in \mathcal{L}^1_{loc}$, $g \geqslant 0$, $g(\xi) = 0$ *if* $\xi < 0$. *Moreover, let* $g$ *have a quasi-asymptotics at infinity with respect to an automodel function* $\rho$ *of order* $\alpha > -1$ *and let the following limit relation be true*

$$\frac{f(\xi)}{g(\xi)} \to C \neq 0, \quad \xi \to \infty. \tag{5.5}$$

*Then* $f$ *has also a quasi-asymptotics at infinity with respect to* $\rho$.

To prove this we use the following lemma which generalizes the lemma of Section 3.4.

**Lemma.** *Let* $g \in \mathcal{S}'_+$ *and let* $\rho$ *be an automodel function of order* $\alpha > -1$. *Then there exists a number* $\beta > 0$ *such that for any functions* $\eta \in \mathcal{D}$ *and* $\gamma$ *such that* $0 < \epsilon \leqslant \gamma(k) \leqslant k^\beta$ *the following limit relation takes place*

$$\frac{1}{\rho(k)}\eta\left(\frac{k\xi}{\gamma(k)}\right)g(k\xi) \to 0, \; k \to \infty \quad \text{in } \mathcal{S}'_+.$$

**Proof.** From the argument of the lemma it follows that there exists a number $C > 0$, and an integer $N \geqslant 0$ such that for any $\phi \in \mathcal{S}_+$

$$\left|\frac{1}{\rho(k)}\left(\eta\left(\frac{k\xi}{\gamma(k}\right)g(k\xi), \phi(\xi)\right)\right| = \left|\left(g(\xi), \frac{1}{k\rho(k)}\eta\left(\frac{\xi}{\gamma(k)}\right)\phi\left(\frac{\xi}{k}\right)\right)\right| \leqslant$$

$$\leqslant \frac{C}{k\rho(k)}\left\|\eta\left(\frac{\xi}{\gamma(k)}\right)\phi\left(\frac{\xi}{k}\right)\right\|_{N, \mathbb{R}^1_+}, \; k \to \infty.$$

As $\eta \in \mathcal{D}$ it is easy to see that

$$\left\|\eta\left(\frac{\xi}{\gamma(k)}\right)\phi\left(\frac{\xi}{k}\right)\right\|_{N, [0,\infty)} \leqslant C_1\gamma^N(k), \; k = 1, 2, \cdots$$

where $C_1$ does not depend on $k$. Thus

$$\left|\frac{1}{\rho(k)}\left(\eta\left(\frac{k\xi}{\gamma(k)}\right)g(k\xi), \phi(\xi)\right)\right| \leqslant C_2\frac{\gamma^N(k)}{k\rho(k)}, \quad k \to \infty.$$

Owing to the condition $\alpha > -1$ and lemma 1 of Section 3.2 it follows that $k^{1-\delta}\rho(k) \to \infty$ ad $k \to \infty$ for a sufficiently small $\delta > 0$. Now we put $\beta = \delta/N$.

Now we prove theorem 3.

**Proof.** Owing to condition (5.5) for some $\xi_0 > 0$ we have

$$f(\xi) = g(\xi)(C + \epsilon(\xi)), \quad \xi \geqslant \xi_0$$

where $\epsilon(\xi) \to 0$ as $\xi \to \infty$. Thus if remains to prove that

$$\frac{1}{\rho(k)}\epsilon(k\xi)g(k\xi) \to 0, \quad k \to \infty \quad \text{in } \mathcal{S}'_+.$$

In turn it will be proved that if for some $N \geqslant 0$ (see theorem 2,* Section 3.4)

$$\frac{1}{k^N\rho(k)}(\epsilon g)^{(-N)}(k) = \frac{1}{\rho(k)}(\epsilon(k\xi)g(k\xi), f_N(1-\xi)) \to 0, \quad k \to \infty. \tag{5.6}$$

Let $\eta$ be a $C^\infty$-function, $0 \leqslant \eta(\xi) \leqslant 1$, $\eta(\xi) \equiv 1$ for $\xi > \xi_0 + 1$ and $\eta(\xi) \equiv 0$ for $\xi < \xi_0$. According to the lemma in order to prove (5.6) it is sufficient to show that for some $\beta \in (0, 1)$

$$\frac{1}{\rho(k)}\left(\eta(k^{1-\beta}\xi)\epsilon(k\xi), f_N(1-\xi)\right) \to 0, \quad k \to \infty. \tag{5.7}$$

But for a sufficiently large $N$ we have

$$\left|\frac{1}{\rho(k)}\int_0^\infty \eta(k^{1-\beta}\xi)\epsilon(k\xi)g(k\xi)f_N(1-\xi)\,d\xi\right| \leqslant$$

$$\leqslant |\epsilon(k\xi_k^*)|\frac{1}{\rho(k)}\int_{\xi_0 k^{\beta-1}}^\infty g(k\xi)f_N(1-\xi)\,d\xi =$$

$$= |\epsilon(k\xi_k^*)|\left[\frac{1}{\rho(k)}(g(k\xi), f_N(1-\xi)) + \epsilon_1(k)\right], \quad k \geqslant k_0 \tag{5.8}$$

where $\epsilon_1(k) = 0$ for $\xi_0 k^{\beta-1} < 1$, $\xi_k^*$ satisfy the inequalities $\xi_0 k^{\beta-1} \leqslant \xi_k^* \leqslant 1$ and thus $\epsilon(k\xi_k^*) \to 0$ as $k \to \infty$; we have also used the properties $g \geqslant 0$ and $f_N \geqslant 0$.

On the other hand, as $g$ has a quasi-asymptotics at infinity with respect to $\rho$ for a sufficiently large $N \geqslant 0$, we have, by virtue of theorem 2,

$$\frac{1}{\rho(k)}(g(k\xi), f_N(1-\xi)) = \frac{1}{k^N\rho(k)}g^{(-N)}(k) \to C_1, \quad k \to \infty.$$

From this and from (5.8) the limit relation (5.7) follows.

---

*This theorem is obviously true also for the case when $C = 0$.

**6. Quasi-asymptotics by selected variable.** Later we shall need the quasi-asymptotic properties of distributions of several variables with respect to one selected variable.

Let $\psi \in \mathcal{S}'(\mathbf{R}_+^1 \times \mathbf{R}^n) = \mathcal{S}'_+(n)$. We say that $\psi(\lambda, u)$, $\lambda \in \mathbf{R}_+^1$, $u \in \mathbf{R}^n$ *has a quasi-asymptotics at infinity by* $\lambda$ *with respect to an automodel function* $\rho$ if there exists the limit relation

$$\frac{\psi(k\lambda, u)}{\rho(k)} \to g(\lambda, u) \not\equiv 0, \ k \to \infty \ \text{ in } \mathcal{S}'_+(n). \tag{6.1}$$

Similarly, if $\Phi \in \mathcal{S}'(\mathbf{R}^{n+1})$ then $\Phi(\kappa, x)$, $\kappa \in \mathbf{R}^1$, $x \in \mathbf{R}^n$, *has a quasi-asymptotics at zero by* $\kappa$ *with respect to an automodel function* $\rho$ if there exists the limit relation

$$\frac{\Phi\left(\frac{\kappa}{k}, x\right)}{\rho(k)} \to g(\kappa, x) \not\equiv 0, \ k \to \infty \ \text{ in } \mathcal{S}'(\mathbf{R}^{n+1}). \tag{6.2}$$

If the order of the automodel function $\rho$ is equal to $\alpha$ then the function $g(\lambda, u)$ in (6.1) has the representation (cf. (4.1), Section 3.4)

$$g(\lambda, u) = f_{\alpha+1}(\lambda) \times \psi_0(u) \tag{6.3}$$

where $\psi_0 \in \mathcal{S}'(\mathbf{R}^n)$ and $\psi_0 \not\equiv 0$. Similarly, if $\Phi \in \mathcal{S}'_+(n)$ then for $g(\kappa, x)$ in (6.2) we have (cf. (4.9), Section 3.4)

$$g(\kappa, x) = f_{-\alpha+1}(\kappa) \times g_0(x), \ g_0 \in \mathcal{S}'(\mathbf{R}^n), \ g_0 \not\equiv 0. \tag{6.4}$$

Most statements of the previous subsections of this section also remain valid in this case. Necessary changes in the formulae are obvious. For instance, we formulate here analogies of theorems 2 and 3 of Section 3.3:

**Theorem 1.** *A distributions* $\psi(\lambda, u)$ *from* $\mathcal{S}'_+(n)$ *has a quasi-asymptotics at infinity by* $\lambda$, *with respect to an automodel function* $\rho$ *of order* $\alpha$, *if and only if there exists a number* $N > -1 - \alpha$ *such that* $\psi^{(-N)}(\lambda, u) = f_N(\lambda) * \psi(\lambda, u)$, *the* $N$*-th primitive by* $\lambda$, *is a local integrable functions in* $\lambda$ *and has the usual asymptotics as* $\lambda \to \infty$ *with respect to* $\lambda^N \rho(\lambda)$. *In other words, for any* $\phi \in \mathcal{S}(\mathbf{R}^n)$ *the distribution*

$$\psi_\phi^{(-N)}(\lambda) = (\psi^{(-N)}(\lambda, u), \phi(u))$$

*is regular and*

$$\frac{\psi_\phi^{(-N)}(\lambda)}{\lambda^N \rho(\lambda)} \to C_\phi^{(N)} \not\equiv 0, \ \lambda \to \infty.$$

**Theorem 2.** *Let* $\psi \in \mathcal{S}'_+(n)$ *and let* $\rho$ *be an automodel function. If for any* $\phi \in \mathcal{S}_+(n)$, *which has a bounded support in* $\lambda$, *the sequence*

$$\frac{1}{\rho(k)}(\psi(k\lambda, u), \phi(\lambda, u)), \ k \to \infty \tag{6.5}$$

*converges, and if at least for one such* $\phi$ *that limit is not equal to zero, then* $\psi$ *has a quasi-asymptotics at infinity by* $\lambda$ *with respect to* $\rho$.

Now we prove theorem 3 which has no one-dimensional analogy. We put

$$T_0 = [(\lambda, u) : \lambda \geqslant 0, |u| \leqslant 1].$$

**Theorem 3.** *Let $\rho$ be an automodel function and let $\psi \in \mathcal{S}'(T_0)$ possess the following properties:*

(i) *For any $\phi \in \mathcal{S}(\mathbf{R}^1)$ and $\nu \in S^{n-1}$ there exists the limit*

$$\lim_{k\to\infty}\frac{1}{\rho(k)}(\psi(k\lambda, u), \phi(\lambda + (\nu, u))) \not\equiv 0; \tag{6.6}$$

(ii) *For $k \geqslant k_0 \geqslant 1$*

$$\frac{1}{\rho(k)}\left|(\psi(k\lambda, u), \phi(\lambda + (\nu, u)))\right| \leqslant C_\phi \tag{6.7}$$

*where $C_\phi$ does not depend on $\nu \in S^{n-1}$ and $k \geqslant k_0$. Then $\psi$ has a quasi-asymptotics at infinity by $\lambda$ with respect to $\rho$.*

**Proof.** Let the order of $\rho$ be equal to $\alpha$. At first we prove the estimate

$$\frac{1}{\rho(k)}\left|(\psi(k\lambda, u), \phi(a\lambda + (\nu, u)))\right| \leqslant C(a^N + a^{-N})\|\phi\|_{N,[-1,\infty)} \tag{6.8}$$

for all $\phi \in \mathcal{S}$, $k \to \infty$, $a > 0$ and for some $C > 0$ and $N > 0$.

Fix $\epsilon > 0$ and choose $k_1 > k_0$ such that for $k > k_1$ the estimates of lemma 1 of Section 3.2 are satisfied. Consider two cases when $a \geqslant k/k_1$ and $a < k/k_1$.

(1) The case $a \geqslant k/k_1 > 1$. If $\psi \in \mathcal{S}'_q(T_0)$ then for some $C_i > 0$ and an integer $q \geqslant 0$ we have

$$\left|\left(\frac{\psi(k\lambda, u)}{\rho(k)}, \phi(a\lambda + (\nu, u))\right)\right| = \frac{1}{k\rho(k)}\left|\left(\psi(\lambda, u), \phi\left(\frac{a\lambda}{k} + (\nu, u)\right)\right)\right| \leqslant$$

$$\leqslant \frac{C_1}{k\rho(k)}\left\|\phi\left(\frac{a\lambda}{k} + (\nu, u)\right)\right\|_{q,T_0} \tag{6.9}$$

where

$$\left\|\phi\left(\frac{a\lambda}{k} + (\nu, u)\right)\right\|_{q,T_0} = \max_{j_0+|j|\leqslant q}\ \sup_{(\lambda,u)\in T_0}(1+\lambda)^q\left|\partial_\lambda^{j_0}\partial_\lambda^{j}\phi\left(\frac{a\lambda}{k} + (\nu, u)\right)\right| \leqslant$$

$$\leqslant C_2 \max_{0\leqslant i,l\leqslant q}\left[\left(\frac{k}{a}\right)^i + \left(\frac{a}{k}\right)^l\right]\|\phi\|_{q,[-1,\infty)} \leqslant \tag{6.10}$$

$$\leqslant C_3 a^q \|\phi\|_{q,[-1,\infty)}.$$

On the other hand, according to lemma 1 of Section 3.2, $\rho(k) \geqslant C_4 k^{\alpha-\epsilon}$ for $k > k_1$ and hence

$$\rho(k) \geqslant C_5 a^{-|\alpha|-\epsilon}. \tag{6.11}$$

Thus from (6.9)-(6.11) we obtain the estimate

$$\left|\left(\frac{\psi(k\lambda,u)}{\rho(k)},\phi(a\lambda+(\nu,u))\right)\right| \leqslant C_6 a^{q+|\alpha|+\epsilon}\|\phi\|_{q,[-1,\infty)}.$$

(2) The case $0 < a < (k/k_1)$, $k > k_1$. We consider the set of tempered distributions $\{f_{\nu,k}, \nu \in S^{n-1}, k > k_1\}$ which is defined by the formula

$$(f_{\nu,k},\phi) = \left(\frac{\psi(k\lambda,u)}{\rho(k)},\phi(\lambda+(\nu,u))\right), \quad \phi \in \mathcal{S}(\mathbf{R}^1).$$

Supports of all the $f_{\nu,k}$ are contained in the set $[-1,\infty)$ and hence $f_{\nu,k} \in \mathcal{S}'([-1,\infty))$. Condition (6.7) shows that this set of distributions is bounded in $\mathcal{S}'([-1,\infty))$ and thus in $\mathcal{S}'_m([-1,\infty))$ for some integer $m \geqslant 0$. This means that the constant $C_\phi$ in (6.7) is equal to $C_7\|\phi\|_{q,[-1,\infty)}$, if $q$ is chosen $\geqslant m$,

$$C_\phi = C_7\|\phi\|_{q,[-1,\infty)}. \tag{6.12}$$

Further, we have

$$\left(\frac{\phi(k\lambda,u)}{\rho(k)},\phi(a\lambda+(\nu,u))\right) = \frac{\rho\left(\frac{k}{a}\right)}{a\,\rho(k)}\left(\frac{\psi\left(\frac{k\lambda}{a},u\right)}{\rho\left(\frac{k}{a}\right)},\phi(\lambda+(\nu,u))\right). \tag{6.13}$$

But according to lemma 1 of Section 3.2,

$$\frac{\rho\left(\frac{k}{a}\right)}{\rho(k)} \leqslant C_8 a^{-\alpha}(a^\epsilon + a^{-\epsilon})$$

and therefore, by (6.7), (6.12) and (6.13), we obtain the estimate (6.8):

$$\left|\left(\frac{\psi(k\lambda,u)}{\rho(k)},\phi(a\lambda+(\nu,u))\right)\right| \leqslant C_9 a^{\alpha+1}(a^\epsilon + a^{-\epsilon})\|\phi\|_{q,[-1,\infty)}.$$

Continuing the proof of theorem 3 we note that the function $\phi(\xi) = e^{iz_0\xi}$ belongs to the space $\mathcal{S}([-1,\infty))$ if $\operatorname{Im} z_0 = y_0 > 0$ and for $N = 0,1,\cdots$,

$$\begin{aligned}\|e^{iz_0\xi}\|_{N,[-1,\infty)} &= \max_{0\leqslant j\leqslant N}\sup_{\xi\geqslant -1}(1+\xi^2)^{N/2}\left|\frac{d^j}{d\xi^j}e^{iz_0\xi}\right| \leqslant \\ &\leqslant \max_{0\leqslant j\leqslant N}\sup_{\xi\geqslant -1}(1+\xi^2)^{N/2}|z_0|^j e^{-y_0\xi} \leqslant \\ &\leqslant C_{10}(1+|z_0|^N)\left(1+\frac{1}{y_0^N}\right)e^{y_0}.\end{aligned} \tag{6.14}$$

Therefore, by (6.8) and (6.14), for any $\xi \in \mathbf{R}^n$, $k > k_1$ and $y_0 > 0$ we have

$$\left|\left(\frac{\psi(k\lambda,u)}{\rho(k)},e^{iz_0(\lambda+(\xi,u))}\right)\right| = \left|\left(\frac{\psi(k\lambda,u)}{\rho(k)},e^{iz_0|\xi|\left(\frac{\lambda}{|\xi|}+\left(\frac{\xi}{|\xi|},u\right)\right)}\right)\right| \leqslant$$

$$\leqslant C_{11}\left(|\xi|^N + |\xi|^{-N} + |z_0|^N + y_0^{-N}\right)e^{y_0|\xi|}. \quad (6.15)$$

By using theorem 2 to prove theorem 3, it remains to show the convergence of the sequence (6.5) only on those test functions $\phi(\lambda, u)$ that have finite supports in $\mathbf{R}^1$ with respect to $\lambda$. But by the hypothesis of theorem 3 the support of $\psi(\lambda, u)$ on $u$ is the unit ball $|u| \leqslant 1$, so it is sufficient to prove theorem 3 for those test functions $\phi(\lambda, u)$ that have compact support with respect to all variables: that is, for $\phi \in D(\mathbf{R}^{n+1})$.

So let $\phi \in D(\mathbf{R}^{n+1})$. Its Laplace transform $\tilde{\phi}(z)$, $z = (z_0, z_1, \ldots, z_n) = (z_0, \mathbf{z}) \in C^{n+1}$ is an entire function which satisfies the estimation

$$(1 + |z|^p)|\tilde{\phi}(z)| \leqslant C_p' e^{R|y|} \quad (6.16)$$

for any $p \geqslant 0$ and some $R > 0$.

Let $\Gamma$ be an $(n+1)$-dimensional surface in $\mathbf{C}^{n+1}$ which is defined by the formula

$$\Gamma = [z : z = z + iy(x), x \in \mathbf{R}^{n+1}]$$

where

$$y_0(x) = \min\left\{1, \left|\frac{x_0}{x_1}\right|, \ldots, \left|\frac{x_0}{x_n}\right|\right\}, \quad \mathbf{y}(x) = \frac{y_0(x)}{x_0}\mathbf{x}. \quad (6.17)$$

Consider the integral

$$J_k[z^\alpha\tilde{\phi}] = \int_\Gamma z^\alpha\tilde{\phi}(z)\left(\frac{\phi(k\lambda, u)}{\rho(k)}, e^{i(z_0\lambda + (\mathbf{z}, u))}\right) dz_0 \cdots dz_n \quad (6.18)$$

where $\alpha = (\alpha_0, \boldsymbol{\alpha})$ and $z^\alpha = z_0^{\alpha_0}\mathbf{z}^{\boldsymbol{\alpha}}$.

According to (6.17), on the surface $\Gamma$ we have

$$z_0\lambda + (\mathbf{z}, u) = z_0[\lambda + (\xi, u)], \quad \xi = \frac{\mathbf{z}}{z_0} = \frac{\mathbf{x}}{x_0} = \frac{\mathbf{y}(x)}{y_0(x)}, \quad (6.19)$$

and, further,

$$|\xi| = \frac{|\mathbf{z}|}{|z_0|} = \frac{(|z_1|^2 + \cdots + |z_n|^2)^{1/2}}{|z_0|} \geqslant \frac{(|z_1| \cdots |z_n|)^{1/n}}{|z_0|}. \quad (6.20)$$

Besides, taking into account that $|y_1| \leqslant 1, \ldots, |y_n| \leqslant 1$ and $|\mathbf{y}| \geqslant 1$ on $\Gamma$, we get

$$y_0|\xi| = |\mathbf{y}| \leqslant \sqrt{n} \quad (6.21)$$

and

$$y_0 \geqslant \frac{|x_0|}{|\mathbf{x}|}|\mathbf{y}| \geqslant \frac{|x_0|}{|\mathbf{x}|} \geqslant \frac{|x_0|}{|z_1| + \cdots + |z_n|} \geqslant \frac{|z_0| - y_0}{|z_1| + \cdots + |z_n|}$$

so that

$$y_0 \geqslant \frac{|z_0|}{1 + |z_1| + \cdots + |z_n|}. \tag{6.22}$$

If we substitute (6.19)-(6.22) into (6.15) we get the estimate

$$\left|\left(\frac{\psi(k\lambda, u)}{\rho(k)}, e^{i(z_0\lambda+(z,u))}\right)\right| \leqslant C_{12}\frac{1 + |z|^{2N}}{|z_0|^N |z_1 \cdots z_n|^{N/n}},$$

$$z \in \Gamma,\ k > k_1,\ y_0 > 0. \tag{6.23}$$

Besides, owing to (6.6) and (6.19) there exists the limit

$$\lim_{k\to\infty}\left(\frac{\psi(k\lambda, u)}{\rho(k)}, e^{i(z_0\lambda+(z,u))}\right) = c(z),\ z \in \Gamma,\ y_0 > 0 \tag{6.24}$$

(that is, for almost all $z \in \Gamma$).

Now we turn our attention to the integral (6.18). We note that the surface of integration $\Gamma$ is piecewise smooth everywhere except the set $[z \in \Gamma: x_0 = 0]$. Represent $\Gamma$ as the $\bigcup$ of pieces $\Gamma_l$, $l = 0, 1, \ldots, n$.

$$\Gamma = \Gamma_0 \bigcup \Gamma_1 \bigcup \cdots \bigcup \Gamma_n$$

where

$$\Gamma_0 = [z \in \Gamma: |x_j| \leqslant |x_0|, j = 1, 2, \ldots, n], \tag{6.25}$$

$$\Gamma_l = [z \in \Gamma: |x_j| \leqslant |x_l|, j = 0, 1, \ldots, n],\ l = 1, 2, \ldots, n. \tag{6.26}$$

On $\Gamma_0$, evidently, we have

$$y_0(x) = 1,\ z_0 = x_0 + i,\ \mathbf{z} = \left(1 + \frac{i}{x_0}\right)\mathbf{x}.$$

Therefore,

$$dz_0\wedge dz_1 \wedge \cdots \wedge dz_n = \left(1 + \frac{i}{x_0}\right)^n dx_0\wedge dx_1 \wedge \cdots \wedge dx_n. \tag{6.27}$$

$$\left|\left(1 + \frac{i}{x_0}\right)^n\right| \leqslant C_{13}\left(1 + \frac{1}{|x|}\right)^n.$$

Similarly, on $\Gamma_l$ we have

$$y_0(x) = \pm\frac{x_0}{x_l},\ z_0 = x_0 \pm i\frac{x_0}{x_l},\ \mathbf{z} = \left(1 + \frac{i}{x_l}\right)\mathbf{x}$$

and hence

$$dz_0 \wedge dz_1 \wedge \cdots \wedge dz_n = \left(1 \pm \frac{i}{x_l}\right)^n dx_0\wedge dx_1 \wedge \cdots \wedge dx_n. \tag{6.29}$$

$$|x_l| \geqslant \frac{|x|}{n+1},\ \left|\left(1 \pm \frac{i}{x_l}\right)^n\right| \leqslant C_{14}\left(1 + \frac{1}{|x|}\right)^n. \tag{6.30}$$

The estimates (6.16), (6.23) and (6.27)-(6.30) show that for sufficiently large $\alpha_j$, $j = 0, 1, \ldots, n$, the integral (6.18) converges absolutely and, by virtue of (6.24) and the Lebesgue theorem, we have

$$J_k[z^\alpha \tilde{\phi}] \to \int_\Gamma z^\alpha \tilde{\phi}(z) c(z)\, dz_0 \wedge dz_1 \wedge \cdots \wedge dz_n, \quad k \to \infty. \tag{6.31}$$

Now we denote

$$\phi_z(\lambda, u) = z^\alpha \tilde{\phi}(z) e^{i(z_0\lambda + (z, u))}.$$

It is clear that $\phi_z \in \mathcal{S}(T_0)$ for $z \in \Gamma$, $x_0 \neq 0$. Besides, owing to (6.22), for any integer $q \geq 0$ there exists a multi-index $\alpha_1$ such that for all $\alpha \geq \alpha_1$ and any $m \geq 0$ we have

$$\|\phi_z\|_{q, T_0} \leq C_m''(1 + |x|)^{-m}, \quad z \in \Gamma, \ x_0 \neq 0. \tag{6.32}$$

From this, denoting $\Gamma_\epsilon = [z \in \Gamma : |x_0| \geq \epsilon]$ and taking into account (6.27)-(6.30), we conclude that

$$\int_{\Gamma_\epsilon} \phi_z(\lambda, u)\, dz \wedge dz_1 \wedge \cdots \wedge dz_n \to \int_\Gamma \phi_z(\lambda, u)\, dz_0 \wedge dz_1 \wedge \cdots \wedge dz_n, \tag{6.33}$$

$\epsilon \to 0$ in $\mathcal{S}_q(T_0)$ for some $q \geq 0$ and also if $\psi \in \mathcal{S}_q'(T_0)$ then

$$\int_{\Gamma_\epsilon} z^\alpha \tilde{\phi}(z) \left( \frac{\psi(k\lambda, u)}{\rho(k)}, e^{i(z_0\lambda + (z, u))} \right) dz_0 \wedge dz_1 \wedge \cdots \wedge dz_n =$$

$$= \left( \frac{\psi(k\lambda, u)}{\rho(k)}, \int_{\Gamma_\epsilon} \phi_z(\lambda, u) dz_0 \wedge dz_1 \wedge \cdots \wedge dz_n \right).$$

For $\alpha \geq \alpha_1$ and $k > k_1$ it follows from this and (6.33) that

$$J_k[z^\alpha \tilde{\phi}] = \left( \frac{\psi(k\lambda, u)}{\rho(k)}, \lim_{\epsilon \to 0} \int_{\Gamma_\epsilon} \phi_z(\lambda, u) dz_0 \wedge dz_1 \wedge \cdots \wedge dz_n \right).$$

From this and (6.31) it follows that the family of distributions

$$\left\{ \frac{\psi(k\lambda, u)}{\rho(k)}, k \to \infty \right\} \tag{6.34}$$

converges on all test functions of the form

$$\lim_{\epsilon \to 0} \int_{\Gamma_\epsilon} z^\alpha \tilde{\phi}(z) e^{i(z_0\lambda + (z, u))} dz_0 \wedge dz_1 \wedge \cdots \wedge dz_n, \quad \phi \in \mathcal{D}(\mathbf{R}^{n+1}). \tag{6.35}$$

Now we deform the surface of integration $\Gamma_\epsilon$ in (6.35). By the Cauchy-Poincaré theorem we have

$$\int_{\Gamma_\epsilon} \phi_z(\lambda, u) dz_0 \wedge dz_1 \wedge \cdots \wedge dz_n = \left( \int_{D_\epsilon} + \int_{\Gamma_\epsilon(0)} \right) \phi_z(\lambda, u) dz_0 \wedge dz_1 \wedge \cdots \wedge dz_n \tag{6.36}$$

where

$$\Gamma_\epsilon(t) = [z \in \mathbf{C}^{n+1} : z = x + ity(x), \ |x_0| > \epsilon, \mathbf{x} \in \mathbf{R}^n],$$

$$D_\epsilon = [z \in \mathbf{C}^{n+1} : z = x + ity(x), \ |x_0| = \epsilon, \mathbf{x} \in \mathbf{R}^n, 0 \leq t \leq 1].$$

We consider at first the integral (6.36) over $D_\epsilon$. For this we divide $D_\epsilon$ into $n+1$ pieces, $D_\epsilon^0, D_\epsilon^1, \ldots, D_\epsilon^n$, $D_\epsilon^j \bigcap D_\epsilon^l = \varnothing$ if $j \neq l$, where

$$D_\epsilon^0 = [z \in D_\epsilon : |x_j| \leqslant \epsilon, j = 1,2,\ldots,n],$$

$$D_\epsilon^l = [z \in D_\epsilon : |x_l| > \epsilon, |x_j| < |x_l|, j = 1,2,\ldots,n, j \neq l], l = 1,2,\ldots,n.$$

On $D_\epsilon^0$ we have

$$z_0 = \pm\epsilon + it, \quad \mathbf{z} = \left(1 \pm \frac{i}{\epsilon}\right)\mathbf{x}.$$

Therefore,

$$dz_0 \wedge dz_1 \wedge \cdots \wedge dz_n = i\left(1 \pm \frac{i}{\epsilon}\right)^n dt \wedge dx_1 \wedge \cdots \wedge dx_n,$$

$$\lim_{\epsilon\to 0}\int_{D_\epsilon^0}\phi_z(\lambda,u)dz_0\wedge dz_1\wedge\cdots\wedge dz_n = \lim_{\epsilon\to 0} i\sum_{+,-}\pm\int_0^1\int_{-\epsilon}^{\epsilon}\cdots\int_{-\epsilon}^{\epsilon}\phi_{(\pm\epsilon+it,(1\pm i/\epsilon)\mathbf{x})}(\lambda,u)$$

$$\left(1\pm\frac{i}{\epsilon}\right)^n dt\, d\mathbf{x} =$$

$$= i\sum_{+,-}\pm(\pm i)^n\int_{S_0}\phi_{(it,i\mathbf{x})}(\lambda,u)\,dt\,d\mathbf{x} =$$

$$= \sum_{+,-}\pm(\pm i)^n I_0^{\pm}(\lambda,u)$$

where $\pm$ is defined by orientations of the pieces of $D_\epsilon^0$ which correspond to the values $x_0 = \pm\epsilon$,

$$S_0 = [(t,\mathbf{x}) \in \mathbf{R}^{n+1}, 0 \leqslant t \leqslant 1, |x_j| \leqslant 1, j = 1,2,\ldots,n].$$

$$I_0^{\pm}(\lambda,u) = \int_{S_0}\phi_{(it,i\mathbf{x})}(\lambda,u)\,dt\,d\mathbf{x}. \tag{6.38}$$

Similarly, on $D_\epsilon^l$, $l = 1,2,\ldots,n$, we have

$$x_0 = \pm\epsilon, \quad y_0 = \pm\frac{\epsilon}{x_l}. \tag{6.39}$$

Denote by $D_\epsilon^l(\pm,\pm)$ these four pieces of $D_\epsilon^l$ which correspond to various combinations of signs in (6.39), and consider, for instance, the integral (6.36) over $D_\epsilon^l(+,+)$; that is, when $x_0 = \epsilon$ and $x_l \geqslant \epsilon$. (Other cases are considered analogously.) Hence

$$z_0 = \epsilon + \frac{i\epsilon t}{x_l}, \quad \mathbf{z} = \left(1 + \frac{it}{x_l}\right)\mathbf{x}, \quad z \in D_\epsilon^l(+,+)$$

so that after some calculations we obtain

$$dz_0\wedge dz_1\wedge\cdots\wedge dz_n = \frac{i\epsilon}{x_e}\left(1+\frac{it}{x_e}\right)^n dt\,d\mathbf{x}.$$

Therefore, by (6.32), we have

$$\int_{D_\epsilon^l(+,+)} \psi_z(\lambda,u)\,dz_0\wedge dz_1\wedge\ldots\wedge dz_n =$$

$$= \int_0^1 dt \int_\epsilon^\infty dx_\epsilon \int_{|x_j|\leqslant x_l} dx_1 \cdots \widehat{dx_l} \cdots dx_n \times$$

$$\times \frac{i\epsilon}{x_l}\left(1+\frac{it}{x_l}\right)^n \phi_{(\epsilon+i\epsilon\frac{t}{x_l},\,(1+\frac{it}{x_l})x)}(\lambda,u).$$

From this, after performing the change of variables of integration $\mathbf{x} \to \mathbf{x}/\epsilon$, we obtain

$$\lim_{\epsilon\to 0}\int_{D_\epsilon^l(+,+)} \phi_z(\lambda,u)dz_0\wedge dz_1\wedge\cdots\wedge dz_n =$$

$$\lim_{\epsilon\to 0}\int_0^1 dt\int_1^\infty dx_l \int_{|x_j|\leqslant x_l} dx_1\cdots\widehat{dx_l}\cdots dx_n \times$$

$$\times\frac{i}{x_l}\left(\epsilon+\frac{it}{x_l}\right)^n \phi_{(\epsilon+\frac{it}{x_l},\,(\epsilon+\frac{it}{x_l})x)}(\lambda,u) =$$

$$= i^{n+1}\int_0^1 t^n\,dt\int_1^\infty \frac{dx_l}{x_l^{n+1}}\int_{|x_j|\leqslant x_l} dx_1\cdots\widehat{dx_l}\cdots dx_n\,\phi_{(\frac{it}{x_l},\,\frac{it}{x_l}x)}(\lambda,u)$$

where the symbol $\widehat{dx_l}$ means that $dx_l$ must be omitted. Again, by the change of variables of integration $t \to (t/x_l)$, $\mathbf{x} \to (t/x_l)\mathbf{x}$ we obtain

$$\lim_{\epsilon\to 0}\int_{D_\epsilon^l(+,+)} \phi_z(\lambda,u)\,dz_0\wedge dz_1\wedge\cdots\wedge dz_n = i^{n+1}I_l^{+,+}(\lambda,u) \tag{6.40}$$

where

$$I_l^{+,+}(\lambda,u) = \int_{S_l}\phi_{(it,\,i\mathbf{x})}(\lambda,u)\,dt\,dx_1\cdots dx_n.$$

$$S_l = [(t,\mathbf{x}) \in \mathbb{R}^{n+1}: 0\leqslant t\leqslant x_l\leqslant 1,\ |x_j|\leqslant x_l,\ j=1,2,\ldots,n]. \tag{6.41}$$

Thus, by (6.33), (6.36), (6.37), (6.40) and similar equalities for the integrals over other parts $D_\epsilon^l(\pm,\pm)$ of $D_\epsilon^l$ - which can be expressed as $\epsilon\to 0$ by the integrals $I_l^{+,-}$, $I^{l^{-,+}}$ and $I_l^{-,-}$, similar to (6.40)-(6.41) - we conclude that

$$\lim_{\epsilon\to 0}\int_{\Gamma_\epsilon}\phi_z(\lambda,u)\,dz_0\wedge dz_1\wedge\cdots\wedge dz_n = I(\lambda,u)+\int x^\alpha\tilde{\phi}(x)e^{i(x_0\lambda+(x,u))}dx =$$

$$= I(\lambda,u)+(2\pi)^n i^{-|\alpha|}\partial_\lambda^{\alpha_0}\partial_u^\alpha\phi(-\lambda,-u) \tag{6.42}$$

where $I$ is a linear combination (with constant complex coefficients) of the quantities $I_0^\pm$, $I_l^{\pm,\pm}$, $l=1,2,\ldots,n$.

Now we prove that there exists the limit

$$\left(\frac{\psi(k\lambda,u)}{\rho(k)},I(\lambda,u)\right) \to C_I,\ k\to\infty \tag{6.43}$$

for a sufficiently large $\alpha$. We verify this statement, for instance, for the term

$I_0^+(\lambda, u)$. By virtue of (6.38) we have

$$I_0^+(\lambda, u) = \int_{S_0} e^{-t\lambda-(\mathbf{x},u)}(it)^{\alpha_0}(i\mathbf{x})^{\boldsymbol{\alpha}}\tilde{\phi}(it, i\mathbf{x})\,dt\,d\mathbf{x}.$$

By the arguments of theorem 3 (cf. 6.24))

$$\left(\frac{\psi(k\lambda, u)}{\rho(k)}, e^{-t\lambda-(\mathbf{x},u)}\right) \to C(t, \mathbf{x}),\ k \to \infty,\ t > 0 \tag{6.44}$$

and, besides, by (6.15) (cf. (6.23))

$$\left|\left(\frac{\psi(k\lambda, u)}{\rho(k)}, e^{-t\lambda-(\mathbf{x},u)}\right)\right| \leqslant C_{16}\left(t^{-N} + |\mathbf{x}|^{-N}\right) \tag{6.45}$$

for $0 < t \leqslant 1$, $|\mathbf{x}| \leqslant 1$, $k > k_1$. On the other hand, for any $q > 0$ we have

$$\|e^{-\lambda t-(\mathbf{x},u)}\|_{q,T_0} \leqslant C_q'' t^{-q},\ t > 0,\ |\mathbf{x}| \leqslant 1. \tag{6.46}$$

Now, we choose a multi-index $\alpha = (\alpha_0, \boldsymbol{\alpha})$ such that $\alpha_j \geqslant N$, $j = 0, 1, \dots, N$. Taking into account the limit relation (6.44) and estimation (6.45), from the Lebesgue theorem we conclude that

$$\int_{S_0} (it)^{\alpha_0}(i\mathbf{x})^{\boldsymbol{\alpha}}\tilde{\phi}(it, i\mathbf{x})\left(\frac{\psi(k\lambda, u)}{\rho(k)}, e^{-\lambda t-(\mathbf{x},u)}\right) dt\,d\mathbf{x} \to$$

$$\int_{S_0} (it)^{\alpha_0}(i\mathbf{x})^{\boldsymbol{\alpha}}\tilde{\phi}(it, i\mathbf{x})C(t, \mathbf{x})\,dt\,d\mathbf{x}. \tag{6.47}$$

On the other hand, owing to (6.46) for $a_j \geqslant q$, $j = 0, 1, \dots, n$, where $q$ is the order of $\psi$ in the left-hand side of the limit relation (6.47), we can insert the factor $(it)^{\alpha_0}(i\mathbf{x})^{\boldsymbol{\alpha}}\tilde{\phi}(it, i\mathbf{x})$ into the large brackets so that

$$\left(\frac{\psi(k\lambda, u)}{\rho(k)}, I_0^+(\lambda, u)\right) \to \int_{S_0} (it)^{\alpha_0}(i\mathbf{x})^{\boldsymbol{\alpha}}\tilde{\phi}(it, i\mathbf{x})C(t, \mathbf{x})\,dt\,d\mathbf{x}$$

and thus we have obtained the desired result (6.43).

Now we note that the family of distributions (6.34) converges on all test functions of the form (6.35) and hence, owing to (6.42), it converges on all test functions of the form

$$\left\{\partial_\lambda^{\alpha_0}\partial_u^{\boldsymbol{\alpha}}\phi(-\lambda, -u), \phi \in \mathcal{D}(\mathbf{R}^{n+1})\right\}$$

for some fixed multi-index $\alpha = (\alpha_0, \boldsymbol{\alpha})$. Since $\operatorname{supp}\psi$ is contained in $T_0$ which is semi-bounded in $\lambda$ and bounded in $u$, then the family (6.34) converges on all test functions from $\mathcal{D}(\mathbf{R}^{n+1})$ and, hence, on all test function from $\mathcal{S}(T_0)$. Thus, theorem 3 has been completely proved.

**Corollary.** *If $f \in \mathcal{S}'_+$ such that for some even $h \in \mathcal{E}'(\mathbf{R}^1)$ and for all $\phi \in \mathcal{S}$ there exists the limit*

$$\lim_{k\to\infty}\left(\frac{f(k\lambda)}{\rho(k)}, h * \phi\right) \not\equiv 0$$

*then $f$ has a quasi-asymptotics at infinity with respect to $\rho$.*

**Proof.** Introduce the distribution $\psi \in \mathcal{S}'(T_0)$, $T_0 \subset \mathbf{R}^2$ by the formula $\psi(\lambda, u) = f(\lambda) \times h(u)$ (without loss of generality we can suppose that $\operatorname{supp} h \subset [-1, 1]$). By the hypothesis for any $\phi \in \mathcal{S}$ there exist the limits

$$\lim_{k \to \infty} \left( \frac{\psi(k\lambda, u)}{\rho(k)}, \phi(\lambda \pm u) \right) = C_\phi^\pm \not\equiv 0.$$

Thus, owing to theorem 3 for $n = 1$ the distribution $\psi(\lambda, u) = f(\lambda) \times h(u)$ has a quasi-asymptotics at infinity by $\lambda$ with respect to $\rho$. But it is possible if and only if the distribution $f(\lambda)$ has this property.

Chapter 2

# MANY-DIMENSIONAL TAUBERIAN THEOREMS

This chapter is devoted to general Tauberian theorems for the distribution of several variables.

## 4. The general Tauberian theorem and its consequences.

In this section we prove a theorem which establishes the connection between the quasi-asymptotics of a distributions in a cone over a family of linear transformations and the asymptotic behaviour of its Laplace transform.

**1. The Tauberian theorem for a family of linear transformations.** Let $\Gamma$ be a regular cone in $\mathbf{R}^n$, with its vertex at the origin, which has a family of linear automorphisms $\{U_k, k \in I\}$ (see Section 3.1), such that

$$U_k \Gamma = \Gamma, \ V_k \Gamma^* = \Gamma^*, \ V_k = (U_k^T)^{-1}, \ J_k = \det U_k > 0, \ k \in I.$$

Let $f \in \mathcal{S}'(\Gamma)$. Its Laplace transform

$$L[f(\xi)] = \tilde{f}(z), \ \tilde{f}(z) \in H(C), \ C = \operatorname{int} \Gamma^*$$

satisfies the estimation

$$|\tilde{f}(z)| \leqslant M \frac{(1+|z|)^\alpha}{\Delta_C^\beta(y)}, \ z = x + iy \in T^C \tag{1.1}$$

for some $M$, $\alpha$ and $\beta$ (see Section 2.5).

**Theorem 1.** *A distribution $f \in \mathcal{S}'(\Gamma)$ has a quasi-asymptotics in the cone $\Gamma$ over the family $\{U_k, k \in I\}$ with respect to a function $\rho_k$, $k \in I$, such that*

$$\frac{1}{\rho(k)} f(U_k \xi) \to g(\xi) \not\equiv 0, \ k \to \infty, \ k \in I \ \text{ in } \mathcal{S}'(\Gamma), \tag{1.2}$$

*only if the following conditions hold: it is necessary that*

$$(1) \ \frac{1}{J_k \rho_k} \tilde{f}(V_k z) \overset{z \in K}{\rightrightarrows} h(z), \ k \to \infty, \ k \in I,$$

*where $K$ is any compact from $T^C$;*

(2) *there exist numbers $M$, $\alpha$, $\beta$ and $k_1$ such that*

$$\frac{1}{J_k \rho_k} |\tilde{f}(V_k z)| \leqslant M \frac{(1+|z|)^\alpha}{\Delta_C^\beta(y)}, \ k > k_1, \ z \in T^C.$$

*and it is sufficient that*

(A) *there is an open set* $\Omega \subset C$ *such that there exists the limit*

$$\frac{1}{J_k \rho_k} \tilde{f}(iV_k y) \to h(iy),\ k \to \infty,\ k \in I,\ y \in \Omega;$$

(B) *there exists numbers* $M'$, $\alpha'$, $\beta'$, $k_2$ *and a vector* $e \in C$ *such that*

$$\frac{1}{J_k \rho_k} |\tilde{f}(V_k(x + i\delta e))| \leqslant M' \frac{(1 + |x|)^{\alpha'}}{\delta^{\beta'}},\ k > k_2,\ k \in I,\ 0 < \delta \leqslant 1.$$

*In this case*

$$\tilde{g}(z) = h(z),\ z \in T^C, \tag{1.3}$$

*and for any* $m > \alpha' + n$ *the primitive* $f^{(-m)}$ *has an asymptotics in the cone* $\Gamma$ *over the family* $\{U_k, k \in I\}$ *with respect to* $J_k^m \rho_k$, $k \in I$. *In other words,* $f^{(-m)}(\xi)$ *satisfies the conditions and there exists the limit*

$$\frac{1}{J_k^m \rho_k} f^{(-m)}(U_k \xi) \xrightarrow{\xi \in K} g^{(-m)}(\xi),\ k \to \infty,\ k \in I, \tag{1.4}$$

*where* $K$ *is any compact in* $\mathbf{R}^n$.

**Proof. Necessity.** Let (1.2) hold. The sequence of distributions $\{(1/\rho(k))f(U_k \xi), k \to \infty, k \in I\}$ converges in $\mathcal{S}'(\Gamma)$ and therefore it converges in the norm in some $\mathcal{S}_p'(\Gamma)$, so that the following inequality holds

$$\left| \left( \frac{1}{\rho k} f(U_k \xi), \phi(\xi) \right) \right| \leqslant M \|\phi\|_{p,\Gamma},\ \phi \in \mathcal{S}_p(\Gamma),\ k \in I \tag{1.5}$$

(see (3.1) and (1.1), Section 1). Nothing that $\phi(\xi) = e^{i(z,\xi)}$ belongs to $\mathcal{S}_p(\Gamma)$ for every $p$ and $z \in T^C$, we have

$$\left( \frac{1}{\rho k} f(U_k \xi), e^{i(z,\xi)} \right) = \frac{1}{J_k \rho_k} \tilde{f}(V_k z) \to (g(\xi), e^{i(z,\xi)}) =$$

$$= \tilde{g}(z),\ k \to \infty,\ k \in I,$$

the convergence being uniform on any compact set from $T^C$. This means that condition (1) holds.

From (1.5) it follows that

$$\left| \frac{1}{J_k \rho_k} \tilde{f}(V_k z) \right| = \left| \left( \frac{1}{\rho_k} f(U_k \xi), e^{i(z,\xi)} \right) \right| \leqslant M \|e^{i(z,\xi)}\|_{p,\Gamma}. \tag{1.6}$$

For the right-hand side in (1.6) we get

$$\| e^{i(z,\xi)} \|_{p,\Gamma} = \sup_{\xi \in \Gamma, |\alpha| \leqslant p} (1 + |\xi|^2)^{p/2} |\partial^\alpha e^{i(z,\xi)}| =$$

$$= \sup_{\xi \in \Gamma, |\alpha| \leqslant p} (1 + |\xi|^2)^{p/2} |(iz)^\alpha| e^{-(y,\xi)} \leqslant$$

$$\leqslant M_1 (1 + |z|)^p \sup_{\xi \in \Gamma, |\xi| > 1} |\xi|^p e^{-\Delta_C(y)|\xi|} \leqslant \tag{1.7}$$

$$\leqslant M_2 \frac{(1+|z|)^p}{\Delta_C^p(y)}, \quad z \in T^C.$$

Here we used the estimate (compare with Section 1.9)

$$(y,\xi) \geqslant \Delta_C(y)|\xi|, \quad \xi \in \Gamma, \; y \in \operatorname{int}\Gamma^* = C.$$

From (1.6) and (1.7) condition (2) follows.

**Sufficiency.** At first let us prove that the set of distributions $\{(1/\rho_k)f(U_k\xi), k > k_2, k \in I\}$ is bounded in $\mathcal{S}'(\Gamma)$.

Indeed, inversing the Laplace transform we have for a sufficiently large $m$ and $|\xi| \geqslant 1$

$$\frac{1}{J_k^m\rho_k} f^{(-m)}(U_k\xi) = \int\limits_{\mathbf{R}^n + i(e/|\xi|)} \frac{1}{J_k^{m+1}\rho_k} e^{-i(z,\xi)} \tilde{f}(V_k z) K_C^m(V_k z)\, dz = \tag{1.8}$$

$$= \frac{1}{J_k\rho_k} \int\limits_{\mathbf{R}^n} \tilde{f}\left(V_k\left(x + i\frac{e}{|\xi|}\right)\right) K_C^m\left(x + i\frac{e}{|\xi|}\right) e^{((\xi/|\xi|),e)-i(x,\xi)}\, dx.$$

Taking into account (6.4) of Section 2.6 and condition (B) we get from (1.8)

$$\frac{1}{J_k^m\rho_k}\left|f^{(-m)}(U_k\xi)\right| \leqslant M_3 \exp\left[\left(\frac{\xi}{|\xi|}, e\right)\right] \int \frac{(1+|x|)^{\alpha'}|\xi|^{mn+\beta'}}{(|\xi|^2|x|^2 + |e|^2)^{m/2}}\, dx \leqslant$$

$$\leqslant M_4 \int \frac{1+|\eta|)^{\alpha'}}{(|\eta|^2 + |e|^2)^{m/2}} |\xi|^{(m-1)(n+\beta')}\, d\eta \leqslant \tag{1.9}$$

$$\leqslant P(|\xi|), \quad k \in I.$$

Here $P(|\xi|)$ is a polynomial. We have also taken into account that $m > \alpha' + n$ and made the change of variables $|\xi|x = \eta$.

For $|\xi| < 1$ we have

$$\frac{1}{J_k^m\rho_k}|f^{(-m)}(U_k\xi)| = \left|\int\limits_{\mathbf{R}^n + ie} \frac{1}{J_k^{m+1}\rho_k} e^{-i(z,\xi)} \tilde{f}(V_k z) K_C^m(V_k z)\, dz\right| =$$

$$= \frac{1}{J_k\rho_k}\left|\int \tilde{f}(V_k(x+ie)) K_C^m(x+ie) e^{-i(x,\xi)+(e,\xi)}\, dx\right| \leqslant$$

$$\leqslant M_4 \int \frac{(2+|x|)^{\alpha'}}{(|x|^2 + |e|^2)^{m/2}}\, dx \leqslant M_6, \quad k \in I. \tag{1.10}$$

Here we have used condition (B) for $\delta = 1$, the etimate (6.4) of Section 2.6 and the fact that $m > \alpha' + n$.

For each $\phi \in \mathcal{S}$ the following relation holds (see Section 2.8)

$$\left(\frac{1}{\rho_k} f(U_k\xi), \phi(\xi)\right) = \left(\theta_\Gamma^{-m}(\xi) * \theta_\Gamma^m(x) * \frac{1}{\rho_k} f(U_k\xi), \phi(\xi)\right) =$$

$$= \left(\theta_\Gamma^m * \frac{1}{\rho_k} f(U_k\xi), (\theta_\Gamma^m(\eta), \phi(\eta+\xi))\right) =$$

$$= \left(\frac{1}{J_k^m \rho_k} f^{(-m)}(U_k \xi), \phi_m(\xi)\right), \quad k \in I, \tag{1.11}$$

where $\phi_m(\xi) = (\theta_\Gamma^{-m}(\eta), \phi(\eta + \xi))$ is a test function from $\mathcal{S}$. From the estimates (1.9), (1.10) and relation (1.11) it follows that the sequence $\{(1/\rho_k) f(U_k \xi), k \to \infty, k \in I\}$ is bounded in $\mathcal{S}'(\Gamma)$.

Let us show that it converges on a dense set of $\mathcal{S}(\Gamma)$. Indeed, by condition (A)

$$\left(\frac{1}{\rho_k} f(U_k \xi), e^{-(y,\xi)}\right) = \frac{1}{J_k \rho_k} \tilde{f}(iV_k y) \to h(iy), \quad k \to \infty, \quad k \in I, \quad y \in \Omega.$$

So, this sequence converges on the family of test functions $\{e^{-(y,\xi)}, y \in \Omega\}$. But the linear hull of this family is dense in $\mathcal{S}(\Gamma)$. To prove this let us consider an arbitrary functional $F \in \mathcal{S}'(\Gamma)$ which vanishes on all the functions from this family. Let $\tilde{F}(z) \in H(C)$ be the Laplace transform of $F$. By the condition $\tilde{F}(iy) = (F(\xi), e^{-(y,\xi)}) = 0$, $y \in \Omega$ and by the uniqueness theorem for holomorphic functions we conclude that $\tilde{F}(z) = 0$ for $z \in T^C$. Now reference to the well-known consequence of the Hahn-Banach theorem completes the proof of our assertion.

Thus, the sequence of functionals $\{(1/\rho_k) f(U_k \xi), k \to \infty, k \in I\}$ is bounded and converges on a dense set. Thus, it converges in $\mathcal{S}'(\Gamma)$, and (1.2) is proved.

From (1.8) and (1.10) we get, for $k \in I$,

$$\frac{1}{J_k^m \rho_k} f^{(-m)}(U_k \xi) = \begin{cases} \displaystyle\int \frac{1}{J_k \rho_k} \tilde{f}\left(V_k\left(x + i\frac{e}{|\xi|}\right)\right) K_C^m\left(x + i\frac{e}{|\xi|}\right) e^{(\xi/|\xi|, e) - i(x,\xi)} dx, & |\xi| \geqslant 1, \\ \displaystyle\int \frac{1}{J_k \rho_k} \tilde{f}(V_k(x + ie)) K_C^m(x + ie) e^{(\xi, e) - i(x,\xi)} dx, & |\xi| < 1. \end{cases} \tag{1.12}$$

If now $\xi \in K \subset\subset \mathbf{R}^n$, then using (6.4) of Section 2.6 and condition (B) we get for the integrand in (1.12) (for $|\xi| \geqslant 1$,

$$\left|\frac{1}{J_k \rho_k} \tilde{f}\left(V_k\left(x + i\frac{e}{|\xi|}\right)\right) K_C^m\left(x + i\frac{e}{|\xi|}\right) e^{(\xi/|\xi|, e) - i(x,\xi)}\right| \leqslant$$

$$\leqslant M'(1 + |x|)^{\alpha'} |\xi|^{\beta'} M_7 \frac{|\xi|^{mn} e^{(\xi/|\xi|, e)}}{[|\xi|^2 |x|^2 + \Delta_C^2(e)]^{m/2}} \leqslant$$

$$\leqslant M_8 \frac{(1 + |x|)^{\alpha'}}{[|x|^2 + \Delta_C^2(e)]^{m/2}}, \quad k \in I. \tag{1.13}$$

A similar estimation holds for $|\xi| < 1$ (cf. (1.10)). This shows that the integrand in (1.12) is bounded by a function which is summable over $x$ and independent of $k$, if $k > k_2$ and $\xi \in K \subset\subset \mathbf{R}^n$. Therefore, using the necessity of the theorem that has just been proved, we conclude, by condition (1) and the Lebesgue theorem, that one can take the limit in (1.12) when $k \to \infty$, $k \in I$ before the integration. This proves (1.4) and (1.5). From the estimates (1.9) and (1.10) it follows also that $f^{(-m)}$ has an asymptotics in the cone $\Gamma$ over the family of transformations $\{U_k, k \in I\}$ with respect to $J_k^m \rho_k$, $k \in I$.

**2. The general Tauberian theorem for the dilatation group.** Any cone $\Gamma$ has a one-parameter group of linear transformations with respect to which it is invariant. This is the dilatation group $\{kI, k > 0\}$; where $I$ is the unit $n \times n$ matrix. From the results of Section 3.3 it follows that the quasi-asymptotics over a one-parameter group makes sense if it is considered with respect to an automodel function. For the Laplace tranform $\tilde{f}(z)$ of a distribution $f \in \mathcal{S}'(\Gamma)$ the following estimate holds. There are numbers $M$, $a$ and $b$ such that

$$|\tilde{f}(x+iy)| \leqslant M\frac{(1+|x|)^a}{\Delta_C^b(y)}, \quad 0 < y \leqslant 1 \tag{2.1}$$

(see estimate (1.1), Section 4.1).

Let $\rho$ be an automodel function of order $\alpha$.

**Theorem 1.** *For a distribution $f \in \mathcal{S}'(\Gamma)$ to have a quasi- asymptotics in the cone $\Gamma$ with respect to an automodel function of order $\alpha$, so that*

$$\frac{1}{\rho(k)} f(k\xi) \to g(\xi) \not\equiv 0, \ k \to \infty \quad \text{in } \mathcal{S}'(\Gamma), \tag{2.2}$$

*the following conditions are necessary:*

(1) $\dfrac{r^n}{\rho\left(\frac{1}{r}\right)} \tilde{f}(rz) \xrightarrow{z \in K} h(z), \ r \to +0,$

*where $K$ is any compact set in $T^C$,*

(2) *there are numbers* M, $c$ *and* $d$ *such that*

$$\left|\frac{r^n}{\rho\left(\frac{1}{r}\right)} \tilde{f}(rz)\right| \leqslant M\frac{(1+|z|)^c}{\Delta_C^d(y)}, \quad 0 < r \leqslant 1, \ z \in T^C;$$

*and the following conditions are sufficient:*

(A) *there exists a solid subcone $C' \subset C$ ($\operatorname{int} C' \neq \varnothing$) such that*

$$\frac{r^n}{\rho\left(\frac{1}{r}\right)} \tilde{f}(iry) \to h(iy), \ r \to +0, \ y \in C'. \tag{2.3}$$

(B) *there are numbers $M'$, $\beta$, $0 \leqslant \gamma < 1$ and a vector $e \in C$, $\Delta_C(e) = 1$ such that*

$$\frac{r^n}{\rho\left(\frac{1}{r}\right)} |\tilde{f}(r(x+i\delta e))| \leqslant \frac{M'}{\delta^\beta}, \quad 0 < r \leqslant 1 \tag{2.4}$$

*for $|x| = \delta^\gamma \leqslant 1$ and $\delta = 1$, $|x| \leqslant 1$* (See Fig. 8).

*In this case*

$$h(z) = \tilde{g}(z), \ z \in T^C \tag{2.5}$$

*and for any $m$ satisfying the inequality*

$$m > m_0 = \max\left\{a+n, \frac{b-\alpha-n\gamma}{1-\gamma}, \frac{\beta-\alpha-n\gamma}{1-\gamma}\right\}, \tag{2.6}$$

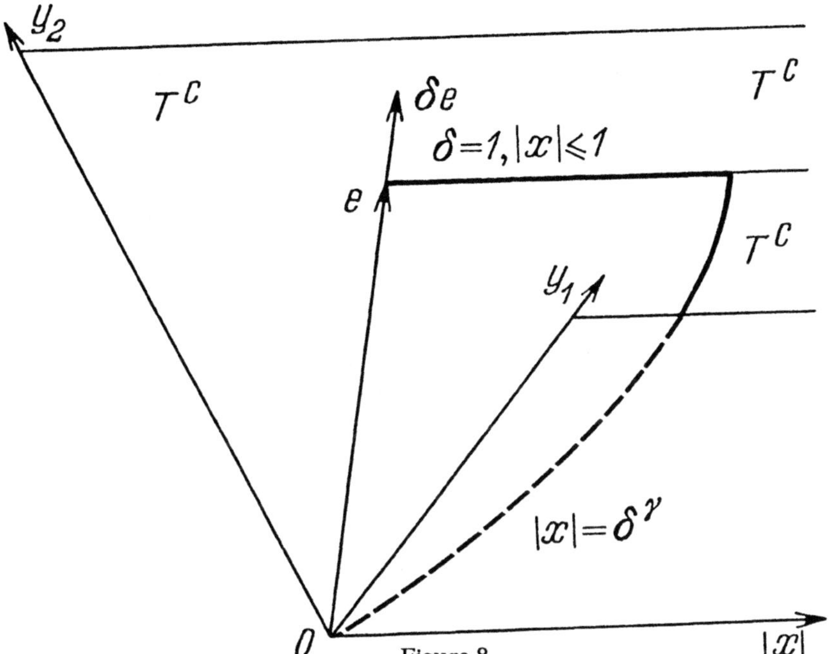


Figure 8.

*where $a$ and $b$ are taken from the estimate (2.1), the primitive $f^{(-m)}(\xi)$ is continuous and has a quasi-asymptotics in $\Gamma$ with respect to an automodel function $k^{nm}\rho(k)$, so that there exists the limit*

$$\frac{1}{k^{nm}\rho(k)} f^{(-m)}(k\xi) \xrightarrow{\xi\in K} g^{(-m)}(\xi), \; k \to \infty \tag{2.7}$$

*where $K$ is any compact set in $\mathbf{R}^n$. In addition, the following equality holds*

$$\mathcal{K}_C^m(z)h(z) = \Gamma(\alpha + mn + n)\int_{\mathrm{pr}\,\Gamma} \frac{g^{(-m)}(\sigma)\,d\sigma}{(-iz,\sigma)^{\alpha+mn+n}}. \tag{2.8}$$

**Proof. Necessity.** This directly follows from the theory of Section 4.1 if one takes into account that $U_k\xi = k\xi$, $V_k(\xi) = k^{-1}\xi$, $J_k = k^n$ and puts $1/k = r$.

**Sufficiency.** Let us show at first that for $m > m_0$ and some $k_1 \geqslant 1$ the following estimate holds

$$\left|\frac{1}{k^{mn}\rho(k)} f^{(-m)}(k\xi)\right| \leqslant P(|\xi|), \quad \xi \in \Gamma, k > k_1, \tag{2.9}$$

where $P$ is some polynomial.

Let $\epsilon > 0$ be so small that $m - m_0 - \epsilon/1{-}\gamma > 0$ and $1/2 \geqslant \epsilon/1{-}\gamma > 0$. According to lemma 1 of Section 3.2 there exists a number $k_\epsilon$ such that for $k > k_\epsilon$ all the estimates of this lemma hold for this particular $\epsilon$. If $k$ is sufficiently large, then from thislemma it also follows that there is an $M_1$ such that

$$M_1 k^{\alpha-\epsilon} < \rho(k), \; k > k_1 \geqslant 1. \tag{2.10}$$

Indeed, to prove this it is sufficient to put $a = 2k_\epsilon k^{-1}$ in the second inequality of the lemma and to assume that $k > 2k_\epsilon = k_1$. Then one can take $M_1$ to be equal to $\rho(k_1)$.

Let $|\xi| \geqslant 1 + \epsilon$. Inversing the Laplace transform we have, for a sufficiently large $m$,

$$\frac{1}{k^{mn}\rho(k)} f^{(-m)}(k\xi) = \frac{2}{k^{n(m+1)}\rho(k)} \int_{\mathbb{R}^n + i(e/|\xi|)} e^{-i(z,\xi)} \tilde{f}\left(\frac{z}{k}\right) K_C^m\left(\frac{z}{k}\right) dz = \tag{2.11}$$
$$= \int \frac{1}{k^n \rho(k)} \tilde{f}\left(\frac{x + i(e/|\xi|)}{k}\right) K_C^m\left(x + i\frac{e}{|\xi|}\right) e^{(\xi/|\xi|, e) - i(x,\xi)} dx.$$

Making the substitution $|\xi| x = \eta$ and taking into account the estimate for the Cauchy kernel (6.4) of Section 2.6, from (2.11) we get

$$\frac{1}{k^{mn}\rho(k)} \left| f^{(-m)}(k\xi) \right| \leqslant \int \frac{|\xi|}{k^n \rho(k)} \left| f\left(\frac{\eta + ie}{k|\xi|}\right) \right| \left| \frac{|\xi|^{nm} M_0 \Delta_C^{(1-n)m}(e)}{[|\eta|^2 + \Delta_C^2(e)]^{m/2}} \right| d\eta \leqslant$$
$$\leqslant M_0 |\xi|^{nm} \int \frac{\rho(k|\xi|)}{\rho(k)} \left| \frac{(k|\xi|)^{-n}}{\rho(k|\xi|)} \tilde{f}\left(\frac{\eta + ie}{k|\xi|}\right) \right| \frac{d\eta}{(1 + |\eta|^2)^{m/2}} \leqslant$$
$$\leqslant M_0 |\xi|^{nm + \alpha + \epsilon} \int \frac{r^n}{\rho\left(\frac{1}{r}\right)} |\tilde{f}(r(\eta + ie))| \frac{d\eta}{(1 + |\eta|^2)^{m/2}}. \tag{2.12}$$

Here we have also used lemma 1 of Section 3.2 and denoted $r = (k|\xi|)^{-1}$, so that $0 < r < r_2 = (k_1|\xi|)^{-1} < 1$.

Let $|\xi| < 1 + \epsilon$. Inversing the Laplace transform and making the estimate similar to (2.12), for a sufficiently large $m$ we have

$$\frac{1}{k^{nm}\rho(k)} |f^{(-m)}(k\xi)| = \frac{1}{k^{n(m+1)}\rho(k)} \left| \int_{\mathbb{R}^n + ie} e^{-i(z,\xi)} \tilde{f}\left(\frac{z}{k}\right) K_C^m\left(\frac{z}{k}\right) dz \right| \leqslant$$
$$\leqslant M_0 e^{1+\epsilon} \int \frac{r^n}{\rho\left(\frac{1}{r}\right)} |\tilde{f}(r(x + ie))| \frac{dx}{(1 + |x|^2)^{m/2}}. \tag{2.13}$$

Here $r = k^{-1}$, so that $0 < r \leqslant k_1^{-1} = r_1 < 1$.

Let us now estimate the function

$$I(r, x) = \frac{r^n}{\rho\left(\frac{1}{r}\right)} \tilde{f}(r(x + ie)) \tag{2.14}$$

for $0 < r \leqslant r_1 < 1$. From estimate (2.4), for $|x| < 1$ and $\delta = 1$ we have

$$|I(r, x)| \leqslant M', \quad 0 < r \leqslant r_1 < 1. \tag{2.15}$$

Assuming now $|x| \geqslant 1$, let us consider two cases:

(1) $0 < r < |x|^{-(1/1-\gamma)} r_1$. Denoting $r' = r|x|^{(1/1-\gamma)}$ we get

$$|I(r,x)| = \left| \frac{(r')^n}{\rho\left(\frac{1}{r'}\right)} \tilde{f}\,(r'(x|x|^{-(1/1-\gamma)} + i|x|^{-(1/1-\gamma)}) \right| \times$$

$$\times \frac{\rho\left(\frac{1}{r}|x|^{-(1/1-\gamma)}\right)}{\rho\left(\frac{1}{r}\right)} \cdot \frac{1}{|x|^{n/1-\gamma}}. \tag{2.16}$$

The first factor in the right-hand side of (2.16) can be estimated according to formula (2.4), in which $r$ should be replaced by $r'$. The second factor for $1-\epsilon < |x|^{-(1/1-\gamma)} \leqslant 1$ and $0 < r < r_1$ is estimated by some constant $M_1$. For $|x|^{-(1/1-\gamma)} < 1-\epsilon$ it is estimated according to the second inequality of lemma 1 of Section 3.2 (in which one should put $a = |x|^{-(1/1-\gamma)}$). Concluding, we obtain the estimate

$$|I(r,x)| \leqslant M_2 |x|^{\beta/1-\gamma} |x|^{-(n/1-\gamma)} |x|^{-(\alpha-\epsilon/1-\gamma)} = M_2 |x|^{\beta-\alpha-n+\epsilon/1-\gamma}. \tag{2.17}$$

(2) $r_1 |x|^{-(1/1-\gamma)} \leqslant r \leqslant r_1 < 1$. Using estimate (2.1) and taking into account that $\Delta(re) \sim r$, we get

$$|I(r,x)| \leqslant \frac{r^n}{\rho\left(\frac{1}{r}\right)} M_3 \frac{(1+|x|r)^a}{r^b} \leqslant \frac{M_3 r^{n-b}}{M_1 r^{-\alpha+\epsilon}} (1+r|x|)^a \leqslant$$

$$\leqslant M_4 |x|^{\epsilon/1-\gamma}$$

Here we have used estimate (2.10) for $k = r^{-1}$. Continuing estimatious further, we obtian

$$|I(r,x)| \leqslant M_5 |x|^{m_1} |x|^{\epsilon/1-\gamma}, \tag{2.18}$$

where

$$m_1 = \max\left\{a, \frac{b-\alpha-n}{1-\gamma}\right\}.$$

From (2.15), (2.17) and (2.18) we get the estimate

$$|I(r,x)| \leqslant M_6 (1+|x|)^{m_2 + (\epsilon/1-\gamma)}, \tag{2.19}$$

where

$$m_2 = \max\left\{a, \frac{b-\alpha-n}{1-\gamma}, \frac{\beta-\alpha-n}{1-\gamma}\right\}.$$

From (2.12) and (2.19), estimate (2.9) follows. We have proved that for $m > m_0$ and for $\epsilon > 0$ there exist numbers $k_1 = k_1(\epsilon)$ and $M = M(\epsilon)$ such that

$$\left| \frac{1}{k^{mn} \rho(k)} f^{(-m)}(k\xi) \right| \leqslant M(1+|\xi|)^{mn+\alpha+\epsilon}, \quad k > k_1. \tag{2.20}$$

So it follows that the sequence $\{1/\rho(k) f(k\xi), k \to \infty\}$ is bounded in $\mathcal{S}'(\Gamma)$, and condition (A) means that this sequence converges on a dense set in $\mathcal{S}(\Gamma)$ (the proof is similar to that of the theorem of Section 4.1). Thus, (2.2) is proved. By necessity, which we have already proved, conditions (1) and (2) hold. If $k = r^{-1}$ we have for $z \in T^C$

$$\frac{r^n}{\rho\left(\frac{1}{r}\right)}\tilde{f}(rz) = \left(\frac{f(k\xi)}{\rho(k)}, e^{i(z,\xi)}\right) \to (g(\xi), e^{i(z,\xi)}) = \tilde{g}(z), \ k \to \infty.$$

Comparing this with (1) we conclude that (2.5) is true. From (2.11), (2.12) and (2.13), and from estimate (2.19), it follows that for $m > m_0$ the function $f^{(-m)}(\xi)$ is continuous. By the same estimation one can take a limit in (2.11), as $k \to \infty$, before the integration (by the Lebesgue theorem) and obtain (2.7). This fact, together with estimate (2.20), proves our assertion that the function $f^{(-m)}(\xi)$ has an asymptotics in $\Gamma$ with respect to the automodel function $k^{mn}\phi(k)$. From (2.7) the continuity of $g^{(-m)}(\xi)$ follows and, therefore, we have for $z \in T^C$

$$\mathcal{K}_C^m(z)h(z) = \mathcal{K}_C^m(z)\tilde{g}(z) = (g^{(-m)}(\xi), e^{i(z,\xi)}) =$$

$$= \int_\Gamma |\xi|^{\alpha+mn} g^{(-m)}\left(\frac{\xi}{|\xi|}\right) e^{i(z,\xi)}\, d\xi =$$

$$= \int_{\mathrm{pr}\,\Gamma} g^{(-m)}(\sigma) \int_0^\infty |\xi|^{\alpha+mn+n-1} e^{|\xi|(iz,\sigma)}\, d|\xi|\, d\sigma =$$

$$= \Gamma(\alpha + mn + n) \int_{\mathrm{pr}\,\Gamma} \frac{g^{(-m)}(\sigma)\, d\sigma}{(-iz,\sigma)^{\alpha+mn+n}}.$$

Here we have used the homogeneity of the quasi-asymtotics (see (3.4), Section 3.3). Thus, (2.8) is proved.

**Remark.** In the hypothesis of the sufficiency of the theory the following limit relation holds for $m > m_0$:

$$\frac{1}{|\xi|^{mn}\rho(|\xi|)} f^{(-m)}(\xi) \underset{\to}{\overset{\xi}{\to}} g^{(-m)}\left(\frac{\xi}{|\xi|}\right), \quad |\xi| \to \infty. \tag{2.21}$$

This follows directly from (2.7) for $|\xi| = k$.

**3. Tauberian theorems for nonnegative measures.** The theorems which have been considered in the previous subsections need the verification of condition (B). This is sometimes a very difficult task. Therefore, it is very desirable to avoid this condition at least in some subclasses of the algebra $H(C)$. In this subsection we consider a class of functions which are the Laplace transforms of nonnegative measures. In the next subsection we shall consider the class of holomorphic functions which have bounded argument in the tube domain $T^C$.

As usual $\Gamma$ is a regular cone which has a family of linear automorphisms $\{U_k, k \in I\}$ (see Section 3.1).

**Theorem 1.** *Let $f = [d\mu]$, where $d\mu$ is a nonnegative tempered measure with support in the cone $\Gamma$. The distribution $f$ has a quasi-asymptotics $g$ in the cone $\Gamma$ over the family $\{U_k, k \in I\}$ with respect to a function $\rho_k$, $k \in I$ if and only if the following conditions hold:*

(A) *there is an open set $\Omega \subset C$ such that there exists the limit*

$$\frac{1}{J_k \rho_k}\tilde{f}(iV_k y) \to h(iy),\ k \to \infty,\ k \in I,\ y \in \Omega;$$

(B) *there exist numbers $M$, $\beta$, $k_0$ and a vector $e \in C$ such that*

$$\frac{1}{J_k \rho_k}\tilde{f}(iV_k \delta e) \leqslant \frac{M}{\delta^\beta},\ k > k_0,\ k \in I,\ 0 < \delta \leqslant 1.$$

*In this case, for any compact set $K$, such that $g^{(-1)}(\xi)$ is continuous in some neighbourhood of $K$, the first primitive of the measure $d\mu$ has an asymptotics over the family of transformations $\{U_k, k \in I\}$ with respect to the function $J_k \rho_k$, $k \in I$ (that is, relation* (1.8) *and estimate* (1.9) of Section 3.1 *hold).*

**Proof.** The statement of the theorem follows directly from theorem 1 of Section 4.1, theorem 3 of Section 3.1, and the following estimation:

$$\frac{1}{J_k \rho_k}|\tilde{f}(V_k(x + i\delta e))| = \frac{1}{J_k \rho_k}\left|\int e^{i(V_k(x+i\delta e),\xi)} d\mu(x)\right| \leqslant$$

$$\leqslant \frac{1}{J_k \rho_k}\int e^{-\delta(V_k e,\xi)} d\mu(\xi) = \frac{1}{J_k \rho_k}\tilde{f}(iV_k \delta e) \leqslant$$

$$\leqslant \frac{M}{\delta^\beta},\ k > k_0,\ k \in I,\ 0 < \delta \leqslant 1. \qquad \square$$

If the family of transformations is the dilatation group, the following theorem holds.

**Theorem 2.** *Let $[d\mu] \in \mathcal{S}'(\Gamma)$ where $d\mu$ is a nonnegative measure. Its primitive has an asymptotics in $\Gamma$ with respect to $k^n \rho(k)$ where $\rho$ is an automodel function of order $\alpha$ if and only if the following condition holds:*

(A) *there is an open subcone $C' \subset C$ such that*

$$\frac{r^n}{\rho\left(\frac{1}{r}\right)}[\tilde{d\mu}](iry) \to h(iy) \not\equiv 0,\ r \to +0,\ y \in C'. \tag{3.1}$$

*In this case the measure $d\mu$ has a quasi-asymptotics in $\Gamma$ with respect to $\rho$:*

$$\frac{1}{\rho(k)}[d\mu](k\xi) \to g(\xi),\ k \to \infty \ \text{ in } \mathcal{S}'(\Gamma) \tag{3.2}$$

*so that*

$$h(z) = \tilde{g}(z),\ z \in T^C,$$

*and*

$$\frac{1}{|\xi|^n \rho(|\xi|)}[d\mu^{(-1)}(\xi) \xrightarrow{\xi \in \Gamma'} g^{(-1)}\left(\frac{\xi}{|\xi|}\right),\ |\xi| \to \infty,\quad \Gamma' \subset\subset \Gamma. \tag{3.3}$$

*The following formula is true*

$$\kappa_C(iy)\tilde{g}(iy) = \Gamma(2n+\alpha)\int_{\mathrm{pr}\,\Gamma}\frac{g^{(-1)}(\sigma)\,d\sigma}{(y,\sigma)^{2n+\alpha}}. \tag{3.4}$$

**Proof. Necessity.** Let $\mu(\xi) = [d\mu]^{(-1)}(\xi)$ have an asymptotics in $\Gamma$ with respect to $k^n\rho(k)$. Using theorem 1 of Section 3.1 we conclude that $d\mu$ has a quasi-asymptotics in $\Gamma$ with respect to $\rho$. Thus (3.2) holds. Now reference to theorem 1 of Section 4.2 completes the proof of the necessity.

**Sufficiency.** Let a vector $e$ belong to $C'$. By the fact that $d\mu$ is nonnegative and by the estimation

$$\rho\left(\frac{1}{\delta}\frac{1}{r}\right) \leqslant M\left(\frac{1}{\delta}\right)^{\alpha+\epsilon}\rho\left(\frac{1}{r}\right), \quad 0<\delta\leqslant 1$$

we have

$$\left|\frac{r^n}{\rho\left(\frac{1}{r}\right)}[d\tilde{\mu}](r(x+i\delta e))\right| = \left|\frac{r^n}{\rho\left(\frac{1}{r}\right)}\int e^{ir(x+\delta e,\xi)}\,d\mu(\xi)\right| \leqslant$$

$$\leqslant \frac{r^n}{\rho\left(\frac{1}{r}\right)}\int e^{-r\delta(e,\xi)}\,d\mu(\xi) \leqslant$$

$$\leqslant M\left(\frac{1}{\delta}\right)^{\alpha+\epsilon}\left(\frac{1}{\delta}\right)^{n}\frac{\delta^n r^n}{\rho\left(\frac{1}{\delta r}\right)}[d\tilde{\mu}](ir\delta e) \leqslant$$

$$\leqslant \frac{M_1}{\delta^{\alpha+n+\epsilon}}.$$

To obtain the last inequality we have used relation (3.1). Thus, conditions (A) and (B) of theorem 1 of Section 4.2 hold (with $\gamma = 0$) and we conclude that (3.2) is true. Now (3.3) follows from (3.2) by theorem 4 of Section 3.3.Formula (3.4) directly follows from (2.8) of Section 4.2 because, in this case, $m = 1$. □

**Remark 1.** *Let an automodel function $\rho$ satisfy the condition that there are numbers $M_0$ and $k_0$ such that*

$$k^n\rho(k) \leqslant M_0 \quad \text{if } k > k_0. \tag{3.5}$$

Let condition (A) of Theorem 2 hold in the following weakened form:

(A1) *there exists a vector $y_0 \in \mathrm{pr}\,C$, such that*

$$\frac{r^n}{\rho\left(\frac{1}{r}\right)}[d\tilde{\mu}](iry_0) \to h(iy_0), \quad r \to +0. \tag{3.6}$$

*Then the measure $d\mu$ has finite mass:*

$$\int d\mu(\xi) = b < \infty, \tag{3.7}$$

*there exists the limit*

$$\lim_{k \to \infty} k^n \rho(k) = d, \tag{3.8}$$

*and for any $z \in T^C$*

$$\frac{r^n}{\rho\left(\frac{1}{r}\right)}[d\tilde{\mu}](rz) \to h(z) \equiv h(iy_0) = bd, \ r \to +\infty. \tag{3.9}$$

Proof. By (3.5) we have

$$\frac{1}{M_0}[d\tilde{\mu}](iry_0) = \frac{1}{M_0}\int e^{-r(y_0,\xi)} d\mu(\xi) \leqslant \frac{r^n}{\rho\left(\frac{1}{r}\right)}\int e^{-r(y_0,\xi)} d\mu(\xi) =$$

$$= \frac{r^n}{\rho\left(\frac{1}{r}\right)}[d\tilde{\mu}](iry_0) \to h(iy_0), \ r \to +0. \tag{3.10}$$

Hence, the expression on the right-hand side of inequality (3.10) has a limit when $r \to +0$. In other words, (3.7) is true and so (3.8) is true also. Hence, for $z \in T^C$ we have

$$\frac{r^n}{\rho\left(\frac{1}{r}\right)}\int e^{ir(y_0,\xi)} d\mu(\xi) \to \tilde{g}(z) = h(iy_0) = db, \ r \to +0.$$

□

For the case of a nonnegative measure with support in the cone $\mathbf{R}^n_+$, K.A. Bukin (1981) has proved the following Tauberian remainder theorem which generalizes the well-known Freud theorem (1951) for $n = 1$.

Let a nonnegative tempered measure $d\mu$ with support in the cone $\mathbf{R}^n_+$ be such that

$$\int_{\mathbf{R}^n_+} e^{-(y,\xi)} d\mu(\xi) = \frac{1}{|y|^\alpha}\left[h\left(\frac{y}{|y|}\right) + r(y)\right], \ h \not\equiv 0, \ y \in \mathbf{R}^n_+$$

for some $\alpha \geqslant 0$, and for $r(y)$ satisfying the estimate

$$|r(y)| \leqslant C_1 \exp[-f(|y|)], \ 0 < |y| \leqslant \delta, \ y \in \mathbf{R}^n_+$$

for some $\delta > 0$ and $C_1$ which do not depend on $y$. Besides, suppose that the functional $f$ has the following properties: it is continuous and monotonically decreases on $(0, \delta]$; moreover, $f(t) \to +\infty$, $t \to +0$.

Then the reproducing function $\mu = [d\mu]^{(-1)}$ of the measure $d\mu$ can be written in the form

$$\mu(\xi) = |\xi|^\alpha\left[H\left(\frac{\xi}{|\xi|}\right) + \nu(\xi)\right].$$

where $\nu(\xi)$ satisfies the condition: there exists a $C_2 > 0$ such that for any cone $\Gamma \subset\subset \mathbf{R}_+^n$

$$|\nu(\xi)| \leqslant C_2\sigma^2 \frac{\psi\left(\frac{|\xi|}{\sigma}\right)}{|\xi|}, \quad \xi \in \Gamma, \quad |\xi| \geqslant \kappa = \kappa(\Gamma) > 0.$$

Here $\psi(\tau)$ is a function which is inverse of the function $tf\,(1/t)$ and $\sigma = \sigma(\Gamma)$ is given by the formula

$$\sigma^2 = \sup_{s \in \operatorname{pr}\Gamma} \left( \frac{1}{s_1^2} + \frac{1}{s_2^2} + \cdots + \frac{1}{s_n^2} \right).$$

**4. Tauberian theorems for holomorphic functions of bounded argument.**

**Definition.** A function $h(z)$, which is holomorphic in a domain $G \subset \mathbf{C}^n$, is said to be a function of *bounded argument* in $G$ if $h(z) \neq 0$, $z \in G$ and there is a number $M$ such that

$$|\arg h(z)| \leqslant M, \quad z \in G,$$

and the function $\arg h(z)$ is assumed to be continuous in $G$.

Note that a holomorphic function with a nonnegative imaginary part in $G$ is a function of bounded argument in $G$.

**Theorem 1.** *Let $f \in \mathcal{S}'(\Gamma)$, and let its Laplace transform $\tilde{f}(z)$ be a function of bounded argument in $T^C$. It has a quasi-asymptotics $g$ in the cone $\Gamma$ over the family $\{U_k, k \in I\}$ with respect to a function $\rho_k$ if and only if the following condition is satisfied:*

(A) *there exists an open set $\Omega \subset C$ such that*

$$\frac{1}{J_k \rho_k} \tilde{f}(iV_k y) \to h(iy), \quad k \to \infty, \quad k \in I, \quad y \in \Omega.$$

*In this case* $\tilde{g}(z) = h(z)$, $z \in T^C$.

**Proof.** According to the general Tauberian theorem it is sufficient to prove that for a function of bounded argument condition (A) implies condition (B) of theorem 1 in Section 4.1. First, let us note that we can restrict ourselves to the case when the function $\tilde{f}$ has a nonnegative imaginary part.

Indeed, suppose that for functions with a nonnegative imaginary part we proved that condition (A) implies condition (B). Since $\tilde{f}$ is a function of bounded argument there exists numbers $\kappa$ and $\phi_0$ such that the function

$$\omega(z) = e^{i\phi_0}[\tilde{f}(z)]^{\kappa}$$

has the arguments which belong to the interval $(0, \pi)$. For this function we have, by hypothesis,

$$e^{i\phi_0}\left[\frac{1}{J_k \rho_k} \tilde{f}(iV_k y)\right]^{\kappa} \to e^{i\phi_0} h^{\kappa}(iy), \quad k \to \infty, \quad k \in I, \quad y \in \Omega.$$

Thus, by our assumption we get the following estimate

$$\left|\frac{1}{J_k \rho_k}\right|^\kappa \left|\tilde{f}(V_k(x+i\delta e))\right|^\kappa \leqslant M'\frac{(1+|x|)^{\alpha'}}{\delta^{\beta'}},\ k>k_2,\ k\in I,\ 0<\delta\leqslant 1$$

for some numbers $M'$, $\alpha'$, $\beta'$ and a vector $e \in C$. Taking the $(1/\kappa)$th power of the both sides of this inequality we obtain condition (B) for $\tilde{f}$.

Thus, without loss of generality we can assume that $\tilde{f}$ has a nonnegative imaginary part.

We shall need the following lemma which is also of interest in its own right.

**Lemma 1.** *Let $U$ be a linear operator in $\mathbf{R}^n$ which leaves the cone $\Gamma$ invariant, and let a function $\tilde{f}(z)$ be holomorphic in $T^C$, and contain a nonnegative imaginary part. Then for any $y \in C$ there exists a constant $M = M(y)$, which does not depend on $U$, such that*

$$|\tilde{f}(V(x+i\delta y))| \leqslant M\frac{(1+|x|)^2}{\delta}|\tilde{f}(iVy)|,\ x\in\mathbf{R}^n,\ y\in C,\ 0<\delta\leqslant 1 \quad (4.1)$$

(see (6.6), Section 2.6).

**Proof.** Let us fix $n$ vectors $y^\nu \in C$, $\nu = 1,\dots,n$, $|y^\nu| = 1$, such that

$$y = \sum_{\nu=1}^n l_\nu y^\nu,\ l_\nu > 0,\ \nu = 1,\dots,n$$

and let

$$x = \sum_{\nu=1}^n q_\nu y^\nu,\quad (q_\nu = q_\nu(x),\ |x|^2 \approx \sum_{\nu=1}^n q_\nu^2)$$

be the decomposition of $x$ with respect to the basis $\{y^\nu, \nu = 1,\dots,n\}$. Let us part

$$z = V(x+i\delta y) = \sum_{\nu=1}^n (q_\nu + i\delta l_\nu)Vy^\nu = \sum_{\nu=1}^n z_\nu V y^\nu.$$

$$\zeta = iVy = \sum_{\nu=1}^n z_\nu V y^\nu.$$

where $z_\nu = q_\nu + i\delta l_\nu$, $\zeta_\nu = il_\nu$, $\nu = 1,\dots,n$, $0<\delta\leqslant 1$. The function

$$f_V(z) = \tilde{f}\left(\sum_{\nu=1}^n z_\nu V y^\nu\right)$$

is holomorphic in the tube domain $\{\operatorname{Im} z_\nu > 0, \nu = 1,\dots,n\} = T^{\mathbf{R}_+^n}$ and has there a nonnegative imaginary part. The distance between the points $z$ and $\zeta$ in the Caratheodory metric is given by the formula

$$C_{R_+^n}(z,\zeta) = \ln\frac{1+\gamma(z,\zeta)}{1-\gamma(z,\zeta)},\quad \gamma(z,\zeta) = \max_\nu\left|\frac{\zeta_\nu - z_\nu}{\zeta_\nu - \bar{z}_\nu}\right|.$$

Now we introduce the holomorhic mapping $F(z) = (f_V(z),\dots,f_V(z)) \in C^n$. Since $\operatorname{Im} f_V(z) > 0$, then

$$F : T^{\mathbf{R}_+^n} \to T^{\mathbf{R}_+^n}.$$

Because the Caratheodory metric does not increase under a holomorphic mapping (see B.V. Shabat, 1976), we have:

$$C_{R_+^n}(\digamma(z),\digamma(\zeta)) = \ln\frac{1+\gamma(\digamma(z),\digamma(\zeta))}{1-\gamma(\digamma(z),\digamma(\zeta))} \leqslant C_{R_+^n}(z,\zeta).$$

Thus, it follows that $\gamma(\digamma(z),\digamma(\zeta)) \leqslant \gamma(z,\zeta)$. In other words

$$\left|\frac{f_V(z)-f_V(\zeta)}{f_V(z)-\overline{f_V(\zeta)}}\right| \leqslant \max_\nu\left|\frac{z_\nu-\zeta_\nu}{z_\nu-\overline{\zeta}_\nu}\right| =$$

$$= \max_\nu\left|\frac{q_\nu+i(1-\delta)l_\nu}{q_\nu+i(1+\delta)l_\nu}\right|. \tag{4.2}$$

Denoting $c_1 = 4\min_\nu l_\nu^2$, $c_2 = 4\max_\nu l_\nu^2$ we have, from (4.2),

$$\left|\frac{\tilde f(z)-\tilde f(\zeta)}{\tilde f(z)-\overline{\tilde f(\zeta)}}\right| \leqslant \max_\nu\left|\frac{\dfrac{q_\nu}{l_\nu}+i(1-\delta)}{\dfrac{q_\nu}{l_\nu}+i(1+\delta)}\right| = \max_\nu\sqrt{1-\frac{4\delta}{\left(\dfrac{q_\nu}{l_\nu}\right)^2+(1+\delta)^2}} \leqslant$$

$$\leqslant \max_\nu\left(1-\frac{4\delta}{\left(\dfrac{q_\nu}{l_\nu}\right)^2+4}\right) \leqslant 1-\frac{c_1\delta}{c_2+|x|^2}. \tag{4.3}$$

Denoting the right-hand side of inequality (4.3) by $c_3$, we conclude that

$$|\tilde f(z)| = |\tilde f(V(x+i\delta y))| \leqslant \frac{c_3+1}{1-c_3}|\tilde f(\zeta)| \leqslant$$

$$\leqslant \frac{2(c_2+|x|^2)}{c_1\delta}|\tilde f(iVy)|.$$

Thus, inequality (4.1) follows. □

To complete the proof of the theorem let us fix a vector $y = e \in \Omega$. Using the lemma we get

$$\frac{1}{J_k\rho_k}|\tilde f(V_k(x+i\delta e))| \leqslant M\frac{(1+|x|)^2}{\delta}\frac{1}{J_k\rho_k}|\tilde f(iV_k e)| \leqslant$$

$$\leqslant M_2\frac{(1+|x|)^2}{\delta}, \tag{4.4}$$

where $k > k_2$, $k \in I$, $0 < \delta \leqslant 1$. In the last inequality of (4.4) we have used condition (A) for $y = e$. Now we use complete the proof by theorem 1 of Section 4.1. □

**Corollary.** *Let the hypothesis of theorem 1 holds and let* $\operatorname{Im}\tilde f(z) \geqslant 0$, $z \in T^C$. *Then it follows from (A) that for any* $m > n+2$ *the primitive* $f^{(-m)}$ *has an asymptotics in* $\Gamma$ *over the family* $\{U_k, k \in I\}$ *with respect to* $J_k^m\rho_k$, $k \in I$.

**Proof.** This statement directly follows from estimate (4.1) of lemma 1 for $y = e \in \Omega$ and from theorem 1 of Section 4.1 where one can put $\alpha' = 2$. □

Later we shall need a local version of theorem 1, adapted for the dilatation group of the cone $\Gamma$. The neighbourhood of zero in $T^C$ is the set $B_\epsilon \cap T^C$.

**Theorem 2.** *Let* $f \in \mathcal{S}'(\Gamma)$, *and let its Laplace transform* $\tilde f$ *have a bounded argument in some neighbourhood of zero in* $T^C$. *For the distribution* $f$ *has a quasi-*

*asymptotics in* $\Gamma$ *with respect to an automodel function* $\rho$.

$$\frac{1}{\rho(k)} f(k\xi) \to g(\xi),\ k \to \infty \quad \text{in } \mathcal{S}'(\Gamma), \tag{4.5}$$

*if and only if the following condition holds:*

(A) *there exists a solid subcone* $C' \subset C$ ($\operatorname{int} C' \neq \emptyset$) *such that*

$$\frac{r^n}{\rho\left(\frac{1}{r}\right)} \tilde{f}(iry) \to h(iy),\ r \to +0,\ y \in C'. \tag{4.6}$$

*In this case* $h(z) = \tilde{g}(z)$, $z \in T^C$.

**Proof.** Let us prove that estimate (2.4) of theorem 1 of Section 4.2, with $\gamma = 0$, follows from (4.6) in some neighbourhood of zero. As in the proof of theorem 1, it is sufficient to consider only the case when $\tilde{f}(z)$ has a nonnegative imaginary part. Note further that, without loss of generality, one can assume that $C = \mathbf{R}^{+^n}$. Indeed, let

$$C_1 = \{y \in \mathbf{R}^n : (e_j, y) > 0,\ j = 1, \ldots, n\},$$

where $e_j \in C^*$, $j = 1, \ldots, n$ are linearly independent vectors such that $C_1 \subset\subset C' \subset C$. Let us consider the linear automorphism

$$y \to Wy = y' = [(e_1, y), \ldots, (e_n, y)]$$

This transforms the cone $C_1$ into the cone $\mathbf{R}^n_+$, and the rays $\{iry, r > 0\}$ into the rays $\{irWy, r > 0\}$. Therefore, if $\tilde{f}(z)$ satisfies condition (4.6) for $y \in C_1 \subset\subset C'$, then $\tilde{F}(z') = \tilde{f}(W^{-1}z')$ satisfies condition (4.6) for $y' \in \mathbf{R}^n_+$. Now if theorem 2 is prove for $C = \mathbf{R}^{+^n}$ then estimate (2.4) of theorem 1 of Section 4.2 holds for some vector $e' \in \mathbf{R}^n_+$, and therefore this estimation holds for $\tilde{f}(z)$ and for some vector $e = W^{-1}e' \in C_1$.

Thus, we assume that $C = \mathbf{R}^n_+$ and $\operatorname{Im} \tilde{f}(z) > 0$ in some neighbourhood of zero in $T^{\mathbf{R}^n_+}$. Let us take a polydisc $\{|z_j - i\epsilon| < \epsilon,\ j = 1, \ldots, n\}$ which lies both in $T^{\mathbf{R}^n_+}$ and in this neighbourhood of zero (it exists if $\epsilon$ is rather small). Let us map it on $T^{\mathbf{R}^n_+}$ by the biholomorphic mapping $\zeta_j = (iz_j / 2\epsilon i - z_j)$, $j = 1, \ldots, n$. Under this mapping each disc $|z_j - i\epsilon| < \epsilon$ is transformed into an upper half-plane $\operatorname{Im} \zeta_j > 0$, the diameter $[iy_j, 0 < y_j < 2\epsilon]$ being transformed into an imaginary axis. The function $\tilde{f}(z)$ becomes a holomorphic function on $\zeta \in T^{\mathbf{R}^n_+}$ which has the limits of (4.6). Using lemma 1 and theorem 1 of Section 4.2 we conclude that the angular limit exists (condtion (1) of theorem 1 of Section 4.2) for all $\zeta \in T^{\mathbf{R}^n_+}$, and that estimate (B) holds for any vector $e \in \mathbf{R}^n_+$ with $\gamma = 0$. Taking the inverse mapping we get estimate (2.4) of theorem 1 of Section 4.2 for $\tilde{f}(z)$, for any $e \in C_1$, and some $\gamma \in (1/2, 1)$. By theorem 1 of Section 4.2 we obtain (4.5). □

**Corollary.** *Let* $f \in$ '$(\Gamma)$ *and* $\operatorname{Im} \tilde{f}(z) > 0$, $z \in T^C$. *If* $f$ *has a quasi-asymptotics in the cone* $\Gamma$ *with respect to an automodel function* $\rho$ *of order* $\alpha$, *then the following inequalities hold:*

$$-1 - n \leqslant \alpha \leqslant 1 - n. \tag{4.7}$$

**Proof.** The proof follows from the fact that for any subcone $C' \subset\subset C$ there is a number $M = M(C')$ such that

$$|\tilde{f}(z)| \leqslant M\frac{1+|z|^2}{|y|}, \quad z \in T^{C'}$$

(see V.S. Valdimirov, 1979) and from the fact that $\mathrm{Im}(-1/\tilde{f}(z)) > 0$, $z \in T^C$, and from condition (A) of theorem 2. □

## 5. Admissible and strictly admissible functions.

From the previous sections we know that the quasi-asymptotics over a dilatation group make sense only if it is considered with respect to automodel (regularly varying) functions. If one considers the quasi- asymptotics over an arbitrary family of linear transformations $\{U_k, k \in I\}$, under which the cone $\Gamma$ is invariant, one has to consider some special classes of functions which are most suitable for this family. This section is devoted to these classes and to some of their generalizations.

1. **Families of linear transformations under which a cone is invariant.** Let $\Gamma$ be a closed, solid, acute cone in $\mathbf{R}^n$ with its vertex at the origin and let $U$ be an automorphism of $\Gamma$ (non-singuluar linear operator in $\mathbf{R}^n$) such that $U\Gamma = \Gamma$. The set of all automorphisms of $\Gamma$ is denoted by $A(\Gamma)$. Each linear operator $U \in A(\Gamma)$ defines two numbers

$$\Lambda = \|U\| = \sup_{|e|=1} |Ue|; \quad \lambda = \inf_{|e|=1} |Ue|. \tag{1.1}$$

For a linear operator $V = (U^T)^{-1}$ under which the cone $\Gamma^*$ is invariant, $V \in A(\Gamma^*)$, the corresponding numbers are

$$\lambda_1 = \frac{1}{\Lambda} = \operatorname{int}_{|e|=1} |Ve|; \quad \Lambda_1 = \frac{1}{\lambda} = \sup_{|e|=1} |Ve|. \tag{1.2}$$

Later we shall require some statements which we now formulate in the form of lemmas.

**Lemma 1.** *For any $\xi \in \operatorname{int}\Gamma$ there is a constant $M = M(\xi)$ which does not depend on $U \in A(\Gamma)$, such that*

$$\frac{|U\xi|}{|\xi|} \leqslant \|U\| = \Lambda \leqslant M\,|U\xi|, \quad U \in A(\Gamma). \tag{1.3}$$

**Proof.** The first inequality of (1.3) is obvious. Let us prove the second inequality. Suppose that it is not satisfied. Then there is a sequence of operators $U_k \in A(\Gamma)$, and a sequence of unit vectors $e_k$, $|e_k| = 1$ such that

$$|U_k e_k| \geqslant k\,|U_k\xi|, \quad k = 1, 2, \cdots$$

Since $\xi \in \operatorname{int}\Gamma$ there exists an $\epsilon > 0$ such that

$$B_\epsilon(\xi) = \xi + B_\epsilon \subset \Gamma.$$

For this $\epsilon$, denoting $\epsilon e_k = \xi_k$, we have

$$|U_k \xi_k| \geqslant k\epsilon |U_k \xi|. \tag{1.4}$$

Since $\xi + \xi_k \in \Gamma$, the sequence $\{U_k(\xi + \xi_k), k = 1, 2, \cdots\}$ lies in $\Gamma$ and therefore

$$\frac{U_k(\xi + \xi_k)}{|U_k(\xi + \xi_k)|} \in \operatorname{pr}\Gamma. \tag{1.5}$$

One can choose a subsequence which converges to some vector $e_0 \in \operatorname{pr}\Gamma$. Without loss of generality we shall assume that sequence (1.5) converges to $e_0$:

$$\frac{U_k\xi + U_k\xi_k}{|U_k\xi + U_k\xi_k|} \to e_0 \in \operatorname{pr}\Gamma,\ k \to \infty.$$

By (1.4) it follows that

$$\frac{U_k\xi_k}{|U_k\xi_k|} \to e_0 \in \operatorname{pr}\Gamma,\ k \to \infty.$$

In a similar way, for the sequence $\{U_k(\xi - \xi_k), k = 1, 2, \cdots\} \subset \Gamma$ it follows that

$$\frac{U_k(\xi - \xi_k)}{|U_k(\xi - \xi_k)|} \in \operatorname{pr}\Gamma. \tag{1.6}$$

Since

$$\frac{U_k(-\xi_k)}{|U_k(-\xi_k)|} = \frac{-U_k\xi_k}{|U_k\xi_k|} \to -e_0,\ k \to \infty$$

then, taking into account (1.4), we have from (1.6)

$$\frac{U_k\xi - U_k\xi_k}{|U_k\xi - U_k\xi_k|} \to -e_0 \in \operatorname{pr}\Gamma,\ k \to \infty.$$

Thus, both vectors $e_0$ and $-e_0$ are shown to belong to $\operatorname{pr}\Gamma$. This fact contradicts the acuteness of the cone $\Gamma$.

**Lemma 2.** *For any $\xi \in \operatorname{int}\Gamma$ there is a number $p = p(\xi)$ which does not depend on $U \in A(\Gamma)$, such that*

$$p\,\Delta_\Gamma(U\xi) \leqslant \lambda \leqslant \frac{\Delta_\Gamma(U\xi)}{\Delta_\Gamma(\xi)},\quad U \in A(\Gamma) \tag{1.7}$$

*where $\Delta_\Gamma(U\xi)$ is the distance between the points $U\xi$ and the boundary of the cone $\Gamma$* (see Section 0.2).

**Proof.** From (1.1) we have $B_\lambda \subset U(B_1)$, where $U(B_1)$ is the image of the unit ball under the action of the operator $U$; therefore, $B_{\lambda\epsilon} \subset U(B_\epsilon)$.

Let $\xi \in \operatorname{int}\Gamma$. There exists an $\epsilon > 0$ such that $\xi + B_\epsilon \subset \Gamma$. Hence,

$$U(\xi + B_\epsilon) = U\xi + U(B_\epsilon) \subset \Gamma.$$

Thus, it follows that $U\xi + B_{\lambda\epsilon} \subset \Gamma$ and, therefore, $\Delta_\Gamma(U\xi) \geqslant \lambda\epsilon$ where one can take $\epsilon = \Delta_\Gamma(\xi)$. Thus, the first inequality in (1.7) has been proved. To prove the second inequality in (1.7) let us suppose that it is not satisfied. Then for any $k > 0$ there exists a linear operator $U_k \in A(\Gamma)$, such that

$$\Delta_\Gamma(U_k \xi) \geqslant k \lambda_k , \tag{1.8}$$

where

$$\lambda_k = \inf_{|e|=1} |U_k e| .$$

In other words, for any $k > 0$, $U_k \xi + B_{k\lambda_k} \subset \Gamma$.

Let us prove that there exists a sequence $\{e_k, |e_k| = 1, k \to \infty\}$ such that $\xi + \alpha e_k \in \Gamma$ for $|\alpha| < 1/2k$. Indeed, it follows from (1.8) that for any $k > 0$ there is $e_k$, $|e_k| = 1$ such that $|U_k e_k| \leqslant 2\lambda_k$. Now for any $\alpha$, $|\alpha| \leqslant 1/2k$ we have

$$|U_k(\alpha e_k)| \leqslant |\alpha||U_k e_k| \leqslant \frac{1}{2}k \cdot 2\lambda_k = k\lambda_k .$$

Hence,

$$U_k \xi + U_k(\alpha e_k) = U_k(\xi + \alpha e_k) \in \Gamma.$$

Therefore, $\xi + \alpha e_k \in \Gamma$.

Since $|e_k| = 1$, one can choose a subsequence $\{e_{k_1}, k_1 \to \infty\}$ which converges to some vector $e_0$, $|e_0| = 1$. Taking into account that the cone $\Gamma$ is closed we conclude that

$$\xi + te_0 \in \Gamma, \quad -\infty < t < \infty.$$

This contradicts the assumption that the cone $\Gamma$ is acute. □

**Remark 1.** It is easy to see from the proof of lemma 2 that the constants $M(\xi)$ and $p(\xi)$ in (1.3) and (1.7) can be assumed to be bounded on each compact set $K \subset\subset \Gamma$. (That is, for any $K \subset\subset \Gamma$ there are constants $M(K)$ and $p(K)$ which depend only on the set $K$ such that (1.3) and (1.7) hold with these constants for any $\xi \in K$.)

**Lemma 3.** *Let $e_1$ and $e_2$ be arbitrary vector from a compact subcone $\Gamma' \subset\subset \Gamma$ such that $|e_1| = |e_2| = 1$. There is a constant $C = C(\Gamma')$ which does not depend on $U \in A(\Gamma)$ such that*

$$\phi(Ue_1, Ue_2) \leqslant C\phi(e_1, e_2), \quad e_1, e_2 \in \Gamma', \tag{1.9}$$

*where $\phi(e_1, e_2)$ is the angle between the vectors $e_1$ and $e_2$.*

**Proof.** Using the fact that there is a number $\delta = \delta(\Gamma') < 1$ (see Section 1.8) such that $(e_1, e_2) \geqslant -\delta$ and

$$\left| \left( \frac{Ue_1}{|Ue_1|}, \frac{Ue_2}{|Ue_2|} \right) \right| \leqslant 1$$

we have

$$\frac{\sin \phi(Ue_1, Ue_2)}{\sin \phi(e_1, e_2)} = \sqrt{\frac{1 - \left( \frac{Ue_1}{|Ue_1|}, \frac{Ue_2}{|Ue_2|} \right)^2}{1 - (e_1, e_2)^2}} =$$

$$\leqslant \sqrt{\frac{\left[ 1 + \left( \frac{Ue_1}{|Ue_1|}, \frac{Ue_2}{|Ue_2|} \right) \right] \left[ 1 - \left( \frac{Ue_1}{|Ue_1|}, \frac{Ue_2}{|Ue_2|} \right) \right]}{[1 + (e_1, e_2)][1 - (e_1, e_2)]}} \leqslant$$

$$\leq \sqrt{\frac{2}{1-\delta}} \sqrt{\frac{1-\left(\frac{Ue_1}{|Ue_1|}, \frac{Ue_2}{|Ue_2|}\right)}{1-(e_1, e_2)}}.$$

Using the identity $1 - 1/2|e_1 - e_2|^2 = (e_1, e_2)$, which holds for unit vectors, we obtain

$$\sqrt{\frac{1-\left(\frac{Ue_1}{|Ue_1|}, \frac{Ue_2}{|Ue_2|}\right)}{1-(e_1, e_2)}} = \frac{\left|\frac{Ue_1}{|Ue_1} - \frac{Ue_2}{|Ue_2|}\right|}{|e_1 - e_2|} =$$

$$= \frac{\left|U\left(\frac{|Ue_2|}{|Ue_1|}e_1 - e_2\right)\right|}{|Ue_2|\,|e_1 - e_2|} \leqslant$$

$$\leqslant \frac{\|U\|}{|Ue_2|} \frac{\left|\frac{|Ue_2|}{|Ue_1|}e_1 - e_2\right|}{|e_1 - e_2|} \leqslant$$

$$\leqslant M(e_2) \frac{\left|\frac{|Ue_2| - |Ue_1|}{|Ue_1|}\right| + |e_1 - e_2|}{|e_1 - e_2|} \leqslant$$

$$\leqslant \frac{M(e_2)}{|e_1 - e_2|}\left[\frac{|U(e_1 - e_2)|}{|Ue_1|} + |e_1 - e_2|\right] \leqslant$$

$$\leqslant M(e_2) + \frac{M(e_2)}{|e_1 - e_2|} \frac{\|U\|\,|e_1 - e_2|}{|Ue_1|} \leqslant$$

$$\leqslant M(e_2)[1 + M(e_1)].$$

Here we have twice used inequality (1.3). Taking into account remark 1 we conclude that

$$\frac{\sin\phi(Ue_1, Ue_2)}{\sin\phi(e_1, e_2)} \leqslant \sqrt{\frac{2}{1-\delta}}\left[M(e_2) + M(e_1)M(e_2)\right] \leqslant C,$$

where $C$ does not depend on $U \in A(\Gamma)$. Thus, (1.9) has been proved. □

Let us know classify some families of nonsingular linear operators $\{U_k, k \in I\}$ under which the cone $\Gamma$ is invariant, assuming that infinity is the limit point of the set $I$. Each such a family is characterized by two functions, $\Lambda(k)$ and $\lambda(k)$, which are defined by formulae (1.1) for any $k \in I$ and $U = U_k$.

**Definition.** We say that the family $\{U_k, k \in I\}$, $U_k\Gamma = \Gamma$, $k \in I$ is

(1) of the *first type* if

$$\Lambda(k) \to \infty,\ \lambda(k) \to \infty,\ k \to \infty,\ k \in I;$$

(2) of the *second type* if there is a constant $b > 0$ such that

$$\Lambda(k) \to \infty,\ k \to \infty,\ \lambda(k) \geqslant b > 0,\ k \in I;$$

(3) of the *third type* if

$$\Lambda(k) \to \infty,\ k \to \infty,\ k \in I;$$

(4) of the *fourth type* if

$$\lambda(k) \to 0,\ \Lambda(k) \to 0,\ k \to \infty,\ k \in I:$$

(5) of the *fifth type* if there is a constant $B > 0$ such that

$$\lambda(k) \to 0,\ k \to \infty,\ \Lambda(k) \leqslant B,\ k \in I;$$

(6) of the *sixth type* if

$$\lambda(k) \to 0,\ k \to \infty,\ k \in I.$$

If the family $\{U_k, k \in I\}$ is given, one can construct a family of linear operators $\{V_k, k \in I\}$, $V_k = (U_k^T)^{-1}$ under which the cone $\Gamma^*$ is invariant. It follows from formulae (1.2) that if the family $\{U_k, k \in I\}$ is of the first type then the corresponding family $\{V_k, k \in I\}$ is of the fourth type. (Similarly, if the family $\{U_k, k \in I\}$ is of the second type the family $\{V_k, k \in I\}$ is of the fifth type, and so on.)

It is clear that any family of the first type is a family of the second type; a family of the second type is in turn a family of the third type; and so on.

Any cone $\Gamma$ has at least one family of the first type (and one family of the fourth type); that is, the family of dilatations $\xi \to k\xi$, $0 < k < \infty$. But not every cone has a family of the second type which is not a family of the first type and a family of the third type which is not a family of the second type. Nevertheless, it is easy to see that the positive octant and the future light cone have a family of the second type which is not a family of the first type, and a family of the third type which is not a family of the second type.

Let $y \in \operatorname{int}\Gamma^* = C$. Consider the $\{\delta y, 0 < \delta \leqslant 1\} \subset C$. If $\{U_k, k \in I\}$ is a family of the first type then by lemma 1 we have

$$|V_k(\delta y)| = \delta |V_k y| \leqslant \delta |y| \, \|V_k\| \leqslant \tag{1.10}$$

$$\leqslant \frac{\delta |y|}{\lambda(k)} \leqslant \frac{|y|}{\lambda(k)} \to 0,\ k \to \infty,\ k \in I.$$

If $\{U_k, k \in I\}$ is a family of the second type then there exists $b > 0$ such that

$$|V_k(\delta y)| \leqslant \frac{\delta |y|}{\lambda(k)} \leqslant \frac{|y|}{b},\ 0 < \delta \leqslant 1,$$

$$\Delta_C(V_k(\delta y)) = \delta\Delta_C(V_k y) \leqslant \frac{\delta p(y)}{\Lambda(k)} \leqslant \frac{p(y)}{\Lambda(k)} \to 0,\ k \to \infty,\ k \in I. \tag{1.11}$$

To obtain (1.11) we have used lemma 1 and lemma 2 for the operator $V_k$ under which the cone $\Gamma^*$ is invariant.

If $\{U_k, k \in I\}$ is a family of the third type then

$$\Delta_C(V_k(\delta y)) \leqslant \frac{p(y)}{\Lambda(k)} \to 0, \ k \to \infty, \ k \in I, \ 0 < \delta \leqslant 1. \tag{1.12}$$

**2. Strictly admissible functions for a family of linear transformations.** Let $\{U_k, k \in I\}$ be a family of nonsingular linear operators in $\mathbf{R}^n$ under which the regular cone $\Gamma$ is invariant. As before we assume that infinity is the limit point of the set $I$.

**Definition.** A distribution $u \in \mathcal{S}'(\Gamma)$ is called *q-strictly admissible for the family* $\{U_k, k \in I\}$ if the following conditons hold:

(1) the primitive of $u$ of order $q$ is a locally summable positive function:

$$u^{(-q)}(\xi) > 0, \ \xi \in \operatorname{int}\Gamma, \ u^{(-q)} \in \mathcal{L}^1_{loc}; \tag{2.1}$$

(2) there exists a vector $\xi_0 \in \operatorname{int} \operatorname{pr}\Gamma$ such that

$$\frac{u^{(-q)}(U_k\xi)}{u^{(-q)}(U_k\xi_0)} \xrightarrow[\longrightarrow]{\xi \in K} g(\xi), \ k \to \infty, \ k \in I, \tag{2.2}$$

where $K$ is any compact set in $\operatorname{int}\Gamma$ and $g(\xi)$ is a continuous positive function in $\operatorname{int}\Gamma$.

(3) there exists $k_0$ such that

$$\frac{u^{(-q)}(U_k\xi)}{u^{(-q)}(U_k\xi_0)} \leqslant \psi(\xi), \ k > k_0, \ k \in I, \ \xi \in \operatorname{int}\Gamma, \tag{2.3}$$

where $\psi(\xi)$ is some tempered function in $\Gamma$ (see Section 0.3).

In the case of one variable for the cone $\Gamma = [0, \infty)$ and for the family of dilatations $\{U_k = k, 0 < k < \infty\}$, 0-strictly admissible functions are nothing but automodel functions (see Section 3.2). In the case of several variables for the family of dilatations of the cone $\Gamma$ and $\{U_k = kI, 0 < k < \infty\}$, where $I$ is the unit matrix, the description of the class of 0-strictly admissible functions was given in A.L. Yakymiv (1981). He called such functions 'regularly varying'. Any such a function has the representation

$$u(\xi) = \Phi(\xi)\exp\left[\eta(\xi) + \int_a^{|\xi|} \frac{\epsilon(\tau)}{\tau} d\tau\right], \ |\xi| \geqslant a, \ \xi \in \operatorname{int}\Gamma, \tag{2.4}$$

where $\Phi(\xi)$ is a continuous, homogeneous function for $\xi \in \operatorname{int}\Gamma$; $\eta(\xi)$ is a measurable function such that $\eta(\xi) \to c$, $|\xi| \to \infty$; and $\epsilon(\tau)$ is an infinitely differentiable function such that $\epsilon(\tau) \to 0$, $\tau \to \infty$.

Thus, the $q$-strictly admissible functions are generalizations of automodel (regularly varying) functions of an arbitrary family $\{U_k, k \in I\}$ of linear transformations under which the cone $\Gamma$ is invariant.

Let us point out some properties of the $q$-strictly admissible functions which follow directly from their definition. It is clear that $g(\xi_0) = 1$. Furthermore, condition (2) holds for any $\xi_1 \in \operatorname{int}\Gamma$, the function $g$ being multiplied by the factor

$$\lim_{k\to\infty} \frac{u^{(-q)}(U_k\xi_0)}{u^{(-q)}(U_k\xi_1)}.$$

*If $u(\xi)$ is a $q$-strictly admissible function for the family $\{U_k, k \in I\}$ then it is $(q+p)$-strictly admissible for the same family if $p = 0, 1, 2, \cdots$.*

**Proof.** Indeed, it is sufficient to check this statement for $p = 1$. Taking the primitive of $u^{(-q)}(\xi)$ with respect to the cone $\Gamma$ and having in mind (2.1) we get that $u^{(-q-1)}(\xi) > 0$ when $\xi \in \operatorname{int}\Gamma$. Denoting

$$\rho_k = u^{(-q)}(U_k\xi_0)$$

we have from (2.2)

$$\frac{1}{\rho_k}\int_{\xi-\Gamma} u^{(-q)}(U_k\eta)\,d\eta \to \int_{\xi-\Gamma} g(\eta)\,d\eta,\ k\to\infty,\ k\in I.$$

The convergence here is uniform in $\xi$ on any compact set $K$ from $\Gamma$. The integral on the right-hand side is a continuous nonvanishing function of $\xi$. Therefore,

$$\frac{1}{\rho_k}\int_{\xi-\Gamma} u^{(-q)}(U_k\xi_0) \to g^{(-1)}(\xi_0) \neq 0,\ k\to\infty,\ k\in I.$$

Hence,

$$\frac{u^{(-q-1)}(U_k\xi)}{u^{(-q-1)}(U_k\xi_0} \xrightarrow[\longrightarrow]{\xi\in K} \frac{g^{(-1)}(\xi)}{g^{(-1)}(\xi_0)} = g_1(\xi) > 0,\ k\to\infty,\ k\in I,\ K \subset\subset \operatorname{int}\Gamma,$$

and the function $g_1(\xi)$ is continuous in $\operatorname{int}\Gamma$. Taking the first primitives with respect to the cone $\Gamma$ from both sides of (2.3) we obtain

$$\frac{1}{J_k\rho_k}u^{(-q-1)}(U_k\xi) \leqslant \psi^{(-1)}(\xi),\ k > k_0,\ k\in I,$$

where $\psi^{(-1)}$ is a tempered function on $\Gamma$. □

**Lemma 1.** *Let $u(\xi)$ be a $q$-strictly admissible function for the family $\{U_k, k\in I\}$. Then there exists the limit*

$$\frac{J_k^{q-1}\tilde{u}(iV_k y)}{u^{(-q)}(U_k\xi_0)} \xrightarrow[\longrightarrow]{y\in Q} \mathcal{K}_{C^*}^{-q}(iy)\tilde{g}(iy),\ k\to\infty,\ k\in I,\ Q\subset\subset C. \tag{2.5}$$

*Besides, there exist numbers $m > 0$ and $M \geqslant m$ which depend on $Q$, such that*

$$m \leqslant \mathcal{K}_{C^*}^{-q}(iy)\tilde{g}(iy) \leqslant M,\ y\in Q. \tag{2.6}$$

**Proof.** Taking into account the regularity of the cone $\Gamma$, we have

$$\frac{J_k^{q-1}\tilde{u}(iV_k y)}{u^{(-q)}(U_k\xi_0)} = \frac{J_k^q}{u^{(-q)}(U_k\xi_0)}(u(U_k,\xi), e^{-(y,\xi)}) =$$

$$= \frac{J_k^q}{u^{(-q)}(U_k\xi_0)}\left(\theta_\Gamma^{-q}(\xi) * \theta_\Gamma^q(\xi) * u(U_k\xi), e^{-(y,\xi)}\right) =$$

$$= \frac{J_k^q}{u^{(-q)}(U_k\xi_0)}\left(\theta_\Gamma^q(\xi) * u(U_k\xi), \left(\theta_\Gamma^{-q}(\eta), e^{-(y,\eta)-(y,\xi)}\right)\right) =$$

$$= K_{\bar{C}}^{q}(iy)\int \frac{u^{(-q)}(U_k\xi)}{u^{(-q)}(U_k\xi_0)} e^{-(y,\xi)}\, d\xi. \tag{2.7}$$

By estimate (2.3) one can take a limit in (2.7) before integrating (by the Lebesgue theorem). Taking into account (2.2) we get (2.5). Now, let $Q$ be any compact set in $\operatorname{int}\Gamma^* = C$. Since the set of the test functions $\{e^{-(y,\xi)}, y \in Q\}$ is bounded in $\mathcal{S}(\Gamma)$, the convergence in (2.5) is uniform in $y \in Q$. From estimates (2.3), estimate (2.6) follows. □

3. **Admissible functions of a cone.** As before we assume that infinity is the limit of the set of indices $I$.

**Definition 1.** A distribution $u \in \mathcal{S}'(\Gamma)$ is called *$q$-admissible for the cone* $\Gamma$, if for any family of nonsingular linear transformations $\{U_k, k \in I\}$, under which the cone $\Gamma$ is invariant, there exists a subsequence $\{U_{k_m}, k_m \to \infty, m \to \infty\}$ such that $u(\xi)$ is $q$ - strictly admissible for this subsequence.

The set of all $q$-admissible functions for the cone $\Gamma$ we denote by $\Delta_q(\Gamma)$; $\Delta_0(\Gamma) = \Delta(\Gamma)$.

**Definition 2.** A distribution $u \in \mathcal{S}'(\Gamma)$ is called *$q$-admissible of the type $N$ ($N = 1, 2, \ldots, 6$) for the cone* $\Gamma$ if for any family of nonsingular linear transformations of the $N$the type, $\{U_k, k \in I\}$, under which the cone $\Gamma$ is invariant, there exists a subsequence $\{U_{k_m}, k_m \to \infty, m \to \infty\}$ such that $u$ is $q$-strictly admissible for this subsequence.

The set of all $q$-admissible functions of type $N$ for the cone $\Gamma$ is denoted by $\Delta_q(\Gamma, N)$, $N = 1, 2, \ldots, 6$; $\Delta_0(\Gamma, N) = \Delta(\Gamma, N)$.

The following embeddings obviously hold

$$\Delta_q(\Gamma, 1) \supset \Delta_q(\Gamma, 2) \supset \Delta_q(\Gamma, 3) \supset \Delta_q(\Gamma) \supset \Delta_q(\Gamma, 6) \supset \Delta_q(\Gamma, 5) \supset \Delta_q(\Gamma, 4).$$

Note that a 0-admissible function of some type for the cone is simultaneously a one-, two-, etc., admissible functions of the same type. That is, the following embeddings hold

$$\cdots \supset \Delta_2(\Gamma, n) \supset \Delta_1(\Gamma, n) \supset \Delta(\Gamma, N), \quad N = 1, 2, \ldots, 6.$$

Now we shall give some simple conditions allowing a function to be 0- admissible for some cones which very often occur in mathematical physics. We shall also give examples of 0-admissible functions for arbitrary acute, convex cones.

Let $\Gamma$ be an acute, convex, closed, solid and selfdual ($\Gamma = \Gamma^*$) cone in $\mathbf{R}^n$, with its vertex at zero; let $\{U_k, k \in I\}$ be a given family of nonsingular linear operators under which the cone $\Gamma$ is invariant. Let $\mathcal{O} \subset \Gamma$ be a set which satisfies the conditions

$$U_k \quad \subset \mathcal{O}, \ k \in I; \ \lambda \ \subset \mathcal{O}, \ \lambda \geqslant 0; \ L(\mathcal{O}) = \mathbf{R}^n, \tag{3.1}$$

where $L(\mathcal{O})$ is the linear hull of $\mathcal{O}$.

Besides, let $u \in \mathcal{S}'(\Gamma)$ be a positive ontinuously differentiable function satisfying the condition: there are numbers $a$ and $b$ such that

$$a \leqslant \frac{(l,\xi)(l,\partial u(\xi))}{u(\xi)} \leqslant b, \quad \forall l \in \mathbb{O}, \quad |l| = 1, \quad \forall \xi \in \operatorname{int}\Gamma, \tag{3.2}$$

where $\partial = (\partial_1, \ldots, \partial_n)$ (see Section 0.4).

Under these hypotheses the following three lemmas are true.

**Lemma 1.** *For any points* $\xi, \eta \in \operatorname{int}\Gamma$ *such that* $\xi - \eta \in \mathbb{O}$ *and any* $U_k$, $k \in I$ *the following inequalities hold:*

$$\left.\begin{matrix} \dfrac{\Delta_\Gamma^a(\xi)}{|\eta|^a}, & a > 0 \\ \dfrac{|\xi|^a}{\Delta_\Gamma^a(\eta)}, & a < 0 \end{matrix}\right\} \leqslant \frac{u(U_k\xi)}{u(U_k\eta)} \leqslant \left\{\begin{matrix} \dfrac{|\xi|^b}{\Delta_\Gamma^b(\eta)}, & b > 0, \\ \dfrac{\Delta^b(\xi)}{|\eta|^b}, & b < 0. \end{matrix}\right. \tag{3.3}$$

**Proof.** Let us consider a segment connecting the points $U_k\xi$ and $U_k\eta$. By (3.1), $U_k(\xi-\eta) \in \mathbb{O}$. Denoting

$$l = \frac{U_k(\xi-\eta)}{|U_\kappa(\xi-\eta)|}, \quad l \in \mathbb{O}, \quad |l| = 1,$$

let us represent this segment in the following parametric form:

$$\tau = \tau(\omega) = U_k\eta + [\omega - (l, U_k\eta)]l, \quad (l, U_k\eta) \leqslant \omega \leqslant (l, U_k\xi),$$

where $\omega = (l,\tau)$. Denoting also $u(\tau(\omega)) = \phi(\omega)$, we have from estimate (3.2) for $\xi = \tau(\omega)$ that $a \leqslant \omega\phi'(\omega)/\phi(\omega) \leqslant b$.

Integrating this inequality over $\omega$ from $(l, U_k\eta)$ to $(l, U_k\xi)$ we obtain

$$a \ln\frac{(l,U_k\xi)}{(l,U_k\eta)} = \int_{(l,U_k\eta)}^{(l\,U_k\xi)} \frac{a}{\omega}\,d\omega \leqslant$$

$$\leqslant \ln\frac{\phi((l,U_k\xi))}{\phi((l,U_k\eta))} \leqslant$$

$$\leqslant \int_{(l,U_k\eta)}^{(l,U_k\xi)} \frac{b}{\omega}\,d\omega = b\ln\frac{(l,U_k\xi)}{(l,U_k\eta)}.$$

Thus, taking into account that

$$\phi((l,U_k\xi)) = u(U_k\xi), \quad \phi((l,U_k\eta)) = u(U_k\eta),$$

we get

$$\frac{(U_k^Tl,\xi)^a}{(U_k^Tl,\eta)^a} = \frac{(l,U_k\xi)^a}{(l,U_k\eta)^a} \leqslant \frac{u(U_k\xi)}{u(U_k\eta)} \leqslant$$

$$\leqslant \frac{(l,U_k\xi)^b}{(l,U_k\eta)^b} = \frac{(U_k^Tl,\xi)^b}{(U_k^Tl,\eta)^b}. \tag{3.4}$$

Dividing the numerator and denominator in the wings of inequality (3.4) by $|U_k^Tl|$, and having in mind the selfduality of $\Gamma$ and the estimate

$$\Delta_\Gamma(\tau) \leqslant \left(\frac{U_k^Tl)}{|U_k^Tl|}, \tau\right) \leqslant |\tau|, \quad \tau \in \Gamma, \tag{3.5}$$

we obtain estimate (3.3). □

**Lemma 2.** *For any compact set $K \subset\subset \operatorname{int}\Gamma$ and for any vector $e \in \operatorname{int}\Gamma$ there are positive constants $C_1$ and $C_2$ such that for any $U_k$, $k \in I$*

$$C_1 \leqslant \frac{u(U_k \xi)}{u(U_k e)} \leqslant C_2, \quad \xi \in K. \tag{3.6}$$

**Proof.** Let $\{e_j, j = 1, \dots, n\}$, $l_j \in \mathbb{O}$ be a basis in $\mathbf{R}^n$. By (3.1) such vectors exist. We have

$$\xi - e = \sum_{j=1}^{n} \alpha_j e_j, \quad \xi \in K, \; e \in \operatorname{int}\Gamma.$$

Let us denote $\xi^1 = e + \alpha_1 l_1, \dots, \xi^s = \xi^{s-1} + \alpha_s l_s, \dots, \xi^n = \xi^{n-1} + \alpha_n l_n = \xi$. Note that either $\xi^s - \xi^{s-1} \in \mathbb{O}$ or $\xi^{s-1} - \xi^s \in \mathbb{O}$. One can assume without loss of generality that for some $s_0$ $(1 \leqslant s_0 \leqslant n)$, $\xi^s - \xi^{s-1} \in \mathbb{O}$, $s \leqslant s_0$ and $\xi^{s-1} - \xi^s \in \mathbb{O}$, $s > s_0$. In this case we have $\xi^s \in \operatorname{int}\Gamma$, $s = 1, \dots, n$. From lemma 1 it follows that for any $U_k$, $k \in I$

$$\frac{|\xi^s|^a}{\Delta_\Gamma^a(\xi^{s-1})} \leqslant \frac{u(U_k \xi^s)}{u(U_k \xi^{s-1})} \leqslant \frac{|\xi^s|^b}{\Delta_\Gamma^b(\xi^{s-1})}, \tag{3.7}$$

if $s \leqslant s_0$ and

$$\frac{\Delta_\Gamma^b(\xi^s)}{|\xi^{s-1}|^b} \leqslant \frac{u(U_k \xi^s)}{u(U_k \xi^{s-1})} \leqslant \frac{\Delta_\Gamma^a(\xi^s)}{|\xi^{s-1}|^a}, \tag{3.8}$$

if $s > s_0$. (Without loss of generality we assume here that $a < 0$, $b > 0$.) Since

$$\frac{u(U_k \xi)}{u(U_k e)} = \frac{u(U_k \xi^n)}{u(U_k \xi^{n-1})} \cdot \frac{u(U_k \xi^{n-1})}{u(U_k \xi^{n-2})} \cdot \dots \cdot \frac{u(U_k \xi^1)}{u(U_k e)},$$

then using estimates (3.7) and (3.8), and taking into account that each $\xi^s$ lies in some compact set $K_s \subset\subset \operatorname{int}\Gamma$ if $\xi$ is in a compact set $K \subset\subset \Gamma$, we get estimate (3.6).

**Lemma 3.** *For any compact set and for any vector $e \in \operatorname{int}\Gamma$ there is a constant $C_3 > 0$ such that for any $U_k$, $k \in I$,*

$$\left| \frac{\partial u(U_k \xi)}{u(U_k e)} \right| \leqslant C_3, \quad \xi \in K \subset\subset \operatorname{int}\Gamma. \tag{3.9}$$

**Proof.** Taking into account that

$$\partial u(U_k \xi) = U_k^T \partial u(\tau)|_{\tau = U_k \xi},$$

we have from estimate (3.2)

$$a \leqslant \frac{(l, U_k \xi)\left(l, \partial u(\tau)|_{\tau=U_k \xi}\right)}{u(U_k \xi)} \leqslant \frac{(l, U_k \xi)(l, (U_k^T)^{-1} \partial u(U_k \xi))}{u(U_k \xi)} \leqslant$$

$$\leqslant (U_k^T l, \xi)\left(U_k^{-1} l, \frac{\partial u(U_k \xi)}{u(U_k e)}\right) \frac{u(U_k e)}{u(U_k \xi)} \leqslant b$$

for any $l \in \mathbb{O}$, $|l| = 1$ and $\xi \in \operatorname{int}\Gamma$. Now, let $\xi \in K$. Using lemma 2 we see that there are constants $a_1$ and $b_1$ such that

$$a_1 \leqslant (U_k^T l\ \xi)\left|U_k^{-1} l, \frac{\partial u(U_k \xi)}{u(U_k e)}\right| \leqslant b_1, \quad l \in \quad , \quad |l| = 1. \tag{3.10}$$

Let us take vectors $\{l_j, j = 1, \ldots, n\}$ in which are basic vectors in $\mathbf{R}^n$, and let us put in (3.10)

$$l = \frac{U_k l_j}{|U_k l_j|}, \quad j = 1, \ldots, n.$$

We get, as the result,

$$a_1 \leqslant \frac{(U_k^T U_k l_j, \xi)}{|U_k l_j|^2}\left|l_j, \frac{\partial u(U_k \xi)}{u(U_k e)}\right| \leqslant b_1, \quad j = 1, \ldots, n. \tag{3.11}$$

Since $\xi \in K$ and a compact set $K$ belongs to some solid subcone $C'' \subset\subset C = \operatorname{int}\Gamma^* = \operatorname{int}\Gamma$, then using formula (9.1) of Section 1.9 and lemma 1 of Section 5.1 we get the estimate

$$\frac{(U_k^T U_k l_j, \xi)}{|U_k l_j|^2} \geqslant \frac{\sigma|\xi|\,|U_k^T U_k l_j|}{|U_k l_j|^2} \geqslant$$

$$\geqslant \frac{\sigma|\xi|\,\|U_k^T U_k\|}{M(l_j)|l_j|^2\|U_k\|^2} \geqslant M_1. \tag{3.12}$$

Here we have used the selfduality of the cone $\Gamma$ and the equality $\|U_k^T U_k\| = \|U_k\|^2$. Estimate (3.9) follows from (3.11) and (3.12). □

From lemma 1 and 2 and the Arzela-Ascoly lemma it follows that if we are given a family of linear operators $\{U_k, k \in I\}$, under which the selfdual cone $\Gamma$ is invariant, and if there is a set $\mathbb{O}$ with properties (3.1) such that inequality (3.2) holds for the function $u(\xi)$, then there exists a subsequence $\{U_{k_m}, k_m \in I, k \to \infty, m \to \infty\}$ (depending on the family of operators) such that

$$\frac{u(U_{k_m}\xi)}{u(U_{k_m}e)} \xrightarrow{\xi \in K} g(\xi), \quad m \to \infty, \quad K \subset\subset \operatorname{int}\Gamma = C, \tag{3.13}$$

where $g$ is a positive continuous function. In other words, condition (3.2) guarantees condition (2.2) of Section 5.2 for the function $u$ to be 0-strictly admissible for the family $\{U_{k_m}, k_m \in I, k_m \to \infty, m \to \infty\}$.

Now we shall restrict ourselves to the cases of positive octant $\mathbf{R}_+^n$ and the future light cone $V_n^+$ and give sufficient conditions for $u \in \Delta(\mathbf{R}_+^n)$ and $u \in \Delta(\overline{V}_n^+)$.

**Theorem 1.** *Let $u$ be a positive and continuously differentiable function in $\mathbf{R}_+^n$ satisfying the condition*

$$-1 < a \leqslant \frac{\xi_j \partial_j u(\xi)}{u(\xi)} \leqslant b, \quad \xi_j > 0, \quad j = 1, \ldots, n. \tag{3.14}$$

*Then $u \in \Delta(\mathbf{R}_+^n)$.*

**Proof.** As the set $\mathbb{O}$ we take the set of vectors of the form $\lambda l_j$, $j = 1, \ldots, n$, where $\lambda > 0$ and $l_j$ are unit coordinate vectors: $l_j = (0, \ldots, 0, 1, 0, \ldots, 0)$ (unity is on the $j$th place). Condition (3.14) is in fact condition (3.2) and, as we noted before, it guarantees by (3.13) condition (2.2) of Section 5.2; that if, condition (2) for the function $u$ to be 0-strictly admissible for a proper subfamily of linear transformations. The theorem will be proved if we prove the estimate

$$\frac{u(U\xi)}{u(U\xi_o)} \leqslant \psi(\xi), \quad \xi \in \mathbf{R}^n_+, \ U \in A(\mathbf{R}^n_+) \tag{3.15}$$

for some $\xi_0 \in \mathbf{R}^n_+$, where $\psi$ is some tempered function in $\mathbf{R}^n_+$ which does not depend on $U \in A(\mathbf{R}^n_+)$.

Let us fix $\xi_0 = (\xi_1^0, \dots, \xi_n^0)$, $\xi_j^0 > 0$, $j = 1, \dots, n$ and let $\xi = (\xi_1, \dots, \xi_n)$ be arbitrary point in $\mathbf{R}^n_+$. We assume that $\xi_1 \leqslant \xi_2 \leqslant \cdots \leqslant \xi_n$. (If this is not the case, one can renumber the coordinates.) Let us take the intermediate points:

$$\xi^0 = \xi = (\xi_1, \dots, \xi_n), \dots, \xi^s =$$
$$= (\xi_1^0, \dots, \xi_s^0, \xi_{s+1}, \dots, \xi_n), \dots, \xi^n = (\xi_1^0, \dots, \xi_n^0) = \xi_0.$$

We have

$$\frac{u(U\xi)}{u(U\xi_0)} = \frac{u(U\xi^0)}{u(U\xi^1)} \cdot \frac{u(U\xi^1)}{u(U\xi^2)} \cdot \dots \cdot \frac{u(U\xi^{n-1})}{u(U\xi_0)}. \tag{3.16}$$

Let us estimate each fraction on the right-hand side of (3.16) by lemma 1 assuming (without loss of generality) that $a < 0$, $b > 0$. Taking into account that

$$|\xi^2| \leqslant \sqrt{|\xi|^2 + |\xi_0|^2},$$
$$\Delta_{\mathbf{R}^n_+}(\xi^s) = \min(\xi_1^0, \dots, \xi_s^0, \xi_{s+1}, \dots, \xi_n)$$

we get

$$\frac{u(U\xi^s)}{u(U\xi^{s+1})} \leqslant \begin{cases} \dfrac{\Delta^a_{\mathbf{R}^n_+}(\xi^s)}{|\xi^{s+1}|^a} \leqslant \dfrac{[\min(\xi_1^0, \dots, \xi_s^0, \xi_{s+1})]^a}{(|\xi|^2 + |\xi_0|^2)^{a/2}}, & \xi^{s+1} - \xi^s \in \mathbf{R}^n_+, \\ \dfrac{|\xi^2|^b}{\Delta^b_{\mathbf{R}^n_+}(\xi^{s+1})} \leqslant \dfrac{(|\xi|^2 + |\xi_0|^2)^{b/2}}{[\min(\xi_1^0, \dots, \xi_s^0, \xi_{s+1}^0)]^b}, & \xi^s - \xi^{s+1} \in \mathbf{R}^n_+. \end{cases} \tag{3.17}$$

Here we have considered that if $\xi^s - \xi^{s+1} \in \mathbf{R}^n_+$ then $\xi_{s+1}^0 \leqslant \xi_{s+1}$, and therefore $\xi_{s+1}^0 \leqslant \xi_{s+1} \leqslant \cdots \leqslant \xi_n$. Hence,

$$\min(\xi_1^0, \dots, \xi_{s+1}^0, \xi_{s+2}, \dots, \xi_n) = \min(\xi_1^0, \dots, \xi_{s+1}^0).$$

From (3.16) and (3.17) it follows that there are numbers $M$ and $m$ such that

$$\frac{u(U\xi)}{u(U\xi_0)} \leqslant M(1 + |\xi|)^m (\xi_1 \cdot \xi_2 \cdot \cdots \cdot \xi_n)^a = \psi(\xi), \quad U \in A(\mathbf{R}^n_+).$$

Now it is clear that if $a > -1$, the function $\psi$ is tempered in $\mathbf{R}^n_+$.

**Theorem 2.** *Let $u$ be a positive, continuously differentiable function in $V_n^+$ satisfying the condition*

$$-1 < a \leqslant \frac{(l,\xi)(l,\partial u(\xi))}{u(\xi)} \leqslant b, \quad \xi \in V_n^+, \ l \in \partial V_n^+, \ |l| = 1, \tag{3.18}$$

*for some $a > -1$ and $b$. Then $u \in \Delta(\overline{V}_n^+)$.*

**Proof.** As in the proof of theorem 1, it is sufficient to prove the estimate

$$\frac{u(U\xi)}{u(U\xi_0)} \leqslant \psi(\xi), \quad \xi \in V_n^+, \quad U \in A(\overline{V}_n^+),$$

where $\psi$ is a tempered function in $\overline{V}_n^+$ which does not depend on $U$. Without loss of generality we assume that $a < 0$, $b > 0$. Let $\xi \in V_n^+$ and $\xi_0 = (\sqrt{2}, 0, \ldots, 0)$. Let us draw a two-dimensional plane through the points $\xi$, $\xi_0$, and zero. It intersects the cone $V_n^+$ at a two-dimensional octant, so for the unit vectors of its boundary, denoted by $e_1$ and $e_2$, we have

$$\xi_0 = e_1 + e_2, \quad \xi = \tau_1 e_2 + \tau_2 e_2, \quad \tau_1 \leqslant \tau_2$$

(see Figure 9).

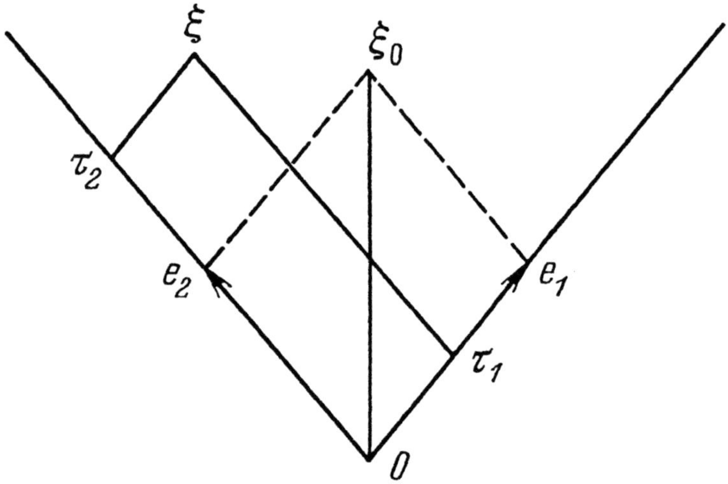


Figure 9.

Let us introduce a vector $\xi^1 = e_1 + \tau_2 e_2$. We have

$$\frac{u(U\xi)}{u(U\xi_0)} = \frac{u(U\xi)}{u(U\xi^1)} \cdot \frac{u(U\xi^1)}{u(U\xi_0)}. \tag{3.19}$$

Making the estimates by lemma 1 we get

$$\frac{u(U\xi)}{u(U\xi^1)} \leqslant \begin{cases} \dfrac{\Delta^a_{V_n^+}(\xi)}{|\xi^1|^a} \leqslant \dfrac{\Delta^a_{V_n^+}(\xi)}{(|\xi|^2 + |\xi_0|^2)^{a/2}}, & \tau_1 < 1, \\[2ex] \dfrac{|\xi|^b}{\Delta^b_{V_n^+}(\xi^1)} \leqslant \dfrac{|\xi|^b}{[\min(1, \tau_2)]^b} \leqslant |\xi|^b, & \tau_1 \geqslant 1; \end{cases}$$

$$\frac{u(U\xi^1)}{u(U\xi_0)} \leqslant \begin{cases} \dfrac{\Delta^a_{V_n^+}(\xi^1)}{|\xi_0|} \leqslant \left|\dfrac{\min(1, \tau_2)}{\sqrt{2}}\right|^a \leqslant \left|\dfrac{\min\left(1, \dfrac{|\xi|}{\sqrt{2}}\right)}{\sqrt{2}}\right|^a, & \tau_2 < 1, \\[2ex] \dfrac{|\xi^1|^b}{\Delta^b_{V_n^+}(\xi_0)} \leqslant |\xi^1|^b \leqslant (|\xi|^2 + |\xi_0|^2)^{b/2}, & \tau_2 \geqslant 1. \end{cases}$$

Here we have taken into account that $|\xi^1| \leqslant \sqrt{|\xi|^2 + |\xi_0|^2}$ if $\tau_1 \geqslant 1$, $1 \leqslant \tau_1 \leqslant \tau_2$ and therefore $\min(1, \tau_2) = 1$; if $\tau_2 < 1$, $\tau_1 \leqslant \tau_2 < 1$ and therefore

$$|\xi| = \sqrt{\tau_1^2 + \tau_2^2} \leqslant \sqrt{2}\tau_2;$$

$$\Delta_{V_n^+}(\xi^1) = \min(1, \tau_2) \geqslant \min\left(1, \frac{|\xi|}{\sqrt{2}}\right).$$

Estimating (3.19) we conclude that there are numbers $M$ and $m$ such that

$$\frac{u(U\xi)}{u(U\xi_0)} \leqslant M(1 + |\xi|)^m \Delta^a_{V_n^+}(\xi)|\xi|^a = \psi(\xi).$$

It is clear that if $a > -1$ and $n \geqslant 1$ the function $\psi$ is tempered in $\bar{V}_n^+$. □

**4. Some examples of admissible functions of a cone.** Let $\Gamma$ be a closed, acute, solid cone in $\mathbf{R}^n$ with its vertex at the origin. Let us consider a function $\chi$ which can be represented in the form

$$\chi(t) = \eta(t)\phi(t), \quad 0 < t < \infty, \tag{4.1}$$

where the functions $\eta(t)$ and $\phi(t)$ satisfy the conditions
(1) there are positive numbers $C_1$ and $C_2$ such that

$$C_1 \leqslant \eta(t) \leqslant C_2, \quad t \in (0, \infty); \tag{4.2}$$

(2) $$\omega(\lambda, t) = \frac{\eta(\lambda t)}{\eta(t)} \xrightarrow{t \in (0, +\infty)} 1, \quad \lambda \to 1; \tag{4.3}$$

(3) the function $\phi$ is a positive, continuously differentiable and satisfies the inequality

$$a \leqslant \frac{t\phi'(t)}{\phi(t)} \leqslant b, \quad 0 < t < \infty \tag{4.4}$$

for some $a$ and $b$.

From condition (4.3) it follows that $\eta(t)$ is a continuous function in $(0, \infty)$ and the family $\{\omega(\lambda, t), \lambda \to 1\}$ is uniformly continuous in $\lambda$ when $t \in (0, \infty)$.

From properties (4.2) and (4.3) we obtain the following lemmas.

**Lemma 1.** *Let $\{r_k, k = 1, 2, \cdots\}$ be any sequence of positive numbers and let $t_k \to t^0 > 0$, $k \to \infty$. Then there exists a subsequence $\{r_{k_m}, m = 1, 2, \cdots\}$ such that*

$$\frac{\eta(r_{k_m} t)}{\eta(r_{k_m} t_{k_m})} \overset{t \in K}{\rightrightarrows} g_1(t), \quad m \to \infty,$$

*where $K$ is any compact set in $(0, \infty)$. Besides, there are positive numbers $C_3$ and $C_4$ such that*

$$c_3 \leqslant g_1(t) \leqslant C_4, \quad t \in (0, \infty)$$

**Lemma 2.** *Let $\{r_k, k = 1, 2, \cdots\}$ be any sequence of positive numbers and let $t_k \to t^0 > 0$, $k \to \infty$. Then there exists a subsequence such that*

$$\frac{\phi(r_{k_m} t)}{\phi(r_{k_m} t_{k_m})} \overset{t \in K}{\rightrightarrows} g_2(t), \quad m \to \infty,$$

*where $K$ is any compact set in $(0,\infty)$; the function $g_2(t)$ is continuous and positive on $(0,\infty)$. Besides, there are positive numbers $C_5$, $C_6$ and $M$ such that*

$$\frac{\phi(r_{k_m}t)}{\phi(r_{k_m}t_{k_m})} \leqslant \begin{cases} C_5 t^b & t \geqslant 1, \\ C_6 t^a, & t < m \end{cases} \quad m > M. \tag{4.5}$$

**Proof.** The cone $\Gamma = [0,\infty)$ is selfdual, $\operatorname{int}\Gamma^* = (0,\infty)$. Any nonsingular operator, under which the cone $\Gamma$ is invariant, is defined by a positive number $r_k$. One can take the whole cone $\Gamma$ as the set in (3.1) of Section 5.3. Then condition (4.4) is nothing but condition (3.2) of Section 5.3. Using lemmas 1, 2, 3 of the previous subsection for our one-dimensional case we get the statement of lemma 2. Estimate (4.5) is a direct consequence of estimate (3.3) of Section 5.3 (see also (3.13), Section 5.3).

**Theorem.** *Let the function $\chi$ satisfy conditions (4.1)-(4.4) and let vectors $e_i$, $i = 1,\dots,p+q$ be fixed. Let the numbers $A_i > 0$ and $\alpha_i$, $i = 1,\dots,p+q$ be given and*

$$\begin{aligned} -\alpha_i a &< 1 \quad \text{if } i = 1,\dots,p, \\ \alpha_i b &< 1 \quad \text{if } i = p+1,\dots,p+q. \end{aligned} \tag{4.6}$$

*Then the function*

$$u(\xi) = \chi(\omega(\xi)), \quad \omega(\xi) = \frac{\sum_{i=1}^{p} A_i (e_i,\xi)^{\alpha_i}}{\sum_{i=p+1}^{p+q} A_i (e_i,\xi)^{\alpha_i}}, \quad \xi \in \Gamma \tag{4.7}$$

*is 0-admissible for the cone $\Gamma$.*

**Proof.** Let $\{U_{k''}, k'' \in I''\}$ be any family of nonsingular linear operators under which the cone $\Gamma$ is invariant. The cone $\Gamma^*$ is invariant under the family $\{U_{k''}^T, k'' \in I''\}$. Let us take a subsequence $\{U_{k'}^T, k' \in I'\} \subset \{U_{k''}^T, k'' \in I''\}$ in such a way that

$$\frac{U_{k'}^T e_i}{|U_{k'}^T e_i|} \to e_i^0 \in \operatorname{pr}\Gamma^*, \quad k' \to \infty, \quad k' \in I', \quad i = 1,\dots,p+q.$$

From this sequence let us select a subsequence $\{U_k^T, k \in I\}$ such that

$$\frac{|U_k^T e_i|^{\alpha_i}}{|U_k^T e_j|^{\alpha_j}} \to B_{ij}, \quad k \to \infty, \quad k \in I, \quad i,j = 1,\dots,p+q \quad (0 \leqslant B_{ij} \leqslant \infty).$$

Changing the numeration if necessary, one can assume that

$$0 \leqslant B_{i1} < \infty, \quad i = 1,\dots,p; \quad 0 \leqslant B_{i,p+1} < \infty, \quad i = p+1,\dots,p+q.$$

Taking into account that

$$\omega(U_k\xi) = \frac{\sum_{i=1}^{p} A_i (U_k^T e_i,\xi)^{\alpha_i}}{\sum_{i=p+1}^{p+q} A_i (U_k^T e_i,\xi)^{\alpha_i}} = \frac{|U_k^T e_i|^{\alpha_1}}{|U_k^T e_{p+1}|^{\alpha_{p+1}}} t_k(\xi).$$

where

$$t_k(\xi) = \frac{\sum_{i=1}^{p} A_i \left( \frac{U_k^T e_i}{|U_k^T e_i|}, \xi \right)^{\alpha_i} \frac{|U_k^T e_i|^{\alpha_i}}{|U_k^T e_i|^{\alpha_1}}}{\sum_{i=p+1}^{p+q} A_i \left( \frac{U_k^T e_i}{|U_k^T e_i|}, \xi \right)^{\alpha_i} \frac{|U_k^T e_i|^{\alpha_i}}{|U_k^T e_{p+1}|^{\alpha_{p+1}}}} =$$

$$= \frac{A_1 \left( \frac{U_k^T e_1}{|U_k^T e_1|}, \xi \right)^{\alpha_1} + \sum_{i=2}^{p} A_i \left( \frac{U_k^T e_i}{|U_k^T e_i|}, \xi \right)^{\alpha_i} \frac{|U_k^T e_i|^{\alpha_i}}{|U_k^T e_1|^{\alpha_1}}}{A_{p+1} \left( \frac{U_k^T e_{p+1}}{|U_k^T e_{p+1}|}, \xi \right)^{\alpha_{p+1}} + \sum_{i=p+2}^{p+q} A_i \left( \frac{U_k^T e_i}{|U_k^T e_i|}, \xi \right)^{\alpha_i} \frac{|U_k^T e_i|^{\alpha_i}}{|U_k^T e_{p+1}|^{\alpha_{p+1}}}}, \quad k \in I,$$

we have

$$t_k(\xi) \xrightarrow{\xi \in K} t(\xi) = \frac{A_1(e_1^0, \xi)^{\alpha_1} + \sum_{i=2}^{p} A_i B_{i,1}(e_i^0, \xi)^{\alpha_i}}{A_{p+1}(e_{p+1}^0, \xi)^{\alpha_{p+1}} + \sum_{i=p+1}^{p+q} A_i B_{i,p+1}(e_i^0, \xi)^{\alpha_i}}, \quad k \to \infty, \tag{4.8}$$

where $K$ is any compact set in $\operatorname{int}\Gamma$. It is clear that there exists $k_0$ such that for $\xi \in \operatorname{int}\Gamma$ and $k > k_0$, $k \in I$ the following estimates hold:

$$M_2 \frac{\Delta_\Gamma^{\alpha_1}(\xi)}{|\xi|^{\alpha_{p+1}} + \cdots + |\xi|^{p+q}} \leqslant t_k(\xi) \leqslant M_1 \frac{|\xi|^{\alpha_1} + \cdots + |\xi|^{\alpha_p}}{\Delta_\Gamma^{\alpha_{p+1}}(\xi)}, \tag{4.9}$$

where $M_1$ and $M_2$ are some positive numbers (see (3.5), Section 5.3). Now, denoting

$$r_k = \frac{|U_k^T e_1|^{\alpha_1}}{|U_k^T e_{p+1}|^{\alpha_{p+1}}},$$

we have, for some $\xi_0 \in \operatorname{int}\Gamma$,

$$\frac{u(U_k \xi)}{u(U_k \xi_0)} = \frac{\chi(r_k t_k(\xi))}{\chi(r_k t_k(\xi_0))}.$$

Now taking into account (4.8), (4.9) and (4.10) we infer from lemma 1 and lemma 2 that there is a subsequence $\{k_m \in I, m = 1, 2, \cdots\}$ such that

$$\frac{u(U_{k_m} \xi)}{u(U_{k_m} \xi_0)} \xrightarrow{\xi \in K} g(\xi), \quad m \to \infty,$$

where $K$ is any compact set in $\operatorname{int}\Gamma$ and $g$ is a continuous positive function in $\operatorname{int}\Gamma$.

Moreover, there are some numbers $M_3$, $m_0$ such that for $m \geqslant m_0$

$$\frac{u(U_{k_m} \xi)}{u(U_{k_m} \xi_0)} \leqslant M_3 \max \left[ \frac{\Delta_\Gamma^{a\alpha_1}(\xi)}{(|\xi|^{\alpha_{p+1}} + \cdots + |\xi|^{\alpha_{p+q}})^a}, \frac{(|\xi|^{\alpha_1} + \cdots + |\xi|^{\alpha_p})^b}{\Delta_\Gamma^{b\alpha_{p+1}}(\xi)} \right].$$

By (4.6) the function on the right-hand side is tempered in $\Gamma$. □

One can construct other examples of 0-admissible functions in a similar way. In particular, one can take linear combinations of the functions of the form (4.7) and take products of them; it is necessary only that the estimating function of this product should be tempered in $\Gamma$.

## 6. Comparison Tauberian theorems.

Comparison Tauberian theorems are usually assumed to be theorem about the asymptotic behaviour of the ratio of some integral transforms of two functions (distributions) if the asymptotic behaviour of the ratio of some other integral transforms is given. In this section we consider some comparison Tauberian theorems in which the integral transform under the consideration is the Laplace transform.

**1. Preliminary theorems.** Let $\Gamma$ be a regular cone and let $\{U_k, k \in I\}$ be a family of linear transformations under which the cone $\Gamma$ is invariant.

**Theorem 1.** *Let $f \in \mathcal{S}'(\Gamma)$ and let $u(\xi)$ be a $q$-strictly admissible function for the family $\{U_k, k \in I\}$. Further, let the following conditions hold:*

(1) *there exists an open set $\Omega \subset C$ such that*

$$\frac{\tilde{f}(iV_k y)}{\tilde{u}(iV_k y)} \to 1, \; k \to \infty, \; k \in I, \; y \in \Omega;$$

(2) *there exists numbers $M$, $\alpha$, $\beta$, $k_0$ and a vector $e \in C$ such that*

$$\left| \frac{\tilde{f}(V_k(x + i\delta e))}{\tilde{u}(iV_k e)} \right| \leqslant M \frac{(1 + |x|)^{\alpha}}{\delta^{\beta}}, \; x \in \mathbf{R}^n, \; 0 < \delta \leqslant 1, \; k > k_0, \; k \in I.$$

*Then for any $m > \alpha + n$, $m \geqslant q$ there exists the limit*

$$\frac{f^{(-m)}(U_k \xi)}{u^{(-m)}(U_k \xi)} \overset{\xi \in K}{\Longrightarrow} 1, \; k \to \infty, \; k \in I, \tag{1.1}$$

*where $K$ is any compact set in* $\operatorname{int} \Gamma$.

**Proof.** Let us put

$$\rho_k = \frac{1}{J_k^q} u^{(-q)}(U_k \xi_0), \; k \in I,$$

where $\xi_0 \in \operatorname{int} \Gamma$ is taken as in (2.2) of Section 5.2. Let us show that for any $m > \alpha + n$ the following limit relation holds:

$$\frac{1}{J_k^m \rho_k} f^{(-m)}(U_k \xi) \overset{\xi \in K}{\Longrightarrow} \theta_{\Gamma}^{m-q} * g, \; k \to \infty, \; k \in I, \; K \subset\subset \mathbf{R}^n, \tag{1.2}$$

where $g$ is a continuous, positive function from the definition of the $q$-strictly admissible function $u$ (see (2.2), Section 5.2). Taking (1.1) into account and using lemma 1 of Section 5.2 we have, for $y \in \Omega$,

$$\frac{\tilde f(iV_k y)}{J_k \rho_k} = \frac{\tilde f(iV_k y)}{\tilde u(iV_k y)} \cdot \frac{J_k^{q-1}\tilde u(iV_k y)}{u^{(-q)}(U_k \xi_0)} \to$$

$$\to \mathcal{K}_C^{-q}(iy)\tilde g(iy),\ k \to \infty,\ k \in I. \tag{1.3}$$

The function $u^{(-q)}$ is 0-strictly admissible, so it has an asymptotics in the cone $\Gamma$ over the family $\{U_k, k \in I\}$ with respect to $J_k^q \rho_k = u^{(-q)}(U_k \xi_0)$, $k \in I$ (see Section 5.2). Therefore, it is a quasi-asymptotics. By theorem 1 of Section 3.1, $u$ has a quasi-asymptotics in $\Gamma$ over the family $\{U_k . k \in I\}$ with respect to $\rho_k$, $k \in I$ and therefore by theorem 1 of Section 4.1 we have the estimate

$$\frac{J_k^{q-1}}{u^{(-q)}(U_k \xi_0)} |\tilde u(V_k z)| \leqslant M_1 \frac{(1+|z|)^{\alpha_1}}{\Delta_C^{\beta_1}(y)}, \tag{1.4}$$

$$k > k_1,\ k \in I,\ z \in T^C,$$

for some $M$, $\alpha_1$, $\beta_1$ and $k_1$. Since

$$\frac{\tilde f(V_k z)}{J_k \rho_k} = \frac{\tilde f(V_k z)}{\tilde u(V_k \zeta)} \cdot \frac{J_k^{q-1}\tilde u(V_k \zeta)}{u^{(-q)}(U_k \xi_0)},\ k \in I,$$

then putting $z = x + i\delta e$, $\zeta = ie$, $0 < \delta \leqslant 1$, $e \in C$ we get the estimate

$$\left|\frac{\tilde f(V_k(x+i\delta e))}{J_k \rho_k}\right| \leqslant M \frac{(1+|x|)^{\alpha}}{\delta^{\beta}} M_1 \frac{(1+|e|)^{\alpha_1}}{\Delta_C^{\beta_1}(e)},\ k \in I.$$

Here we have taken into account condition (2) of theorem 1 for the first factor and for the second we have taken estimate (1.4). So we have shown that $f$ satisfies conditions (A) and (B) of theorem 1 in Section 4.1. Hence, we conclude that for any $m > \alpha + n$ the limit relation (1.2) holds, as

$$\frac{1}{J_k^m \rho_k} f^{(-m)}(U_k \xi) \xrightarrow{\xi \in K} (\theta_\Gamma^{m-q} * g)(\xi),\ k \to \infty,\ k \in I,$$

where $K$ is any compact set in $\mathbf{R}^n$ (see (1.4), Section 4.1).

By (2.2) of Section 5.2 we have

$$\frac{u^{(-q)}(U_k \xi)}{\rho_k} = \frac{J_k^q \theta_\Gamma^q(\xi) * u(U_k \xi)}{u^{(-q)}(U_k \xi_0)} =$$

$$= \frac{u^{(-q)}(U_k \xi)}{u^{(-q)}(U_k \xi_0)} \xrightarrow{\xi \in K} g(\xi),\ k \to \infty,\ k \in I, \tag{1.5}$$

where $K$ is any compact set in $\operatorname{int}\Gamma$. Taking into account estimates (2.3) of Section 5.2 we get, from (1.5),

$$\frac{u(U_k \xi)}{\rho_k} \to \theta_\Gamma^{-q} * g,\ k \to \infty,\ k \in I \text{ in } \ '(\Gamma). \tag{1.6}$$

By the properties of smoothness of kernels of fractional integration $\theta_\Gamma(\xi)$ with respect to the cone $\Gamma$ for large (see corollary, Section 2.8) it follows from (1.6) and (1.2) that for $m > \alpha + n$, $m \geqslant q$

$$\frac{f^{(-m)}(U_k\xi)}{J_k^m\rho_k} \xrightarrow{\xi\in K} g^{(-m+q)}(\xi) \xleftarrow{\xi\in K} \frac{u^{(-m)}(U_k\xi)}{J_k^m\rho_k}, \ k\to\infty,\ k\in I \tag{1.7}$$

where $K$ is any compact set in $\operatorname{int}\Gamma$. By the fact that $g$ is positive and continuous in $\operatorname{int}\Gamma$ we get (1.7) from (1.1). □

**Theorem 2.** *Let $f\in\mathcal{S}'(\Gamma)$ and let $\operatorname{Im}\tilde f(z)>0$, $z\in T^C$. Moreover, let $u(\xi)$ be a $q$-strictly admissible function for the family $\{U_k, k\in I\}$. If condition (1) of theorem 1 holds, then for any $m>2+n$, $m\geqslant q$, there exists the limit (1.1) in theorem 1.*

**Proof.** It is sufficient to show that condition (2) of theorem 1 follows from condition (1). Let us fix a vextor $e\in\Omega\subset C$. Using lemma 1 of Section 4.4 we have

$$\left|\frac{\tilde f(V_k(x+i\delta e))}{\tilde u(iV_k e)}\right| \leqslant M\frac{(1+|x|)^2}{\delta}\left|\frac{\tilde f(iV_k e)}{\tilde u(iV_k e)}\right|,\ k\in I.$$

Thus, we get condition (2) of theorem 1 from condition (1) for a sufficiently large $k, k\in I$. □

**Remark 1.** The statement of the theorem will remain true if one assumes only that $\tilde f(z)$ has a bounded argument in $T^C$ instead of the condition of the non-negativity of its imaginary part. But in this case the number $m$ will probably greater than $n+2$ because one may taken into account the variation of $\arg\tilde f(z)$ in $T^C$.

**Theorem 3.** *Let $f=[d\mu]\in\mathcal{S}'(\Gamma)$, where $d\mu$ is a nonnegative measure with support in $\Gamma$, and let $u$ be a 1- strictly admissible function for the family $\{U_k, k\in I\}$. Besides, let the following conditions hold:*

(1) *there is an open set $\Omega\subset C$ such that*

$$\frac{\tilde f(iV_k y)}{\tilde u(iV_k y)}\to 1,\ k\to\infty,\ k\in I,\ y\in\Omega;$$

(2) *there are numbers $M$, $\beta$, $k_0$ and a vector $e\in C$ such that*

$$\frac{\tilde f(iV_k\delta e)}{\tilde u(iV_k\delta e)}\leqslant\frac{M}{\delta^\beta},\ 0<\delta\leqslant 1,\ k>k_0,\ k\in I.$$

*Then there exists the limit*

$$\frac{\mu(U_k\xi)}{u^{(-1)}(U_k\xi)} \xrightarrow{\xi\in K} 1,\ k\to\infty,\ k\in I, \tag{1.8}$$

*where $K$ is any compact set in* $\operatorname{int}\Gamma$, $\mu(\xi)=[d\mu]^{(-1)}$.

**Proof.** Since $u$ is 1-strictly admissible we have

$$\frac{u^{(-1)}(U_k\xi)}{u^{(-1)}(U_k\xi_0)} \xrightarrow{\xi\in K} g(\xi),\ k\to\infty,\ k\in I,\ K\subset\subset\operatorname{int}\Gamma \tag{1.9}$$

(see (2.2), Section 5.2). Hence $\xi_0\in\operatorname{int}\Gamma$ and $g$ is taken from (2.2) of Section 5.2.

Putting $\rho_k=J_k^{-1}u^{(-1)}(U_k\xi_0)$ let us prove that

$$\frac{f(U_k\xi)}{\rho_k}\to\theta_\Gamma 1 * g = g^{(1)}(\xi),\ k\to\infty,\ k\in I \text{ in } \mathcal{S}'(\Gamma). \tag{1.10}$$

Taking into account condition (1) and using lemma 1 of Section 5.2 we have, for any $y \in \Omega$,

$$\frac{\tilde{f}(iV_k y)}{J_k \rho_k} = \frac{\tilde{f}(iV_k y)}{\tilde{u}(iV_k y)} \frac{\tilde{u}(iV_k y)}{u^{(-1)}(U_k \xi_0)} \to K_C^{-1}(iy)\tilde{g}(iy), \; k \to \infty, \; k \in I. \tag{1.11}$$

Since $f = [d\mu]$ is nonnegative the following estimate holds

$$\begin{aligned} |\tilde{f}(V_k(x+i\delta y))| &= \left|\int e^{i(V_k x,\xi)-\delta(V_k y,\xi)} d\mu(\xi)\right| \leqslant \\ &\leqslant \int e^{-(V_k \delta y,\xi)} d\mu(\xi) = \\ &= \tilde{f}(iV_k \delta y). \end{aligned}$$

Using this estimate for $y = e$ and condition (2) for $k > k_0$, $k \in I$, we get

$$\begin{aligned} \left|\frac{\tilde{f}(V_k(x+i\delta e))}{J_k \rho_k}\right| &\leqslant \frac{\tilde{f}(iV_k \delta e)}{u^{(-1)}(U_k \xi_0)} = \\ &= \frac{\tilde{f}(iV_k \delta e)}{\tilde{u}(iV_k \delta e)} \frac{\tilde{u}(iV_k \delta e)}{u^{(-1)}(U_k \xi_0)} \leqslant \\ &\leqslant \frac{M}{\delta^\beta} M_1 \frac{(1+|\delta e|)^{\alpha_1}}{\Delta_C^{\beta_1}(\delta e)} \leqslant \\ &\leqslant \frac{M_2}{\delta^{\beta+\beta_1}}, \quad 0 < \delta \leqslant 1. \end{aligned} \tag{1.12}$$

To obtain (1.12) we have used estimate (1.4) of theorem 1 for $q = 1$. Formulae (1.11) and (1.12) show that conditions (A) and (B) of theorem 1 in Section 4.1 are valid. Therefore, the limit relation (1.10) holds. Since the first primitive of $g^{(1)}$ is the function $g$, which is continuous (see (2.2), Section 5.2), and since $f = [d\mu]$ is nonnegative, then by theorem 3 of Section 3.1 we have

$$\frac{\mu(U_k \xi)}{J_k \rho_k} = \frac{\mu(U_k \xi)}{u^{(-1)}(U_k \xi_0)} \xrightarrow{\xi \in K} g(\xi), \; k \to \infty, \; k \in I, \; K \subset\subset \operatorname{int}\Gamma. \tag{1.13}$$

Comparing (1.9) and (1.13), and taking into account that $g(\xi) > 0$ for $\xi \in \operatorname{int}\Gamma$, we derive the limit relation (1.8). □

**2. The comparison Tauberian theorems for measures and for holomorphic functions with a nonnegative imaginary part.** Let $\Gamma$ be a regular homogeneous cone in $\mathbf{R}^n$ with its vertex at the origin (see Section 2.6).

**Definition.** Let $\phi$ be a function defined on the cone $\Gamma$. We say that $\phi$ tends to the unit in $\Gamma$ as $\xi \to \infty$, if for any $\epsilon > 0$ there is $\delta = \delta(\epsilon)$ such that

$$|\phi(\xi - 1| < \epsilon \quad \text{if} \quad \Delta_\Gamma(\xi) > \delta. \tag{2.1}$$

In this case we write $\phi(\xi) \to 1$, $\xi \to \infty$ in $\Gamma$.

**Theorem 1.** *Let $f = [d\mu]$, where $d\mu$ is a nonnegative tempered measure with support in $\Gamma$. The following statements are valid:*

(a) *If* $u \in \Delta_1(\Gamma, 1)$ *and there exists the limit*

$$\frac{\tilde f(iy)}{\tilde u(iy)} \to 1, \ y \to 0, \ y \in C, \tag{2.2}$$

*then*

$$\frac{\mu(\xi)}{u^{(-1)}(\xi)} \to 1, \ \xi \to \infty \ \text{in} \ \Gamma. \tag{2.3}$$

(b) *If* $u \in \Delta_1(\Gamma, 2)$ *and for any* $b > 0$ *there exists the limit*

$$\frac{\tilde f(iy)}{\tilde u(iy)} \to 1, \ \Delta_C(y) \to 0, \ y \in C, \ |y| < b, \tag{2.4}$$

*then for any* $\delta > 0$

$$\frac{\mu(\xi)}{u^{(-1)}(\xi)} \to 1, \ |\xi| \to \infty, \ \Delta_\Gamma(\xi) > \delta. \tag{2.5}$$

(c) *If* $u \in \Delta_1(\Gamma, 3)$ *and there exists the limit*

$$\frac{\tilde f(iy)}{\tilde u(iy)} \to 1, \ \Delta_C(y) \to 0, \ y \in C, \tag{2.6}$$

*then*

$$\frac{\mu(\xi)}{u^{(-1)}(\xi)} \to 1, \ |\xi| \to \infty, \ \xi \in \operatorname{int}\Gamma. \tag{2.7}$$

Theorem 2. *Let* $f \in \mathcal{S}'(\Gamma)$ *and let* $\operatorname{Im} \tilde f(z) > 0$, $z \in T^C$. *Then*

(a) *If* $u \in \Delta_q(\Gamma, 1)$ *and there exists the limit*

$$\frac{\tilde f(iy)}{\tilde u(iy)} \to 1, \ y \to 0, \ y \in C, \tag{2.8}$$

*then for any* $m > n + 2$, $m \geqslant q$

$$\frac{f^{(-m)}(\xi)}{u^{(-m)}(\xi)} \to 1, \ \xi \to \infty \ \text{in} \ \Gamma; \tag{2.9}$$

(b) *If* $u \in \Delta_q(\Gamma, 2)$ *and for any* $b > 0$ *there exists the limit*

$$\frac{\tilde f(iy)}{\tilde u(iy)} \to 1, \ \Delta_C(y) \to 0, \ y \in C, \ |y| < b, \tag{2.10}$$

*then for any* $\delta > 0$ *and any* $m > n + 2$, $m \geqslant q$

$$\frac{f^{(-m)}(\xi)}{u^{(-m)}(\xi)} \to 1, \ |\xi| \to \infty, \ \Delta_\Gamma(\xi) > \delta; \tag{2.11}$$

(c) *If* $u \in \Delta_q(\Gamma, 3)$ *and there exists the limit*

$$\frac{\tilde f(iy)}{\tilde u(iy)} \to 1, \ \Delta_C(y) \to 0, \ y \in C, \tag{2.12}$$

*then for any* $m > n + 2$, $m \geqslant q$

$$\frac{f^{(-m)}(\xi)}{u^{(-m)}(\xi)} \to 1, \ |\xi| \to \infty, \ \xi \in \operatorname{int}\Gamma. \tag{2.13}$$

*Here everywhere* $\mu = [d\,\mu]^{-1}$; $\Delta_C(y) = \Delta(y, \partial C)$, $\Delta_\Gamma(\xi) = \Delta(\xi, \partial\Gamma)$ (see Section 0.2).

**Proof.** Let us consider case (a) of theorem 1. Suppose that the statement is false in other words, suppose that (2.2) holds while (2.3) fails. Then there exists an $\epsilon > 0$ and a sequence $\xi_k \in \operatorname{int}\Gamma$, $\Delta_\Gamma(\xi_k) \geqslant k$, $k = 1, 2, \cdots$ such that

$$\left| \frac{\mu(\xi_k)}{u^{(-1)}(\xi_k)} - 1 \right| > \epsilon. \tag{2.14}$$

Let us fix a vector $e \in \operatorname{int}\Gamma$, $|e| = 1$ and consider a family of linear operators $\{U_k, k = 1, 2, \cdots\}$ under which the cone $\Gamma$ is invariant, such that $U_k e = \xi_k$. Such operators exists because the cone $\Gamma$ is homogeneous. Since $|U_k e| = |\xi_k| \geqslant k$, then

$$k \leqslant \Delta_\Gamma(\xi_k) = \Delta_\Gamma(U_k e) \leqslant \frac{\lambda(k)}{p(e)},$$

where $\lambda(k) = \inf_{|e|=1} |U_k e|$ and $p(e)$ is a positive number from lemma 2 of Section 5.1. So it follows that $\lambda(k) \to \infty$, $k \to \infty$ and therefore $\Lambda(k) = \|U_k\| \to \infty$, $k \to \infty$. Hence, the family $\{U_k, k = 1, 2, \cdots\}$ is a family of the first type. Since $u \in \Delta_1(\Gamma, 1)$ there is a subsequence $\{U_{k_m}, m \to \infty\}$ such that $u$ is 1-strictly admissible for this subsequence. From (2.14) it follows that

$$\left| \frac{\mu(\xi_{k_m})}{u^{(-1)}(\xi_{k_m})} - 1 \right| > \epsilon, \ m \to \infty. \tag{2.15}$$

On the other hand, by (1.10) of Section 5.1 we have

$$V_{k_m}(\delta y) \to 0, \ m \to \infty, \ 0 < \delta \leqslant 1, \ y \in \Omega \tag{2.16}$$

where $\Omega$ is any open set in $C$. By condition (2.2)

$$\frac{\tilde{f}(iV_{k_m} y)}{\tilde{u}(iV_{k_m} y)} \to 1, \ m \to \infty, \ y \in \Omega.$$

Besides, fixing a vector $y = e \in \Omega$ and taking into account (2.2), we see from (2.16) that conditions (1) and (2) of theorem 3 in Section 6.1 are valid. Therefore (see (1.8), Section 6.1),

$$\frac{\mu(U_{k_m} e)}{u^{(-1)}(U_{k_m} e)} = \frac{\mu(\xi_{k_m})}{u^{(-1)}(\xi_{k_m})} \to 1, \ m \to \infty. \tag{2.17}$$

Comparing (2.17) and (2.15) we get a contradiction. So case (a) of theorem 1 has been proved.

The proof of case (a) of theorem 2 can be carried out in a similar way. The only difference is that we have to use theorem 2 of Section 6.1 to obtain a contradiction.

The proofs of other cases are similar. We should only use formula (1.11), (1.12) of Section 5.1 and theorems 2 and 3 of Section 6.1, respectively. □

**Remark.** Theorem 2 will be valid if one assumes only that $\tilde{f}(z)$ has a bounded argument in $T^C$ instead of the condition $\operatorname{Im} \tilde{f}(z) > 0$. But in this case the number $m$ in (2.9), (2.11) and (2.13) should be enlarged in order to take into account the variation of $\arg \tilde{f}(z)$ for $z \in T^C$.

**Comments on Chapter 2.**

Results on many-dimensional Tauberian theorems can be found in the works of various authors. We mention here the works of G. Hardy and J. Littlewood (1913), K. Knopp (1939), T. Ganelius (1971), H. Delange (1963a, b) and others. The remainder therms of many-dimensional Tauberian theorems were studied in the works of M.A. Subhankulov (1976) and by his disciples, and also in the works of L. Frennemo (1965, 1966). In these works the results of one-dimensional Tauberian theory were directly extended to the many- dimensional case (namely, to the case of many-dimensional domains which were the direct products of one-dimensional domains). Also studied were nonnegative measures with supports in $\mathbb{R}_+^n$ and their integral transforms, multiple power series and trigonometric series.

In 1976 V.S. Vladimirov extended the one-dimensional Hardy-Littlewood theorem for nonnegative measures to the case of many-dimensional acute cones, or, in other words, to the case when the corresponding Laplace transform was a holomorphic function in a tube domain over an arbitrary acute cone. The remainder term in the Vladimirov theorem was tudied by K.A. Bukin (1981).

The many-dimensional Tauberian theory was further developed inthe works of Yu.N. Drozzinov and B.I. Zavialov (1977, 1980, 1984, 1985a), Yu.N. Drozzinov (1981, 1983a), B.I. Zavialov (1973a, b, 1974, 1977), by V.S. Vladimirov and B.I. Zavialov (1979, 1980, 1981, 1982) (for tempered distributions with support in an arbitrary acute cone), by V.V. Zharinov (1980) (for Fourier hyperfunctions); and also in the works of A.L. Yakymiv (1981, 1983), B. Stankovic (1982, 1985a, b), S.M. Kozlov (1983), E. Omey (1982), L. de Haan and E. Omey (1983), U. Stadtmüller (1981) and others (for nonnegative measures with support in the cone $\mathbb{R}_+^n$).

These works stated some applications of many-dimensional Tauberian theory in mathematical physics, in quantum field theory, in probability theory, in spectral theory, and so on.

The works of B. Bajsanski and I. Karamata (1968, 1969), E. Omey (1982), by L. de Haan and E. Omey (1983), A.L. Yakymiv (1981, 1983), C.M. Kozlov (1983), by Yu.N. Drozzinov and B.I. Zavialov (1985a) were devoted to a many-dimensional generalization of regularly varying (automodel) functions.

The main results of Chapter 2 are contained in the works of V.S. Vladimirov (1976), Yu.N. Drozzinov and B.I. Zavialov (1980, 1984, 1985a) and Yu.N. Drozzinov (1081, 1983a).

## Chapter 3

# ONE-DIMENSIONAL TAUBERIAN THEOREMS

In the case of one variable and for the usual functions or measures, Tauberian theory is highly advanced and has many applications in number theory, in probability theory, in harmonic analysis and in mathematical physics. Among many publications on this subject we mention G.H. Hardy (1949), A.G. Postnikov (1979), T.H. Ganelius (1971), M.A. Subhankulov (1976) and N.H. Bingham, C.M. Goldie and J.L. Tengels (1900), where one can also find a bibliography.

In this chapter we consider the distributions of one variable. One of the main topics that we study here is the relation between the quasi-asymptotic properties of the real imaginary parts of a distribution, which is the Fourier transform of a distribution from $\mathcal{S}'_+$ (see Section 1.3). In certain instances we give more precise information about the order of the primitive of a distribution with quasi-asymptotics which has a normal asymptotics.

We use the results obtained to find the relations between the Cesaro and Abel summation of a divergent series with respect to an automodel weight. In conclusion, we consider a quasi-asymptotic expansion of distributions from $\mathcal{S}'_+$.

This chapter is mainly based on the works of Yu.N. Drozzinov and B.I. Zavialov (1977, 1985b) and V.S. Vladimirov, Yu.N. Drozziniv and B.I. Zavialov (1984a, b).

To emphasize that we consider functions of one variable we denote this variable by $t$, its dual variable under the Fourier transformation by $x$, and under the Laplace transformation by $z = x + iy$.

## 7. The General Tauberian Theorem and its Consequences.

In this section we consider the general Tauberian theorem of Section 4 in the one-dimensional case and we present some of its consequences.

**1. The general Tauberian theorem and its particular cases.** Theorem 1, Section 4.2, leads to the following theorem.

**Theorem 1.** *Let $f \in \mathcal{S}'_+$, let $\tilde{f}(z)$ be its Laplace transform and let $\rho(t)$ be an automodel function of order $\beta$. Then $f$ has a quasi-asumptotics at infinity with respect to $\rho$:*

$$\frac{1}{\rho(k)} f(kt) \to C f_{\beta+1}(t), \; k \to \infty \; \text{ in } \mathcal{S}'_+, \; C \neq 0, \tag{1.1}$$

*if and only if the following conditions are satisfied:*

(A) *there exists the limit*

$$\lim_{y\to+0}\frac{y}{\rho\left(\frac{1}{y}\right)}\tilde{f}(iy)=C; \tag{1.2}$$

(B) *there exist numbers* $M$, $m$ *and* $r_0$ *such that*

$$\left|\frac{r}{\rho\left(\frac{1}{r}\right)}\tilde{f}(re^{i\phi})\right|\leqslant\frac{M}{\sin^m\phi},\quad 0<r\leqslant r_0,\quad 0<\phi<\pi. \tag{1.3}$$

Later we shall use only theorem 1, but one can deduce a stronger result from theorem 1, Section 4.2:

**Theorem 1'.** *Let* $f\in\mathcal{S}'_+$, *let* $\tilde{f}(z)$ *be its Laplace transform and let* $\rho(t)$ *be an automodel function of order* $\beta$. *In order that the distribution* $f$ *has a quasi-asymptotics at infinity with respect to* $\rho$ *the following conditions are necessary:*

$$(1)\ \frac{r}{\rho\left(\frac{1}{r}\right)}\tilde{f}(rz)\ \underset{\longrightarrow}{\overset{z\in K}{\longrightarrow}}\ \frac{C}{(-iz)^{\beta+1}},\quad r\to+0,\quad z=x+iy,$$

*where* $K$ *is any compact set from the upper half-plane* ($y>0$);

(2) *there exist numbers* $M$, $p$ *and* $\alpha$ *such that*

$$\left|\frac{r}{\rho\left(\frac{1}{r}\right)}\tilde{f}(rz)\right|\leqslant M\frac{(1+|z|)^p}{y^\alpha},\quad 0<r\leqslant 1,\ y>0;$$

*and the following conditions are sufficient: condition* (A) *from theorem 1 and*

(B') *there exists numbers* $M_1$, $m$, $r_0$ *and* $0\leqslant\gamma<1$ *such that*

$$\left|\frac{r}{\rho\left(\frac{1}{r}\right)}\tilde{f}(re^{i\phi})\right|\leqslant\frac{M_1}{\sin^m\phi},\quad 0<r\leqslant r_0,$$

$$|x|\leqslant y^\gamma,\quad z=re^{i\phi}=x+iy,\quad y>0.$$

*In this case for any* $q>q_0$,

$$q_0=\max\left\{a+1,\ \frac{b-\beta-\gamma}{1-\gamma},\ \frac{m-\beta-\gamma}{1-\gamma}\right\}$$

*the function* $f^{(-q)}(t)$ *is continuous for* $t>0$ *and has the following asymptotics with respect to* $t^q\rho(t)$:

$$\lim_{t\to+\infty}\frac{f^{(-q)}(t)}{t^q\rho(t)}=\frac{C}{\Gamma(\beta+1+q)}.$$

*Here* $a$ *and* $b$ *are taken from the estimate*

$$|\tilde{f}(x+iy)|\leqslant M_2\frac{(1+|x|)^a}{y^b},\quad x\in\mathbf{R}^1,\quad 0<y\leqslant 1$$

(Estimate (1.1) of Section 4.1 makes it possible to find such an $a$ and $b$.)

Let us remark that in some particular cases the number $q_0$ can be taken in a more precise way (for the case $\gamma = 0$ see, for example, Yu.N. Drozzinov and B.I. Zavialov, 1985b).

The following examples show that condition (B) of theorem 1 cannot be essentially weakened. For instance, it cannot be replaced by the condition for the existence of a limit (1.2) along all the straight lines from some sector of the upper half-plane (example 1) or even along all the straight lines from the upper half-plane to the origin (example 2).

**Example 1.** Let us consider the function $\tilde{f}(z) = z^{\epsilon}\exp[1/2 \ln^2(-iz)]$, $\epsilon > 0$. For $0 < \phi < \pi$, we have

$$|\tilde{f}(re^{i\phi})| = r^{\epsilon}e^{(\pi/2-\phi)\ln r} = r^{\pi/2+\epsilon-\phi}.$$

It is easy to see that $\tilde{f}(z) \in H(\mathbf{R}_+^1)$. If $\epsilon = \pi/4$, then

$$\tilde{f}(re^{i\phi}) \to 0, \ r \to +0, \ 0 < \phi < \frac{4}{4}\pi$$

$$\tilde{f}(re^{i\phi}) \to \infty, \ r \to +0, \ \frac{3}{4}\pi < \phi < \pi.$$

**Example 2.** In the half-disc $\{0 < \phi < \pi, 0 < r < 1\}$ we consider the function

$$\Phi(z) = \frac{1}{z^2}e^{-i(1/4)\ln^4 z}.$$

We have

$$|\Phi(re^{i\phi})| = r^{-2-\phi^3+\phi\ln^2 r}.$$

When $\phi = \arg z$, $0 < \phi < \pi$, is fixed the following limit relation holds

$$-2 - y^3 + \phi\ln^2 r \to +\infty, \ r \to +0.$$

Therefore, if $\phi$ is fixed, then

$$\Phi(re^{i\phi}) \to 0, \ r \to +0, \ 0 < \phi < \pi.$$

But on the line

$$r = e^{-(1/\sqrt{\phi})}, \ (\phi\ln^2 r = 1),$$

which comes to the origin, the inequality

$$|\Phi(z)| = r^{-1-\phi^3} > \frac{1}{r} = e^{1/\sqrt{\phi}}$$

holds, which shows that estimate (B) is impossible for such a function. Using the linear fractional transformation (which transroms origin into origin, the point $i$ into infinity and the imaginary half-axis into itself), one can transform the function $\Phi(z)$ into the function $\tilde{f}(z)$, for which the limit (1.2) exists, but estimate (B) does not hold. It is clear that $\tilde{f} \in H(\mathbf{R}_+^1)$.

In the next chapter we shall need some version of theorem 1, where only the values of $\tilde{f}(z)$ on the imaginary axis are considered.

**Theorem 2.** *The distribution $f \in \mathcal{S}'_+$ has a quasi-asymptotics at infinity with respect to an automodel function $\rho(t)$ if and only if the following necessary and sufficient conditions are satisfied:*

$$\lim_{y \to +0} \frac{y}{\rho\left(\frac{1}{y}\right)} \tilde{f}(iy) = C \neq 0 \tag{1.4}$$

*and*

$$\left| \frac{y^{n+1}}{\rho\left(\frac{1}{y}\right)} \frac{\partial^n \tilde{f}(iy)}{\partial y^n} \right| \leqslant M n \ln^{\sigma} n, \quad 0 < y \leqslant y_0, \; n = 1, 2, \cdots \tag{1.5}$$

*for some $M$, $\sigma$, $y_0$.*

**Proof.** At first, let us prove the *necessity*. According to theorem 1 it is sufficient to prove only estimate (1.5). Because of the condition $\rho^{-1}(k) f(kt) \to C$, $k \to \infty$ in $\mathcal{S}'_+$ the family of distributions $\{1/\rho(k) f(kt)\}$ is bounded when $k \to \infty$. Therefore (see Section 1.3) there exists numbers $N$, $k_0$ and $A$ such that

$$\left\| \left( \frac{1}{\rho(k)} f(kt), \phi(t) \right) \right\| \leqslant A \|\phi\|_{N, \mathbb{R}^1_+}, \phi \in {}_+, \; k \geqslant k_0.$$

Taking into account that $y^n \tilde{f}^{(n)}(iy) = (f(t), y^n t^n e^{-yt})$ we have

$$\left| \frac{y^{n+1}}{\rho\left(\frac{1}{y}\right)} \tilde{f}^{(n)}(iy) \right| = \left\| \left( \frac{1}{\rho\left(\frac{1}{y}\right)} f\left(\frac{1}{y} t\right), t^n e^{-t} \right) \right\| \leqslant$$

$$\leqslant A \|t^n e^{-t}\|_{N, \mathbb{R}^1_+}$$

when $y \leqslant y_0$. From the definition of the norm $\|\cdot\|_{N, \mathbb{R}^1_+}$ (see Section 1.1) it follows that $\| |t^n e^{-t}\|_{N, \mathbb{R}^1_+} \leqslant A_1 n \ln^{N+1} n$ and this proves estimate (1.5) if we put $\sigma = N + 1$.

Let us now prove the *sufficiency*. Assuming that estimate (1.5) holds we shall prove estimate (1.3). Let us draw a straight line, from the point $i$, which is perpendicular to the line connecting the point $e^{i\phi}$ and the origin, and let the point $z_1 = r_1 e^{i\phi}$ be the intersection point of these two lines. Let $d$ denote the distance between $i$ and $z_1$ (see Figure 10).

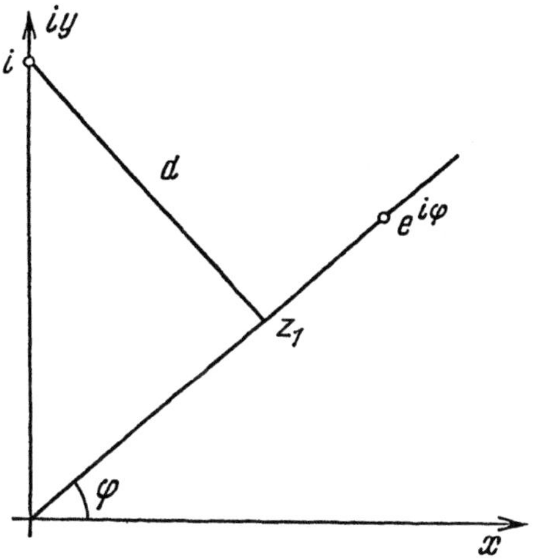


Figure 10.

Expanding the function $\tilde{f}_r(z) = \tilde{f}(rz)$ in Taylor series with the centre at the point $i$, we get

$$\left| \frac{r}{\rho\left(\frac{1}{r}\right)} \tilde{f}(rz_1) \right| = \left| \sum_{n=0}^{\infty} \frac{1}{n!} \frac{r}{\rho\left(\frac{1}{r}\right)} r^n \tilde{f}^{(n)}(ri)(z_1 - i)^n \right| \leqslant$$

$$\leqslant M \sum_{n=0}^{\infty} n^{\sigma} d^n \leqslant \frac{M_1}{(1-d)^{\sigma+1}} \leqslant \frac{M_2}{(\sin\phi)^{2\sigma+2}} \tag{1.6}$$

when $0 < r < y_0$. Since $|z_1| = \sin\phi$, we have, if $0 < r < y_0 \sin\phi$,

$$\left| \frac{r}{\rho\left(\frac{1}{r}\right)} \tilde{f}(re^{i\phi}) \right| = \left| \frac{r}{\rho\left(\frac{1}{r}\right)} \tilde{f}\left(\frac{r}{\sin\phi} z_1\right) \right| =$$

$$= \sin\phi \left| \frac{\frac{r}{\sin\phi}}{\rho\left(\frac{1}{\sin\phi}\frac{\sin\phi}{r}\right)} \tilde{f}\left(\frac{r}{\sin\rho} z_1\right) \right|.$$

By lemma 1, Section 3.2, for a sufficiently small $y_0$ and $0 < r < y_0 \sin\phi$

$$\rho\left(\frac{1}{\sin\phi}\frac{\sin\phi}{r}\right) \geqslant \frac{M_3}{(\sin\phi)^{\beta-\epsilon}} \rho\left(\frac{\sin\phi}{r}\right).$$

where $\beta$ is the order of the function $\rho$. From this and (1.6) we get (1.3) if $0 < r < y_0 \sin\phi$.

For $y_0 \sin\phi < r < r_0$, estimate (1.3) follows from lemma 1 of Section 3.2 and from the following general estimate for the elements of the algebra $H(\mathbf{R}^1_+)$ (see Section 2.5):

$$|\tilde{f}(z)| \leqslant M(1 + |z|^2)^{a/2}(1 + y^{-b}), \; z = x + iy, \; y > 0,$$

with some constants $M = M(f)$, $a$ and $b$, if $r_0$ is sufficiently small. □

The one-dimensional version of theorem 2 in Section 4.3 leads to the following well-known Hardy-Littlewood-Karamata theorem.

**Theorem 3.** *Let $d\mu$ be a nonnegative measure with support on the positive half-axis $\mathbf{R}^1_+$ and let $\rho$ be some automodel function of order $\beta$. Then*

$$\frac{1}{\rho\left(\frac{1}{y}\right)}\int_0^\infty e^{-yt}\, d\mu(t) \to C \neq 0, \; y \to +0$$

*if and only if*

$$\frac{\mu(k)}{\rho(k)} = \frac{1}{\rho(k)}\int_0^k d\mu(t) \to \frac{C}{\Gamma(\beta + 1)}, \; k \to \infty.$$

Let us now formulate another version of theorem 2 of Section 4.4.

**Theorem 4.** *Let $f \in \mathcal{S}'_+$ and in some neighbourhood of the origin let its Laplace transform $\tilde{f}(z)$ have a nonnegative real part. $f$ has a quasi-asymptotics at infinity with respect to an automodel function $\rho$ if and only if*

$$\frac{y}{\rho\left(\frac{1}{y}\right)}\tilde{f}(iy) \to C \neq 0, \; y \to +0.$$

**Remark.** Under the hypothesis of theorem 4, the order $\beta$ of the function $\rho$ automatically satisfies the restriction

$$-2 \leqslant \beta \leqslant 0 \tag{1.7}$$

(see remark 2, Section 4.4).

**Definitions.** (1) A class of functions from $H(\mathbf{R}^1_+)$ which have nonnegative real parts in the upper half-plane are denoted by $H_+$.

(2) $H^+$ denotes a class of functions $\tilde{f} \in H(\mathbf{R}^1_+)$ such that real parts of their buondary values are nonnegative measures in some neighbourhood of the origin. In other words, $\tilde{f}(z) \in H^+$ if and only if there exists $\delta > 0$ such that

$$\operatorname{Re}\tilde{f}(x) \equiv u(x)d\mu(x) + u_1(x), \tag{1.8}$$

where $d\mu$ is a nonnegative measure and

$$\operatorname{supp} u_1 \subset [x \in \mathbf{R}^1 \colon |x| \geqslant \delta].$$

(3) $M_+$ denotes a class of functions $\tilde{f}(z) \in H(\mathbf{R}^1_+)$ such that the real parts of their boundary values are measures on the whole axis and they are nonnegative

measures in some neighbourhood of the origin. In other words, $\tilde{f}(z) \in M_+$ if and only if

$$\operatorname{Re} \tilde{f}(z) \equiv u(x) = d\mu(x),$$

and there exists a $\delta > 0$ such that

$$d\mu(x) \geqslant 0 \quad \text{if} \quad |x| < \delta.$$

It is obvious that

$$H_+ \subset M_+ \subset H^+ \subset H(\mathbf{R}^1_+).$$

The set of all boundary values of functions from $H_+$, $M_+$, $H^+$ is denoted by $H_+$, $M_+$, $H^+$, respectively.

**2. Quasi-asymptotics of a distribution $f$ from $\mathcal{S}'_+$ and a function $\arg \tilde{f}$.** In the previous theorems the asymptotic behaviour of a distribution was studied with respect to an automodel function that was given before. The following question is of some interest. If a distribution is given, does it have a quasi-asymptotics with respect to some automodel function? in certain cases it is possible to answer this question.

**Theorem 1.** *Let $f \in \mathcal{S}'_+$ and $\tilde{f} \in H_+$. If $f$ has a quasi-asymptotics at infinity then it is necessary that there is a limit*

$$\lim_{r \to +0} \arg \tilde{f}(rz) = \omega(z), \quad z = x + iy, \ y > 0;$$

*and conversely, for this it is sufficient if there exist two points $z_1 = x_1 + iy_1$, $y_1 > 0$ and $z_2 = x_2 + iy_2$, $y_2 > 0$, which do not lie on the same ray going through the origin and such that*

$$\arg \tilde{f}(rz_j) \to \omega(z_j), \quad r \to +0, \quad j = 1, 2.$$

*In this case a proper automodel function $\rho$ can be chosen as follows*

$$\rho(t) = \frac{1}{t} \left| \tilde{f}\left(i\frac{1}{t}\right) \right|.$$

**Proof. Necessity.** Suppose that $f$ has a quasi-asymptotics with respect to an automodel function $\rho$. It follows from theorem 1 of Section 7.1 that

$$\frac{r\tilde{f}(rz)}{\rho\left(\dfrac{1}{r}\right)} \to \frac{C}{(-iz)^{\beta+1}}, \quad C \neq 0, \ r \to +0$$

where $\beta$ is the order of the automodel function $\rho$. Therefore,

$$\arg \tilde{f}(rz) \to \arg \frac{C}{(iz)^{\beta+1}}, \quad r \to +0.$$

**Sufficiency.** Let us consider the harmonic function $\arg \tilde{f}(z) = V(x, y)$. According to the hypothesis of the theorem, it has limits along the two rays $\{rz_j, \ j = 1, 2; r > 0\}$. It follows from the Loomis theorem (see L. Loomis, 1943) that there is a limit along any ray coming to the origin from the upper half-plane.

Let us consider the sequence

$$\{V_r(x,y) = V(rx,ry), r \to +0\}.$$

This sequence is bounded and tends uniformly to some harmonic function on every compact set, together with all its derivatives. (Note that $\arg \tilde{f}(z)$ is bounded since $\tilde{f} \in H_+$.)

In particular,

$$\partial_x V_r(x,y) = r\,\partial_{rx} V(rx,ry) \to \partial_x V_0(x,y),\ r \to +0.$$

Hence, putting $x = 0$, $y = 1$ and $y' = r$, we get

$$y'\partial_x V(x,y')\Big|_{x=0} \to \partial_x V_0(x,1)\Big|_{x=0} = \alpha,\ y' \to +0. \tag{2.2}$$

For the conjugated function $u(x,y)$ we have

$$\ln \tilde{f}(z) = u(x,y) + iV(x,y).$$

Using the Cauchy-Riemann equations we derive from (2.2)

$$y\frac{\partial}{\partial y}u(0,y) \to \alpha,\ y \to +0.$$

Thus, it follows that

$$y\frac{\partial}{\partial y}\ln|\tilde{f}(iy)| \to \alpha,\ y \to +0.$$

Putting $\rho_1(t) = |\tilde{f}(i(1/t))|$, we get

$$\frac{y\frac{\partial}{\partial y}|\tilde{f}(iy)|}{|\tilde{f}(iy)|} = \frac{t\rho_1'(t)}{\rho_1(t)} \to -\alpha,\ t \to +\infty,$$

where $y = t^{-1}$. Hence, according to lemma 2 of Section 3.2, $\rho_1(t)$ is an automodel function of order $-\alpha$, and

$$\frac{1}{\rho_1(t)}\tilde{f}(iy) = \frac{y}{y\rho_1(t)}\tilde{f}(iy) \to \lim_{y \to +0} \arg \tilde{f}(iy) =: \omega(i),\ y \to +0.$$

Theorem 4 of Section 7.1 implies that the distribution $f$ has a quasi- asymptotics with respect to the automodel function

$$\rho = \frac{1}{t}\rho_1(t) = \frac{1}{t}|\tilde{f}(i(1/t))|.$$ □

**Remark 1.** The sufficient condition in the theorem can be reformulated in the following way:

Let $\operatorname{Re}\tilde{f}(z) \geqslant 0$ when $y > 0$ and let there exist points $z_1$ and $z_2$ in the upper half-plane ($\operatorname{Im}(z_1/z_2) \neq 0$), and positive functions $\rho_1(t)$ and $\rho_2(t)$ such that

$$\frac{r\tilde{f}(rz_j)}{\rho_j\left(\frac{1}{r}\right)} \to C_j \neq 0,\ r \to +0,\ j = 1,2.$$

Then $\rho_1$ and $\rho_2$ are equivalent automodel functions and $f$ has a quasi-asymptotics at infinity with respect to each of them.

**Remark 2.** Theorems analogous to the last one are valid for those classes of distributions in which condition (B) of theorem 1 in Section 7.1 follows from condition (A). For instance, if $[d\mu] \in \mathcal{S}'_+$ is a nonnegative measure and there exists numbers $\phi_1 < \phi_2$ from an interval $[0, \pi]$, such that the function $\arg[d\tilde{\mu}](z)$ has a limit along any ray $\{z = re^{i\phi}, r \to +0, \phi_1 < \phi < \phi_2\}$, uniform in $\phi$, then $d\mu$ has a quasi-asymptotics at infinity with respect to any automodel function of the form

$$\rho(t) = \frac{1}{t}\left|[d\tilde{\mu}]\left(\frac{1}{t}e^{i\phi_0}\right)\right|$$

where $\phi_0$ is an arbitrary number from the interval $(0, \pi)$.

**3. Tauberian theorem for distributions from the class $H^+$.** In this subsection we extend theorem 4 of Section 7.1 to distributions from the class $H^+$. We begin with some preliminary remarks of a general character.

Let $\mathcal{D}_{\mathcal{L}^\infty}$ denote the space of infinitely differentiable functions on $\mathbf{R}^1$ for which

$$\|\phi\|_{\mathcal{L}^\infty, N} = \max_{0 \leqslant j \leqslant N} \sup_{\xi \in \mathbf{R}^1} |\phi^{(j)}(\xi)| < \infty, \ N = 0, 1, \ldots, \phi^{(j)}(\xi) \to 0, \quad |\xi| \to \infty. \quad (3.1)$$

(These norms are the norms in the spaces $C^N(\mathbf{R}^1)$, see Section 0.4.) The countable family of norms (3.1) defines a natural topology on $\mathcal{D}_{\mathcal{L}^\infty}$. It is clear that $\mathcal{S}(\mathbf{R}^1)$ is topologically embedded in $\mathcal{D}_{\mathcal{L}^\infty}$. Therefore, for the dual spaces we have the embedding $\mathcal{D}'_{\mathcal{L}^\infty} \subset \mathcal{S}'(\mathbf{R}^1)$. Every distribution with compact support clearly belongs to $\mathcal{D}'_{\mathcal{L}^\infty}$. If $\omega(\xi)$ is an arbitrary distribution from $\mathcal{S}'(\mathbf{R}^1)$, then there is a nonnegative integer $N$ such that $(1 + \xi^{2N})^{-1}\omega(\xi) \in \mathcal{D}'_{\mathcal{L}^\infty}$.

For any $z = x + iy$, $y > 0$, the function $\phi_z(\xi) = (1/z - \xi)$ belongs to $\mathcal{D}_{\mathcal{L}^\infty}$, so the expression $(\omega(\xi), (1/z - \xi))$ is well defined and is an analytic functions in the upper half-plane from $H(\mathbf{R}^1_+)$.

The following lemma holds.

**Lemma 1.** *Let the distribution $\omega$ belong to $\mathcal{D}'_{\mathcal{L}^\infty}$ and be real. Let us put*

$$\tilde{f}(z) = \left(\omega(\xi), \frac{i}{\pi(z - \xi)}\right).$$

*In addition, let $\tilde{f}(x)$ be the boundary value of $\tilde{f}(z)$ (in the sense of distributions) on the real axis, and let $u(x)$, and $v(x)$ be its real and imaginary parts, respectively, so that $\tilde{f}(x) = u(x) + iv(x)$. Then $u(x) = \omega(x)$.*

**Proof.** Since the real part of a boundary value is equal to the boundary value of the real part it is sufficient to verify that for any $\phi \in \mathcal{S}(\mathbf{R}^1)$

$$\lim_{y \to +0} \int \operatorname{Re} \tilde{f}(x + iy)\phi(x)\,dx = (\omega, \phi). \quad (3.2)$$

For the left-hand side of (3.2) we have

$$\int \operatorname{Re} \tilde{f}(x + iy)\phi(x)\,dx = \int \left(\omega(\xi), \frac{y}{\pi[(x - \xi)^2 + y^2]}\right)\phi(x)\,dx =$$

$$= \left( \omega(\xi), \frac{1}{\pi} \int \frac{y}{(x-\xi)^2 + y^2} \phi(x)\, dx \right). \tag{3.3}$$

The possibility of integration before the functional action follows from the fact that the Darboux sums of the integral

$$\frac{1}{\pi} \int \frac{y}{(x-\xi)^2 + y^2} \phi(x)\, dx$$

tend to this integral in the topology of the space $\mathcal{D}_{\mathcal{L}_\infty}$ (over $\xi$). Let us notice also that this integral is simply the convolution of the function $\phi$ with a Poisson kernel, and it is known that it tends to $\phi(\xi)$ when $y \to +0$ in the topology of $\mathcal{D}_{\mathcal{L}_\infty}$. Now (3.3) implies (3.2). □

**Corollary 1.** *Let $\tilde{f} \in H(\mathbf{R}^1_+)$, and let $u(x)$ be the real part of its boundary value. Take an integer $N_0 \geqslant 0$, such that*

$$\frac{u(\xi)}{1+\xi_0^{2N}} \in \mathcal{D}'_{\mathcal{L}_\infty}.$$

*Then there exists a polynomial $(z)$ with purely imaginary coefficients such that for any integer $N \geqslant N_0$*

$$\tilde{f}(z) = (1+z^{2N}) \left( u(\xi), \frac{i}{\pi(z-\xi)(1+\xi^{2N})} \right) + P(z),\ y > 0. \tag{3.4}$$

**Proof.** According to lemma 1 the real parts of the boundary values of the functions $\tilde{f}(z)$ and

$$\tilde{f}_1(z) = (1+z^{2N}) \left( u(\xi), \frac{i}{\pi(z-\xi)(1+\xi^{2N})} \right)$$

are the same. Hence, by the Bogolybov 'edge of the wedge' theorem (see V.S. Vladimirov, 1966) these two functions may differ only by an entire function which is purely imaginary on the real axis. As tempered, this function can be but a polynomial with purely imaginary coefficients. □

For a more general representation in the case of several variables see V.S. Vladimorov (1979). Representation (3.4) can be found in H. Bremermann's work (1965).

**Remark.** In some particular cases the arbitariness of representation (3.4) can be reduced. For instance, let $\tilde{f} \in H_+$. Then the following well-known Herglotz-Nevanlinna representation (see V.S. Vladimorov, 1979) can be applied

$$\tilde{f}(z) = (1+z^2) \int \frac{i d\sigma(\xi)}{x - \xi + iy} - iz \operatorname{Re} \tilde{f}(i) + i \operatorname{Im} \tilde{f}(i),$$

where $d\sigma$ is a nonnegative finite measure, $\int d\sigma(\xi) < \infty$.

Now we prove the following theorem.

**Theorem.** *Let $f \in \mathcal{S}'_+$, and let its Laplace transform be $\tilde{f} \in H^+$. $f$ has a quasi-asymptotics at infinity with respect to an automodel function $\rho(t)$ of order $\beta$ if and only if condition (A) of theorem 1 in Section 7.1 holds:*

$$\lim_{y \to +0} \frac{y}{\rho\left(\frac{1}{y}\right)} \tilde{f}(iy) = C,\ C \neq 0. \tag{3.5}$$

*In this case* $\beta \leqslant 0$.

**Proof.** The *necessity* follows from the general Tauberian theorem 1, Section 7.1. Let us prove the *sufficiency*. Since $\tilde{f}(z) \in H^+$ we have

$$\operatorname{Re} \tilde{f}(x) \equiv u(x) = d\mu(x) + u_1(x), \tag{3.6}$$

where $d\mu$ is a nonnegative measure, $\operatorname{supp} d\mu \subset [-\delta, \delta]$ for some $\delta > 0$, and the support of the distribution $u_1$ is contained in the set $[\,|x| \geqslant \delta]$. According to Corollary 1 there exists a number $N_0$ such that for any integer $N > N_0$ representation (3.4) holds, where the polynomial $(z)$ has purely imaginary coefficients ($\operatorname{Re} \mathcal{P}(x) \equiv 0$). Using (3.6) we have

$$\tilde{f}(z) = \left(d\mu(\xi), \frac{i}{\pi(z-\xi)(1+\xi^{2N})}\right) + \left(u_1(\xi), \frac{i}{\pi(z-\xi)(1+\xi^{2N})}\right) +$$

$$+ z^{2N}\left(u(\xi), \frac{i}{\pi(z-\xi)(1+\xi^{2N})}\right) + \mathcal{P}(z) = \tag{3.7}$$

$$= f_1(z) + f_2(z) + f_3(z) + \mathcal{P}(z).$$

Since $u_1(x) = 0$ in some neighbourhood of the origin, $f_2(z)$ is analytic in a neighbourhood of the origin and can be expanded into a Taylor series with purely imaginary coefficients. Considering only terms up to the order of $2N - 1$, and taking into account that the remainder tends to zero when $y \to +0$ faster than $y^{2N-1}$ (yet the limit in (3.5) does exist), we have

$$f_2(z) = f_{2,1}(z) + Q_{2N-1}(z), \quad \operatorname{Re} Q_{2N-1}(x) \equiv 0, \tag{3.8}$$

$$|f_{2,1}(z)| \leqslant M_1 |z|^{2N},$$

when $|z| < (\delta/2)$. Let us estimate $f_3(z)$. Since $u \in \mathcal{S}'$, there eixsts a norm $\|\cdot\|_p$ in $\mathcal{S}(\mathbf{R}^1)$ (see Section 1.1. and Section 1.3) such that

$$|(u, \phi)| \leqslant M_2 \|\phi\|_p, \quad \phi \in \mathcal{S}$$

and, therefore,

$$\left| z^{2N}\left(u(\xi), \frac{i}{\pi(1+\xi^{2N})(z-\xi)}\right)\right| \leqslant |z|^{2N} M_2 \left\| \frac{i}{\pi(1+\xi^{2N})(z-\xi)} \right\|_p \leqslant$$

$$\leqslant M_3 \frac{|z|^{2N}}{y^{p+1}}, \quad |z| < \frac{\delta}{2}. \tag{3.9}$$

Let us remark that from (1.7), (3.7), (3.8) and (3.9) it follows that the order of the automodel function $\rho$ in the hypothesis of the theorem satisfies the condition $\beta \leqslant 0$.

We shall prove the theorem by induction over $\beta$.

At first, let us consider the case $-2 \leqslant \beta \leqslant 0$. The theorem will be proved if, for $\tilde{f}(z)$, we prove condition (B) of the general Tauberian theorem (Theorem 1, Section 7.1).

If $N$ is sufficiently large, the angular estimate (B) for $f_3(z)$ follows from (3.9). Taking the terms of degree zero and one from the polynomials $P(z)$ and $Q_{2N-1}(z)$, from (3.7) and (3.8) we get

$$\tilde{f}(z) = f_1(z) + a_0 i + a_i iz + f_4(z) \equiv f_{1,}1(z) + f_4(z),$$

$$f_{1,1}(z) = f_1(z) + a_0 i + a_1 iz;$$

$$f_4(z) = f_{2,1}(z) + f_3(z) + P(z) + Q_{2N-1}(z) - a_0 i - a_1 iz;$$

where $a_0$ and $a_1$ are real numbers, and $f_4(z)$ satisfies the angular estimate (B) and the estimate

$$|f_4(iy)| \leqslant M_4 |y|^2,$$

when $|y| \leqslant \delta/2$. Therefore, it is sufficient to prove the theorem only for the function

$$f_{1,1}(z) = i \int \frac{d\mu_1(\xi)}{z - \xi} + a_0 i + a_1 iz, \tag{3.10}$$

where

$$d\mu_1(\xi) = \frac{d\mu(\xi)}{\pi(1 + \xi^{2N})}$$

is a nonnegative finite measure over $\mathbf{R}^1$.

First, let us assume that $-2 < \beta \leqslant 0$. In this case we have

$$\frac{y}{\rho\left(\frac{1}{y}\right)} ia_1(iy) \to 0, \; y \to +0.$$

Hence, by (3.5)

$$\frac{y}{\rho\left(\frac{1}{y}\right)} (f_1(iy) + a_0 i) \to C, \; y \to +0. \tag{3.11}$$

But $f_1(z) + a_0 i \in H_+$ and the angular estimate for $f_1(z) + a_0 i$ follows from (3.11) and theorem 4 of Section 7.1. Since the term $a_1 iz$ does not effect the angular estimate (B), when $\beta > -2$, this estimate also holds for $f_{1,1}$ (see (3.10)).

Let us now assume that $\beta = -2$. According to (3.5) we get

$$\frac{y}{\rho\left(\frac{1}{y}\right)} [f_1(iy) + a_0 i - a_1 y] \to C \neq 0, \; y \to +0. \tag{3.12}$$

Denoting

$$\operatorname{Re} f_1(iy) \equiv W(y) = \int \frac{y \, d\mu_1(\xi)}{\xi^2 + y^2}$$

we have

$$\frac{y^2}{\rho\left(\frac{1}{y}\right)} \left(\frac{W(y)}{y} - a_1\right) \to \operatorname{Re} C, \; y \to +0.$$

Note that $(1/y)W(y)$ is nondecreasing when $y \to +0$. The following three cases are possible:

(i) $(1/y)W(y) \to +\infty$, $y \to +0$ and therefore $(1/y^2)\rho(1/y) \to +\infty$. In this case $a_1 y^2(1/\rho(1/y)) \to 0$, $y \to +0$, and therefore the term $a_1 iz$ adds nothing to the limit relation (3.12). Now the angular estimate (B) follows from theorem 4 of Section 7.1.

(ii) $(1/y)W(y) - a_1 \to A \neq 0$, $y \to +0$. In this case either again $(1/y^2)\rho(1/y) \to +\infty$, $y \to +0$ (if $\operatorname{Re} C = 0$), or $(1/y^2)\rho(1/y) \to A_1 \neq 0$, $y \to +0$. The first case has been already considered. In the second case there exists the limit

$$\lim_{y \to +0} \frac{y}{\rho\left(\frac{1}{y}\right)}[f_1(iy) + a_0 i] = C_2 \neq 0.$$

Thus, the term $a_1 iz$, as before, does not change the angular estimate (B).

(iii) $(1/y)W(y) - a_1 \to 0$, $y \to +0$. This means that

$$\int \frac{d\mu_1(\xi)}{\xi^2 + y^2} \to a_1, \quad y \to +0$$

so that

$$\int \frac{d\mu_1(\xi)}{\xi^2} < \infty,$$

and therefore $d\mu_2(\xi) = (1/\xi^2)d\mu_1(\xi)$ is a nonnegative measure which is continuous at zero. Let us prove that

$$iy \int \frac{d\mu_2(\xi)}{iy - \xi} \to 0, \quad y \to +0. \tag{3.13}$$

Indeed, taking into account that the measure $d\mu_2(\xi)$ has a compact support, we have

$$\begin{aligned} iy \int \frac{d\mu_2(\xi)}{iy - \xi} &= y^2 \int \frac{d\mu_2(\xi)}{y^2 + \xi^2} - iy \int \frac{\xi\, d\mu_2(\xi)}{y^2 + \xi^2} = \\ &= \frac{y(y - i\xi)\mu_2(\xi)}{y^2 + \xi^2}\Bigg|_{-\infty}^{+\infty} - y^2 \int \frac{2\xi\mu_2(\xi)\, d\xi}{(y^2 + \xi^2)^2} + \\ &+ iy \int \frac{\mu_2(\xi)(y^2 - \xi^2)}{(y^2 + \xi^2)^2}\, d\xi = \\ &= -\int \frac{\mu_2(y\eta)2\eta\, d\eta}{(1 + \eta^2)^2} + i \int \frac{\mu_2(y\eta)(1 - \eta^2)}{(1 + \eta^2)^2}\, d\eta. \end{aligned}$$

Passing to the limit when $y \to +0$, and taking into account that the function $\mu_2(\xi)$ is continuous at zero and bounded, and having in mind the equality

$$\int \frac{2\eta\, d\eta}{(1 + \eta^2)^2} = 0 = \int \frac{1 - \eta^2}{(1 + \eta^2)^2}\, d\eta$$

we get the limit relation (3.13). Let us now consider the function

$$f_{1,2}(z) = \frac{1}{-z^2} f_{1,1}(z).$$

It is sufficient to prove that its boundary value $f_{1,2}(x)$ has a quasi-asymptotics at the origin with respect to the automodel function $k^2\rho(k)$. Indeed, if $f_{1,2}(x)$ has a quasi-asymptotics at the origin then, according to theorem 4 of Section 3.4, its inverse Fourier transform has a quasi-asymptotics at infinity with respect to the automodel function $k\rho(k)$ and therefore the angular estimate (B) for $f_{1,2}(z)$ holds by theorem 1 of Section 7.1.

By (3.10) and (3.12) we have

$$\frac{y}{\frac{1}{y^2}\rho\left(\frac{1}{y}\right)} f_{1,2}(iy) = \frac{y}{\rho\left(\frac{1}{y}\right)} f_{1,1}(iy) \to C \neq 0, \; y \to +0. \tag{3.14}$$

Obviously, the boundary value of the real part of $f_{1,2}(z)$ is equal to $d\mu_2(x)$ if $x \neq 0$. Denoting now Re $f_{1,2}(x)$ by $u_2(x)$ we have

$$u_2(x) = \frac{d\mu_1(x)}{x^2} + \sum_{j=0}^{p} C_j \delta^{(j)}(x).$$

By Corollary 1, the following representation takes place

$$f_{1,2}(z) = i\int \frac{d\mu_2(\xi)}{z-\xi} + \sum_{j=0}^{p} \frac{C_j}{(-iz)^{j+1}} + Q(z) =$$

$$= f_{2,2}(z) + \sum_{j=0}^{p} \frac{C_j}{(-iz)^{j+1}} + Q(z),$$

where $f_{2,2}(z) = i\int(d\mu_2(\xi)/z-\xi)$ and $Q(z)$ is a polynomial with purely imaginary coefficients. From (3.13) it follows that

$$yf_{2,2}(iy) \to 0, \; y \to +0. \tag{3.15}$$

The automodel function $(1/y^2)\rho(1/y)$ is of zero order, when $y \to +0$. Comparing (3.14) and (3.15) we conclude that all the coefficients $C_j$, $j = 1, 2, \ldots, p$, are equal to zero. In other words,

$$f_{1,2}(z) = f_{2,2}(z) + \frac{C_0}{-iz} + Q(z).$$

If now $C_0 = 0$ then the real part of the boundary value of $f_{1,2}(z)$ is a nonnegative measure $d\mu_2(x)$ and all previous considerations are valid. In particular, $f_{1,2}(z)$ satisfies the angular estimate (B) and therefore $f_{1,1}(z)$ also satisfies the same estimate. If $C_0 \neq 0$, then from (3.14) we get

$$\frac{1}{\frac{1}{y^2}\rho\left(\frac{1}{y}\right)}[yf_{2,2}(iy) + C_0] \to C \neq 0, \; y \to +0.$$

By (3.15) this is possible only if

$$\frac{1}{y^2}\rho\left(\frac{1}{y}\right) \to \frac{C_0}{C} \neq 0, \; y \to +0$$

and the proof can be completed in the same way as in case (ii). The theorem has been proved for $-2 \leqslant \beta \leqslant 0$.

Let $-4 \leqslant \beta < -2$ and (3.5) hold. Using the representation (3.4), estimates (3.8) and (3.9), and taking a sufficiently large $N$, we can restrict ourselves to the case

$$\tilde{f}(z) = i\int\frac{d\mu(\xi)}{z-\xi} + {}_1(z) = f_5(z) + P_1(z),$$

where $P_1(z)$ is a polynomial with purely imaginary coefficients. Since $\beta < -2$, we have

$$\frac{1}{y}\tilde{f}(iy) \to 0, \; y \to +0.$$

As before, from this it follows that

$$\int\frac{d\mu(\xi)}{\xi^2} < \infty.$$

In particular, the measure $d\mu_3(x) = x^{-2}d\mu(x)$ is continuous at zero. Let us consider the function

$$f_6(z) = \frac{1}{-z^2}\tilde{f}(z),$$

with its boundary value denoted by $u_3(x) + iv_3(x)$. Observe that $u_3(x) = x^{-2}d\mu(x)$ if $x \neq 0$ and therefore

$$u_3(x) = \frac{d\mu(x)}{x^2} + \sum_{j=0}^{p} C_j^1\delta^{(j)}(x).$$

Hence, by Corollary 1,

$$f_6(z) = i\int\frac{d\mu_3(\xi)}{z-\xi} + \sum_{j=0}^{p}\frac{C_j^1}{(-iz)^{j+1}} + \mathrm{P}_2(z), \tag{3.16}$$

where ${}_2(z)$ is a polynomial with the purely imaginary coefficients. As before it follows from (3.5) that

$$\frac{y}{\frac{1}{y^2}\rho\left(\frac{1}{y}\right)}f_6(iy) = \frac{y}{\rho\left(\frac{1}{y}\right)}\tilde{f}(iy) \to C \neq 0, \; y \to +0. \tag{3.17}$$

But now the order of the automodel function $(1/y^2)\rho(1/y)$ when $y \to +0$ belongs to the half-open interval $[-2, 0)$. Thus,

$$yf_6(iy) \to 0, \; y \to +0. \tag{3.18}$$

Since the measure $d\mu_3(x)$ is continuous at zero we have, as in (3.13),

$$iy\int\frac{d\mu_3(\xi)}{iy-\xi} \to 0, \; y \to +0. \tag{3.19}$$

Comparing (3.17), (3.18) and (3.19) we get from (3.16) that $C_j^1 = 0$, $j = 0, 1, \ldots, p$. Thus, we have arrived at the case $2 \leqslant \beta < 0$, which we have already considered. Continuing in a similar way, we complete the proof for an arbitrary $\beta \leqslant 0$. □

**4. The decomposition theorem.** The following theorem is essential for our further considerations. It describes the decomposition of a distribution, under which the quasi- asymptotic properties of this distribution are preserved.

**Theorem.** *Let* $f \in \mathcal{S}'(\mathbf{R}^1)$ *and let* $\rho$ *be an automodel function of order* $\beta \neq -1, -2, \cdots$ *If there exists the limit*

$$\frac{1}{\rho(k)} f(kt) \to g(t) 0, \; k \to \infty \tag{4.1}$$

*in* $\mathcal{S}'$*, then* $f$ *can be represented as*

$$f = f^+ + f^-; \; f^+ \in \mathcal{S}'_+, \; f^- \in \mathcal{S}'_-, \tag{4.2}$$

*where* $f^+$ *and* $f^-$ *have quasi-asymptotics at plus and minus infinity, respectively (one of them might be trivial).*

**Remark 1.** If $\beta = -1, -2, \cdots$ then the assertion of the theorem does not hold. One can consider the distribution $f(t) = \; (1/t)$ as the counter- example. If we put $\rho(k) = k^{-1}$, then the limit (4.1) obviously exists. On the other hand, for any decomposition (4.2) there exist constants $C_0, C_1, ..., C_m$, such that for any $\phi \in \mathcal{S}$

$$(f^+, \phi) = \int_0^1 \frac{\phi(t) - \phi(0)}{t} dt + \int_1^\infty \frac{\phi(t)}{t} dt + \sum_{j=0}^m C_j \phi^{(j)}(0).$$

It is easy to see that if $\phi(0) \neq 0$, then for any choice of constants $C_j$, $j = 0, ..., m$

$$(kf^+(kt), \phi(t)) \to \infty, \; k \to \infty.$$

We shall prove the theorem using the following two lemmas.

**Lemma 1.** *Let* $f \in \mathcal{S}'(\mathbf{R}^1)$ *and suppose that for some automodel function of order* $\beta \neq -1, -2, \cdots$ *there exists the limit*

$$\frac{1}{\rho(k)} f(kt) \to g(t) \not\equiv 0, \; k \to \infty \tag{4.3}$$

*in* $\mathcal{S}'$*. Also, let*

$$f = f^+ + f^-; \; f^+ \in \mathcal{S}'_+, \; f^- \in \mathcal{S}'_-. \tag{4.4}$$

*Then there is an integer* $N$ *such that for the distributions*

$$f_N^+(t) = t^N f^+(t), \; f_N^-(t) = t^N f^-(t)$$

*the following limits exists*

$$\frac{1}{k^N \rho(k)} f^{\pm}(kt) \to C_{\pm} f_{\beta+N+1}(\pm t), \; k \to +\infty, \tag{4.5}$$

*where the constants* $C_+$ *and* $C_-$ *do not vanish simultaneously.*

**Proof.** Indeed, by (4.3) there is an $m_1$ such that the convergence in (4.3) is the convergence with respect to the norm $\|\cdot\|_{m_1}$ (see Section 1.3). Let $\phi(t)$ be an arbitrary test function from $\mathcal{S}_+$. Obviously, the function

$$\phi_N(t) = \begin{cases} t^N \phi(t), & t \geq 0, \\ 0, & t < 0. \end{cases}$$

belongs to $\mathcal{S}_{m_1}(\mathbf{R}^1)$ for any itneger $N \geq m_1$ (see Section 1.1). On the other hand, for any decomposition (4.4) there exists $m_2$ such that for any $N \geqslant m_2$ and $k > 0$

$$(f(kt), \phi_N(t)) = (f^+(kt), \phi_N(t)).$$

Thus, taking $n \geqslant \max\{m_1, m_2\}$ we obtain the limit relation

$$\frac{1}{k^N \rho(k)}(f_N^+(kt), \phi(t)) = \frac{1}{\rho(k)}(f^+(kt), t^N \phi(t)) =$$
$$= \frac{1}{\rho(k)}(f(kt), \phi_N(t)) \to (g(t), \phi_N(t)), \ k \to \infty.$$

A similar relationholds for $f_N^-$. Hence, the left-hand sides of (4.5) tend to limits which by formula (4.1) of Section 3.4 have exactly the same form as in (4.5). The constants $C_+$ and $C_-$ cannot vanish simulatenously because, by the condition on $\beta$, $g(t)$ is not the distribution with support at the point $t = 0$. Hence, the lemma has been proved.

**Lemma 2.** *Let $f \in \mathcal{S}'_+$, and let $\rho$ be an automodel function of order $\beta \neq -1, -2, \cdots$. Assume that for some integer $n \geqslant 0$ the distribution $f_N(t) = t^N f(t)$ has a quasi-asymptotics at infinity with respect to $k^N \rho(k)$. Then:*

if $\beta > -1$, $f(t)$ has a quasi-asymptotics at infinity with respect to $\rho(t)$;

if $\beta < -1$, there exists numbers $a_j$, $j = 0, 1, \ldots, p$ such that the distribution

$$g(t) = f(t) + \sum_{j=0}^{p} a_j \delta^{(j)}(t)$$

*has a quasi-asymptotics at infinity with respect to $\rho(t)$.*

**Proof.** Passing to the Laplace transform we get

$$i^N \tilde{f}_N(z) = \frac{d^N}{dz^N} \tilde{f}(z), \ z = x + iy, \ y > 0.$$

Let us prove by induction that for any integer $n$, $0 \leqslant n \leqslant N$, the functions $i^n \tilde{f}_n(z) = (d^n/dz^n)\tilde{f}(z)$ can be represented in the form

$$i^n \tilde{f}_n(z) = \tilde{g}_n(z) + \sum_{j=0}^{p_n} b_j^{(n)} z^j, \tag{4.6}$$

where $\tilde{g}_n(z) = (-i)^n \tilde{g}^n(z)$ satisfy the conditions

$$\frac{1}{\left(\frac{1}{y}\right)^{n+1} \rho\left(\frac{1}{y}\right)} \tilde{g}_n(iy) \to C_n \neq 0, \ y \to +0, \tag{4.7}$$

$$\left| \frac{1}{\left(\frac{1}{r}\right)^{n+1} \rho\left(\frac{1}{r}\right)} \tilde{g}_n(re^{i\phi}) \right| \leqslant \frac{M_n}{(\sin\phi)^{\alpha_n}}, \ 0 < r \leqslant r_0. \tag{4.8}$$

Indeed, according to the hypothesis of the theorem, and by theorem 1 of Section 7.1, these relations hold for $n = N$. Let us suppose that they are valid for some $n = l > 0$. We shall prove that they hold for $n = l - 1$. We consider two cases: $l > -\beta$, and $l < -\beta$ (the case $l = -\beta$ is excluded by the hypothesis).

(1) $l > -\beta$. As $\tilde{g}_{l-1}(z)$ we take the function

$$i\int_\gamma \tilde{g}_l(z')\,dz',$$

where $\gamma$ is any curve connecting the points $ir_0$ and $z$ in the upper half-plane. In particular,

$$\frac{1}{\left(\frac{1}{y}\right)^l \rho\left(\frac{1}{y}\right)}\tilde{g}_{l-1}(iy) = \frac{i}{\left(\frac{1}{y}\right)^l \rho\left(\frac{1}{y}\right)}\int_y^{r_0} \tilde{g}_l(iy')\,dy'. \tag{4.9}$$

Using the well-known property of automodel functions (see Section 3.2)

$$\frac{1}{k\,\rho(k)}\int_{k_0}^{k} \rho(t)\,dt \to C \quad 0,\ k \to \infty,\ \beta > -1, \tag{4.10}$$

we get from (4.7) that, for $n = l$, expressions (4.9) tend to a nonnegative limit when $y \to +0$.

To prove estimate (4.8) for $n = l - 1$ we take the curve $\gamma$ consisting of an arc connecting the points $ir_0$ and $r_0e^{i\phi}$, and a segment connecting the points $r_0e^{i\phi}$ and $e^{i\phi}$ (see Figure 11).

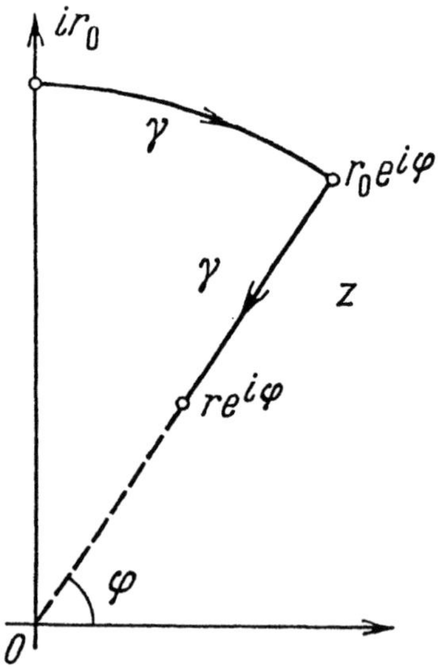


Figure 11.

After this, using (4.8) for $n = l$ and (4.10), it is easy to prove estimate (4.8) for $n = l - 1$.

(2) $l < -\beta$. We now put

$$\tilde{g}_{l-1}(z) = i\int_0^z \tilde{g}_l(z')\,dz'.$$

where the integration is taken over the segment connection the points 0 and $z$. The proof can be carried out as in the previous case, only instead of (4.10) one should use the relation

$$\frac{1}{k\,\rho(k)}\int_k^\infty \rho(t)\,dt \to C \neq 0,\ k \to \infty,\ \beta < -1.$$

Because the primitives of order $N$ of one function may differ only by a polynomial of degree $N-1$, the relations (4.6), (4.7) and (4.8) are proved. Considering again the function $f(t)$ and using (4.7), (4.8) and theorem 1 of Section 7.1 we conclude that the function $f$ differs from a function which has a quasi-asymptotics at infinity with respect to $\rho$, but only by a linear combination of the $\delta$-function and its derivatives. (If $\beta > -1$ this means that the function $f$ itself has a quasi-asymptotics with respect to $\rho$.) □

**Proof.** Now let us prove our theorem. Let $f \in \mathcal{S}'$ and let

$$f = f^+ + f^-;\ f^+ \in \mathcal{S}'_+,\ f^- \in \mathcal{S}'_-$$

be some decomposition of the function $f$. Then by lemma 1 and 2

$$f(t) = f_1(t) + \sum_{j=0}^{p} a_j^{\pm}\delta^{(j)}(t),$$

where

$$\frac{1}{\rho(k)} f_1^{\pm}(kt) \to C_{\pm} f_{\beta+1}(\pm t),\ k \to +\infty,$$

and the constants $C_+$ and $C_-$ do not vanish simultaneously (say, for instance, $C_+ \neq 0$). Putting

$$f_2^+ = f_1^+(t),\ f_2^- = f^-(t) + \sum_j a_j^+\delta^{(j)}(t) = f(t) - f_2^+$$

we obtain

$$f(t) = f_2^+ + f_2^-;\ f_2^+ \in {}'_+,\ f_2^- \in {}'_-,$$

where $f_2^+$ and $f_2^-$ have quasi-asymptotics at infinity with respect to the automodel function$\rho(t)$ (because of our assumption, $f_2^-$ may have trivial quasi-asymptotics). □

Using this theorem and theorems 1 of Section 7.1 and 4 of Section 3.4 one can give an another interpretation of the quasi-asymptotics of a distribution at the origin if the order of an automodel function is not equal to $0, -1, -2, \cdots$. Namely, the following statement holds.

**Corollary 1.** *A distribution $\tilde{f}(x)$ from $\mathcal{S}(\mathbf{R}^1)$ has a quasi-asymptotics at the origin with respect to an automodel function $\rho$ of order $\alpha \neq 0, -1, -2, \cdots$ if and only if*

$$\tilde{f}(x) = \tilde{f}^+(x) + \tilde{f}^-(x),$$

*where $\tilde{f}^+(x)$ and $\tilde{f}^-(x)$ are the boundary values of some functions $\tilde{f}^+(z)$ and $\tilde{f}^-(z)$, from $H(\mathbf{R}^1_+)$ and $H(\mathbf{R}^1_-)$, respectively, such that*

$$\frac{1}{\rho\left(\frac{1}{y}\right)}\tilde{f}^{\pm}(\pm iy) \to C_{\pm}, \ y \to +0,$$

*where* $(C_+, C_-)$ 0 *and the following estimate holds*

$$\left|\frac{1}{\rho\left(\frac{1}{r}\right)}\tilde{f}^{\pm}(re^{\pm i\phi})\right| \leqslant \frac{M_{\pm}}{(\sin\phi)^{\alpha_{\pm}}}, \quad 0 < r \leqslant r_0, \ 0 < \phi < \pi,$$

*for some constants* $M_+$, $M_-$, $\alpha_+$, $\alpha_-$, *and* $r_0$.

As we mentioned before, this statement is not true if $\alpha = 0, -1, -2, \cdots$.

## 8. Quasi-asymptotics properties of distributions at the origin.

In this section we study the connection between the quasi-asymptotics of a distribution $\tilde{f}(x)$ $(f \in \mathcal{S}'_+)$ at the origin and the asymptotic properties of $\mathrm{Re}\,\tilde{f}(x)$.

1. **The general case.** First, let us write representations (3.4) of Section 7.3 in a more convenient form. We put

$$\omega(\xi) = \frac{u(\xi)}{\pi(1+\xi^{2N})}, \quad \omega \in \mathcal{D}'_{L^\infty}. \tag{1.1}$$

$$\tilde{\Phi}(z) = \left(\omega(\xi), \frac{i}{x-\xi+iy}\right)\tilde{\Phi} \in H(\mathbf{R}^1_+), \ z = x+iy. \tag{1.2}$$

Let us introduce the distribution $\Phi(t)$ from $\mathcal{S}'_+$ by the formula

$$\tilde{\Phi}(z) = L[\Phi] = (\Phi(t), e^{izt}), \ z = x+iy, \ y > 0; \tag{1.3}$$

$$\left(\Phi(t) = L^{-1}[\tilde{\Phi}]\right),$$

and put

$$\Phi_1(t) = 2\pi F^{-1}[\omega] = (\omega(\xi), e^{-i\xi t}). \tag{1.4}$$

Note that $\Phi_1(t)$ is continuous. Hence, the distribution $\theta(t)\Phi_1(t)$ is well defined. Noting that

$$F^{-1}\left[\frac{i}{x+i0}\right] = \theta(t)$$

(see Section 2.1), we obtain from (1.2)-(1.4)

$$\Phi(t) = \theta(t)\Phi_1(t). \tag{1.5}$$

Thus, by the reality of $\omega$ we have

$$\Phi_1(t) = \Phi(t) + \overline{\Phi(-t)}; \quad \Phi_1(t) = \overline{\Phi_1(-t)}. \tag{1.6}$$

Now representation (3.4) takes the form

$$\tilde{f}(z) = (1 + z^{2N})\tilde{\Phi}(z) + \mathfrak{P}(z), \ y > 0. \tag{1.7}$$

where $\mathfrak{P}(z)$ is a polynomial with purely imaginary coefficients.

The next two theorems state a connection between the quasi-asymptotics of the boundary value of a holomorphic function in the upper half-plane from $H(\mathbf{R}_+^1)$ and the asymptotic properties of the real part of this boundary value (the latter is in fact the weight funciton in representation (3.4)).

Let $\tilde{f} \in H(\mathbf{R}_+^1)$, and let $\tilde{f}(x)$ be its boundary values:

$$\tilde{f}(x+iy) \to \tilde{f}(x) = u(x) + iv(x), \ y \to +0$$

in $\mathcal{S}'$.

**Theorem 1.** *If $\tilde{f} \in \tilde{\mathcal{S}}'_+$ has a quasi-asymptotics at the origin with respect to an automodel function $\rho$ of order $\alpha \neq 0, -1, -2, \cdots$, then its real part $u(x)$ also has a quasi- asymptotics at the origin with respect to $\rho$.*

*Conversely, if $u(x)$ has a quasi-asymptotics at the origin with respect to an automodel function $\rho$ of order $\alpha$ $0, -1, -2, \cdots$, then there exists a polynomial $Q(x)$, with purely imaginary coefficients, such that the distribution $\tilde{f}(x) - Q(x)$ has a quasi-asymptotics at the origin with respect to $\rho$.*

**Proof.** Let $\tilde{f}$ have a quasi-asymptotics at the origin with respect to $\rho$. According to (4.8) of Section 3.4 we have (see (1.4), Section 2.1)

$$\frac{1}{\rho(k)}\tilde{f}\left(\frac{x}{k}\right) \to \frac{Ci^{\alpha}}{(x+i0)^{\alpha}} =$$
$$= C\Gamma(1-\alpha)\left[e^{i(\pi/2)\alpha} f_{-\alpha+1}(x) + e^{-i(\pi/2)\alpha} f_{-\alpha+1}(-x)\right], \ k \to \infty$$

in $\mathcal{S}'$, where $C = a + ib \neq 0$. Taking the real parts from both sides we get

$$\frac{1}{\rho(k)}u\left(\frac{x}{k}\right) \to \Gamma(1-\alpha)\left[\left(a\cos\frac{\pi}{2}\alpha - b\sin\frac{\pi}{2}\alpha\right) f_{-\alpha+1}(x) + \right.$$
$$\left. + \left(a\cos\frac{\pi}{2}\alpha + b\sin\frac{\pi}{2}\alpha\right) f_{-\alpha+1}(-x)\right], \ k \to \infty \ \text{ in } \mathcal{S}'. \tag{1.8}$$

Since $\alpha \neq 0, -1, -2, \cdots$, it is easy to see that the right-hand side of (1.8) is not equal to zero.

Let us prove the converse statement. Let $u(x)$ have a quasi-asymptotics at the origin with respect to $\rho$. Then for any $N \geqslant 0$ the function

$$\omega(x) = \frac{u(x)}{\pi(1+x^{2N})}$$

also has a quasi-asymptotics at the origin with respect to $\rho$. By (1.4), $\Phi_1(t)$ has a quasi-asumptotics at infinity:

$$\frac{k}{\rho(k)}\Phi_1(kt) \to \kappa(t) \ 0, \ k \to \infty \ \text{ in } \mathcal{S}'.$$

By the theorem in Section 7.4 there are distributions $\Phi_{\pm} \in S'_{\pm}$, such that

$$\Phi_1(t) = \Phi_+(t) + \Phi_-(t),$$

where $\Phi_\pm(t)$ has a quasi-asymptotics at infinity with respect to $\rho_1(k) = k^{-1}\rho(k)$. Without loss of generality, one can assume that

$$\Phi_+(t) = \overline{\phi_-(-t)}. \tag{1.9}$$

(If this is not the case one can take the functions

$$\frac{1}{2}[\Phi_+(t) + \overline{\Phi_-(-t)}]$$

and

$$\frac{1}{2}[\Phi_-(t) + \overline{\Phi_+(-t)}]$$

instead of $\Phi_+$ and $\Phi_-$, which already have property (1.9).) Comparing (1.9) with (1.6) we get

$$\Phi_+(t) = \Phi(t) - \Delta(t), \quad \Delta(t) = -\overline{\Delta(-t)}, \tag{1.10}$$

where

$$\Delta(t) = \sum_{k=0}^{p} C_k\, \delta^{(k)}(t); \quad \Phi(t) = \theta(t)\Phi_1(t).$$

Performing a Fourier transform and taking into account that $\Phi_+(t)$ has a quasi-asymptotics at infinity with respect to an automodel function $\rho_1(k)$ of order $\beta = \alpha - 1 \neq -1, -2, \cdots$, we conclude by (1.10) that

$$\tilde{\Phi}(x) - \sum_{k=0}^{p} C_k\,(-ix)^k$$

has a quasi-asymptotics at the origin with respect to $\rho(k)$ and the polynomial

$$P_1(x) = \sum_{k=0}^{p} C_k\,(-ix)^k$$

has purely imaginary coefficients.

By representation (1.7), $\tilde{f}(x) = (1 + x^{2N})\tilde{\Phi}(x) + P(x)$, and therefore $\tilde{f}(x) - Q(x)$ has a quasi- asymptotics at the origin with respect to $\rho$. Here $Q(x) \equiv P(x) + (1 + x^{2N})P_1(x)$. □

**Theorem 2.** *If $\tilde{f} \in \tilde{\mathcal{S}}_+$ has a quasi-asymptotics at the origin with respect to an automodel function $\rho$ of order $\alpha = -n$, $n = 0, 1, 2, \cdots$, so that*

$$\frac{1}{\rho(k)}\tilde{f}\left(\frac{x}{k}\right) \to \tilde{g}(x) = \frac{C}{(x + i0)^\alpha}, \quad k \to \infty \tag{1.11}$$

*in $\mathcal{S}'$, $C = a + ib \neq 0$, then there exists a solution $h \in \mathcal{S}'$ of the equation*

$$x^{n+1}h(x) = u(x); \quad u(x) = \operatorname{Re}\tilde{f}(x), \tag{1.12}$$

*which has a quasi-asymptotics at the origin with respect to the automodel function*

$$\rho_1(k) = k^{n+1}\rho(k) \tag{1.13}$$

*of order 1, so that*

$$\frac{1}{\rho_1(k)} h\left(\frac{x}{k}\right) \to g_n(x), \; k \to \infty \tag{1.14}$$

*in $\mathcal{S}'$, where*

$$g_n(x) = \begin{cases} a\,\mathcal{P}\frac{1}{x} + \pi b\,\delta(x), & n = 0, \\ b\,\mathcal{P}\frac{1}{x} - \pi a\,\delta(x), & n = 1, \\ -a\,\mathcal{P}\frac{1}{x} - \pi b\,\delta(x), & n = 2, \\ -b\,\mathcal{P}\frac{1}{x} + \pi a\,\delta(x), & n = 3, \end{cases} \pmod 4.$$

*Conversely, if there exists a solution $h \in \mathcal{S}'$ of equation (1.12) which has a quasi-asymptotics at the origin with respect to the function $\rho_1$ of order 1, then there exists a polynomial $Q(x)$ with purely imaginary coefficients such that $\tilde{f}(x) - Q(x)$ has a quasi- asymptotics at the origin with respect to the automodel function $\rho(k) = k^{-n-1}\rho_1(k)$, so that*

$$\frac{1}{\rho(k)}\left[\tilde{f}\left(\frac{x}{k}\right) - Q\left(\frac{x}{k}\right)\right] \to \tilde{g}_1(x) = \frac{C_1}{(x + i\,0)^{\alpha}}, \; k \to \infty \tag{1.15}$$

*in $\mathcal{S}'$, $C_1 = a_1 + b_1 i \neq 0$.*

**Proof.** Let $\tilde{f}$ have a quasi-asymptotics at the origin with respect to $\rho$ of order $\alpha = -n$, $n = 0, 1, \cdots$; that is, (1.11) holds. Then the distribution $(x + i\,0)^{-n-1}\tilde{f}(x)$, which is the boundary value of the function $z^{-n-1}\tilde{f}(z)$, has a quasi- asymptotics at the origin with respect to the automodel function (1.13) $\rho_1 = k^{n+1}\rho(k)$ (see formulae (4.2) and (4.8), Section 3.4):

$$\frac{\tilde{f}\left(\frac{x}{k}\right)}{k^{n+1}\rho(k)\left(\frac{x + i\,0}{k}\right)^{n+1}} \to \frac{(-i)^n}{x + i\,0} C, \; k \to \infty$$

in $\mathcal{S}'$, $C = a + ib \neq 0$.

Denoting the real part of the distribution $(x + i\,0)^{-n-1}\tilde{f}(x)$ by $h(x)$, we get the limit relation (1.14) (see (1.5), Section 2.1). It remains only to note that $h(x)$ satisfies equation (1.12).

Indeed,

$$x^{n+1}h(x) = x^{n+1}\,\mathrm{Re}\frac{\tilde{f}(x)}{(x + i\,0)^{n+1}} = \mathrm{Re}\left[x^{n+1}\frac{\tilde{f}(x)}{(x + i\,0)^{n+1}}\right] =$$

$$= \mathrm{Re}\,\tilde{f}(x) = u(x).$$

Let us now prove the converse statement. Let $h(x)$ be a solution of equation (1.12), which has a quasi-asymptotics at the origin (1.14) with respect to the automodel function $\rho_1$ of order 1. Then the distribution

$$h_1(x) = \frac{h(x)}{\pi(1+x^{2N})}$$

satisfies the equation

$$x^{n+1}h_1(x) = \omega(x), \quad \omega(x) = \frac{u(x)}{\pi(1+x^{2N})}, \tag{1.16}$$

and has a quasi-asymptotics at the origin with respect to $\rho_1$. By the theorem 1 Section 7.4 the distribution $H_1 = F^{-1}[h_1]$ can be decomposed in the following way:

$$H_1 = H_1^+ + H_1^-,$$

where $H_1^{\pm} \in \mathcal{S}'_{\pm}$ has quasi-asymptotics at infinity with respect to the automodel function $k^{-1}\rho_1(k)$ of order 0. By (1.16) the distribution $\Phi_1 = F^{-1}[\omega]$ has the form

$$\Phi_1 = (H_1^+)^{(n+1)} + (H_1^-)^{(n+1)},$$

where $(H_1^{\pm})^{(n+1)} \in \mathcal{S}'_{\pm}$ has a quasi-asymptotics at infinity with respect to the automodel function $k^{-n-2}\rho_1(k)$.

Due to the reality of $\omega$ one can assume (without loss of generality) that

$$(H_1^+)^{(n+1)}(t) = \overline{(H_1^-)^{(n+1)}(-t)}$$

(see the proof of theorem 1). Comparing this with (1.6), we obtain

$$(H_1^+)^{(n+1)} = \Phi - \Delta, \quad \Delta(t) = \sum_{k=0}^{p} C_k \delta^{(k)}(t).$$

Introducing an automodel function $\rho(k)$ according to formula (1.13), and proceeding in the same manner as in the proof of theorem 1, we complete the proof of our theorem. □

**Remark 1.** In formulae (1.14) and (1.15) the coefficients $C$ and $C_1$ are in fact equal ($a = a_1, b = b_1$), with the possible exception of the case when

$$\lim_{k\to\infty} k^n \rho(k) = d, \quad d \neq 0.$$

But even in this case the coefficient preceding $(1/x)$ in (1.14) ($a$ or $b$ dependently on $n$) is equal to a corresponding coefficient in (1.15).

Let us now prove a lemma that we shall require later.

**Lemma.** *Let $v \in \mathcal{S}'$ be even and have a quasi-asymptotics at the origin with respect to an automodel function $\rho$ of order $\alpha = 0$ so that*

$$\frac{1}{\rho(k)} v\left(\frac{x}{k}\right) \to A \neq 0, \quad k \to \infty \tag{1.17}$$

*in $\mathcal{S}'$. Then there is a unique odd solution $h(x)$ of the equation*

$$xh(x) = v(x) \tag{1.18}$$

*which has a quasi-asymptotics at the origin with respect to the automodel function $k\rho(k)$, so that*

$$\frac{1}{k\rho(k)} h\left(\frac{x}{k}\right) \to A \frac{1}{x}, \quad k \to \infty \tag{1.19}$$

in $\mathcal{S}'$.

**Proof.** It is easy to see that the unique odd solution of equation (1.18) is given by the formula

$$(h(\xi),\phi(\xi)) = \frac{1}{2}\left(\nu(\xi), \frac{\phi(\xi)-\phi(-\xi)}{\xi}\right), \quad \phi \in S.$$

Thus, by (1.17) we have

$$\frac{1}{k\rho(k)}\left(h\left(\frac{\xi}{k}\right),\phi(\xi)\right) = \frac{1}{\rho(k)}(h(\xi),\phi(k\xi)) =$$

$$= \frac{1}{2\rho(k)}\left(\nu(\xi), \frac{\phi(k\xi)-\phi(-k\xi)}{\xi}\right) =$$

$$= \frac{k}{\rho(k)}\left(\nu(\xi), \frac{\phi(k\xi)-\phi(-k\xi)}{2k\xi}\right) =$$

$$= \frac{1}{\rho(k)}\left(\nu\left(\frac{\xi}{k}\right), \frac{\phi(\xi)-\phi(-\xi)}{2\xi}\right) \to \frac{A}{2}\int\frac{\phi(\xi)-\phi(-\xi)}{\xi}\,d\xi =$$

$$= A\left(\mathrm{P}\frac{1}{\xi},\phi(\xi)\right), \quad k \to \infty,$$

and the limit equation (1.12) is proved. □

In conclusion we give a simple example which shows that theorem 1 fails for the order of automodelity $\alpha = 0, -1, -2, \cdots$. On the other hand, it illustrates theorem 2.

Let us put $\tilde{f}(z) = -i\ \ln(-z)$. Then

$$\tilde{f}(x) = \pi\theta(x) - i\ \ln|x| \equiv u(x) + iv(x).$$

It is obvious that the distribution $u(x) = \pi\theta(x)$ has a quasi- asymptotics at the origin with respect to the automodel function $\rho \equiv 1$ of order 0. Nevertheless, $\tilde{f}$ has no such proprty, regardless of the polynomial added to it.

On the other hand, one of the solutions of the equation

$$xh(x) = u(x), \quad u(x) = \pi\theta(x),$$

is given by the formula

$$(h(x),\phi(x)) = \pi\int_0^1\frac{\phi(x)-\phi(0)}{x}\,dx + \pi\int_1^\infty\frac{\phi(x)}{x}\,dx, \quad \phi \in S,$$

and has a quasi-asymptotics at the origin with respect to the automodel function $\rho = k\ \ln k$.

Indeed, for any $\phi \in \mathcal{S}$ we have

$$\frac{1}{k\ \ln k}\left(h\left(\frac{x}{k}\right),\phi(x)\right) = \frac{\pi}{\ln k}\left[\int_0^1\frac{\phi(kx)-\phi(0)}{x}\,dx + \int_1^\infty\frac{\phi(kx)}{x}\,dx\right] =$$

$$= \frac{\pi}{\ln k}\left[\int_0^k \frac{\phi(x)-\phi(0)}{x}\,dx + \int_k^\infty \frac{\phi(x)}{x}\,dx\right] =$$

$$= \frac{\pi}{\ln k}\left[\int_0^1 \frac{\phi(x)-\phi(0)}{x}\,dx + \int_1^\infty \frac{\phi(x)}{x}\,dx - \int_1^k \frac{\phi(0)}{x}\,dx\right] =$$

$$= -\pi\phi(0) + \frac{\pi}{\ln k}\left[\int_0^1 \frac{\phi(x)-\phi(0)}{x}\,dx + \int_1^\infty \frac{\phi(x)}{x}\,dx\right].$$

Now it is obvious that

$$\frac{1}{k\ln k}h\left(\frac{x}{k}\right) \to -\pi\delta(x),\ k \to \infty \ \text{ in } \mathcal{S}'.$$

According to theorem 2, $\tilde{f}(x)$ has a quasi-asymptotics at the origin with rspect to $\ln k$. (It is clear that the polynomial ambiguity is not essential here.) One can easily check that this is indeed the case.

**2. Quasi-asymptotics of distributions from $H^+$ and asymptotic properties of the reproducing functions of measures.** We consider distributions $f \in \mathcal{S}'_+$. Their Laplace transforms $\tilde{f} \in H(\mathbf{R}^1_+)$ have boundary values $\tilde{f}(x) = u(x) + iv(x)$ in $\mathcal{S}'$, when $y \to +0$. Restricting the class of distributions under consideration, we assume that $\tilde{f} \in H^+$, so that in some neighbourhood of zero, $|x| < \delta$, $u(x) = d\mu(x)$ is a nonnegative measure. The function $\mu(x)$ is called the reproducing function of this measure. As before, $\rho(t)$ is an automodel function of order $\alpha$. We use also the automodel function $\rho_1(t) = t^{-1}\rho(t)$ of order $\beta = \alpha - 1$.

**Theorem 1.** *Let $f \in \mathcal{S}'_+$, and let $\rho$ be an automodel function of order $\alpha \neq 1, 0, -1, -2, \cdots$. The following statements are equivalent:*

(1) *There exists a distribution*

$$\Delta(t) = \sum_{k=0}^{p} i^k C_k \delta^{(k)}(t),\ \bar{C}_k = -C_k,$$

*such that $f(t) - \Delta(t)$ has a quasi-asymptotics at infinity with respect to the automodel function $\rho_1(H) = t^{-1}\rho(t)$ of order $\beta = \alpha - 1$, so that*

$$\frac{1}{\rho_1(k)}(f(kt) - \Delta(kt)) \to Cf_{\beta+1}(t),\ k \to \infty \tag{2.1}$$

*in $\mathcal{S}'$, $C = a + ib \neq 0$.*

(2) *There exists a polynomial purely imaginary coefficients*

$$\mathcal{P}(z) = \sum_{k=0}^{p} C_k z^k$$

*such that*

$$\frac{1}{\rho\left(\frac{1}{y}\right)}[\tilde{f}(iy) - \mathcal{P}(iy)] \to C = a + ib \neq 0,\ y \to +0. \tag{2.2}$$

(3) *The reproducing function* $\mu(x)$ *is continuous at the point* $x = 0$. *If one assumes that* $\mu(0) = 0$, *then there exist the limits*

$$\frac{\pm\mu(\pm x)}{\rho_1\left(\frac{1}{x}\right)} \to C_\pm, \quad x \to +0, \quad (C_+ . C_-) \neq 0, \tag{2.3}$$

*where*

$$C_\pm = \frac{a \cos\frac{\pi}{2}\alpha \mp b \sin\frac{\pi}{2}\alpha}{1-\alpha}. \tag{2.4}$$

*In this case the following inequalities hold*

$$\alpha < 1; \ a \cos\frac{\pi}{2}\alpha \geqslant \left| b \sin\frac{\pi}{2}\alpha \right|. \tag{2.5}$$

**Remark 1.** In the hypothesis if theorem 1, of $0 < \alpha < 1$ the subtracted polynomial does not influence the existence of the limit (2.2). In this case theorem 1 can be also deduced from the Keldysh type of Tauberian theorem. (For the formulation and proof of the Keldysh type of Tauberian theorem see, for instance, A.G. Kostjuchenko and I.S. Sargs'jan (1979).)

**Theorem 2.** *Let* $f \in \mathcal{S}'_+$ *and let its Laplace transform be* $\tilde{f}(z) \in H^+$. *Let* $\rho$ *be an automodel function of order* $\alpha = 1, 0, -1, -2, \cdots$ *and, as before,* $\rho_1(t) = t^{-1}\rho(t)$, $\beta = \alpha - 1$. *Then the following statements are equivalent:*

(1') *There exists a distribution*

$$\Delta(t) = \sum_{k=0}^{p} i^k C_k \delta^{(k)}(t), \quad \bar{C}_k = -C_k$$

*such that* (2.1) *holds.*

(2') *There exists a polynomial*

$$P(z) = \sum_{k=0}^{p} C_k z^k, \quad \bar{C}_k = -C_k$$

*such that* (2.2) *holds.*

(3') *If* $\alpha = 1$ $(\beta = 0)$ *then there exist the limits*

$$\frac{1}{\rho_1(k)} \int_{-1/k}^{1/k} d\mu(x) \to \pi a, \ a > 0; \tag{2.6}$$

$$\frac{1}{\rho_1(k)} \int_{-1/k}^{1/k} x d\mu(x) \to b = 0, \ k \to \infty.$$

*If* $\alpha = -n$, $n = 0, 1, \cdots$, *there are numbers* $A$ *and* $B$ *such that the following limits exist:*

$$\frac{1}{k^n \rho(k)} \left( \int_{\Delta_k} \frac{d\mu(\xi)}{\xi^{n+1}} - A \right) \to -q_n', \ k \to \infty,$$

$$\frac{1}{k^{n-1}\rho(k)} \left( \int_{\Delta_k} \frac{d\mu(\xi)}{\xi^n} - B \right) \to -q_n'', \ k \to \infty, \tag{2.7}$$

*where*

$$\Delta_k = [-\delta, -\frac{1}{k}) \bigcup (\frac{1}{k}, \delta], \ \delta > 0,$$

$$q_n' = \begin{cases} \pi b \\ -\pi a \\ -\pi b \\ \pi a \end{cases} q_n'' = \begin{matrix} 2a \\ 2b \\ -2a \\ -2b \end{matrix} \ \text{if} \ \begin{matrix} n = 0, \\ n = 1, \\ n = 2, \\ n = 3, \end{matrix} \ (\text{mod } 4). \tag{2.8}$$

(4') *In some neighbourhood of zero the distribution* $[d\mu]$ *can be represented as follows* ($\alpha = -n$; $n = -1, 0, 1, 2, \ldots$):

$$d\mu(x) = x^{n+1} d\left(\frac{t_1(x)}{|x|^n}\right) + x^n \, d\left(\frac{t_2(x)}{|x|^{n-1}}\right), \tag{2.9}$$

*where* $(1/|x|^n)t_1(x)$ *and* $(1/|x|^{n-1})t_2(x)$ *are locally summable odd functions such that*

$$\frac{t_1(x)}{\rho\left(\frac{1}{x}\right)} \to \frac{1}{2} q_n', \ \frac{t_2(x)}{\rho\left(\frac{1}{x}\right)} \to \frac{1}{2} q_n'', \ x \to +0. \tag{2.10}$$

(*Here* $q_{-1}' = \pi a > 0$, $q_{-1}'' = 0$. *For* $n = 0, 1, \cdots$, $q_n'$ *and* $q_n''$ *are given by formulae* (2.8).)

*In this case the following inequality holds* (cf. (2.5)):

$$a \cos\frac{\pi}{2}\alpha \geqslant \left|b \sin\frac{\pi}{2}\alpha\right|.$$

*Formula* (2.9) *is understood in the sence of distributions:*

$$([d\mu](x), \phi(x)) = -\int\left[\frac{t_1(x)}{|x|^n}\frac{d}{dx}(x^{n+1}\phi(x)) + \frac{t_2(x)}{|x|^{n-1}}\frac{d}{dx}(x^n\phi(x))\right]dx,$$

$$\phi \in D(-\delta, \delta), \ \delta > 0.$$

*If* $n = 1$, *then the second term in* (2.9) *is an odd distribution. Its values on even test functions are equal to zero, but if* $\phi(x)$ *is an odd test function from* $D(-\delta, \delta)$ *then* $x^{-1}\phi(x) \in D(-\delta, \delta)$.

**Remark 2.** Under the hypothesis of theorem 2 for $\alpha < 1$ the limit relations (2.3)-(2.4) hold. The converse statement is not true.

**Proof.** Let us now turn to the proof of theorem 1. The equivalence of conditions (1) and (2) follows from theorem in Section 7.3 where we also have noted that $\beta \leqslant 0$ and, therefore, $\alpha = \beta + 1 \leqslant 1$.

Let us prove the implication (1) → (3). According to theorem 4 of Section 3.4, condition (1) is equivalent to the condition that the distribution $\tilde{f}(x) - \mathcal{P}(x)$ has a quasi-asymptotics at the origin with respect to $\rho$. But then by theorem 1 of Section 8.1 its real part $u(x)$ also has a quasi-asymptotics at the origin with respect to $\rho$.

Relation (1.8) implies, for $\alpha < 1$, that

$$\frac{1}{\rho(k)}u\left(\frac{x}{k}\right) = \frac{1}{\rho(k)}[d\mu]\left(\frac{x}{k}\right) \to \gamma_\alpha(x),\ k \to \infty$$

in $\mathcal{S}'$, where

$$\gamma_\alpha(x) = \left(a\cos\frac{\pi}{2}\alpha - b\sin\frac{\pi}{2}\alpha\right)\theta(x)x^{-\alpha} + \tag{2.11}$$

$$+\left(a\cos\frac{\pi}{2}\alpha + b\sin\frac{\pi}{2}\alpha\right)\theta(-x)(-x)^{-\alpha} \tag{2.11}$$

is a locally summable function. As the function $\gamma_\alpha(x)$ is the limit of a sequence of nonnegative measures, it is nonnegative; that is,

$$a\cos\frac{\pi}{2} \pm b\sin\frac{\pi}{2}\alpha \geqslant 0$$

and this is equivalent to inequality (2.5). Let $\chi(x) = \theta_{[0,1]}(x)$ be the characteristic function of the interval [0, 1]. By (2.11) we conclude that

$$\int\gamma_\alpha(x)\chi(x)\,dx = \int_0^1\left(a\cos\frac{\pi}{2}\alpha - b\sin\frac{\pi}{2}\alpha\right)x^{-\alpha}\,dx =$$

$$= \frac{a\cos\frac{\pi}{2}\alpha - b\sin\frac{\pi}{2}\alpha}{1-\alpha} = C_+.$$

Let us take two test functions $\phi_1, \phi_2$ from $\mathcal{S}$ with the properties (see Figure 12).

$$\phi_1(x) \leqslant \chi(x) \leqslant \phi_2(x);$$

$$\left|\int\gamma_\alpha(x)\phi_j(x)\,dx - C_+\right| < \epsilon,\ j = 1, 2.$$

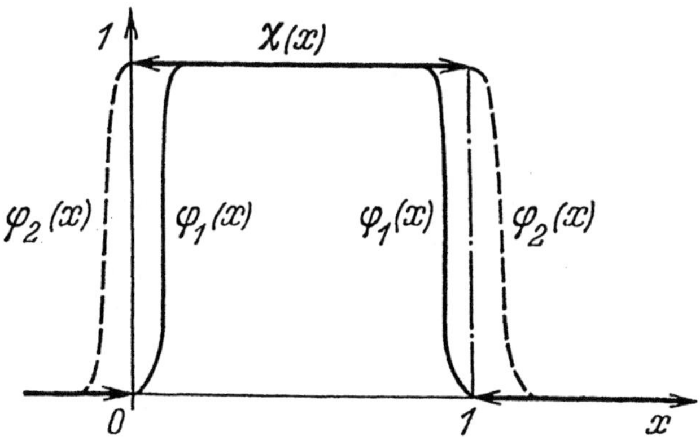


Figure 12.

For a sufficiently large $k$ we have

$$\frac{1}{\rho(k)}\left([d\mu]\left(\frac{x}{k}\right), \phi_1(x)\right) \leqslant \frac{1}{\rho(k)}\left([d\mu]\left(\frac{x}{k}\right), \chi(x)\right) \leqslant$$

$$\leqslant \frac{1}{\rho(k)} \left( [d\,\mu] \left( \frac{x}{k} \right), \phi_2(x) \right)$$

where $[d\,\mu](x)$ denotes the corresponding distribution (see Section 1.2). Taking the liminf and limsup we have

$$C_+ - \epsilon \leqslant \int \gamma_\alpha(x)\phi_1(x)\,dx = \lim_{k\to\infty} \frac{1}{\rho(k)} \left( [d\,\mu] \left( \frac{x}{k} \right), \phi_1(x) \right) \leqslant$$

$$\leqslant \varliminf_{k\to\infty} \frac{1}{\rho(k)} \left( [d\,\mu] \left( \frac{x}{k} \right), \chi(x) \right) \leqslant$$

$$\leqslant \varlimsup_{k\to\infty} \frac{1}{\rho(k)} \left( [d\,\mu] \left( \frac{x}{k} \right), \chi(x) \right) \leqslant$$

$$\leqslant \lim_{k\to\infty} \frac{1}{\rho(k)} \left( [d\,\mu] \left( \frac{x}{k} \right), \phi_2(x) \right) = \int \gamma_\alpha(x)\phi_2(x)\,dx \leqslant$$

$$\leqslant C_+ + \epsilon. \tag{2.12}$$

Since $\epsilon > 0$ is arbitrarily small, we conclude from (2.12) that

$$\frac{1}{\rho(k)} \left( [d\,\mu] \left( \frac{x}{k} \right), \chi(x) \right) \to C_+,\ k \to \infty,$$

so that

$$\frac{k}{\rho(k)}(d\,\mu(x), \chi(kx)) = \frac{k}{\rho(k)}\mu\left( \frac{1}{k} \right) \to C_+,\ k \to \infty.$$

The second part of (2.3) can be proved in a similar way. Thus, the implication (1) $\to$ (3) has been proved.

Let us now prove the impliction (3) $\to$ (1). Since $C_+$ and $C_-$ do not simultaneously vanish and $\mu(x) \sim C_\pm \rho_1(1/|x|)$, $x \to \pm 0$, and as the order of the automodel function $\rho_1$ is in the interval $(-\infty, 0)$, we conclude by theorem 3 of Section 3.4 that the distribution $\mu(x)$ has a quasi-asymptotics at the origin with respect to $\rho_1(k)$. Differentiating the limit relation

$$\frac{k}{\rho(k)}\mu\left( \frac{x}{k} \right) \to C_+\theta(x)x^{1-\alpha} + C_-\theta(-x)(-x)^{1-\alpha},\ k \to \infty$$

in $\mathcal{S}'$, we obtain that $u(x)$ has a quasi-asymptotics at the origin with respect to $\rho(k)$. By theorem 1 of Section 8.1 there is a polynomial $P(x)$ with purely imaginary coefficients such that $\tilde{f}(x) - P(x)$ has a quasi-asymptotics at the origin with respect to $\rho(k)$. Using theorem 4 of Section 3.4, we complete the proof of the implication (3) $\to$ (1).

Let us now prove theorem 2. The equivalence of conditions (1') and (2') follows directly from the theorem in Section 7.3. In particular, $\beta \leqslant 0$, so $\alpha \leqslant 1$.

Let us prove the implication (1') → (3'). According to theorem 4 of Section 3.4, condition (1') is equivalent to the condition that the distribution $\tilde{f}(x) - \mathcal{P}(x)$ has a quasi-asymptotics at the origin with respect to the automodel function $\rho$ of order $\alpha$.

Let us first consider the case $\alpha = 1$. We have

$$\frac{1}{\rho(k)}\left[\tilde{f}\left(\frac{x}{k}\right) - \mathcal{P}\left(\frac{x}{k}\right)\right] \to (a + ib)\frac{i}{x + i0}, \quad a + ib \neq 0, \ k \to \infty \ \text{ in } \mathcal{S}'.$$

Taking into account that $\operatorname{Re} \mathcal{P}(x) \equiv 0$, for the real part $u(x) = d\mu(x)$ we have

$$\frac{1}{\rho(k)}[d\mu]\left(\frac{x}{k}\right) \to \pi a\,\delta(x) - b\mathcal{P}\frac{1}{x}, \ k \to \infty.$$

Since the measure is nonnegative, we have $b = 0$, $a > 0$. Taking as $\chi(x)$ in (2.12) the characteristic function of the interval $[-1, 1]$, and applying the same construction, we have, for a sufficiently large $k$,

$$\frac{1}{\rho(k)}\left([d\mu]\left(\frac{x}{k}\right), \chi(x)\right) = \frac{k}{\rho(k)}\int_{-1/k}^{1/k} d\mu(x) \to \pi a, \ k \to \infty.$$

Similarly, putting in (2.12)

$$\chi(x) = \begin{cases} x & \text{when } |x| \leqslant 1, \\ 0 & \text{when } |x| > 1, \end{cases}$$

we have

$$\frac{k^2}{\rho(k)}\int_{-1/k}^{1/k} x\, d\mu(x) \to 0, \ k \to \infty.$$

(we recall here that $b = 0$). Thus, the limit relation (2.6) is proved.

Now consider the case $\alpha = -n$, $n = 0, 1, \cdots$. From theorem 2 of Section 8.1. it follows that there exists a distribution $h(x)$ which satisfies the equation

$$x^{n+1}h(x) = u(x), \ u(x) = \operatorname{Re}[\tilde{f}(x) - P(x)] = \operatorname{Re}\tilde{f}(x) \tag{2.13}$$

and the limit relation in theorem 2 of Section 8.1 hold. Due to the fact that $u(x) = d\mu(x)$ in some neighbourhood of zero (say $|x| < \delta$), the solution of equation (2.4) is equal in this neighbourhood to the solution of the equation

$$x^{n+1}h(x) = [d\mu](x). \tag{2.14}$$

Let $\chi_{\delta/2}$ be the characteristic function of the segment $[-(\delta/2), \delta/2]$, then the general solution equation (2.14) is given by the formula

$$(h(\xi), \phi(\xi)) = \left([d\mu](\xi), \frac{\phi(\xi) - \sum_{j=0}^{n}\frac{1}{j!}\xi^j\phi^{(j)}(0)\chi_{\frac{\delta}{2}}(\xi)}{\xi^{n+1}}\right) +$$

$$+\sum_{j=0}^{n} d_j \phi^{(j)}(0), \quad \phi \in \mathcal{D}(-\delta, \delta). \tag{2.15}$$

Note that if $\phi_1$ and $\phi_2$ satisfy the conditions

$$\phi_1^{(j)}(0) = \phi_2^{(j)}(0), \ j = 0, 1, \ldots, n; \ x^{n+1}\phi_1(x) \leqslant x^{n+1}\phi_2(x),$$

then for sufficiently large $k$ we have, by (2.15),

$$\left(h\left(\frac{\xi}{k}\right), \phi_1(\xi) - \phi_2(\xi)\right) = \left([d\mu]\left(\frac{\xi}{k}\right), \frac{\phi_1(\xi) - \phi_2(\xi)}{\xi^{n+1}}\right) \leqslant 0.$$

It is clear from (1.14) of Section 8.1 that

$$\int g_n(x)\chi_{\frac{\delta}{2}}(x)\,dx = q_n',$$

where $q_n'$ is given by formula (2.8). Now let us take $\phi_1$ and $\phi_2$ from $\mathcal{D}(-\delta, \delta)$, with the properties mentioned above, in such a way that

$$\left|\int g_n(x)\phi_l(x)\,dx - q_n'\right| < \epsilon, \ l = 1, 2,$$

$$x^{n+1}\phi_1(x) \leqslant x^{n+1}\chi_{\frac{\delta}{2}}(x) \leqslant x^{n+1}\phi_2(x).$$

(See Figure 13 for the case $n = 2$.)

From these properties it follows that for a sufficiently large $k$

$$\frac{1}{k^{n+1}\rho(k)}\left(h\left(\frac{\xi}{k}\right), \phi_1(\xi)\right) \leqslant \frac{1}{k^{n+1}\rho(k)}\left(h\left(\frac{\xi}{k}\right), \chi_{\frac{\delta}{2}}(\xi)\right) \leqslant$$

$$\leqslant \frac{1}{k^{n+1}\rho(k)}\left(h\left(\frac{\xi}{k}\right), \phi_2(\xi)\right).$$

Taking the liminf and limsup, by (1.14) of Section 8.1 we get

$$q_n' - \epsilon \leqslant \int g_n(\xi)\phi_1(\xi)\,d\xi = \lim_{k\to\infty}\frac{1}{k^{n+1}\rho(k)}\left(h\left(\frac{\xi}{k}\right), \phi_1(\xi)\right) \leqslant$$

$$\leqslant \varliminf_{k\to\infty}\frac{1}{k^{n+1}\rho(k)}\left(h\left(\frac{\xi}{k}\right), \chi_{\frac{\delta}{2}}(\xi)\right) \leqslant$$

$$\leqslant \varlimsup_{k\to\infty}\frac{1}{k^{n+1}\rho(k)}\left(h\left(\frac{\xi}{k}\right), \chi_{\frac{\delta}{2}}(\xi)\right) \leqslant$$

$$\leqslant \lim_{k\to\infty}\frac{1}{k^{n+1}\rho(k)}\left(h\left(\frac{\xi}{k}\right), \phi_2(\xi)\right) =$$

$$= \int g_n(\xi)\phi_2(\xi)\,d\xi \leqslant g_n' + \epsilon. \tag{2.16}$$

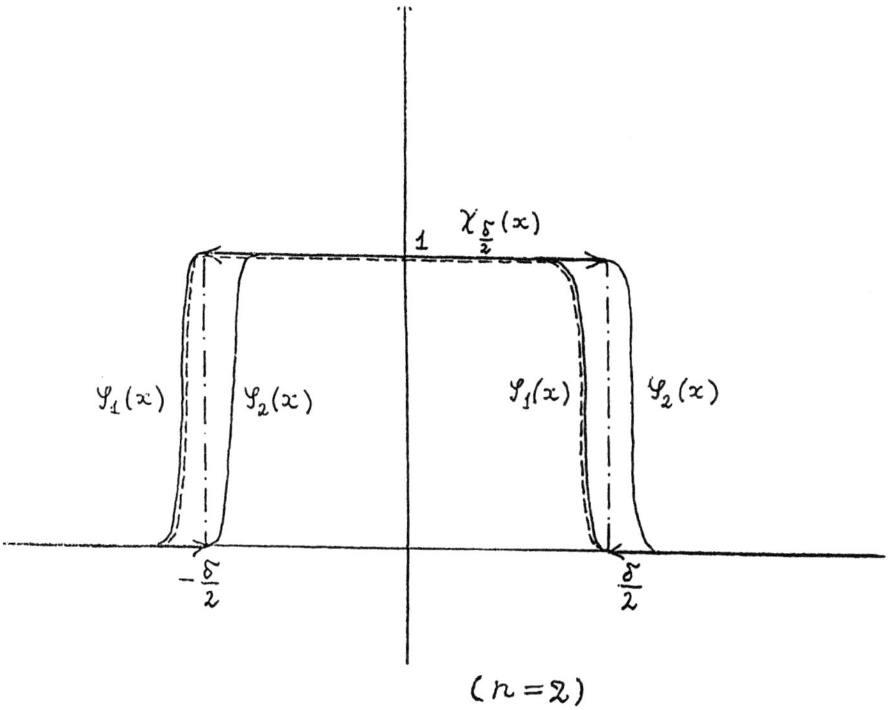


Figure 13.

Since $\epsilon > 0$ is arbitrarily small we conclude that

$$\frac{1}{k^{n+1}\rho(k)}\left(h\left(\frac{\xi}{k}\right).\chi_{\frac{\delta}{2}}(\xi)\right) \to q_n^{'},\ k \to \infty.$$

Using (2.15), from this we get

$$\frac{1}{k^{n+1}\rho(k)}\left(h\left(\frac{\xi}{k}\right).\chi_{\delta/2}(\xi)\right) = \frac{1}{k^n\rho(k)}\left(h(\xi).\chi_{\delta/2}(k\xi)\right) =$$

$$= \frac{1}{k^n\rho(k)}\left[\left(d\mu(\xi).\ \frac{\chi_{\frac{\delta}{2}}(k\xi)-\chi_{\frac{\delta}{2}}(\xi)}{\xi^{n+1}}\right)+d_0\right] =$$

$$= \frac{1}{k^n\rho(k)}\left[-\int_{\Delta_k^{'}}\frac{d\mu(\xi)}{\xi^{n+1}}+d_0\right] \to q_n^{'},\ k \to \infty,$$

where

$$\Delta_k^{'} = \left[-\frac{\delta}{2},-\frac{\delta}{2k}\right) \cup \left(\frac{\delta}{2k},\frac{\delta}{2}\right].$$

Putting $k' = (\delta/2)k$ and remembering that the function $k^n \rho(k)$ is an automodel function of order 0, we get the first formula (2.7).

The second formula (2.7) can be proved in a similar way, but the function $\chi_{\delta/2}(x)$ in (2.16) whould be replaced by the function $x\chi_{\delta/2}(x)$. So the implication (1') → (3') has been proved.

Let us now prove the implication (3') → (4'). At first let $\alpha = -n$, $n = 0, 1, \cdots$, and let conditions (2.7) hold. Let us introduce two odd functions $t_1(x)$ and $t_2(x)$ such that

$$t_1(x) = -\frac{x^n}{2}\left[\int_{\Delta_{1/x}} \frac{d\mu(\eta)}{\eta^{n+1}} - A\right], \quad 0 < x < \delta,$$

$$t_2(x) = -\frac{x^{n-1}}{2}\left[\int_{\Delta_{1/x}} \frac{d\mu(\eta)}{\eta^{n}} - B\right], \quad 0 < x < \delta. \tag{2.17}$$

By (2.7) we have

$$\frac{t_1(x)}{\rho\left(\frac{1}{x}\right)} \to \frac{1}{2}q_n', \quad \frac{t_2(x)}{\rho\left(\frac{1}{x}\right)} \to \frac{1}{2}q_n'', \quad x \to +0,$$

where $q_n'$ and $q_n''$ are given by formulae (2.8). Note that the functions

$$\gamma_1(x) = \frac{1}{|x|^n} t_1(x)$$

and

$$\gamma_2(x) = \frac{1}{|x|^{n-1}} t_2(x)$$

are locally summable, and one of them is non-decreasing when $x > 0$, because the measure $d\mu$ is nonnegative (namely, it is the function $\gamma_2$ if $n$ is even, and the function $\gamma_1$ if $n$ is odd). We show that the distribution

$$\omega(x) = \omega_1(x) + \omega_2(x);$$

$$\omega_1(x) = x^{n+1} d\left(\frac{t_1(x)}{|x|^n}\right),$$

$$\omega_2(x) = x^n \left[d\frac{t_2(x)}{|x|^{n-1}}\right]$$

is equal to the measure $d\mu$ in some neighbourhood of the point $x = 0$. Indeed, denoting

$$d\mu_1(x) = \frac{1}{2}[d\mu(x) + (-1)^{n+1} d\mu(-x)];$$

$$d\mu_2(x) = \frac{1}{2}[d\mu(x) + (-1)^{n} d\mu(-x)],$$

for any $\phi(x) \in D(-\delta, \delta)$ we have

$$(\omega_1(x), \phi(x)) = -\left(\frac{1}{|x|^n} t_1(x), \frac{d}{dx}(x^{n+1}\phi(x))\right) =$$

$$= \lim_{\epsilon \to +0} \left\{ \int_{\epsilon}^{\infty} \left[ \int_{\Delta_{1/x}} \frac{d\mu(\eta)}{\eta^{n+1}} - A \right] \frac{d(x^{n+1}\phi(x))}{2} + \right.$$

$$\left. + \int_{-\infty}^{-\epsilon} \left[ -\int_{\Delta_{1/|x|}} \frac{d\mu(\eta)}{\eta^{n+1}} - A \right] \frac{d(x^{n+1}\phi(x))}{2} \right\} =$$

$$= \lim_{\epsilon \to +0} \left\{ -\frac{\epsilon^{n+1}\phi(\epsilon)}{2} \left( \int_{\Delta_{1/\epsilon}} \frac{d\mu(\eta)}{\eta^{n+1}} - A \right) + \right.$$

$$\left. + \frac{(-\epsilon)^{n+1}\phi(-\epsilon)}{2} \int_{\Delta_{1/\epsilon}} \frac{d\mu(\eta)}{\eta^{n+1}} - A + \left( \int_{\epsilon}^{\infty} + \int_{-\infty}^{-\epsilon} \right) \phi(x)\, d\mu_1(x) \right\}$$

$$= \int \phi(x)\, d\mu_1(x).$$

We have used the fact that the measure $d\mu_1(x)$ is continuous at zero, and the existence of the limit

$$\lim_{\epsilon \to +0} \epsilon^n \left( \int_{\Delta_{1/\epsilon}} \frac{d\mu(\eta)}{\eta^{n+1}} - A \right),$$

which easily follows from (2.7). In a similar way one can show that

$$(\omega_2, \phi) = \int \phi(x)\, d\mu_2(x).$$

Therefore,

$$(\omega, \phi) = \int \phi(x)(d\mu_1(x) + d\mu_2(x)) = \int \phi(x)\, d\mu(x).$$

Now let $\alpha = 1$ ($n = -1$) and conditions (2.6) hold. Let us introduce two odd functions

$$t_1(x) = \frac{1}{2|x|} \int_{-x}^{x} d\mu(\xi); \quad t_2(x) = \frac{1}{2|x|^2} \int_{-x}^{x} \xi\, d\mu(\xi).$$

By (2.6) we have

$$\frac{t_1(x)}{\rho\left(\frac{1}{x}\right)} \to \frac{1}{2}\pi a, \quad \frac{t_2(x)}{\rho\left(\frac{1}{x}\right)} \to 0, \quad x \to +0.$$

Let us put

$$\omega = \omega_1 + \omega_2;$$

$$\omega_1(x) = \frac{1}{2}\frac{d}{dx}\int_{-x}^{x} d\mu(\xi),$$

$$\omega_2(x) = \frac{1}{2x}\frac{d}{dx}\int_{-x}^{x} \xi\, d\mu(\xi).$$

We assume that $\omega_2(x)$ is an odd distribution, vanishing on the even test funcitons. Thus, $\omega_2(x)$ is well defined, and so is $\omega_1(x)$. Let us verify that in the neighbourhood of zero $(-\delta, \delta)$ $\omega = [d\mu](x)$. Indeed, for any odd $\phi \in D(-\delta, \delta)$ we have

$$(\omega_2, \phi) = -\frac{1}{2}\int \int_{-x}^{x} \xi\, d\mu(\xi)\frac{d}{dx}\left(\frac{\phi(x)}{x}\right) dx =$$

$$= \frac{1}{2}\int \frac{\phi(x)}{x}[x\, d\mu(x) - (-x)\, d\mu(-x)] =$$

$$= \int \phi(x)\, d\mu_2(x) = ([d\mu_2](x), \phi(x)),$$

where $[d\mu_2](x) = 1/2([d\mu](x) - [d\mu](-x))$. Similarely, for any $\phi \in D(-\delta, \delta)$

$$(\omega_1, \phi) = \int \phi(x)\, d\mu_1(x) = ([d\mu_1](x), \phi(x)),$$

where

$$[d\mu_1](x) = ([d\mu](x) + [d\mu](-x)).$$

Hence,

$$(\omega, \phi) = (\omega_1, \phi) + (\omega_2, \phi) = ([d\mu_1] + [d\mu_2], \phi) = ([d\mu], \phi)$$

and the implications (3') → (4') is proved.

Let us prove the implication (4') → (1'). Let conditions (2.9) and (2.10) hold in some neighbourhood of zero. We shall construct a distribution $h \in \mathcal{S}$ such that

$$x^{n+1}h(x) = u(x)$$

$$\frac{1}{k^{n+1}\rho(k)}h\left(\frac{x}{k}\right) \to q_n'\delta(x) + \frac{1}{2}q_n''\mathcal{P}\frac{1}{x} = g_n(x),\ k \to \infty \tag{2.18}$$

in $\mathcal{S}$, where $u(x) = d\mu(x)$ in some neighbourhood of zero. In this case let us put

$$h_1(x) = \frac{d}{dx}\left[\frac{1}{|x|^n}t_1(x)\right].$$

By the hypothesis, the function

$$\gamma_1(x) = \frac{1}{|x|^n}t_1(x)$$

has an asymptotics of order $\alpha = 0$:

$$\gamma_1(x) \sim \pm\frac{1}{2}q_n'\frac{1}{|x|^n}\rho\left(\frac{1}{|x|}\right),\quad x \to +0.$$

By theorem 3 of Section 3.4 it also has a quasi-asymptotics:

$$\frac{1}{k^n \rho(k)} \gamma_1\left(\frac{x}{k}\right) \to \frac{1}{2} q_n' \theta(x) - \frac{1}{2} q_n' \theta(-x), \; k \to \infty$$

in $\mathcal{S}'$. Differentiating (see Section 1.4) we get

$$\frac{1}{k^{n+1} \rho(k)} h_1\left(\frac{x}{k}\right) \to q_n' \delta(x), \; k \to \infty$$

in $\mathcal{S}'$. By (2.10) we have

$$\gamma_2(x) = \frac{1}{|x|^{n-1}} t_2(x) \sim \pm \frac{1}{2} q_n'' \frac{1}{|x|^{n-1}} \rho\left(\frac{1}{|x|}\right), \; x \to \pm 0.$$

Thus, the distribution $\gamma_2(x)$ has the quasi-asymptotics:

$$\frac{1}{k^{n-1} \rho(k)} \gamma_2\left(\frac{x}{k}\right) \to \frac{1}{2} q_n'' x, \; k \to \infty$$

in $\mathcal{S}'$. Differentiating, we get

$$\frac{1}{k^n \rho(k)} \left(\frac{d}{dx} \gamma_2\right)\left(\frac{x}{k}\right) \to \frac{1}{2} q_n'', \; k \to \infty \tag{2.19}$$

in $\mathcal{S}'$. The distribution $\gamma_2(x)$ is odd, hence $\gamma_2'$ is even, and by (2.19) it has a quasi-asymptotics of order 0. Thus, by the lemma in Section 8.1 there is a unique odd solution of the equation

$$x h_2(x) = \frac{d}{dx} \gamma_2(x)$$

which has a quasi-asymptotics:

$$\frac{1}{k(k^n \rho(k))} h_2\left(\frac{x}{k}\right) \to \frac{1}{2} q_n'' \mathcal{P}\frac{1}{x}, \; k \to \infty$$

in $\mathcal{S}'$. Putting $h(x) = h_1(x) + h_2(x)$ we obtain a proper solution which satisfies (2.18).

For $n = 0, 1, 2, \cdots$, by theorem 2 of Section 8.1 there exists a polynomial $\mathcal{P}(x)$ with purely imaginary coefficients such that $\tilde{f}(x) - \mathcal{P}(x)$ has a quasi-asymptotics at the origin with respect to $\rho$. Now by means of theorem 4 in Section 3.4 we complete the proof of the implication (4') → (1') for $n = 0, 1, 2, \cdots$.

If $n = -1$, then $h(x) = u(x)$, and using theorem 1 of Section 8.1 and theorem 4 of Section 3.4 we prove our statement. □

Let us prove the statement of remark 2. From condition (1'), repeating all the arguments of the proof of the implication (1) → (3), one can see that condition (3) holds. In other words, the limit relations (2.3) are valid even if $\alpha = 0, -1, -2, \cdots$. But both constants $C_+$ and $C_-$ may now be zero simultaneously. The converse implication (3) → (2'), do not hold. One can see this from the example that has been considered as an illustration of theorem 2 in Section 8.1. Indeed, let $\tilde{f}(x) = \pi\theta(x) - i \ln |x|$. Then condition (2.3) holds with $\rho_1(k) = k^{-1}$, but condition

(2.2) (and therefore (2')) fails, for any polynomial $P(x)$.

Later we need some properties of the functions $t_1(x)$ and $t_2(x)$, representing by (2.9) a measure which is finite on the whole axis.

Let the measure $d\mu$ be represented in the form

$$d\mu(x) = x^{n+1} d\left[\frac{1}{|x|^n} t(x)\right], \quad n = -2, -1, 0, 1, 2, \ldots, \tag{2.20}$$

where $(1/|x|^n)t(x)$ is a locally summable odd function.

**Lemma.** *The measure $d\mu$, represented in the form (2.20) is finite on the whole axis if and only if for $n = -2, 0, 1, 2, \cdots$*

$$t(x) = Cx^n + \eta(x), \quad x > 0,$$

$$\operatorname*{Var}_0^\infty(x\,\eta(x)) < \infty, \quad \int_0^\infty |\eta(x)|\,dx < \infty; \tag{2.21}$$

*and for $n = -1$,*

$$\operatorname*{Var}_0^\infty(x(t(x)) < \infty. \tag{2.22}$$

**Proof.** First, let $n \neq -2, -1$. We have

$$\int_1^\tau \frac{d\mu(\xi)}{\xi^{n+1}} = -t(1) + \frac{t(\tau)}{\tau^n}.$$

By the finiteness of the measure, the integral on the left-hand side tends to a finite limit when $\tau \to +\infty$. So we can conclude that $\tau^{-n} t(\tau) \to C$, $\tau \to +\infty$. Hence, we can put

$$t(x) = Cx^n + \eta(x),$$

where $x^{-n}\eta(x) \to 0$, $x \to +\infty$. Thus, by (2.20) the measure $d\mu$ has the representaiton

$$d\mu(x) = x^{n+1} d\left[\frac{1}{x^n}\eta(x)\right], \quad x > 0.$$

Dividing both sides of this equality by $x^{n+1}$, and integrating from $\tau$ to $\tau_1$, we have

$$\int_\tau^{\tau_1} \frac{d\mu(x)}{x^{n+1}} = -\frac{\eta(\tau)}{\tau^n} + \frac{\eta(\tau_1)}{\tau_1^n}.$$

Taking the limit when $\tau_1 \to \infty$, we get

$$\int_\tau^\infty \frac{d\mu(x)}{x^{n+1}} = -\frac{\eta(\tau)}{\tau^n}.$$

So,

$$|\eta(\tau)| \leqslant \tau^n \int_\tau^\infty \frac{|d\mu(x)|}{x^{n+1}}.$$

By integrating of this inequality over $(0, \infty)$ we obtain

$$\int_0^\infty |\eta(\tau)|\, d\tau \leqslant \int_0^\infty \tau^n\, d\tau \int_\tau^\infty \frac{|d\mu(x)|}{x^{n+1}} = \frac{1}{n+1}\int_0^\infty |d\mu(x)| < \infty. \tag{2.23}$$

Using the identity

$$\frac{d}{dx}(x\eta(x)) = (n+1)\eta(x) + x^{n+1}\frac{d}{dx}\left(\frac{1}{x^n}\eta(x)\right),$$

we get

$$d(x\eta(x)) = (n+1)\eta(x)\,dx + d\mu(x). \tag{2.24}$$

Now (2.21) follows from (2.23).

The converse statement obviously follows from (2.24).

If $n = -1$ then (2.22) follows directly from the finiteness of the measure $d\mu$, because in this case $d\mu(x) = d(xt(x))$.

Let now $n = -2$. Putting $t(x) = (C/x^2) + \eta(x)$, $x^2\eta(x) \to 0$, $x \to +0$, we have

$$d\mu(x) = \frac{1}{x}d(x^2t(x)) = \frac{1}{x}d(x^2\eta(x)).$$

It follows that

$$\int_\epsilon^\tau x\, d\mu(x) = \tau^2\eta(\tau) - \epsilon^2\eta(\epsilon).$$

Letting $\epsilon \to 0$ we get

$$\eta(\tau) = \frac{1}{\tau^2}\int_0^\tau x\, d\mu(x).$$

Integrating this equality over $(0, \infty)$ we see that

$$\int_0^\infty |\eta(\tau)|\, d\tau \leqslant \int_0^\infty \frac{d\delta}{\tau^2}\int_0^\tau x\,|d\mu(x)| = \int_0^\infty |d\mu(x)| < \infty. \tag{2.25}$$

Having in mind (2.25) and the finiteness of the measures $d\mu$ we get (2.21) from (2.24). □

The converse statement follows from (2.24).

## 9. Asymptotic properties of the Fourier transform of distributions from $M_+$.

As we have already stated, a distribution $f \in \mathcal{S}'_+$ has a quasi-asymptotics at infinity with respect to an automodel function $\rho$ if and only if a certain primitive $f^{(-m)}$ has a normal asymptotics with respect to $k^m\rho(k)$ (see Section 3.4). However, as we have already mentioned, in the general case it is rather difficult to establish a criterion which says which primitive has this property. In this section we formulate such a criterion for the Fourier transform of distributions from $M_+$ (see

Section 7.1). We start with the properties of the Fourier transforms of finite measures.

**1. Asymptotic properties of the Fourier transform of finite measures.** Let $d\mu$ be a measure that is finite on the whole axis (not necessarily nonnegative), $|d\mu| < \infty$. Using the notation of Section 8.1. we introduce the function

$$\Phi(t) = 2\pi\theta(t)F^{-1}[\omega] = \frac{\theta(t)}{\pi}\int e^{-ixt}\, d\mu(x). \tag{1.1}$$

(We assume here that $N = 0$, $\omega = (1/\pi)[d\mu]$.) The Laplace transform of this function can be written as follows

$$\tilde{\Phi}(z) = \omega * \frac{i}{z} = \frac{i}{\pi}\int\frac{d\mu(\xi)}{z-\xi}.$$

The primitive of order $\gamma$ of the function $\Phi(t)$ has the form

$$\Phi^{(-\gamma)}(t) = f_\gamma(t) * \frac{\theta(t)}{\pi}\int e^{-ixt}\, d\mu(x). \tag{1.2}$$

The aim of this section is to establish a connection between the asymptotic properties of $\Phi^{(-\gamma)}(t)$, when $t \to +\infty$, and the asymptotic properties of the reproducing function $\mu(x)$ of the measure $d\mu(x)$ at the origin.

Let the measure $d\mu$ be continuous at the origin. Then formula (1.2) can be written as

$$\Phi^{(-\gamma)} = \Phi_+^{(-\gamma)} + \Phi_-^{(-\gamma)},$$

$$\Phi_+^{(-\gamma)} = \frac{1}{\pi}\int_0^\infty \frac{K_\gamma(tx)}{x^\gamma}\, d\mu(x), \quad \gamma \geqslant 0, \tag{1.3}$$

$$\Phi_-^{(-\gamma)} = \frac{1}{\pi}\int_{-\infty}^0 \frac{K_\gamma(tx)}{x^\gamma}\, d\mu(x) = \frac{1}{\pi}\int_0^\infty \frac{\overline{K_\gamma(tx)}}{x^\gamma}\, d(-\mu(x)),$$

where

$$K_\gamma(t) = f_\gamma(t) * [\theta(t)e^{-it}]. \tag{1.4}$$

Let us point out some properties of the function $K_\gamma$ which follow from its definition and from the properties of $f_\gamma$ (see (6.8), Section 1.6). The following recurrent relation hold:

$$iK_\gamma = f_\gamma - K_{\gamma-1}, \quad K_\gamma = f_{\gamma+1} - iK_{\gamma+1}. \tag{1.5}$$

Moreover, (1.6)

if $0 \leqslant \gamma < 1$, then $K_\gamma$ is bounded on $[0, +\infty)$,
if $-1 < \gamma < 0$, then $K_\gamma$ is bounded on $[\epsilon, +\infty)$,
if $0 < \gamma$, then $(1/t^\gamma)K_\gamma(t)$ is bounded on $(0, +\infty)$,
if $-1 < \gamma < 0$, then $(1/t^\gamma)K_\gamma(t)$ is bounded on $(0, R)$.

Here $\epsilon > 0$, $R > 0$ are arbitrary numbers. Decomposing the measure $d\mu$ into the sum of its odd and even parts,

$$d\mu = d\mu_s + d\mu_a; \quad d\mu_s = \frac{1}{2}d(\mu(x) - \mu(-x)).$$

$$d\mu_a = \frac{1}{2} d(\mu(x) + \mu(-x))$$

we write representation (1.3) in the form

$$\Phi^{(-\gamma)}(t) = \frac{2}{\pi}\int_0^\infty \mathrm{Re}\frac{K_\gamma(xt)}{x^\gamma} d\mu_s(x) + i\frac{2}{\pi}\int_0^\infty \mathrm{Im}\frac{K_\gamma(xt)}{x^\gamma} d\mu_a(x), \quad \gamma \geqslant 0. \tag{1.7}$$

We assume, further, that the measure $d\mu$ has the property that there exists $\delta > 0$ such that

$$d\mu(x) \geqslant 0 \quad \text{when} \quad |x| \leqslant \delta. \tag{1.8}$$

As usual, $\rho(t)$ is an automodel function of order $\alpha$: $\rho_1(t) = t^{-1}\rho(t)$ is an automodel function of order $\beta = \alpha - 1$.

If a measure $d\mu$, finite on $\mathbf{R}^1$, satisfies (1.8), and if it has a quasi-asymptotics at the origin with respect to $\rho(k)$, then, as we have already seen for $\beta \leqslant 0$, for $\beta \neq 0, -1, -2, \cdots$

$$\frac{\pm\mu(\pm x)}{\rho_1\left(\frac{1}{x}\right)} \to C_\pm, \; x \to +0 \tag{1.9}$$

(see theorem 1, Section 8.2). It follows from this that the measure $d\mu$ has the moments of

$$M_{-k} = \frac{1}{\pi}\int_{-\infty}^{+\infty} \frac{d\mu(x)}{x^k}, \; k = 0, 1, \ldots, [-\beta], \tag{1.10}$$

where $[-\beta]$ is the greatest integer in $-\beta$ (see Section 0.5).

**Theorem 1.** *Let $\rho$ be an automodel function of order $\alpha < 1$, $\alpha \neq 0, -1, -2, \cdots$, and let a measure $d\mu$, finite on $\mathbf{R}^1$, satisfy conditions* (1.8)-(1.9). *Then for any $\gamma$, $\gamma + \alpha > 1$, there exists the limit*

$$\lim_{t\to+\infty} \frac{1}{t^{\gamma-1}\rho(t)}\left[\Phi(t) + \sum_{k=1}^{[1-\alpha]} i^k M_{-k}\,\delta^{(k-1)}(t)\right]^{(-\gamma)} = C, \tag{1.11}$$

*where*

$$C = \frac{i(\alpha-1)\left[C_+ e^{i(\pi/2)\alpha} - C_- e^{-i(\pi/2)\alpha}\right]}{\Gamma(\gamma+\alpha)\sin\pi\alpha}. \tag{1.12}$$

*The function $\Phi(t)$ is given by formula (1.5) and the numbers $M_{-k}$ are given by* (1.10). *If $[1-\alpha] = 0$, the sum in* (1.11) *is absent.*

**Proof.** Using the recurrent formula (1.5), from representation (1.3) we get

$$\begin{aligned}\Phi^{(-\gamma)}(t) = &-\sum_{k=1}^{[-\beta]} iM_{-k} f_{\gamma-k+1}(t) + \\ &+ \frac{i^{[-\beta]}}{\pi}\int_0^\infty \frac{K_{\gamma-[-\beta]}(\xi t)}{\xi^\gamma} d\mu(\xi) + \\ &+ \frac{(-1)^{[-\beta]}}{\pi}\int_0^\infty \frac{\overline{K_{\gamma-[-\beta]}(\xi t)}}{\xi^\gamma} d(-\mu(-\xi))\end{aligned} \tag{1.13}$$

where $\beta = \alpha - 1$ is the order of the automodel function $\rho_1 = t^{-1}\rho(t)$. Since the first sum in (1.13) is simply the primitive of order $\gamma$ of the sum of the $\delta$-function and its derivatives in (1.11), it is sufficient to prove that there exists the limit

$$\frac{i^{[-\beta]}}{\pi t^{\gamma}\rho_1(t)}\int_0^{\infty}\frac{K_{\gamma-[-\beta]}(\xi t)}{\xi^{\gamma}}d\mu(\xi) \to C_1,\ t \to +\infty \tag{1.14}$$

and similarly for the second limit in (1.13). First, let us note that it is sufficient to prove the theorem for $\gamma + \beta = \epsilon$, where $0 < \epsilon < 1$, since in this case all the primitives of larger order will have proper limits. Thus, without loss of generality, we assume that $1 > \gamma - [-\beta] > 0$.

We shall need the following two lemmas.

**Lemma 1.** *Let $\rho(t)$ be an automodel function of the order $\kappa \leqslant 0$ and let a number $\gamma$ satisfy the conditions $\gamma + \kappa > 0$, $1 \geqslant \gamma - [-\kappa] \geqslant 0$. If $d\mu$ is a measure that is finite over $\mathbf{R}^1$, then for any $\delta > 0$*

$$\frac{1}{t^{\gamma}\rho(t)}\int_{|\xi|>\delta}\frac{K_{\gamma-[-\kappa]}(\xi(t)}{|\xi|^{\gamma}}d\mu(\xi) \to 0,\ t \to +\infty. \tag{1.15}$$

**Proof.** Indeed, since $\gamma + \kappa > 0$, then $t^{\gamma}\rho(t) \to +\infty$, $t \to +\infty$ (see Section 3.2), and since $1 \geqslant \gamma - [-\kappa] \geqslant 0$, then $K_{\gamma-[-\kappa]}(\xi t)$ is bounded (see (1.6)). From the estimate

$$\left|\int_{|\xi|>\delta}\frac{K_{\gamma-[-\kappa]}(\xi t)}{\xi^{\gamma}}d\mu(\xi)\right| \leqslant M\int_{|\xi|>\delta}\frac{|d\mu(\xi)|}{|\xi|^{\gamma}} \leqslant$$
$$\leqslant \frac{M}{\delta^{\gamma}}\int |d\mu(\xi)| \leqslant M_1$$

the limit relation (1.15) follows. □

**Lemma 2.** *Let $d\mu$ be a nonnegative finite measure and let there exists the limit*

$$\lim_{x\to+0}\frac{\mu(x)}{\rho\left(\frac{1}{x}\right)} = C_+.$$

*where $\rho(k)$ is an automodel function of the order $\kappa \leqslant 0$. Then the following conditions hold:*

(1). *For any $x > 0$,*

$$\lim_{t\to+\infty}\frac{\mu\left(\frac{x}{t}\right)}{\rho(t)} = C_+x^{-\kappa}. \tag{1.16}$$

(2) *For any $\epsilon > 0$ there exists numbers $T_\epsilon$ and $M$ such that for $t > T_\epsilon$*

$$\frac{\mu\left(\frac{x}{t}\right)}{\rho(t)} \leqslant \begin{cases} Mx^{-\kappa+\epsilon}, & \text{when } x \geqslant 1, \\ Mx^{-\kappa-\epsilon}, & \text{when } 0< <x < 1^{\cdot} \end{cases} \tag{1.17}$$

**Proof.** Taking the limit for $t \to +\infty$ from both sides of the identity

$$\frac{\mu\left(\frac{x}{t}\right)}{\rho(t)} = \frac{\mu\left(\frac{x}{t}\right)}{\rho\left(\frac{t}{x}\right)}\cdot\frac{\rho\left(\frac{t}{x}\right)}{\rho(t)}$$

and remembering that $\rho$ is an automodel function, we obtain the limit relation (1.16).

Let $(T_\epsilon/t) < 1/x < 1-\epsilon$, where $T_\epsilon$ is taken from lemma 1 of Section 3.2. Then we have

$$\frac{\mu\left(\dfrac{x}{t}\right)}{\rho(t)} = \frac{\rho\left(\dfrac{1}{x}t\right)}{\rho(t)} \cdot \frac{\mu\left(\dfrac{x}{t}\right)}{\rho\left(\dfrac{1}{x}t\right)} \leqslant \left(\frac{1}{x}\right)^{\kappa-\epsilon} M_1 = M_1 x^{-\kappa+\epsilon}.$$

If $1-\epsilon < x < 1+\epsilon$, then because of the automodelity of $\rho$ we have

$$\frac{\mu\left(\dfrac{x}{t}\right)}{\rho(t)} = \frac{\mu\left(\dfrac{x}{t}\right)}{\rho\left(\dfrac{t}{x}\right)} \cdot \frac{\rho\left(\dfrac{t}{x}\right)}{\rho(t)} \leqslant M_2.$$

If $x < 1-\epsilon$ (that is, $1/x > 1+\epsilon$), then by lemma 1 of Section 3.2 for $t > T_\epsilon$, we obtain

$$\frac{\rho\left(\dfrac{t}{x}\right)}{\rho(t)} \leqslant \left(\frac{1}{x}\right)^{\kappa+\epsilon} = x^{-\kappa-\epsilon}.$$

Therefore,

$$\frac{\mu\left(\dfrac{x}{t}\right)}{\rho(t)} \leqslant M_3 x^{-\kappa-\epsilon}.$$

If $x > t/T_\epsilon$, we have

$$\frac{\mu\left(\dfrac{x}{t}\right)}{\rho(t)} \leqslant \frac{\mu(\infty)}{\rho(t)} \leqslant M_4 t^{-\kappa+\epsilon} \leqslant M_4 (T_\epsilon x)^{-\kappa+\epsilon} \leqslant M_5 x^{-\kappa+\epsilon}.$$

Considering all the estimates we obtain (1.17). □

Let us continue the proof of the theorem. Note that by lemma 1 we can assume that the upper limit of the integral in (1.14) is replaced by an arbitrarily small number $\delta > 0$. And since $d\mu(x) \geqslant 0$ if $|x| < \delta$, we shall assume that the measure $d\mu$ is finite and nonnegative on the whole axis.

Hence taking a number $a > 0$ and denoting $[-\beta] = n \geqslant 0$, we rewrite the left-hand side of (1.14) in the form

$$\frac{i^n}{\pi t^\gamma \rho_1(t)} \int_0^\infty \frac{K_{\gamma-n}(\xi t)}{\xi^\gamma} d\mu(\xi) = I_1 + I_2,$$

where

$$I_1 = \frac{i^n}{\pi t^\gamma \rho_1(t)} \int_{a/t}^\infty \frac{K_{\gamma-n}(\xi t)}{\xi^\gamma} d\mu(\xi);$$

$$I_2 = \frac{i^n}{\pi t^\gamma \rho_1(t)} \int_0^{a/t} \frac{K_{\gamma-n}(\xi t)}{\xi^\gamma} d\mu(\xi).$$

Taking into account that $|K_{\gamma-n}(\xi t)| \leqslant M$ because $1 > \gamma - n > 0$ we have

$$\frac{1}{M}|I_1| \leqslant \frac{1}{\pi}\int_{a/t}^{\infty} \frac{1}{(t\xi)^\gamma} \frac{d\mu(\xi)}{\rho_1(t)} = \int_a^\infty \frac{1}{\xi^\gamma} \frac{d\mu\left(\frac{\xi}{t}\right)}{\pi\rho_1(t)} =$$

$$= \frac{\mu\left(\frac{\xi}{t}\right)}{\pi\rho_1(t)\xi^\gamma}\Bigg|_a^\infty + \frac{1}{\pi}\int_a^\infty \frac{\gamma\mu\left(\frac{\xi}{t}\right)}{\rho_1(t)} \frac{d\xi}{\xi^{\gamma+1}} =$$

$$= -\frac{\mu\left(\frac{a}{t}\right)}{\pi\rho_1(t)} \cdot \frac{1}{a^\gamma} + \frac{\gamma}{\pi}\int_a^\infty \frac{\mu\left(\frac{\xi}{t}\right)}{\rho_1(t)} \frac{d\xi}{\xi^{\gamma+1}} = \Psi(t, a). \tag{1.18}$$

Taking into account estimate (1.17), and the limit relation (1.16) of lemma 2 (its hypothesis holds by (1.9)), and taking the limit before integration (which is possible by the Lebesgue theorem), we get

$$\Psi(t, a) \to \left(-\frac{C_+}{\pi}\frac{1}{a^{\gamma+\beta}} + \frac{\gamma}{\pi}\int_a^\infty \frac{C_+ d\xi}{\xi^{\gamma+\beta+1}}\right) =$$

$$= \Psi_\infty(a), \; t \to +\infty.$$

Note that $\Psi_\infty(a) \to 0$ when $a \to +\infty$ because $\gamma + \beta > 0$. Using the recurrent relation (1.5) we infer that

$$I_2 = \frac{i^n}{\pi\rho_1(t)} \int_0^{a/t} \frac{K_{\gamma-n}(\xi t)}{(\xi t)^\gamma} d\mu(\xi) =$$

$$= \frac{i^n}{\pi\rho_1(t)} \int_0^a \frac{K_{\gamma-n}(\xi)}{\xi^\gamma} d\mu\left(\frac{\xi}{t}\right) =$$

$$= \frac{\mu\left(\frac{\xi}{t}\right)}{\pi\rho_1(t)} \frac{iK_{\gamma-n}(\xi)}{\xi^\gamma}\Bigg|_0^a - i^n \int_0^a \frac{\mu\left(\frac{\xi}{t}\right)}{\pi\rho_1(t)} \left[\frac{K_{\gamma-n-1}(\xi)}{\xi^\gamma} - \gamma\frac{K_{\gamma-n}(\xi)}{\xi^{\gamma+1}}\right] d\xi =$$

$$= \frac{i^n K_{\gamma-n}(a)}{a^\gamma} \cdot \frac{\mu\left(\frac{a}{t}\right)}{\pi\rho_1(t)} + i^n \int_0^a \left[\gamma\frac{K_{\gamma-n}(\xi)}{\xi^{\gamma+1}} - \frac{K_{\gamma-n-1}(\xi)}{\xi^\gamma}\right] \frac{\mu\left(\frac{\xi}{t}\right)}{\pi\rho_1(t)} d\xi.$$

Because

$$\frac{K_{\gamma-n}(\xi)}{\xi^\gamma} \sim \frac{1}{\xi^n},$$

$$\mu\left(\frac{\xi}{t}\right) \sim \rho_1\left(\frac{t}{\xi}\right),$$

$$\left|\gamma\frac{K_{\gamma-n}(\xi)}{\xi^{\gamma+1}} - \frac{K_{\gamma-n-1}(\xi)}{\xi^{\gamma}}\right| \sim \frac{1}{\xi^{n+1}},$$

when $\xi \sim 0$, and by estimate (1.17) of lemma 2, we obtain

$$\lim_{t\to+\infty} I_2 = i^n\left[\frac{K_{\gamma-n}(a)}{a^{\gamma+\beta}}\frac{C_+}{\pi} + \frac{C_+}{\pi}\int_0^a\left[\gamma\frac{K_{\gamma-n}(\xi)}{\xi^{\gamma+1}} - \frac{K_{\gamma-n-1}(\xi)}{\xi^{\gamma}}\right]\frac{d\xi}{\xi^{\beta}}\right] =$$

$$= -i^n\beta\frac{C_+}{\pi}\int_0^a\frac{K_{\gamma-n}(\xi)}{\xi^{\gamma+\beta+1}}\,d\xi.$$

Now let $\epsilon > 0$ be given. Let us take $a$ and $t_1$ sufficiently large so that for $t > t_1$, $|I_1| < \epsilon$,

$$\left|I_2 + i^n\beta\frac{C_+}{\pi}\int_0^a\frac{K_{\gamma-n}(\xi)}{\xi^{\gamma+\beta+1}}\,d\xi\right| < \epsilon$$

and

$$\left|\beta\frac{C_+}{\pi}\int_a^\infty\frac{K_{\gamma-n}(\xi)}{\xi^{\gamma+\beta+1}}\,d\xi\right| < \epsilon$$

(which is possible because for a sufficiently large $\xi$

$$\left|\frac{K_{\gamma-n}(\xi)}{\xi^{\gamma+\beta+1}}\right| \leqslant \frac{M}{\xi^{\gamma+\beta+1}}).$$

Thus we obtain

$$\left|\frac{i^n}{\pi t^{\gamma}\rho_1(t)}\int_0^\infty\frac{K_{\gamma-n}(\xi t)}{\xi^{\gamma}}\,d\mu(\xi) + i^n\beta\frac{C_+}{\pi}\int_0^\infty\frac{K_{\gamma-n}(\xi)}{\xi^{\gamma+\beta+1}}\,d\xi\right| \leqslant$$

$$\leqslant |I_1| + \left|I_2 + i^n\beta\frac{C_+}{\pi}\int_0^a\frac{K_{\gamma-n}(\xi)}{\xi^{\gamma+\beta+1}}\,d\xi\right| +$$

$$+\left|i^n\beta\frac{C_+}{\pi}\int_a^\infty\frac{K_{\gamma-n}(\xi)}{\xi^{\gamma+\beta+1}}\,d\xi\right| \leqslant 3\epsilon.$$

In a similar way one can prove that the second intergral in (1.13) when divided by $t^{\gamma}\rho_1(t)$ has also a limit. Hence, the statement (1.11) has been proved. Let us now calculate the constant C in (1.11).

From (1.11) (see theorem 1, Section 3.4) it follows that the primitive of order $\gamma$ of the function

$$\Psi(t) = \Phi(t) + \sum_{m=1}^{[1-\alpha]} i^m M_{-m}\,\delta^{(m-1)}(t)$$

has a quasi-asymptotics with respect to the automodel function $k^{\gamma-1}\rho(k) = k^{\gamma}\rho_1(k)$. Hence by theorem 1 of Section 3.1,

$$\frac{1}{\rho_1(k)}\left(\Phi(kt) + \sum_{m=1}^{[1-\alpha]} i^m M_{-m}\delta^{(m-1)}(kt)\right) \to C\Gamma(\gamma+\alpha)f_{\gamma+\alpha}(t),\ k \to \infty$$

in $\mathcal{S}'_+$. Therefore, condition (1) of theorem 1 in Section 8.1 holds. Hence, by (1.9) condition (3) of this theorem also holds. So, the coefficient $C\Gamma(\gamma+\alpha) = a + ib$ is connected with the coefficients $C_+$ by formula (2.4) of Section 8.2. Reversing formula (2.4) we get

$$C = \frac{a+ib}{\Gamma(\gamma+\alpha)} = \frac{i(\alpha-1)}{\Gamma(\gamma+\alpha)\sin\pi\alpha}\left[C_+e^{i(\pi/2)\alpha} - C_-e^{-i(\pi/2)\alpha}\right]$$

and the proof of formula (1.12) is complete. □

**Remark 1.** From the proof of the theorem - namely, from the formulae (1.12)-(1.14) - we obtain the equality

$$\frac{i}{\pi}\int_0^\infty \frac{K_{\gamma-n}(\xi)}{\xi^{\gamma+\alpha}}d\xi = \frac{-ie^{i(\pi/2)\alpha}}{\Gamma(\gamma+\alpha)\sin\pi\alpha},$$

$$0 \leqslant \gamma - n < 1,$$
$$0 \leqslant \alpha + n < 1,\quad \alpha+\gamma > 1.$$

Suppose that $\rho(t)$ is an automodel function of the order $\alpha = 0, -1, -2, \cdots$ and $\rho_1 = t^{-1}\rho(t)$ is an automodel function of the order $\beta = \alpha - 1 = -n$, $n = 1, 2, 3, \cdots$. Let $d\mu$ be a finite measure on $\mathbf{R}^1$ which is nonnegative in some neighbourhood of the origin (see condition (1.8)) and let

$$d\mu(x) = d\mu_1(x) + d\mu_2(x), \tag{1.19}$$

where (cf. (2.9), Section 8.2)

$$d\mu_1(x) = x^n d\left(\frac{t_1(x)}{|x|^{n-1}}\right),\quad d\mu_2(x) = x^{n-1}d\left(\frac{t_2(x)}{|x|^{n-2}}\right),$$

and where

$$\frac{t_1(x)}{|x|^{n-1}},\quad \frac{t_2(x)}{|x|^{n-2}}$$

are locally summable, odd functions and

$$\frac{t_1(x)}{\rho\left(\frac{1}{x}\right)} \to \frac{1}{2}q'_{n-1},$$

$$\frac{t_2(x)}{\rho\left(\frac{1}{x}\right)} \to \frac{1}{2}q''_{n-1}, \tag{1.20}$$

$$x \to +0,\quad q'^2_{n-1} + q''^2_{n-1} \neq 0.$$

From the results of the lemma in Section 8.2 it follows that $t_1(x)$ can be represented in the form

$$t_1(x) = Dx^{n-1} + \eta(x);\quad x\eta(x) \to 0,\ x \to +\infty. \tag{1.21}$$

**Theorem 2.** *Let a measure $d\mu$, which is continuous at the origin and finite on $\mathbf{R}^1$, satisfy conditions (1.8), (1.19) and (1.20), where $\rho$ is an automodel function of order $\alpha = 0, -1, -2, \cdots$, and let $\rho_1 = t^{-1}\rho(t)$ be an automodel function of order $\beta = \alpha - 1 = -n$, $n = 1, 2, 3, \cdots$. Then, for any $\gamma > n$ there exists the limit*

$$\lim_{t\to+\infty}\frac{1}{t^{\gamma-1}\rho(t)}\left[\Phi(t) + \sum_{k=1}^{n-1} i^k M_{-k}\,\delta^{(k-1)}(t) + i^n\frac{2D}{\pi}\delta^{(n-1)}\right]^{(-\gamma)} = C, \tag{1.22}$$

*where*

$$C = \frac{i^n}{\Gamma(\gamma - n + 1)}\left[\frac{1}{\pi}q_{n-1}' - i\frac{1}{2}q_{n-1}''\right]. \tag{1.23}$$

*The function $\Phi(t)$ is defined by formula (1.1), $M_{-k}$ are defined by (1.10) and $D$ is given by (1.21). If $n = 1$ the sum in (1.22) can be deleted.*

**Proof.** At first we note that as in the proof of theorem 1 we can consider only the case $1 > \gamma - n > 0$. Using the recurrent formula (1.5) from representation (1.7) we get

$$\begin{aligned}
\Phi^{(-\gamma)}(t) &= -\sum_{k=1}^{n-1} i^k \int_{-\infty}^{+\infty}\frac{f_{\gamma-k+1}(t)\xi^{\gamma-k}}{\xi^\gamma\pi}d\mu(\xi) + \\
&+ \frac{2}{\pi}\int_0^\infty \mathrm{Re}\frac{i^{n-1}K_{\gamma-n+1}(\xi t)}{\xi^\gamma}d\mu_s(\xi) + \\
&+ i\frac{2}{\pi}\int_0^\infty \mathrm{Im}\frac{i^{n-1}K_{\gamma-n+1}(\xi t)}{\xi^\gamma}d\mu_a(\xi) = \\
&= \left(-\sum_{k=1}^{n-1} i^k M_{-k}\,\delta^{(k-1)}(t)\right)^{(-\gamma)} + \\
&+ i^{n-1}\frac{2}{\pi}\int_0^\infty \mathrm{Re}\frac{K_{\gamma-n+1}(\xi t)}{\xi^\gamma}d\mu_2(\xi) + \\
&+ i^n\frac{2}{\pi}\int_0^\infty \mathrm{Im}\frac{K_{\gamma-n+1}(\xi t)}{\xi^\gamma}d\mu_1(\xi).
\end{aligned} \tag{1.24}$$

Let us take an $a > 0$ and divide the integral over $(0, +\infty)$ in (1.24) into the integrals over$(0, (a/t))$ and $((a/t), +\infty)$. By (1.5) we get

$$I_1 = \frac{2}{\pi}\int_{a/t}^\infty \mathrm{Re}\frac{K_{\gamma-n+1}(\xi t)}{\xi^\gamma}d\mu_2(\xi) = -\frac{2}{\pi}\int_{a/t}^\infty \mathrm{Im}\frac{K_{\gamma-n}(\xi t)}{\xi^\gamma}d\mu_2(\xi). \tag{1.25}$$

And

$$\begin{aligned}
\frac{2}{\pi}\int_{a/t}^\infty \mathrm{Im}\frac{K_{\gamma-n+1}(\xi t)}{\xi^\gamma}d\mu_1(\xi) &= \frac{2}{\pi}\int_{a/t}^\infty \mathrm{Im}\frac{iK_{\gamma-n}(\xi t) - if_{\gamma-n+1}(\xi t)}{\xi^\gamma}d\mu_1(\xi) = \\
&= \frac{2}{\pi}\int_{a/t}^\infty \mathrm{Re}\frac{K_{\gamma-n}(\xi t)}{\xi^\gamma}d\mu_1(\xi) -
\end{aligned}$$

$$-\frac{2}{\pi}f_{\gamma-n+1}(t)\int_{a/t}^{\infty}\frac{\xi^{\gamma-n}}{\xi^{\gamma}}\xi^{n}\,d\left(\frac{t_1(\xi)}{\xi^{n-1}}\right)=$$

$$=I_2-\frac{2}{\pi}f_{\gamma-n+1}(t)\left[D-\frac{t_1\left(\frac{a}{t}\right)}{\left(\frac{a}{t}\right)^{n-1}}\right].$$

where

$$I_2=\frac{2}{\pi}\int_{a/t}^{\infty}\mathrm{Re}\frac{K_{\gamma-n}(\xi t)}{\xi^{\gamma}}d\,\mu_1(\xi). \tag{1.26}$$

Here we have used formula (1.21). Taking into account (1.24) we have

$$\frac{1}{t^{\gamma}\rho_1(t)}\left[\Phi(t)+\sum_{k=1}^{n-1}i^{k}M_{-k}\,\delta^{(k-1)}(t)+i^{n}\frac{2D}{\pi}\delta^{(n-1)}(t)\right]^{(-\gamma)}=$$

$$=\frac{i^{n-1}}{t^{\gamma}\rho_1(t)}\left[I_1+iI_2+i\frac{2}{\pi}f_{\gamma-n+1}(t)\frac{t_1\left(\frac{a}{t}\right)}{\left(\frac{a}{t}\right)^{n-1}}+I_3+iI_4\right]. \tag{1.27}$$

Where $I_1$ and $I_2$ are given by formulae (1.25) and (1.26),

$$I_3=\frac{2}{\pi}\int_0^{a/t}\mathrm{Re}\frac{K_{\gamma-n+1}(\xi t)}{\xi^{\gamma}}d\,\mu_2(\xi);$$

$$I_4=\frac{2}{\pi}\int_0^{a/t}\mathrm{Im}\frac{K_{\gamma-n+1}(\xi t)}{\xi^{\gamma}}d\,\mu_1(\xi).$$

Let us calculate the limit of $(1/t^{\gamma}\rho_1(t))I_3$ when $t\to+\infty$. We have

$$\frac{1}{t^{\gamma}\rho_1(t)}I_3=\frac{2}{\pi\rho_1(t)}\int_0^{a}\mathrm{Re}\frac{K_{\gamma-n+1}(\xi)}{\xi^{\gamma}}d\,\mu_2\left(\frac{\xi}{t}\right)=$$

$$=\frac{2}{\pi}\int_0^{a}\mathrm{Re}\frac{K_{\gamma-n+1}(\xi)}{\xi^{\gamma-n+1}}\cdot\frac{1}{t\,\rho_1(t)}d\left(\frac{t_2\left(\frac{\xi}{t}\right)}{\xi^{n-2}}\right)=$$

$$=\frac{2}{\pi}\mathrm{Re}\left\{\left.\frac{K_{\gamma-n+1}(\xi)t_2\left(\frac{\xi}{t}\right)}{\xi^{\gamma-n+1}t\,\rho_1(t)\xi^{n-2}}\right|_0^{a}-\right.$$

$$\left.-\int_0^{a}\frac{t_2\left(\frac{\xi}{t}\right)}{t\,\rho_1(t)\xi^{n-2}}\left[\frac{K_{\gamma-n}(\xi)}{\xi^{\gamma-n+1}}-(\gamma-n+1)\frac{K_{\gamma-n+1}(\xi)}{\xi^{\gamma-n+2}}\right]d\,\xi\right\}=$$

$$=\frac{2}{\pi}\mathrm{Re}\left\{\frac{K_{\gamma-n+1}(a)t_2\left(\frac{a}{t}\right)}{a^{\gamma-1}t\,\rho_1(t)}-\int_0^{a}\cdots d\,\xi\right\}\to$$

$$\to \frac{2}{\pi}\mathrm{Re}\left\{\frac{K_{\gamma-n+1}(a)}{a^{\gamma-n}}\frac{1}{2}q''_{n-1} - \int_0^a \frac{1}{2}q''_{n-1}\left[\frac{K_{\gamma-n}(\xi)}{\xi^{\gamma-n+1}} - (\gamma-n+1)\frac{K_{\gamma-n+1}(\xi)}{\xi^{\gamma-n+1}}\right]\xi\, d\xi\right\} =$$

$$= \frac{q''_{n-1}}{\pi}\int_0^a \mathrm{Re}\frac{K_{\gamma-n+1}(\xi)}{\xi^{\gamma-n+1}}d\xi = \Psi_1(a).$$

If $a \to +\infty$, then

$$\Psi_1(a) \to \frac{q''_{n-1}}{\pi}\int_0^\infty \mathrm{Re}\frac{K_{\gamma-n+1}(\xi)}{\xi^{\gamma-n+1}}\, d\xi =$$

$$= \frac{q''_{n-1}}{2\Gamma(\gamma-n+1)} \equiv \Psi_1(+\infty). \tag{1.28}$$

Let us now calculate the limit of $(1/t^\gamma\rho_1(t))I_4$ when $t \to +\infty$. We have

$$\frac{1}{t^\gamma\rho_1(t)}I_4 = \frac{2}{\pi}\int_0^a \mathrm{Im}\frac{K_{\gamma-n+1}(\eta)}{\eta^{\gamma-n}\, t\,\rho_1(t)}\, d\left(\frac{t_1\left(\frac{\eta}{t}\right)}{\eta^{n-1}}\right) =$$

$$= \frac{2}{\pi}\mathrm{Im}\left\{\frac{K_{\gamma-n+1}(\eta)}{\eta^{\gamma-n}\,\eta^{n-1}}\frac{t_1\left(\frac{\eta}{t}\right)}{t\,\rho_1(t)}\Bigg|_0^a - \int_0^a \frac{t_1\left(\frac{\eta}{t}\right)}{t\,\rho_1(t)}\times\right.$$

$$\left.\times\left[\frac{K_{\gamma-n}(\eta)}{\eta^{\gamma-n}} - (\gamma-n)\frac{K_{\gamma-n+1}(\eta)}{\eta^{\gamma-n+1}}\right]\frac{d\eta}{\eta^{n-1}}\right\} =$$

$$= \frac{2}{\pi}\mathrm{Im}\left\{\frac{t_1\left(\frac{a}{t}\right)}{t\,\rho_1(t)}\cdot\frac{K_{\gamma-n+1}(a)}{a^{\gamma-1}} - \int_0^a \frac{t_1\left(\frac{\eta}{t}\right)}{t\,\rho_1(t)}\times\right.$$

$$\left.\times\left[\frac{K_{\gamma-n}(\eta)}{\eta^{\gamma-n}} - (\gamma-n)\frac{K_{\gamma-n+1}(\eta)}{\eta^{\gamma-n+1}}\right]\frac{d\eta}{\eta^{n-1}}\right\} =$$

$$= \Psi_2(a,t) \to \frac{2}{\pi}\mathrm{Im}\left\{\frac{1}{2}q'_{n-1}\frac{K_{\gamma-n+1}(a)}{a^{\gamma-n}} - \int_0^a \frac{1}{2}q'_{n-1}\times\right.$$

$$\times \left[ \frac{K_{\gamma-n}(\eta)}{\eta^{\gamma-n}} - (\gamma-n)\frac{K_{\gamma-n+1}(\eta)}{\eta^{\gamma-n+1}} \right] d\eta \Bigg] \equiv 0. \tag{1.29}$$

Let us consider the behaviour of $(1/t^{\gamma}\rho_1(t))(I_1 + iI_2)$ when $t \to +\infty$. Since the measure $d\mu$ is finite, its even and odd parts $d\mu_1$ and $d\mu_2$ are also finite. By lemma 1 the upper limits of integrals (1.25) and (1.26), which define $I_1$ and $I_2$, can be arbitrarily small, so that if $t$ is sufficiently large the integration in (1.25) and (1.26) will be only over the interval on which $d\mu$ is nonnegative. Hence, without loss of generality we can assume that the measure $d\mu$ is nonnegative on the whole axis. But in this case the even part of the measure is also nonnegative. If we suppose that the even part of the measure $d\mu$ is $d\mu_1$, we obtain

$$\frac{1}{t^{\gamma}\rho_1(t)}\left|I_1 + iI_2\right| = \frac{2}{\pi\rho_1(t)}\left| -\int_{a/t}^{\delta} \operatorname{Im}\frac{K_{\gamma-n}(\xi t)}{(\xi t)^{\gamma}}\, d\mu_2(\xi) + i\int_{a/t}^{\delta} \operatorname{Re}\frac{K_{\gamma-n}(\xi t)}{(\xi t)^{\gamma}}\, d\mu_1(\xi)\right| \leqslant$$

$$\leqslant \frac{2}{\pi\rho_1(t)}\left[\int_{a/t}^{\delta} \frac{|K_{\gamma-n}(\xi t)|}{(\xi t)^{\gamma}} |d\mu(\xi) - d\mu_1(\xi)| + \int_{a/t}^{\delta} \frac{|K_{\gamma-n}(\xi t)|}{(\xi t)^{\gamma}}\, d\mu_1(\xi)\right] \leqslant$$

$$\leqslant \frac{2M}{\pi\rho_1(t)}\left[\int_{a/t}^{\delta} \frac{d\mu(\xi)}{(\xi t)^{\gamma}} + 2\int_{a/t}^{\delta} \frac{d\mu_1(\xi)}{(\xi t)^{\gamma}}\right] =$$

$$= \frac{2M}{\pi\rho_1(t)}\left[3\int_{a/t}^{\delta} \frac{d\mu_1(\xi)}{(\xi t)^{\gamma}} + \int_{a/t}^{\delta} \frac{d\mu_2(\xi)}{(\xi t)^{\gamma}}\right].$$

We have taken into account that $1 > \gamma - n > 0$ and therefore the function $K_{\gamma}(\eta)$ is bounded. Let us continue our estimates. Using (1.19) we see that

$$\frac{\pi}{2Mt^{\gamma}\rho_1(t)}|I_1 + iI_2| \leqslant \frac{3}{\rho_1(t)}\int_a^{\infty} \frac{1}{t^n \xi^{\gamma-n}}\, d\left(\frac{t_1\left(\frac{\xi}{t}\right)}{\left(\frac{\xi}{t}\right)^{n-1}}\right) + \frac{1}{\rho_1(t)}\int_a^{\infty} \frac{d\left(\frac{t_2\left(\frac{\xi}{t}\right)}{\left(\frac{\xi}{t}\right)^{n-2}}\right)}{\xi^{\gamma-n+1}t^{n-1}} =$$

$$= \frac{3t_1\left(\frac{\xi}{t}\right)}{\xi^{\gamma-1}t\,\rho_2(t)}\Bigg|_a^{\infty} + 3\int_a^{\infty} \frac{t_1\left(\frac{\xi}{t}\right)}{t\rho_1(t)}(\gamma-n)\frac{d\xi}{\xi^{\gamma}} + \frac{t_2\left(\frac{\xi}{t}\right)}{\xi^{\gamma-1}t\,\rho_1(t)}\Bigg|_a^{\infty} +$$

$$+ \int_a^{\infty} \frac{t_2\left(\frac{\xi}{t}\right)}{t\rho_1(t)}(\gamma-n+1)\frac{d\xi}{\xi^{\gamma}} =$$

$$= -\frac{3t_1\left(\frac{a}{t}\right)}{a^{\gamma-1}t\,\rho_1(t)} + 3(\gamma-n)\int_a^\infty \frac{t_1\left(\frac{\xi}{t}\right)}{t\,\rho_1(t)}\frac{d\xi}{\xi^\gamma} - \frac{t_2\left(\frac{a}{t}\right)}{a^{\gamma-1}t\,\rho_1(t)} +$$

$$+ (\gamma-n+1)\int_a^\infty \frac{t_2\left(\frac{\xi}{t}\right)}{t\,\rho_1(t)}\frac{d\xi}{\xi^\gamma} = \Psi_3(a,t).$$

We have taken into account that $t_2(\xi) \sim \text{const}\,\xi^{n-2}$, $t_1(\xi) \sim \text{const}\,\xi^{n-1}$ when $\xi \to \infty$ and $\gamma - n > 0$. By (1.20) we have, for $t \to +\infty$,

$$\int_a^\infty \frac{t_1\left(\frac{\xi}{t}\right)}{t\,\rho_1(t)}\frac{d\xi}{\xi^\gamma} \to \frac{1}{2}q_{n-1}\int_a^\infty \frac{d\xi}{\xi^{\gamma-n+1}},$$

and similarly for $t_2(\xi)$. Therefore, if $t \to +\infty$

$$\Psi_3(a,t) \to \Psi_3(a,\infty) = -\frac{3q_{n-1}'}{2a^{\gamma-n}} + \frac{3(\gamma-n)q_{n-1}'}{2}\int_a^\infty \frac{d\xi}{\xi^{\gamma-n+1}} -$$

$$- \frac{q_{n-1}''}{2a^{\gamma-n}} + \frac{(\gamma-n+1)q_{n-1}''}{2}\int_a^\infty \frac{d\xi}{\xi^{\gamma-n+1}}.$$

It is clear that

$$\Psi_3(a,\infty) \to 0 \quad \text{when } a \to +\infty. \tag{1.30}$$

Note that

$$\frac{2i^n}{\pi t^\gamma \rho_1(t)}\frac{t_1\left(\frac{a}{t}\right)}{\left(\frac{a}{t}\right)^{n-1}}f_{\gamma-n+1}(t) = \frac{2i^n t_1\left(\frac{a}{t}\right)}{\pi\Gamma(\gamma-n+1)a^{n-1}t\,\rho_1(t)} \to$$

$$\to \frac{i^n}{\Gamma(\gamma-n+1)}\frac{q_{n-1}'}{\pi}, \quad t \to +\infty. \tag{1.31}$$

Now let a number $\epsilon > 0$ be given. Taking into account (1.28), (1.29) and (1.30) we have, for sufficiently large $t$ and $a$,

$$\left|\frac{i^n}{t^\gamma \rho_1(t)}I_4\right| = |\Psi_2(a,t)| < \epsilon,$$

$$\left|\frac{1}{t^\gamma\rho_1(t)}I_3 - \Psi_1(\infty)\right| = \left|\frac{1}{t^\gamma\rho_1(t)}I_3 - \Psi_1(a) + \Psi_1(a) - \Psi_1(\infty)\right| \leqslant$$

$$\leqslant \left|\frac{1}{t^\gamma\rho_1(t)}I_3 - \Psi_1(z)\right| + |\Psi_1(a) - \Psi_1(\infty)| \leqslant 2\epsilon,$$

$$\left|\frac{1}{t^\gamma\rho_1(t)}(I_1 + iI_2)\right| \leqslant \frac{2M}{\pi}\Psi_3(a,t) = \frac{2M}{\pi}[\Psi_3(a,t) - \Psi_3(a,\infty) + \Psi_3(a,\infty)] \leqslant$$

$$\leqslant \frac{2M}{\pi}|\Psi_3(a,t)+\Psi_3(a,\infty)| + \frac{2M}{\pi}|\Psi_3(a,\infty)| \leqslant \frac{4M}{\pi}\epsilon.$$

From this and from (1.31) we obtain (1.22) and (1.23). □

It remains only to consider the case $\alpha = 1$. Suppose that $\rho$ is an automodel function of order 1 and that $d\mu$ is a finite measure on $\mathbf{R}^1$, such that

$$d\mu(x) = d\mu_s(x) + d\mu_a(x);$$
$$d\mu_s(x) = d(|x|t_1(x)),$$
$$d\mu_a(x) = \frac{1}{x} d(x^2 t_2(x)) \tag{1.32}$$

where $|x|t_1(x)$ and $x^2 t_2(x)$ are locally summable functions, and

$$\lim_{x\to+0}\frac{t_1(x)}{\rho\left(\frac{1}{x}\right)} = \frac{\pi}{2}a, \quad \lim_{x\to+0}\frac{t_2(x)}{\rho\left(\frac{1}{x}\right)} = -b, \quad a^2+b^2 \neq 0. \tag{1.33}$$

Note that the odd part $d\mu_a(x)$ of the measure $d\mu(x)$ vanishes on even test functions, so that

$$(d\mu_a(x), \phi(x)) \equiv \int \frac{\phi(x)-\phi(-x)}{2x} d(x^2 t_2(x)), \quad \phi \in \mathcal{S}.$$

From the results of the lemma in Section 8.2 it follows that

$$\operatorname*{Var}_0^\infty(xt_1(x)) < \infty; \; t_2(x) = \frac{D}{x^2} + \eta(x), \; x > 0 \tag{1.34}$$

where

$$\operatorname*{Var}_0^\infty(x\eta(x)) < \infty, \quad \int_0^\infty |\eta(x)|\,dx < \infty.$$

Now we assume that $d\mu(x) \geqslant 0$ in some neighbourhood of the origin (that is, condition (1.8) holds). In this case, necessarily

$$a > 0, \; b = 0.$$

**Theorem 3.** *Suppose that $d\mu$, a finite measure on $\mathbf{R}^1$, satisfies conditions* (1.8), (1.32) *and* (1.33), *where $\rho(t)$ is an automodel function of order* 1. *Then for any $\gamma > 0$ there exists the limit*

$$\lim_{t\to+\infty}\frac{1}{t^{\gamma-1}\rho(t)}\Phi^{(-\gamma)}(t) = \frac{a}{\Gamma(\gamma+1)} \tag{1.35}$$

*where $\Phi(t)$ is defined by formula* (1.1).

**Proof.** As in the proofs of the previous theorems, together with $\rho(t)$ we consider the automodel function $\rho_1(t) = t^{-1}\rho(t)$ of order 0. First note that one can assume the measure $d\mu$ to be continuous at zero.

Indeed, if this is not the case then

$$d\mu(x) = A\,\delta(x) + d\tilde{\mu}(x), \tag{1.36}$$

where $d\tilde{\mu}(x)$ is continuous at zero. So we have

$$\rho_1\left(\frac{1}{x}\right) \to W \neq 0, \ x \to +0$$

and by representation (1.32)-(1.33) $A = \pi a W$. It is also clear that in the representation (1.32) of the measure $d\tilde{\mu}$

$$d\tilde{\mu} = d(|x|\tilde{t}_1(x)) + \frac{1}{x} d(x^2\tilde{t}_2(x))$$

the corresponding limits (1.33) are equal to zero:

$$\lim_{x\to+0} \frac{x\tilde{t}_1(x)}{\rho_1\left(\frac{1}{x}\right)} = \lim_{x\to+0} \frac{x\tilde{t}_2(x)}{\rho_1\left(\frac{1}{x}\right)} = 0.$$

Using (1.1) and (1.36) we have

$$\frac{1}{t^\gamma\rho_1(t)}\Phi^{(-\gamma)}(t) = \frac{1}{t^\gamma\rho_1(t)}\left[f_\gamma(t) * \int\frac{\theta(t)}{\pi}e^{-i\xi t}\,d\mu(\xi)\right] =$$

$$= \frac{1}{t^\gamma\rho_1(t)}\left[Af_\gamma(t) * \frac{\theta(t)}{\pi} + f_\gamma(t) * \int\frac{\theta(t)}{\pi}e^{-i\xi t}\,d\tilde{\mu}(\xi)\right].$$

Therefore,

$$\lim_{t\to+\infty}\frac{\Phi^{(-\gamma)}(t)}{t^\gamma\rho_1(t)} = \lim_{t\to+\infty}\frac{Af_{\gamma+1}(t)}{\pi t^\gamma\rho_1(t)} +$$

$$+ \lim_{t\to+\infty}\frac{1}{t^\gamma\rho_1(t)}\left[f_\gamma(t) * \frac{\theta(t)}{\pi}\int e^{-i\xi t}\,d\tilde{\mu}(\xi)\right]. \tag{1.37}$$

But

$$\lim_{t\to+\infty}\frac{Af_{\gamma+1}(t)}{t^\gamma\rho_1(t)\pi} = \frac{\pi a W}{W\pi\Gamma(\gamma+1)} = \frac{a}{\Gamma(\gamma+1)}.$$

Hence, the calculation of the limit (1.37) can be reduced to the calculation of a similar limit in the case when the measure $d\mu$ is continuous at zero.

Assuming that the measure $d\mu$ is continuous at zero, we have (see (1.7))

$$\frac{\Phi^{(-\gamma)}(t)}{t^\gamma\rho_1(t)} = \frac{2}{\pi t^\gamma\rho_1(t)}\left[\int_0^\infty \mathrm{Re}\frac{K_\gamma(\xi t)}{\xi^\gamma}\,d\mu_s(\xi) + i\int_0^\infty \mathrm{Im}\frac{K_\gamma(\xi t)}{\xi^\gamma}\,d\mu_a(\xi)\right].$$

As in the proof of theorems 1 and 2 it is sufficient to consider the case $0 < \gamma < 1$. We divide the integrals over $(0,\infty)$ into the sum of the integrals over $(0, d/t)$ and $(d/t, \infty)$. As before, the integral over $(d/t, \infty)$ by lemma 1 can be replaced by the integral over $(d/t, \delta)$, where $\delta > 0$ is arbitrarily small. But on the intervals $(0, d/t)$ and $(d/t, \delta)$ we have $d\mu \geqslant 0$. So without loss of generality we can assume that the measure $d\mu$ is nonnegative on the whole axis.

Taking estimations similar to the estimations in the proof of theorem 2, we get

$$\frac{2}{\pi t^\gamma\rho_1(t)}\left|\int_{d/t}^\infty \mathrm{Re}\frac{K_\gamma(\xi t)}{\xi^\gamma}\,d\mu_1(\xi) + i\int_{d/t}^\infty \mathrm{Im}\frac{K_\gamma(\xi t)}{\xi^\gamma}\,d\mu_2(\xi)\right| \leqslant \Psi_3(d, t),$$

where

$$\Psi_3(d\,,t\,) \to \Psi_3(d\,,\infty),\ t \to +\infty;\ \ \Psi_3(d\,,\infty) \to 0,\ d \to \infty.$$

Let us consider the remaining integrals. We have

$$\frac{2}{\pi t^{\gamma}\rho_1(t\,)}\int_0^{d/t}\mathrm{Re}\frac{K_\gamma(\xi t\,)}{\xi^\gamma}\,d\,\mu_1(\xi) = \frac{2}{\pi}\int_0^{d/t}\mathrm{Re}\frac{K_\gamma(\xi t\,)}{(\xi t\,)^\gamma}\frac{d\,(\xi t_1(\xi))}{\rho_1(t\,)} =$$

$$= \frac{2}{\pi}\int_0^{d}\mathrm{Re}\frac{K_\gamma(\eta)}{\eta^\gamma}\frac{d\left(\eta t_1\left(\frac{\eta}{t}\right)\right)}{t\,\rho_1(t\,)} =$$

$$= \frac{2}{\pi}\mathrm{Re}\left\{\frac{K_\gamma(d\,)}{d^{\gamma-1}}\frac{t_1\left(\frac{d}{t}\right)}{t\,\rho_1(t\,)} - \int_0^d \frac{\eta t_1\left(\frac{\eta}{t}\right)}{t\,\rho_1(t\,)}\left[\frac{K_{\gamma-1}(\eta)}{\eta^\gamma} - \gamma\frac{K_\gamma(\eta)}{\eta^{\gamma+1}}\right]d\,\eta\right\} =$$

$$= \frac{2}{\pi}\mathrm{Re}\left\{\frac{K_\gamma(d\,)}{d^{\gamma}}\frac{t_1\left(\frac{d}{t}\right)d}{t\,\rho_1(t\,)} + i\int_0^d \frac{\eta t_1\left(\frac{\eta}{t}\right)}{t\,\rho_1(t\,)}\left[\frac{K_\gamma(\eta)}{\eta^\gamma} - \gamma\frac{K_{\gamma+1}(\eta)}{\eta^{\gamma+1}}\right]d\,\eta\right\} = \Psi(d\,,t\,).$$

We have taken into account that, according to (1.5),

$$\frac{K_{\gamma-1}(\eta)}{\eta^\gamma} - \gamma\frac{K_\gamma(\eta)}{\eta^{\gamma+1}} = \frac{f_{\,\gamma}(\eta)}{\eta^\gamma} - \gamma\frac{f_{\,\gamma+1}(\eta)}{\eta^{\gamma+1}} - i\left[\frac{K_\gamma(\eta)}{\eta^\gamma} - \gamma\frac{K_{\gamma+1}(\eta)}{\eta^{\gamma+1}}\right]$$

and by the recurrent formula $\gamma\Gamma(\gamma) = \Gamma(\gamma + 1)$

$$\frac{f_{\,\gamma}(\eta)}{\eta^\gamma} - \gamma\frac{f_{\,\gamma+1}(\eta)}{\eta^{\gamma+1}} = 0.$$

Since $(1/\eta^n)K_n(\eta)$ is bounded it follows from lemma 2 that in the expression for $\psi(d\,,t\,)$ one can take a limit when $t \to +\infty$ before integrating. So, by (1.33), we have, for $t \to +\infty$

$$\Psi(d\,,t\,) \to \frac{2}{\pi}\mathrm{Re}\left\{\frac{K_\gamma(d\,)}{d^\gamma}\frac{\pi a}{2} + i\int_0^d \frac{\pi a}{2}\left[\frac{K_\gamma(\eta)}{\eta^\gamma} - \gamma\frac{\mathrm{K}_{\gamma+1}(\eta)}{\eta^{\gamma+1}}\right]d\,\eta\right\} =$$

$$= a\left\{\mathrm{Re}\frac{K_\gamma(d\,)}{d^\gamma} - \int_0^d \mathrm{Im}\left[\frac{K_\gamma(\eta)}{\eta^\gamma} - \gamma\frac{K_{\gamma+1}(\eta)}{\eta^{\gamma+1}}\right]d\,\eta\right\} =$$

$$= a\left\{\mathrm{Re}\frac{K_\gamma(d\,)}{d^\gamma} - \mathrm{Re}\int_0^d \left[\frac{K_{\gamma-1}(\eta)}{\eta^\gamma} - \gamma\frac{K_\gamma(\eta)}{\eta^{\gamma+1}}\right]d\,\eta\right\} =$$

$$= a\left\{\mathrm{Re}\frac{K_\gamma(d\,)}{d^\gamma} - \mathrm{Re}\int_0^d d\left(\frac{K_\gamma(\eta)}{\eta^\gamma}\right)\right\} =$$

$$= a\ \mathrm{Re}\frac{K_\gamma(\eta)}{\eta^\gamma}\Bigg|_{\eta\to+0} = \frac{a}{\Gamma(\gamma+1)}. \tag{1.39}$$

For the other integral we get

$$i\frac{2}{\pi t^\gamma \rho_1(t)}\int_0^{d/t}\mathrm{Im}\frac{K_\gamma(\xi t)}{\xi^\gamma}\,d\mu_2(\xi) = \frac{-2i}{\pi t^\gamma\rho_1(t)}\int_0^{d/t}\mathrm{Re}\frac{K_{\gamma+1}(\xi t)}{\xi^{\gamma+1}}\,d(\xi^2 t_2(\xi)) =$$

$$= \frac{-2i}{\pi}\int_0^d \mathrm{Re}\frac{K_{\gamma+1}(\eta)}{\eta^{\gamma+1}}\cdot\frac{d\left(\left(\frac{\eta}{t}\right)^2 t_2\left(\frac{\eta}{t}\right)\right)}{\rho_1(t)} =$$

$$= \frac{-2i}{\pi}\mathrm{Re}\left\{\frac{K_{\gamma+1}(d)}{d^{\gamma-1}}\frac{t_2\left(\frac{d}{t}\right)}{t\rho_1(t)} - \int_0^d \frac{\eta^2 t_2\left(\frac{\eta}{t}\right)}{t\rho_1(t)}\left[\frac{K_\gamma(\eta)}{\eta^{\gamma+1}} - (\gamma+1)\frac{K_{\gamma+1}(\eta)}{\eta^{\gamma+2}}\right]d\eta\right\} = \omega(d,t).$$

Using the estimates from lemma 2 in the expression for $\omega(d,t)$ one can take the limit when $t\to+\infty$ before the integrating (by the Lebesgue theorem). Taking into account (1.33) we have, for $t\to+\infty$,

$$\omega(d,t)\to\omega(d,\infty) =$$

$$= \frac{-2i}{\pi}\mathrm{Re}\left\{\frac{K_{\gamma+1}(d)}{d^\gamma}(-b) - (-b)\int_0^d\left[\frac{K_\gamma(\eta)}{\eta^{\gamma+1}} - (\gamma+1)\frac{K_{\gamma+1}(\eta)}{\eta^{\gamma+2}}\right]\eta\,d\eta\right\} = 0,$$

since $b = 0$. Now reasoning as before in the proof of the previous theorem, we complete the proof fo theorem 3. □

**2. Asymptotic properties of the Fourier transform of distributions from $M_+$.** In this subsection we study the asymptotic properties of the primitives of the Fourier transforms of distributions from $M_+$.

**Theorem 1.** *Suppose that $f\in\mathcal{S}'_+$, $\tilde f\in M_+$ and for some real $r$*

$$\int\frac{|d\mu(x)|}{(1+|x|)^r} < \infty,\quad \mathrm{Re}\,\tilde f(x) = d\mu(x). \tag{2.1}$$

*Let $f$ have a quasi-asymptotics at infinity with respect to an automodel function $\rho$ of order $\alpha$, so that*

$$\frac{1}{\rho(k)}f(kt)\to cf_{\alpha+1}(t),\ k\to\infty \tag{2.2}$$

*in $\mathcal{S}'_+$, $c\neq 0$. Then for any $\gamma\geqslant\max(\gamma_0,r)$, where $\gamma_0$ is any number such that $\gamma_0>-\alpha$, the primitive $f^{(-\gamma)}(t)$ of the order $\gamma$ of the distribution $f$ is a continuous function for $t>0$, and there exists the limit*

$$\lim_{t\to+\infty}\frac{1}{t^\gamma\rho(t)}f^{(-\gamma)} = \frac{c}{\Gamma(\gamma+\alpha+1)}. \tag{2.3}$$

*(Note that, according to the results of Section 8.2, in this case $\alpha \leqslant 0$ always).*

To prove the theorem we need the following lemma:

**Lemma.** *Let $d\nu$ be a finite measure over $\mathbf{R}^1$ such that for some $\kappa$, $0 \leqslant \kappa < 2$,*

$$\int (1 + |\xi|)^\kappa |d\nu(\xi)| < \infty. \tag{2.4}$$

*Then the derivative of order $\kappa$ of the function*

$$\psi(t) = 2\pi\theta(t) F^{-1}[d\nu] = \theta(t) \int e^{-i\xi t} d\nu(\xi)$$

*is continuous for $t > 0$, and there exist numbers $M > 0$ and $T_0$ such that*

$$|\psi^{(\kappa)}(t)| < M \quad \text{if } t > T_0. \tag{2.5}$$

**Proof.** Let us consider the case $1 < \kappa < 2$ (the case $0 \leqslant \kappa \leqslant 1$ can be considered in the similar way). The function $\psi^{(\kappa)}$ can be represented in the form

$$\begin{aligned}\psi^{(\kappa)}(t) &= f_{-\kappa} * \psi = \\ &= \left[\delta(t)\int d\nu(\xi) + \theta(t)\int e^{-i\xi t}(-i\xi)\, d\nu(\xi)\right]^{(\kappa-1)} = \\ &= f_{\kappa-1}(t)\int d\nu(\xi) - i\int_0^\infty K_{-\kappa+1}(\xi t)\xi^{\kappa-1}\xi\, d\nu(\xi) + i\int_0^\infty \overline{K_{-\kappa+1}(\xi t)}\xi^{\kappa-1}(-\xi)\, d(-\nu(-\xi)),\end{aligned} \tag{2.6}$$

where the $K_\gamma$ are given by formula (1.4): $K_\gamma(t) = f_\gamma(t) * (\theta(t) e^{-it})$. Let us consider the function

$$\begin{aligned}\psi_1(t) &= \int_0^\infty K_{-\kappa+1}(\xi t)\xi^{\gamma-1}\xi\, d\nu(\xi) = \\ &= \left(\int_0^{1/t} + \int_{1/t}^\infty\right) K_{-\kappa+1}(\xi t)\xi^{\kappa-1}\cdot\xi\, d\nu(\xi).\end{aligned}$$

If $0 < \xi < 1/t$, using (1.6) we obtain

$$\begin{aligned}\left|\frac{1}{t^{\kappa-1}}\int_0^{1/t} K_{-\kappa+1}(\xi t)(\xi t)^{\kappa-1}\xi\, d\nu(\xi)\right| &\leqslant \frac{M}{t^{\kappa-1}}\left|\int_0^\infty \xi\, d\nu(\xi)\right| \leqslant \\ &\leqslant \frac{M_1}{t^{\kappa-1}}.\end{aligned}$$

If $\xi > 1/t$, using also by (1.6), we get

$$\left|\int_{1/t}^\infty K_{-\kappa+1}(\xi t)\xi^{\kappa-1}\xi\, d\nu(\xi)\right| \leqslant M_2 \int_{1/t}^\infty |\xi^\kappa\, d\nu(\xi)| < \infty.$$

Estimating the other term in (2.6) in a similar way and nothing that $f_{\kappa-1}(t)$ is continuous and bounded when $\kappa > 1$, we complete the proof. □

**Proof.** Let us now prove the theorem. Let $\gamma$ satisfy the hypothesis of the theorem. Suppose that $N$ is an integer such that $\gamma \leqslant 2N < \gamma + 2$. According to representation (3.4) of Section 7.3 we have

$$\tilde{f}(z) = (1 + z^{2N})\tilde{\Phi}(z) + P(z), \quad \operatorname{Re} P(x) \equiv 0,$$
$$\tilde{\Phi}(z) = \frac{i}{\pi} \int \frac{d\mu(\xi)}{(1 + \xi^{2N})(z - \xi)}, \quad y > 0. \tag{2.7}$$

Thus, we get

$$f(t) = \left[ 1 + \left( -i\frac{\partial}{\partial t} \right)^{2N} \right] \Phi(t) + \sum_{j=0}^{p} C_j \delta^{(j)}(t),$$

where

$$\Phi(t) = \frac{\theta(t)}{\pi} \int e^{-i\xi t} \frac{d\mu(\xi)}{1 + \xi^{2N}}. \tag{2.8}$$

Therefore,

$$f^{(-\gamma)}(t) = \Phi^{(-\gamma)}(t) + (-i)^{2N}\Phi^{(2N-\gamma)} + \left( \sum_{j=0}^{p} C_j \delta^{(j)}(t) \right)^{(-\gamma)}. \tag{2.9}$$

Denoting $(1 + x^{2N})^{-1} d\mu(x) = d\nu(x)$, we see that the hypothesis of the lemma holds since $2 > 2N - \gamma \geqslant 0$. From the property $t^{\gamma}\rho(t) \to \infty$, when $t \to \infty$ (see Section 3.2), it follows that

$$\frac{1}{t^{\gamma}\rho(t)} \Phi^{(2N-\gamma)}(t) \to 0, \quad t \to +\infty. \tag{2.10}$$

Since the measure $d\nu(x)$ is finite, then from (2.8) it follows that $\Phi^{(-\gamma)}(t)$ is continuous for $t > 0$, and therefore the functions $f^{(-\gamma)}(t)$ and

$$f_1^{(-\gamma)}(t) = \Phi^{(-\gamma)}(t) + \left( \sum_{j=0}^{p} C_j \delta^{(j)}(t) \right)^{(-\gamma)}$$

are continuous for $t > 0$. One can see by the hypothesis of the theorem that $f_1(t)$ has a quasi-asymptotics at infinity with respect to $\rho(t)$ of order $\alpha$. From the results of Section 9.1 it follows that there exists a distribution of the form

$$\left( \sum_{j=0}^{p_1} C_j' \delta^{(j)}(t) \right)^{(-\gamma)}$$

such that there exists the limit

$$\frac{1}{t^{\gamma}\rho(t)} \left[ \Phi^{(-\gamma)}(t) + \left( \sum_{j=0}^{p_1} C_j' \delta^{(j)}(t) \right)^{(-\gamma)} \right] \to \hat{C}_1, \quad t \to +\infty. \tag{2.11}$$

Let us rewrite $f^{(-\gamma)}$ in the form

$$f^{(-\gamma)}(t) = \left[ \Phi^{(-\gamma)}(t) + \left( \sum_{j=0}^{p_1} C_j' \delta^{(j)}(t) \right)^{(-\gamma)} \right] + (-i)^{2N}\Phi^{(2N-\gamma)}(t) + h^{(-\gamma)}(t), \tag{2.12}$$

where

$$h(t) = \left[ \sum_{j=0}^{p} C_j \delta^{(j)}(t) - \sum_{j=0}^{p_1} C_j' \delta^{(j)}(t) \right] \equiv \sum_{j=0}^{p_2} C_j'' \delta^{(j)}(t). \tag{2.13}$$

Note that $h(t)$ has a quasi-asymptotics at infinity (which may be trivial) with respect to $\rho(t)$. Indeed, $f^{(-\gamma)}$ has a quasi- asymptotics with respect to $t^\gamma\rho(t)$. From (2.10) and (2.11) it follows that the first two terms in the right-hand side of (2.12) have a quasi-asymptotics at infinity with respect to $t^\gamma\rho(t)$, and therefore $h^{(-\gamma)}(t)$ also has a quasi-asymptotics. Thus, it follows that $h(t)$ has a quasi-asymptotics with respect to $\rho(t)$; that is,

$$\frac{1}{\rho(k)}h(kt) \to h_1(t),\ k \to \infty \ \text{ in } \mathcal{S}',$$

or

$$\frac{1}{\rho(k)}\sum_{j=0}^{p_2}\frac{C_j''}{k^{j+1}}\delta^{(j)}(t) \to h_1(t),\ k \to \infty \ \text{ in } \mathcal{S}'. \tag{2.14}$$

If all the constants $C_j''$ in (2.13) are equal to zero, the theorem is proved. Let us suppose this is not the case, and let $C_n''$ be the first coefficient which is not equal to zero. Acting with (2.14) on a test function $\phi$ such that $\phi^{(k)}(0) = 0$, $k = 0, 1, \ldots, n$, we see that $\rho(k)k^{n+1}$ tends either to a finite limit or to $+\infty$ when $k \to \infty$. In both cases we get $\alpha \geqslant -n$.

Now, in the right-hand side of the relation

$$\frac{1}{t^\gamma\rho(t)}h^{(-\gamma)}(t) = \frac{1}{t^{\gamma-n-1}t^{n+1}\rho(t)}\sum_{j=n}^{p_2}C_j''f_{\gamma-j}(t)$$

the essential term is equal to

$$\frac{C_n''t^{\gamma-n-1}}{t^{\gamma-n-1}t^{n+1}\rho(t)}$$

(or zero, if $\gamma$ is an integer) and therefore the limit

$$\lim_{t\to+\infty}\frac{1}{t^\gamma\rho(t)}h^{(-\gamma)}(t)$$

is finite. □

The next example shows that the theorem is strict (with respect to the order of a primitive).

**Example.** Let us consider the function

$$\tilde{f}(z) = -z^2\left(\frac{i}{z} - \frac{iq}{z-1} - \frac{iq}{z+1}\right),\quad 0 < 2q < 1. \tag{2.15}$$

We have

$$\operatorname{Re}\tilde{f}(z) = (1-2q)y + \frac{qy}{(x-1)^2+y^2} + \frac{qy}{(x+1)^2+y^2} > 0,\ y > 0,$$

$$\tilde{f}(iy) = y^2\left[\frac{1}{y} - 2q\frac{y}{y^2+1}\right].$$

So, $\tilde{f}(iy)$ is real. Therefore, $\tilde{f}(z)$ is a positive real function (that is, $\tilde{f} \in H_+$ and it is real on the imaginary axis). On the other hand, for the automodel function $\rho(t) = t^{-2}$ (of order $\alpha = -2$) we have

$$\frac{y}{\rho\left(\frac{1}{y}\right)}\tilde{f}(iy) \to 1, \quad y \to +0.$$

The corresponding distribution $f(t)$ has the following second primitive

$$f^{(-2)}(t) = \theta(t)[1 - 2q \cos t].$$

But is has no limit when $t \to +\infty$.

**Corollary.** *Suppose that* $f \in \mathcal{S}'_+$ *and* $\tilde{f} \in H_+$. *If there exists the limit*

$$\lim_{y \to +0} \frac{y}{\rho\left(\frac{1}{y}\right)}\tilde{f}(iy) = h,$$

*where* $\rho$ *is an automodel function of order* $\alpha$ *(in this case necessarily* $-2 \leqslant \alpha \leqslant 0$*), then for* $\gamma \geqslant \max(2, \gamma_0)$*, where* $\gamma_0 > -\alpha$*, the function* $f^{(-\gamma)}(t)$ *is continuous for* $t > 0$ *and there exists the limit*

$$\lim_{t \to +\infty} \frac{1}{t^{\gamma}\rho(t)} f^{(-\gamma)}(t) = \frac{h}{\Gamma(\gamma + \alpha + 1)}. \tag{2.16}$$

**Proof.** Indeed, from the Herglotz–Nevanlinna representation (see remark of Section 7.3) it follows that in this case $r = 2$. Now the statement which we were to prove follows directly from theorem 1 and theorem 4 of Section 7.1. □

Let us note that if $-2 < \alpha \leqslant 0$ one can assume that $\gamma = 2$, and if $\alpha = -2$, then $\gamma = 2 + \epsilon$, where $\epsilon > 0$ is arbitrarily small.

**3. The Abel and Čezaro series summation with respect to an automodel weight.** In this subsection we consider the application of the above results to the Abel and Čezaro summation of series. Let us consider the series

$$\sum_{k=0}^{\infty} C_k \tag{3.1}$$

with tempered coefficients. That is, there are numbers $M$ and $l$ such that $|C_k| \leqslant M(1+k)^l$. The corresponding power series

$$F(\omega) = \sum_{k=0}^{\infty} C_k \omega^k, \quad \omega = re^{ix} \tag{3.2}$$

represents an analytic function in the unit disc $B = [|\omega| < 1]$. Let $\rho$ be an automodel function of order $\alpha$.

**Definitions.** (1) We say that the series (3.1) converges in the *sense of Abel with respect to an automodel weight* $\rho$ (and we write $A(\rho)$) if there exists the limit

$$\frac{1-r}{\rho\left(\frac{1}{1-r}\right)} F(r) \to h, \quad r \to 1-0. \tag{3.3}$$

(2) We say that the series (3.1) converges in the *sense of Čezaro with the order* $\gamma$ *with respect to an automodel weight* $\rho$ (and we write $(C, \gamma; \rho)$) if there exists the limit

$$\frac{1}{N^{\gamma+1}\rho(N)}\sum_{k=0}^{N} C_k (N-k)^\gamma \to h,\ N \to \infty. \tag{3.4}$$

Following G.H. Hardy we say that we have a $C$-convergence of series (3.1) with respect to an automodel weight if (3.4) holds for some $\gamma$.

If $\rho(t) = t^{-1}$ the $A(\rho)$-convergence is a convergence in the sense of Abel, and the $(C, \gamma; \rho)$-convergence in the sense of Čezaro of order $\gamma$. In this case (3.3) follows from (3.4) (see G.H. Hardy, 1949). If the converse statement is to be true, some additional conditions on the coefficients $C_k$ are necessary. These are usually known as the Tauberian conditions.

For the given series (3.1) let us introduce a distribution $f \in \mathcal{S}'_+$ by the formula

$$f(t) = \sum_{k=0}^{\infty} C_k \delta(t-k). \tag{3.5}$$

Then

$$\sum_{k=0}^{N} C_k (N-k)^\gamma = \Gamma(\gamma+1) f^{(-\gamma-1)}(N)$$

and (3.4) can be rewritten as

$$\frac{\Gamma(\gamma+1)}{N^{\gamma+1}\rho(N)} f^{(-\gamma-1)}(N) \to h,\ N \to \infty. \tag{3.6}$$

The function $F(\omega)$ has, on the unique circumference, a boundary value $F(e^{ix})$ related to distributions (see, for instance, V.S. Vladimirov, 1979). If, in (3.2), we change the variables

$$\omega = e^{iz},\ z = x + iy,\ y > 0,$$

we get

$$F(\omega) = \sum_{k=0}^{\infty} C_k e^{izk} = (f(t), e^{izt}) = \tilde{f}(z). \tag{3.7}$$

The Laplace transform $\tilde{f}$ of the distribution $f$ defined by (3.5) is an analytic, periodic function in the upper half plane with a period $2\pi$. From the general Tauberian theorem (theorem 1, Section 7.1) there follows directly

**Proposition 1.** *For the series* (3.1) *to C-convergence with respect to an automodel weight* $\rho$, *it is necessary and sufficient that the following conditions hold:*

(A) *The series* (3.1) $A(\rho)$*-converges.*

(B) *There exist numbers* $M, m, p_0$ *such that*

$$\left| \frac{\sqrt{r^2+1-2r\cos x}}{\rho\left(\dfrac{1}{\sqrt{r^2+1-2r\cos x}}\right)} F(re^{ix}) \right| \leqslant \frac{M}{\left(\dfrac{1-r\cos x}{\sqrt{r^2+1-2r\cos x}}\right)^m},$$

$$0 < \sqrt{r^2+1-2r\cos x} \leqslant p_0.$$

*Here*

$$\frac{1 - r\cos x}{\sqrt{r^2 + 1 - 2r\cos x}} = \sin\phi,$$

*where $\phi$ is the angle between the line tangent to the circumference $|\omega| = 1$ at the point $\omega = 1$, and the ray going from the point $\omega = 1$ to the point $rl^{ix}$* (see Figure 14).

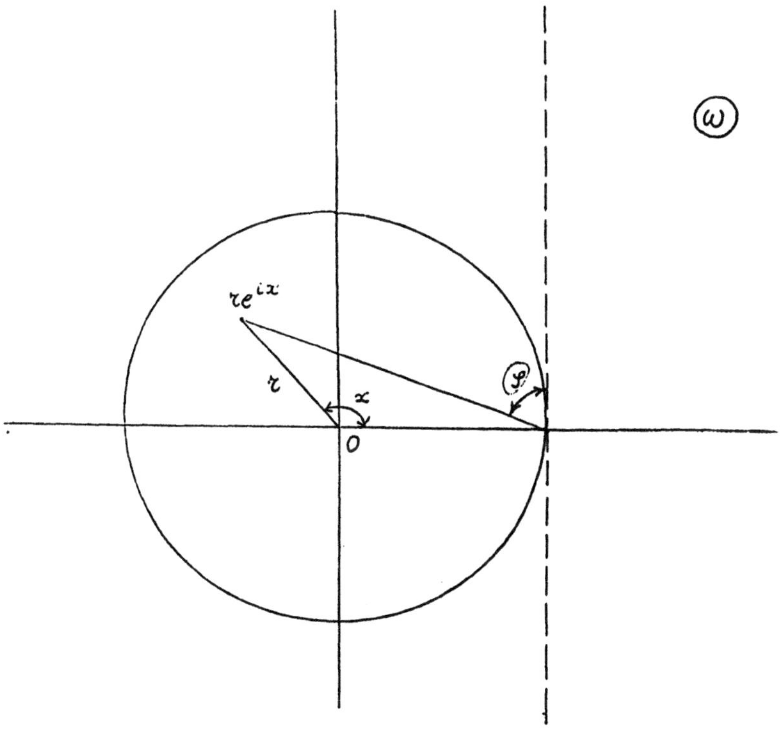


Figure 14.

Note that estimate (B) of proposition 1 is just the same as estimate (B) of theorem 1 of Section 7.1.

**Definition.** By analogy to the class $M_+$ we introduce the class $M_+(B)$ which consists of all the functions which are holomorphic in the unique disc, with the boundary values of their real parts being measures which are nonnegative in some neighbourhoods of the point $\omega = 1$:

$$\operatorname{Re} F(e^{ix}) = d\mu(x); \quad -\pi < x \leqslant \pi; \quad d\mu(x) \geqslant 0$$

for $|x| < \delta$ for some $\delta > 0$ ($\delta$ depends on $F$).

For instance, let

$$c_0 = \frac{1}{2\pi}\int_{-\pi}^{\pi} d\mu(\xi), \quad c_k = \frac{1}{\pi}\int_{-\pi}^{\pi} e^{i\xi k}\, d\mu(\xi), \quad k = \pm 1, \pm 2, \cdots \tag{3.8}$$

where $d\mu$ is a measure on $(-\pi, \pi]$, nonnegative in some neighbourhood of zero. Then the series (3.2) is a complex trigonometric Fourier-Stiltjes series. It is easy to see that it belong to $M_+(\mathbf{B})$. Indeed, taking into account that $c_k = \bar{c}_{-k}$ we have (see Section 2.2)

$$\operatorname{Re} f(re^{ix}) \to c_0 + \frac{1}{2}\sum_{|k|>0} c_k e^{ixk} = d\mu(x), \quad r \to 1-0$$

in $\mathcal{D}'_{2\pi}$ and therefore in $\mathcal{S}'$.

**Theorem 1.** *Let the series* (3.1) *be such that the corresponding function* $F \in M_+(\mathbf{B})$. *If* $A$ *($\rho$)-converges with respect to an automodel weight of order* $\alpha$, *then it converges* $(C, \kappa; \rho)$ *with any*

$$\kappa > \max\{-\alpha - 1, 0\}. \tag{3.9}$$

**Proof.** Since $\operatorname{Re} F(e^{ix})$, the measure $d\mu$ is nonnegative in some neighbourhood of the point $x = 0$, which is finite on $(-\pi, \pi]$, so that

$$\int_{-\pi}^{\pi} d\mu(x) = \operatorname{Re} F(0) = c_0,$$

then by (3.7) we have $\tilde{f}(x) = d\tilde{\mu}(x) + iv(x)$, where $d\tilde{\mu}$ is a periodic measure over $\mathbf{R}^1$ with a period $2\pi$. Therefore, for any $\epsilon > 0$

$$\int \frac{|d\tilde{\mu}(x)|}{(1+|x|)^{1+\epsilon}} < \infty.$$

By the hypothesis we have

$$\frac{y}{\rho\left(\frac{1}{y}\right)} \tilde{f}(iy) \to h, \quad y \to +0.$$

Hence, by the theorem of Section 7.3, $f(t)$ defined by (3.5) has a quasi- asymptotics at infinity with respect to an automodel function $\rho$ of order $\alpha$ (in this case $\alpha \leqslant 0$, as already mentioned). Using the theorem of Section 9.2 we get that (3.6) holds for any $\gamma$, such that

$$\gamma + 1 > \max\{-\alpha, 1\}.$$

So, (3.9) follows. □

**Theorem 2.** *Let the distribution*

$$f(t) = \sum_{k=0}^{\infty} c_k \delta(t-k)$$

*be such that* $\tilde{f} \in M_+$. *If there exists the limit*

$$\frac{y}{\rho\left(\frac{1}{y}\right)} \tilde{f}(iy) \to h \neq 0, \quad y \to +0,$$

*where* $\rho$ *is an automodel function of order* $\alpha$, *then for any* $\gamma > -\alpha$ *the primitive* $f^{(-\gamma)}(t)$ *has an asymptotics by translation with respect to the automodel function*

$t^{\gamma}\rho(t)$. *That is, for any* $\phi \in \mathcal{S}$

$$\frac{1}{t^{\gamma}\rho(t)}(f^{(-\gamma)}(\xi), \phi(t-\xi)) \to \frac{h}{\Gamma(\gamma+\alpha+1)}\int \phi(\xi)\,d\xi, \quad t \to +\infty. \tag{3.10}$$

**Proof.** According to theorem 2 of Section 3.5 and its following remark, it is sufficient to prove relation (3.10) only for the test functions $\phi \in \mathcal{D}(-1/2, 1/2)$. We can assume also that the summation in (3.5) begins from $k = 1$.

Let us consider the function

$$g(t) = f(t) * \phi.$$

Note that

$$\operatorname{supp} g \subset \operatorname{supp} f + \operatorname{supp}\phi \subset [1, \infty) \bigcup [-\frac{1}{2}, \frac{1}{2}] \subset [0, \infty)$$

(see Section 1.8). If $\phi \in (-1/2, 1/2)$ satisfies the conditons

$$(1) \int \phi(\xi)\,d\xi = m \neq 0, \quad (2)\ \phi(\xi) = \overline{\phi(-\xi)}; \tag{3.11}$$

then its Fourier transform $\tilde{\phi}$ is real and $\tilde{\phi}(0) = m \neq 0$. Since $\tilde{g}(z) = \tilde{\phi}(z) \cdot \tilde{f}(z)$ and $\tilde{\phi}(z)$ is an entire function (in particular it is analytic in the neighbourhood of zero) then by theorem 1 of Section 7.1, $g$ has a quasi- asymptotics at infinity with respect to $\rho$. Moreover, we have

$$\operatorname{Re}\tilde{g}(x) = \tilde{\phi}(x)\operatorname{Re}\tilde{f}(x) = \tilde{\phi}(x)\,d\mu(x).$$

Assuming (without loss of generality) that $m$ in (3.11) is positive, we conclude that the measure $\operatorname{Re}\tilde{g}(x)$ is nonnegative in some neighbourhood of zero. From the theorem of Section 9.2 it follows that, for any $\gamma > -\alpha$, $g^{(-\gamma)}$ has the usual asymptotics at infinity with respect to $t^{\gamma}\rho(t)$:

$$\frac{1}{t^{\gamma}\rho(t)}g^{(-\gamma)}(t) = \frac{1}{t^{\gamma}\rho(t)}(f^{(-\gamma)} * \phi) \to \frac{hm}{\Gamma(\gamma+\alpha+1)}, \quad t \to +\infty. \tag{3.12}$$

So, the theorem is proved for the test functions, which have properties (3.11). Similarly, one can prove (3.10) in the case when condition (2) of (3.11) is replaced by

$$(3)\ \phi(\xi) = -\overline{\phi(-\xi)}.$$

In the general case, $\phi(t) = \phi_1(t) + \phi_2(t)$, where $\phi_1(t) = \overline{\phi_1(-t)}$, $\phi_2(t) = -\overline{\phi_2(-t)}$.

Thus, (3.10) holds if condition (1) in (3.11) holds for both $\phi_1$ and $\phi_2$. Let this condition be false. Supposing, for instance, that $\int \phi_1(\xi)\,d\xi = 0$, we consider the function $\phi_1 + \psi$, where $\psi \in \mathcal{D}(-1/2, 1/2)$ satisfies condition (3.11). Now (3.10) holds for $\psi$ and $\phi_1 + \psi$, so it holds for $\phi_1$. Similar arguments are applicable to $\phi_2$ if $\tilde{\phi}_2(-) = 0$. □

**Corollary.** *If, in the hypothesis of theorem 1,* $-1 < \alpha \leqslant 0$, *then*

$$\frac{1}{N\rho(N)}f^{(-1)}(N) \to \frac{h}{\Gamma(2+\alpha)}, \quad N \to \infty.$$

*In other words, the series* (3.1) $(C, 0; \rho)$*-converges.*

**Proof.** Indeed, taking $\phi \in \mathcal{D}(-1/2, 1/2)$ such that $\int \phi(\xi)\,d\xi = 1$, from (3.10) we have when $\gamma = 1$

$$\frac{1}{N\rho(N)}(f^{(-1)}(\xi), \phi(N + \frac{1}{2} - \xi)) = \frac{1}{N\rho(N)} f^{(-1)}(N) \to \frac{h}{\Gamma(2+\alpha)}, \quad N \to \infty. \qquad \square$$

Hence if, for instance, the Fourier-Stiltjes series of the measure $d\mu$, which is nonnegative in some neighbourhood of the point $x = 0$, satisfies condition (3.3), where $\rho$ is an automodel funciton of order $\alpha$, then if $\alpha \leqslant -1$, the series $(C, \gamma; \rho)$-converges, where $\gamma$ is arbitrary number greater than $-\alpha - 1$; and, if $-1 < \alpha \leqslant 0$, the series $(C, 0; \rho)$-converges.

## 10. Quasi-asymptotic expansions.

In this section we consider the quasi-asymptotic expansions of distributions from $\mathcal{S}'_+$ and give some of their applications to the study of the asymptotic behaviour of solutions of convolution equations. In applications it is seldom convenient to use the entire scale of automodel functions, as there are many automodel functions of the same order which are not equivalent at infinity (see Section 3.2). In this section we use only the power scale of automodel funcitons. In other words, we assume that $\rho(t) = t^\alpha$. In this case the order of automodelity completely defines the asymptotic behaviour of a distribution at infinity.

### 1. Open and closed quasi-asymptotic expansions.

**Definition 1.** We say that $g \in \mathcal{S}'_+$ has a *quasi-asymptotic of order* $\alpha$ if it has a quasi-asymptotic at infinity with respect to $\rho = t^\alpha$ (see (4.1), Section 3.4). That is

$$\frac{1}{k^\alpha} g(kt) \to C f_{\alpha+1}(t), \quad C \neq 0, \quad k \to \infty \quad \text{in } \mathcal{S}'_+. \tag{1.1}$$

**Definition 2.** We say that a distribution $g \in \mathcal{S}'_+$ has a *closed quasi-asymptotic expansion of order* $\alpha$ *and of length* $l$, $0 \leqslant l < +\infty$, if there exists an integer $N$, a family of real numbers $\alpha_j$, $j = 1, \ldots, N$, and a corresponding family of complex coefficients $C_j$, $j = 1, \ldots, N$, such that

$$\frac{1}{k^{\alpha-l}} \left| g(kt) - \sum_{j=1}^{N} C_j f_{\alpha_j+1}(kt) \right| \to 0, \quad k \to \infty \quad \text{in } \mathcal{S}'_+. \tag{1.2}$$

In this case we write

$$g(t) \sim \sum_{j=1}^{N} C_j f_{\alpha_j+1}(t), \quad t \to +\infty.$$

**Definition 3.** We say that a distribution $g \in \mathcal{S}'_+$ has an *(open) quasi-asymptotic expansion of order* $\alpha$ *and of length* $l$, $0 < l \leqslant +\infty$, if for any $l_1 < l$ the distribution $g$ has a closed quasi-asymptotic expansion of order $\alpha$ and of length $l_1$.

In this case there exists a sequence (which may be infinite) of real numbers $\alpha_1 = \alpha > \alpha_2 > \cdots > \alpha_N > \cdots$, and a corresponding sequence of complex coefficients $C_j$, $j = 1, 2, \ldots$, such that for any $l_1$, $0 \leqslant l_1 < l$, the expression

$$\sum_{\alpha_j \geqslant \alpha - l_1} C_j f_{\alpha_j+1}(t)$$

gives a proper quasi-asymptotic expansion of the length $l_1$. In this case we write

$$g(t) \sim \sum_j C_j f_{\alpha_j+1}(t) + \cdots, \; t \to +\infty.$$

It is clear that if $g$ has a closed quasi-asymptotic expansion of order $\alpha$ and of length $l$, then it has an open quasi-asymptotic expansion of the same order and length (the converse statement is not true). If the term $C_{\alpha-l} f_{\alpha-l+1}(t)$ belongs to a closed expansion then it does not belong to an open expansion of order $\alpha$ and of length $l$.

**Lemma 1.** *A distribution $g \in \mathcal{S}'_+$ has a closed quasi-asymptotic expansion*

$$\sum_{j=1}^{N} C_j f_{\alpha_j+1}(t)$$

*of order $\alpha$ and of length $l$ if and only if its Laplace transform has the form*

$$\tilde{g}(z) = \sum_{j=1}^{N} C_j' \left(\frac{1}{z}\right)^{\alpha_j+1} + \epsilon(z), \tag{1.3}$$

*where $C_j' = C_j e^{i(\pi/2)(\alpha_j+1)}$, and $\epsilon(z)$ is analytic in the upper half-plane function with the following properties:*

(*a*) $y^{\alpha-l+1}\epsilon(iy) \to 0, \; y \to +0.$

(b) *There exist numbers $M$, $m$, and $r_0$, such that*

$$\left| r^{\alpha-l+1}\epsilon(re^{i\phi}) \right| \leqslant \frac{M}{\sin^m \phi}, \quad 0 < r \leqslant r_0.$$

**Proof.** The lemma directly follows from the general Tauberian theorem (theorem 1, Section 7.1) and from example 2 of Section 2.5. □

Now let us consider some examples.

**Example 1.** The functions

$$f_2(t) + \theta(t)\ln(t+1), \quad \sum_{n=1}^{\infty} \frac{1}{n^2} f_{1+1/n}(t)$$

have open quasi-asymptotic expansions of order 1 and of length 1 but they have no closed quasi-asymptotic expansions of the same order and length.

The function $f_{\alpha+1}(t)$ has a quasi-asymptotic expansion of order $\alpha$ and of infinite length, consisting only of one term.

**Example 2.** Let us show that any distribution $g \in \mathcal{S}'_+$ with compact support has an open quasi-asymptotic expansion of order not greater than -1 ($\alpha \leqslant -1$) and of infinite length.

Indeed, in this case $\tilde{g}(z)$ can be expanded in the Taylor series at the point $z = 0$:

$$\tilde{g}(z) = \sum_{j=1}^{\infty} C_{-j}' \left(\frac{1}{z}\right)^{-j+1}.$$

Hence, by lemma 1

$$g(t) \sim \sum_{j=1}^{\infty} C_{-j} f_{-j+1}(t) + \cdots, \quad t \to +\infty.$$

Thus any distribution with a compact support has a quasi-asymptotic expansion of infinite length which consits of terms of orders $\alpha_j = -j \;\; j = 1, 2, \cdots$.

**Example 3.** Let us consider a distribution $g \in \mathcal{S}'_+$ periodic for $t \to +\infty$. This means that there exist numbers $t_0 \geqslant 0$ and $T > 0$ such that for any $\phi \in \mathcal{S}$, $\operatorname{supp}\phi \subset [t \geqslant t_0]$, we have $(g, \phi) = (g(t), \phi(t - nT))$, $n = 0, 1, 2, \cdots$.

We show that such a function $g$ has a quasi-asymptotic expansion of infinite length, with terms only of the orders $\alpha_j = -j + 1$, $j = 1, 2, \cdots$. At first we prove the following lemma.

**Lemma 2.** *If $f$ is a distribution from $\mathcal{S}'_+$ periodic for $t \to +\infty$, then it can be represented as follows*

$$f(t) = f_1(t) + f_2(t) * \sum_{k=0}^{\infty} \delta(t - kT),$$

*where $f_1, f_2 \in \mathcal{E}'(\mathbf{R}^1)$ and $T$ is the period of $f$.*

**Proof.** Let $e_T(t)$ be a function from $\mathcal{D}(\mathbf{R}^1)$ such that

$$\sum_{k=-\infty}^{+\infty} e_T(t - kT) = 1$$

(see Section 2.2). Then the function

$$\eta(t) = \sum_{k=0}^{\infty} e_T(t - kT)$$

is infinitely differentiable, vanishes for a large negative $t$ and is equal to 1 for a large positive $t$. If $F(t)$ is the periodic continuation of $f$ onto the whole axis, then the distribution

$$f_1(t) = f(t) - \eta(t) F(t)$$

has compact support. By the periodicity of $F(t)$ we have

$$\eta(t)F(t) = \left\{ \sum_{k=0}^{\infty} e_T(t - kT) \right\} \cdot F(t) =$$

$$= \sum_{k=0}^{\infty} e_T(t - kT) F(t) =$$

$$= \sum_{k=0}^{\infty} e_T(t - kT) F(t - kT) =$$

$$= (e_T(t) F(t)) * \sum_{k=0}^{\infty} \delta(t - kT).$$

Denoting now $e_T F$ by $f_2$ we complete the proof. □

Let us continue the proof of the statement of example 3. According to lemma 2 the distribution $g(t)$ and its Laplace transform can be represented in the form

$$g(t) = g_1(t) + g_2(t) * \left[\sum_{k=0}^{\infty} \delta(t - kT)\right].$$

$$\tilde{g}(z) = \tilde{g}_1(z) + \tilde{g}_2(z) \cdot \chi(z)$$

where $g_1(t)$ and $g_2(t)$ have compact supports, and

$$\chi(x) = \sum_{k=0}^{\infty} e^{ikTz} = \frac{1}{1 - e^{iTz}} = \sum_{j=0}^{\infty} C_j \left(\frac{1}{z}\right)^{-j+1}, \quad |z| < \frac{\pi}{T}.$$

Using lemma 1 and theorem 2 we get the needed result.

## 2. Quasi-asymptotic expansion and convolutions.

**Lemma 1.** *Let $f$ and $g$ belong to $\mathcal{S}'_+$ and have closed quasi- asymptotic expansions of order $\alpha$ and $\beta$ and of lengths $l_1$ and $l_2$, respectively:*

$$f(t) \sim \sum_{i=1}^{p} A_i f_{\alpha_i+1}(t), \quad g(t) \sim \sum_{j=1}^{q} B_j f_{\beta_j+1}(t), \quad t \to +\infty. \tag{2.1}$$

*Then the convolution $h(t) = g * f$ has a closed quasi-asymptotic expansion of order $\gamma = \alpha + \beta + 1$ and of length $l = \min\{l_1, l_2\}$*

$$h(t) \sim \sum_{\alpha_i+\beta_j \geqslant \alpha+\beta-l} A_i B_j f_{\alpha_i+\beta_j+2}(t), \quad t \to +\infty. \tag{2.2}$$

**Proof.** According to lemma 1 of Section 10.1

$$\tilde{f}(z) = \sum_{i=1}^{p} A_i' \left(\frac{1}{z}\right)^{\alpha_i+1} + \epsilon_1(z),$$

$$\tilde{g}(z) = \sum_{j=1}^{q} B_j' \left(\frac{1}{z}\right)^{\beta_j+1} + \epsilon_2(z),$$

where the functions $\epsilon_1(z)$ and $\epsilon_2(z)$ have properties (a) and (b) of lemma 1 of Section 10.1. Multiplying $\tilde{f}(z)$ by $\tilde{g}(z)$ we get

$$\tilde{h}(z) = \sum_{\alpha_i+\beta_j \geqslant \alpha+\beta-l} A_i' B_j' \left(\frac{1}{z}\right)^{\alpha_i+\beta_j+2} + \epsilon_2(z),$$

where the function $\epsilon_3(z)$ consists of summands of the four types:

$$\left(\frac{1}{z}\right)^{\gamma+2}, \quad \gamma < \alpha + \beta - l,$$

$$\epsilon_1(z)\left(\frac{1}{z}\right)^{\delta+1}, \quad \gamma \leqslant \beta,$$

$$\epsilon_2(z)\left(\frac{1}{z}\right)^{\kappa+1}, \quad \kappa \leqslant \alpha,$$

$$\epsilon_1(z)\epsilon_2(z).$$

All these expressions show that $\epsilon_3(z)$ satisfies conditions (a) and (b) of lemma 1 of Section 10.1. So, the closed quasi-asymptotic expansion (2.2) follows. □

**Corollary 1.** The statement of lemma 1 holds if everywhere the closed quasi-asymptotic expansions are replaced by open ones.

Let us consider the convolution equation in:

$$\mathcal{K} * u = h, \ \mathcal{K} \in \mathcal{S}'_+, \ h \in \mathcal{S}'_+. \tag{2.3}$$

For its solution $u(t)$ the following statements about its quasi-asymptotic behaviour hold.

**Lemma 2.** *Let the kernel $\mathcal{K}$ have a quasi-asymptotics of the order $\alpha$ and let the distribution $h$ have a quasi-asymptotics of the order $\beta$. Moreover, let there exist $\gamma < \alpha$, such that*

$$\frac{1}{k^{\gamma}}\left[\mathcal{K}(kt) - C_1 f_{\alpha+1}(kt)\right] \to 0, \ k \to \infty \tag{2.4}$$

*in $\mathcal{S}'_+$. Then the solution $u(t) \in \mathcal{S}'_+$ of the convolution equation (2.3) (if it exists) has a quasi-asymptotics of the order $\beta - \alpha - 1$.*

**Proof.** Taking the Laplace transform of (2.3) we obtain

$$\tilde{\mathcal{K}}(z)\tilde{u}(z) = \tilde{h}(z), \ z = x + iy = re^{i\phi}, \ y > 0. \tag{2.5}$$

According to the general Tauberian theorem and to the hypothesis, $z^{\alpha+1}\tilde{\mathcal{K}}(z)$ and $z_{\beta+1}\tilde{h}(z)$ tend, when $y \to +0$, $x = 0$, to nonzero constants. Therefore, $y^{\beta-\alpha}\tilde{u}(iy)$ tends, when $y \to +0$, to a nonezero constant too. So, condition (A) of the general Tauberian theorem (theorem 1, Section 7.1) holds with $\rho = t^{\beta-\alpha-1}$.

Let us prove estimate (B) of the same theorem. By (2.5) we have

$$t^{\alpha+1}\tilde{\mathcal{K}}(z)z^{\beta-\alpha}\tilde{u}(z) = z^{\beta+1}\tilde{h}(z). \tag{2.6}$$

On the other hand, using lemma 1 of Section 10.1 we obtain

$$\tilde{\mathcal{K}}(z) = C_1'\frac{1}{z^{\alpha+1}} + \epsilon(z),$$

where, according to property (b) of lemma 1 of Section 10.1,

$$|r^{\gamma+1}\epsilon(re^{i\phi})| \leqslant \frac{M}{\sin^m \phi}. \tag{2.7}$$

Denoting the difference $\alpha - \gamma$ by $\delta$, we get from (2.7) the inequality:

$$r^{\alpha+1}|\tilde{\mathcal{K}}(re^{i\phi})| \geqslant |C_1'| - \frac{Mr^{\delta}}{\sin^m \phi}.$$

Let

$$r_1 = \left(\frac{|C_1'|}{2M}\right)^{2/\delta}.$$

Then for $r < \min\{r_1, (\sin\phi)^{(2m/\delta)}\} = r_2$ we have

$$r^{\alpha+1}|\tilde{\mathcal{K}}(re^{i\phi})| \geqslant \frac{|C_1'|}{2}.$$

From this and from (2.6), using the fact that $\tilde{h}(z)$ satisfies estimate (B) of theorem 1 of Section 7.1, we get the proper estimate (B) in the domain $[z = re^{i\phi}, r \leqslant r_2]$. Let now $r < r_1$, but $r > (\sin\phi)^{(2m/\delta)}$. By the fact that $\tilde{u}(z)$ belongs to the class $H(\mathbf{R}_+^1)$, we have for $r < r_1$ (see Section 2.5)

$$|z^{\beta-\alpha}\tilde{u}(z)| \leqslant \frac{M_1}{|y|^{m_1}}.$$

So, we obtain

$$|r^{\beta-\alpha}\tilde{u}(re^{i\phi})| \leqslant \frac{M_1}{(r\sin\phi)^{m_1}} \leqslant \frac{M_2}{(\sin\phi)^{m_1(2m/\delta+1)}}$$

and the estimate (B) for $\tilde{u}(z)$ is proved. Now the statement of the lemma follows from the general Tauberian theorem. □

Now we prove a theorem about the quasi-asymptotic expansion of a solution of a convolution equation in the convolution algebra $\mathcal{S}'_+$.

**Theorem.** *Let* $K \in \mathcal{S}'_+$ *have a closed quasi-asymptotic expansion of order* $\alpha$ *and of length* $l_1 > 0$ *and let* $h \in \mathcal{S}'_+$ *have a closed quasi-asymptotic expansion of order* $\gamma$ *and of length* $l_2 \leqslant l_1$. *Then the solution* $u(t)$ *of the convolution equation* (2.3) *(if it exists) has a closed quasi-asymptotic expansion of order* $\beta = \gamma - \alpha - 1$ *and of length* $l_2$.

**Proof.** By the definition of the quasi-asymptotic expansion (1.2) of Section 10.1 we have

$$(t) = \sum_{i=1}^{p} A_i f_{\alpha_i+1}(t) + \epsilon(l_1, t);$$

$$\frac{1}{k^{\alpha-l_1}}\epsilon(l_1, kt) \to 0,\ k \to \infty \ \text{ in } \mathcal{S}_+^1,$$

$$h(t) = \sum_{j=1}^{q} C_j f_{\nu_j+1}(t) + \epsilon(l_2, t);$$

$$\frac{1}{k^{\gamma-l_2}}\epsilon(l_2, kt) \to 0,\ k \to \infty \ \text{ in } \mathcal{S}'_+. \tag{2.8}$$

Note that the expansion for $h(t)$ ends with the term of order $\gamma - l_2$. By lemma 2 (the hypothesis of which holds because $l_1 > 0$) $u(t)$ has a quasi-asymptotics of order $\beta = \gamma - \alpha - 1$, so that

$$u(t) \sim B_1 f_{\beta+1}(t),\ t \to +\infty,\ B_1 = \frac{C_1}{A_1}.$$

Let us consider the distribution $u_1(t) = u(t) - B_1 f_{\beta+1}(t)$. This satisfies the equation

$$K * u_1 = h_1,\ h_1 = h_1' + \epsilon_1,$$

where

$$h_1(t) = \sum_{j=2}^{q} C_j f_{\nu_j+1}(t) + B_1 f_{\beta+1}(t) * \left|\sum_{i=1}^{p} A_i f_{\alpha_i+1}(t) + \epsilon(l_1, t)\right| + \epsilon(l_2, t),$$

where $h_1'$ is a finite linear combination of $f_{\kappa+1}(t)$ with $\gamma - l_2 \leqslant \kappa < \gamma$ and

$$\frac{1}{k^{\gamma-l_2}}\epsilon_1(kt) \to 0, \ k \to \infty \ \text{in} \ \mathcal{S}'_+.$$

Here we have taken into account that $l_1 \geqslant l_2$. So, the expansion for $h_1(t)$ ends with the term of order $\gamma - l_2$ also and has a length smaller than $l_2$.

If $h_1(t) \not\equiv 0$, then by lemma 2 we can select the next quasi-asymptotic term whose order satisfies the inequalities

$$\gamma - \alpha - 1 - l_2 \leqslant \beta_1 < \beta.$$

Let us consider the distribution $u_2(t) = u_1 - \beta_2 f_{\beta_1+1}(t)$ and the equation $K * u_2 = h_2 = h_2' + \epsilon_2$, where $\epsilon_2$ satisfies condition (2.9). After a finite number of stepts we shall come to a distribution $u_n(t)$ which satisfies the equation

$$K * u_n = \epsilon_n(t), \tag{2.10}$$

where $\epsilon_n(t)$ satisfies (2.9) again. Indeed, if (2.10) were not true, that is

$$K * u_n = h_n = h_n' + \epsilon_n, \ h_n' \not\equiv 0, \ n = 2, 3, \cdots$$

then the distribution $u$ would have an open quasi-asymptotic expansion of order $\gamma - \alpha - 1$ and of length $l \leqslant l_2$ consisting of an infinite number of terms. By corollary 1 the distribution $h(t)$ would have an open quasi-asymptotic expansion of length $l \leqslant l_2$, which by (2.2), would consist of an infinite number of terms too. This would contradict the condition that $h(t)$ has a closed quasi-asymptotic expansion of length $l_2$. Using (2.10) and lemma 2 we conclude that

$$\frac{1}{k^{\gamma-\alpha-1-l_2}}u_n(kt) \to 0, \ k \to \infty \ \text{in} \ \mathcal{S}'_+.$$

But this means that $u(t)$ has a closed quasi-asymptotic expansion of the order of $\beta = \gamma - \alpha - 1$ up to $\beta - l_2$. □

**Corollary.** *Theorem 1 holds if the closed quasi-asymptotic expansion is replaced by an open one.*

Chapter 4

# ASYMPTOTIC PROPERTIES OF SOLUTIONS OF CONVOLUTION EQUATIONS

## 11. Quasi-asymptotics of the fundamental solutions of convolution equations.

1. **Quasi-asymptotics and convolutions.** Let $\Gamma$ be a regular cone in $\mathbf{R}^n$ invariant under a family of linear nonsingular transformations $\{U_k, k \in I\}$, $J_k = \det U_k$, $V_k = U_k^{-1T}$ (see Section 3.1). In the convolution algebra $\mathcal{S}'(\Gamma)$ we consider the equation (see Section 1.9)

$$\mathcal{K} * u = f. \tag{1.1}$$

This subsection is devoted to the determination of certain sufficient conditions on the kernel $\mathcal{K} \in \mathcal{S}'(\Gamma)$ and the distribution $f \in \mathcal{S}'(\Gamma)$ under which there exists a solution $u(\xi)$ of the many-dimensional convolution equation (1.1) in the algebra $\mathcal{S}'(\Gamma)$ possessing a quasi-asymptotics in the cone $\Gamma$ over the family $\{U_k, k \in I\}$ with respect to some function $\rho_k$, $k \in I$.

**Lemma 1.** *Let $f_j \in \mathcal{S}'(\Gamma)$, $j = 1, 2, \ldots, p$, and let each distribution $f_j$ have a quasi-asymptotics in the cone $\Gamma$ over the family $\{U_k, k \in I\}$ with respect to $\rho_k^j$, $k \in I$,*

$$\frac{1}{\rho_k^j} f_j(U_k \xi) \to g_j(\xi) \not\equiv 0, \ k \to \infty, \ k \in I \ \text{in} \ \mathcal{S}'(\Gamma). \tag{1.2}$$

*Then the convolution $f = f_1 * f_2 * \cdots * f_p$ has a quasi-asymptotics in the cone $\Gamma$ over the family $\{U_k, k \in I\}$ with respect to the function $J_k^{p-1} \rho_k^1 \cdots \rho_k^p$, $k \in I$, and*

$$\frac{1}{J_k^{p-1} \rho_k^1 \cdots \rho_k^p} f(U_k \xi) \to g = g_1 * \cdots * g_p, \ k \to \infty, \ k \in I \ \text{in} \ \mathcal{S}'(\Gamma). \tag{1.3}$$

**Proof.** In fact, according to theorem 1 of Section 4.1 from (1.2) follow the limit relations

$$\frac{1}{J_k \rho_k^j} \tilde{f}(iV_k y) \to \tilde{g}_k(iy), \ k \to \infty, \ k \in I, \ y \in C, \ j = 1, \ldots, p,$$

and estimate (B) of theorem 1 of Section 4.1 for any $e \in C$. Therefore

$$\frac{1}{J_k^p \rho_k^1 \cdots \rho_k^p} \tilde{f}(iV_k y) = \prod_{j=1}^{p} \frac{1}{J_k \rho_k^j} \tilde{f}_j(iV_k y) \to$$

$$\to \prod_{j=1}^{p} \tilde{g}_j(iy), \ k \to \infty, \ k \in I, \ y \in C.$$

Estimate (B) for $\tilde{f}(z) = f_1(z) \cdots f_p(z)$ is also satisfied. We again use theorem 1 of Section 4.1 and obtain the limit relation (1.3). □

From lemma 1 there evidently follows

**Corollary.** *Let a $(N \times N)$-matrix function $Z(\xi)$ from $\mathcal{S}'(\Gamma)$ have a quasi-asymptotics $G(\xi)$ in the cone $\Gamma$ over the family $\{U_k, k \in I\}$, with respect to a function $\rho_k$, $k \in I$, and let the $N$-vector function $v(\xi)$ from $'(\Gamma)$ have a quasi-asymptotics $v^0(\xi)$ in the cone $\Gamma$ over the family $\{U_k, k \in I\}$ with respect to a function $\eta_k$, $k \in I$:*

$$\frac{z_{ij}(U_k \xi)}{\rho_k} \to g_{ij}(\xi), \; k \to \infty, \; k \in I \text{ in } \mathcal{S}'(\Gamma), \; i, j = 1, \ldots, N,$$

$$\frac{v_j(U_k \xi)}{\eta_k} \to v_j^0(\xi), \; k \to \infty, \; k \in I \text{ in } \mathcal{S}'(\Gamma), \; j = 1, \ldots, N$$

*where*

$$Z(\xi) = (z_{ij}(\xi)), \; z_{ij} \in \mathcal{S}'(\Gamma),$$

$$G(\xi) = (g_{ij}(\xi)), \; v(\xi) = (v_1(\xi), \ldots, v_n(\xi)),$$

$$v_j \in \mathcal{S}'(\Gamma), \; v^0(\xi) = (v_1^0(\xi), \ldots, v_n^0(\xi)).$$

*Then the vector function*

$$u = Z * v$$

has a quasi-asymptotics $u^0(\xi)$ in the cone $\Gamma$ over the family $\{U_k, k \in I\}$ with respect to the function $J_k \rho_k \eta_k$, $k \in I$, and

$$u^0 = G * v^0.$$

We recall that a vector function (a matrix function) possesses some property if each of its components possesses that property (see Section 0.4). By a convolution of two matrix functions (or a matrix function and a vector function) we understand the matrix function (or resp. vector function) obtained in the usual way but under the condition that all multiplications should be changed into convolutions.

Now we consider the following convolution equation

$$\mathcal{K}_1 * \cdots * \mathcal{K}_p * \mathcal{E} = \delta, \; \mathcal{K}_j \in \mathcal{S}'(\Gamma), \; j = 1, \ldots, p. \tag{1.4}$$

Its solution $\mathcal{E}$ from $\mathcal{S}'(\Gamma)$, if it exists, is called the *fundamental solution* of equation (1.4) (see Section 1.8).

**Lemma 2.** *Let $\mathcal{K}_j \in \mathcal{S}(\Gamma)$ have a quasi-asymptotics in the cone $\Gamma$ over a family $\{U_k, k \in I\}$ with respect to a function $\rho_k^j$, $k \in I$:*

$$\frac{1}{\rho_k^j} \mathcal{K}_j(U_k \xi) \to g_j(\xi) \; 0, \; k \to \infty, \; k \in I \text{ in } \mathcal{S}'(\Gamma), \tag{1.5}$$

*and let $\tilde{\mathcal{K}}_j(z)$ be a function of bounded argument in $T^C$, $j = 1, \ldots, p$. Then the fundamental solution $\mathcal{E}(\xi)$ of equation (1.4) exists in $\mathcal{S}'(\Gamma)$ and it has a quasi-asymptotics in the cone $\Gamma$ over the family $\{U_k, k \in I\}$, with respect to the function $J_k^{-p-1}(\rho_k^1 \cdots \rho_k^p)^{-1}$, $k \in I$,*

$$J_k^{p+1} \rho_k^1 \cdots \rho_k^p \mathcal{E}(U_k \xi) \to \mathcal{E}_0(\xi), \; k \to \infty, \; k \in I \text{ in } \mathcal{S}'(\Gamma), \tag{1.6}$$

*and also $\tilde{\mathcal{E}}_0(z)$ is of bounded argument in $T^C$ and*

$$\tilde{\mathcal{E}}_0(z) = [\tilde{g}_1(z) \cdots \tilde{g}_p(z)]^{-1}, \quad z \in T^C. \tag{1.7}$$

**Proof.** According to (1.5), from theorem1 of Section 4.1 it follows that

$$\frac{1}{J_k \rho_k^j} \tilde{\mathcal{K}}_j(V_k z) \to \tilde{g}_j(z), \quad k \to \infty,$$

$$k \in I, \quad z \in T^C, \quad j = 1, \ldots, p,$$

and the convergence is uniform in any compact set in $T^C$. Since $\mathcal{K}_j(z)$ is of bounded argument in $T^C$, then $g_j(z)$ is also of bounded argument in $T^C$. If we apply the Laplace transform to equality (1.4) (see Section 2.5) then we obtain (the cone $\Gamma$ is regular!)

$$\tilde{\mathcal{E}}(z) = \frac{1}{\tilde{\mathcal{K}}_1(z) \cdots \tilde{\mathcal{K}}_p(z)}, \quad z \in T^C$$

so that the function $\tilde{\mathcal{E}}(z)$ is holomorphic and of bounded argument in $T^C$. Hence, $\mathcal{E} \in \mathcal{S}'(\Gamma)$. Further, as we have shown, for $z \in T^C$ and $k \to \infty$ we have

$$J^p \rho_k^1 \cdots \rho_k^p \tilde{\mathcal{E}}(V_k z) = \frac{1}{\dfrac{\tilde{\mathcal{K}}_1(V_k z) \cdots \tilde{\mathcal{K}}_p(V_k z)}{J_k^p \rho_k^1 \cdots \rho_k^p}} \to$$

$$\to \frac{1}{\tilde{g}_1(z) \cdots \tilde{g}_p(z)} = \tilde{\mathcal{E}}(z), \tag{1.8}$$

then according to theorem 1 of Section 4.4 the limit relation (1.6) and equality (1.7) are valid. □

From lemma 1 and 2 we immediately obtain the following:

**Theorem 1.** *Let $\mathcal{K} \in \mathcal{S}'(\Gamma)$ have a quasi-asymptotics $\mathcal{K}_0(\xi)$ in the cone $\Gamma$ over the family $\{U_k, k \in I\}$ with respect to the function $\rho_k$, $k \in I$ and let $\tilde{\mathcal{K}}(z)$ be of bounded argument in $T^C$. Further, let $f \in \mathcal{S}'(\Gamma)$ have a quasi- asymptotics $g_0(\xi)$ in the cone $\Gamma$ over the family $\{U_k, k \in I\}$ with respect to a function $\eta_k$, $k \in I$. Then the convolution equation (1.1) has a solution $u \in \mathcal{S}'(\Gamma)$ which has a quasi- asymptotics $u_0(\xi)$ in the cone $\Gamma$ over the family $\{U_k, k \in I\}$ with respect to the function $(J_k \rho_k)^{-1}\eta_k$, $k \in I$, and also $u_0$ satisfies the equation*

$$\mathcal{K}_0 * u_0 = g_0. \tag{1.9}$$

2. **Quasi-asymptotics of the fundamental solutions of hyperbolic operators with constant coefficients.** Consider a differential operator of the order $m$, with constant coefficients

$$Q(\partial) = \sum_{|\alpha| \leq m} a_\alpha \partial^\alpha, \quad \sum_{|\alpha| = m} |a_\alpha| \neq 0. \tag{2.1}$$

The operator $Q(\partial)$ is called *hyperbolic with respect to a cone C* if there exists a point $y^0 \in \mathbf{R}^n$ such that

$$Q(y^0 - iz) \neq 0, \quad z \in T^C. \tag{2.2}$$

Then the polynomial $Q(-iz)$ is also called *Hyperbolic.*

The fundamental solution $E(\xi)$ of the operator $Q(\partial)$,

$$Q(\partial)E(\xi) = \delta(\xi).$$

exists in the algebra $\mathcal{S}'(\Gamma+)$, it is unique and it can be represented (see Section 2.5) in the form

$$E(\xi) = e^{(y_0,\xi)}\mathcal{E}(\xi), \tag{2.3}$$

where $(\xi)$ is the fundamental solution of the operator

$$P(\partial) = Q(y^0 + \partial),\ P(-iz) = Q(y^0 - iz),\ z \in T^C, \tag{2.4}$$

$$P(\partial)\mathcal{E}(\xi) = \delta(\xi) \tag{2.5}$$

in the algebra $\mathcal{S}'(\Gamma)$.

We represent the polynomial $P(-iz)$ in the form

$$P(-iz) = \sum_{j=l}^{m} P_j(-iz), \tag{2.6}$$

where $P_j(-iz)$ is a homogeneous polynomial of degree $j$, $m$ is the order of the differential operator $Q(\partial)$ and $l$ is the order of the lowest term $P_l(\partial)$ of $P(\partial)$.

**Lemma 1.** *A hyperbolic polynomial $P(-iz)$ which does not vanish in $T^C$ is of bounded argument in $T^C$.*

**Proof.** In fact, for any $z = x + iy \in T^C$ the polynomial $P(-ix - i\lambda y)$ of degree $m$ in $\lambda$ does not vanish in the upper half-plane $\mathrm{Im}\,\lambda > 0$ and therefore the following expansion holds:

$$P(-ix - i\lambda y) = P_m(-iy)[\lambda - \lambda_1(x,y)] \cdots [\lambda - \lambda_m(x,y)],$$

$$\mathrm{Im}\,\lambda_j(x,y) \leqslant 0,\ z = x + iy \in T^C,\ j = 1,2,\dots,m;$$

and also without loss of generality we can assume that all $\lambda_j(x,y)$ are continuous in $T^C$. From this follows the representation

$$P(-iz) = P_m(-iy)[i - \lambda_1(x,y)] \cdots [i - \lambda_m(x,y)]. \tag{2.7}$$

Further, we note that the homogeneous polynomial $P_m(-iz)$ is also hyperbolic and therefore it is proportional to a polynomail with real coefficients (see L. Hörmander, 1963, ch. 5.5). Now the assertion of the lemma follows from representation (2.7). □

**Theorem 1.** *Let $\Gamma$ be a regular cone which is invariant under a family of linear nonsingular transformations $\{U_k, k \in I\}$ and let $P(-iz)$ be a polynomial which does not vanish in $T^C$, $C = \mathrm{int}\,\Gamma^*$. If there exists a function $\rho_k$, $k \in I$ such that*

$$J_k \rho_k P(V_k y) \to h(iy),\ k \to \infty,\ k \in I,\ y \in \Omega \tag{2.8}$$

*for some domain $\Omega \subset C$ then the fundamental solution $\mathcal{E}(\xi)$ of the hypterbolic operators $P(\partial)$ exists and has a quasi-asymptotics $\mathcal{E}_0(\xi)$ in the cone $\Gamma$ over the family $\{U_k, k \in I\}$, with respect to the function $\rho_k$, $k \in I$; that is,*

$$\frac{1}{\rho_k}\mathcal{E}(U_k\xi) \to \mathcal{E}_0(\xi),\ k \to \infty,\ k \in I \tag{2.9}$$

*in $\mathcal{S}'(\Gamma)$, and, in addition, the following formula is valid*

$$\mathcal{E}_0(z) = \frac{1}{h(z)}, \quad z \in T^C, \tag{2.10}$$

*where the function $h(z)$ is of bounded argument in $T^C$.*

**Proof.** From lemma 1 it follows that the polynomial $P(-iz)$ is of bounded argument in $T^C$ and condition (2.8) shows, according to theorem 1 of Section 4.4, that the distribution $P(\partial)\delta(\xi)$ has a quasi- asymptotics in the cone $\Gamma$ over the family $\{U_k, k \in I\}$ with respect to $(J_k^2 \rho_k)^{-1}$, $k \in I$. Now we use lemma 2 of Section 11.1 to accomplish the proof of the theorem. □

**Corollary.** *Let $\Gamma$ be a regular cone and let $P(\partial)$ be a hyperbolic operator with respect to the cone $C = \operatorname{int}\Gamma^*$ of the form (2.6). Then the fundamental solution $\mathcal{E}(\xi)$ of the operator $P(\partial)$ has a quasi-asymptotics in $\Gamma$ with respect to the function $\rho(k) = k^{l-n}$, $k > 0$,*

$$\frac{1}{k^{l-n}}\mathcal{E}(k\xi) \to \mathcal{E}_0(\xi), \quad k \to \infty \tag{2.11}$$

*in $\mathcal{S}'(\Gamma)$ and the quasi-asymptotics $\mathcal{E}_0(\xi)$ is defined completely by the lowest term $P_l(\partial)$ of the operator $P(\partial)$; namely $\mathcal{E}_0(\xi)$ is the fundamental solution of the hyperbolic operator $P_l(\partial)$,*

$$P_l(\partial)\mathcal{E}_0(\xi) = \delta(\xi). \tag{2.12}$$

**Proof.** In fact, any cone is invariant under the one-parameter group of dilatations $\xi \to k\xi$, $k > 0$ ($J_k = k^n$). For this group, $\rho(k) = k^{l-n}$. Using representation (2.6), we have

$$k^n \rho(k) P\left(\frac{y}{k}\right) = k^l \sum_{j=l}^{m} P_j\left(\frac{y}{k}\right) =$$

$$= \sum_{j=l}^{m} k^{-j+l} P_j(y) \to P_l(y), \quad k \to \infty, \ y \in C.$$

Thus, condition (2.8) of theorem 1 is fulfilled and $h(z) = P_l(-iz)$ and therefore $P_l(-iz)$ is of bounded argument in $T^C$. Hence $P_l(\partial)$ is a hyperbolic operator with respect to the same cone $C$. From (2.9) and (2.10) the limit relation (2.11) and equality (2.12) follow.

As an illustration we consider hyperbolic operators of second order with two independent variables, $m = n = 2$. Without loss of generality we can assume that

$$\Gamma = [(\xi_0, \xi_1) \in \mathbf{R}^2 : \xi_0 > |\xi_1|] = \bar{V}_1^+, \quad \operatorname{int}\Gamma^* = C = V_1^+.$$

By a change of variables it is possible to reduce such hyperbolic operators with constant coefficients to the form

$$Q(\partial) = \partial_0^2 - \partial_1^2 \pm c^2 \quad (c \geqslant 0).$$

We consider several cases.

(1) The *wave equation*, $c = 0$. In this case $Q(\partial) = P_2(\partial) = \partial_0^2 - \partial_1^2$ and the fundamental solution $\mathcal{E}(\xi)$ has a quasi-asymptotics with respect to $\rho(k) = 1$

($l = n = 2$); that is, the quasi-limit $\mathcal{E}_0(\xi)$ coincides with $\mathcal{E}(\xi)$:

$$\mathcal{E}(\xi) = \mathcal{E}_0(\xi) = \frac{1}{2}\theta(\xi_0 - |\xi_1|).$$

(2) The *Klein-Gordon equation*,

$$Q(\partial) = \partial_0^2 - \partial_1^2 + c^2 = P_2(\partial) + P_0(\partial),$$

where $l = 0$, $P_0(-iz) = c^2$. The fundamental solution (see V.S. Vladimirov [1])

$$\mathcal{E}(\xi) = \frac{1}{2}\theta(\xi_0 - |\xi_1|)J_0(c\sqrt{\xi_0^2 - \xi_1^2})$$

has a quasi-asymptotics $\mathcal{E}_0(\xi)$ with respect to $\rho(k) = k^2$,

$$k^2\mathcal{E}(k\xi) \to \frac{1}{c^2}\delta(\xi), \quad k \to \infty \tag{2.13}$$

in $\mathcal{S}'(\bar{V}_1^+)$, such that

$$\mathcal{E}_0(\xi) = \frac{1}{c^2}\delta(\xi), \quad \tilde{\mathcal{E}}_0(z) = \frac{1}{c^2},$$

where $J_0$ is the Bessel function.

(3) The *telegraph equation*,

$$Q(\partial) = \partial_0^2 - \partial_1^2 - c^2,$$

where $y^0 = (c, 0)$ and

$$P(\partial) = Q(y^0 + \partial) = \partial_0^2 - \partial_1^2 + 2c\,\partial_0 = P_2(\partial) + P_1(\partial);$$

that is, $P_1(y) = 2cy_0$, $l = 1$. The fundamental solution for the operator $P(\partial)$

$$\mathcal{E}(\xi) = \frac{1}{2}\theta(\xi_0 - |\xi_1|)I_0(c\sqrt{\xi_0^2 - \xi_1^2})$$

has a quasi-asymptotics $\mathcal{E}_0(\xi)$ with respect to $\rho(k) = k^{-1}$,

$$k\mathcal{E}(k\xi) \to \frac{1}{2c}\theta(\xi_0) \times \delta(\xi_1), \quad k \to \infty \tag{2.14}$$

in $\mathcal{S}'(\bar{V}_1^+)$, such that

$$\mathcal{E}_0(\xi) = \frac{1}{2c}\theta(\xi_0) \times \delta(\xi_1), \quad \tilde{\mathcal{E}}_0(z) = \frac{i}{2cz_0},$$

where $I_0$ is the Bessel function of an imaginary argument.

**3. Quasi-asymptotics of the solutions of the Cauchy problem for the heat equation.** We consider the Cauchy problem for the heat equation

$$\partial_t u = \partial_\xi^2 u, \quad u(0, \xi) = u_0(\xi). \tag{3.1}$$

We are interested in the classical problem of the asymptotic behaviour of the solution $u(t, \xi)$ as $t \to +\infty$ at some fixed point $\xi \in \mathbf{R}^1$ (the problem of the stabilization of the solution). It has been known that even for the class of smooth bounded initial functions $u_0(\xi)$ the solution $u(t, \xi)$ cannot have any limit as $t \to +\infty$ (that is, the solutions are not stabilized). In the class of bounded and even half-bounded initial functions it is possible to show that the existence of a spherical limiting mean for the initial function is a necessary and sufficient condition for the stabilization of the solution of the Cauchy problem (3.1). Essentially the proof of those facts is a reformulation of the well-known Tauberian theorem by Wiener (for details see Yu.N. Drozzinov, 1969).

In this section, using the general (real) Tauberian theorem (theorem 2 of Section 7.1), we establish necessary and sufficient conditions for the existence of a quasi-asymptotics of the solution $u(t,\xi)$ in the class of tempered initial functions.

*The generalized Cauchy problem* for the heat equation and for the initial data $u_0 \in \mathcal{S}'$ is called the problem of determination of a generalized solution $u \in \mathcal{S}'_+(1)$ of the equation

$$\partial_t u = \partial_\xi^2 u + u_0(\xi) \times \delta(t). \tag{3.2}$$

This problem has the unique solution $u(t,\xi)$, which by virtue of hypo- ellipticity of the heat conductivity operator, is infinitely differentiable in $t > 0$ and is a real analytic function in $\xi$ belonging to $\mathcal{S}'(\mathbf{R}^1)$ for each fixed $t > 0$. If in equatin (3.2), we perform the Laplace transform with respect to $t$, and then solve the obtained differential equation, we have for the distribution $\tilde{u}(z,\xi) = L_t[u]$

$$\tilde{u}(z,\xi) = \frac{1}{2\sqrt{-z}} e^{-\sqrt{-iz}\cdot|\xi|} * u_0(\xi), \quad z = x+iy, \; x \in \mathbf{R}^1, \; y > 0. \tag{3.3}$$

Henceforth we choose that branch of the square root $\sqrt{-iz}$ which is positive for $z = iy, y > 0$.

**Theorem 1.** *The solution $u(t,\xi)$ of the generalized Cauchy problem* (3.2) *has the following quasi-asymptotics as $t \to +\infty$ with respect to an automodel function $\rho(k)$ of order $\alpha$,*

$$\frac{1}{\rho(k)} u(kt,\xi) \to A(\xi) \times f_{\alpha+1}(t), \; k \to \infty \; in \; \mathcal{S}', \; \xi \in K \tag{3.4}$$

*where $K$ is any compact set in $\mathbf{R}^1$, if and only if the following conditions are fulfilled:*

$$\text{(i)} \lim_{z\to+0} \frac{r}{\rho\left(\frac{1}{r}\right)} e^{-r|\xi|} * u_0(\xi) = A(\xi), \; \xi \in K. \tag{3.5}$$

(ii) *there exists numbers $N \geq 0$, $M > 0$, and $r_0 > 0$, depending only on the compact set $K$, such that*

$$\frac{r^{m+1}}{\rho\left(\frac{1}{r}\right)} \left| |\xi|^m e^{-r|\xi|} * u_0(\xi) \right| \leq M 2^{m/2} m^N m!, \tag{3.6}$$

$$0 < r \leq r_0, \; \xi \in K, \; m = 1,2, \cdots$$

In the proof of the theorem we use the following lemmas.

**Lemma 1.** *Let $g(z)$ be holomorphic in the circle $|z| < d$ and satisfy the estimate: there exist numbers $M > 0$ and $N \geq 0$, such that*

$$|g(z)| \leq M(d - |z|)^{-N}, \quad |z| < d. \tag{3.7}$$

*Then there exists a number $C_n$ which does not depend on $d$, $M$ and $m$, such that the following estimates are valid*

$$|g^{(m)}(0)| \leqslant MC_N \frac{m!}{d^{m+N}} m^N, \quad m = 1, 2, \cdots \tag{3.8}$$

*Conversely, if the estimates*

$$|g^{(m)}(0)| \leqslant M_1 \frac{m!}{N!d^{m+N+1}} m^N, \quad m = 1, 2, \ldots, \quad |g(0)| \leqslant \frac{M_1}{d^{N=1}} \tag{3.8'}$$

*are satisfied then $g(z)$ satisfies the estimate*

$$|g(z)| \leqslant M_1(d - |z|)^{-N-1}, \quad |z| < d. \tag{3.7'}$$

**Proof.** Let estimate (3.7) be satisfied. By the Cauchy formula we have

$$\frac{1}{m!} g^{(m)}(0) = \frac{1}{2\pi i} \int_{|\zeta|=r} \frac{g(\zeta)\, d\zeta}{\zeta^{m+1}}, \quad m = 0, 1, \cdots$$

from which follows estimate (3.8)

$$\left| \frac{1}{m!} g^{(n)}(0) \right| \leqslant \min_{0 \leqslant r \leqslant d} \frac{M}{(d-r)^N r^m} = MC_N m^N d^{-m-N}, \quad m = 1, 2, \cdots$$

for some

$$C_N \leqslant \frac{1}{N^N} e^{N+N^2}.$$

Conversely, let estimate (3.8') be satisfied. Using the Taylor expansion formula we get, for the function $g(z)$, estimate (3.7'):

$$|g(z)| = \left| \sum_{m=0}^{\infty} \frac{g^{(m)}(0)}{m!} z^m \right| \leqslant M_1 \frac{1}{N 1 d^{N+1}} \sum_{m=1}^{\infty} m^N \left( \frac{|z|}{d} \right)^m + \frac{M_1}{d^{N+1}} \leqslant$$

$$\leqslant \frac{M_1}{(d - |z|)^{N+1}}.$$

□

**Lemma 2.** *Let $F(z)$ be holomorphic in the upper half-plane $y > 0$, and let $\rho$ be an automodel function. The function $\Phi(z) = F(i\sqrt{-iz})$ satisfies the estimate:*

(B) *There exists numbers $M > 0$, $N \geqslant 0$ and $y_0 > 0$ such that*

$$\frac{y^{m+1}}{\rho\left(\frac{1}{y}\right)} |\Phi^{(m)}(iy)| \leqslant Mm!m^N, \quad 0 < y \leqslant y_0, \; m = 1, 2, \ldots, \tag{3.9}$$

*if and only if there exist numbers $M_1 > 0$, $N_1 > 0$ and $y_1 > 0$, such that*

$$\frac{y^{m+1}}{\rho\left(\frac{1}{y}\right)} |F^{(m)}(iy)| \leqslant M_1 2^{m/2} m! m^{N_1}, \quad 0 < y \leqslant y_1, \; m = 1, 2, \cdots \tag{3.10}$$

**Proof.** Let condition (B) be satisfied. From this as in the proof of theorem 2 of Section 7.1, we obtain for the function $\Phi(z)$ the estimate

$$\frac{r}{\rho\left(\frac{1}{r}\right)} |\Phi(re^{i\phi})| \leqslant \frac{M_2}{\sin^{N_1}\phi}, \quad 0 < r \leqslant r_0 = y_0 \tag{3.11}$$

for some $M_2 > 0$ and $N_1 \geqslant 0$. Hence, for the function $F(z)$ we have

$$\frac{r}{\rho\left(\frac{1}{r}\right)} |F(re^{i\phi})| \leqslant \frac{M_3}{\left[\left(\phi - \frac{\pi}{4}\right)\left(\frac{3\pi}{4} - \phi\right)\right]^{N_1}}, 0 < r \leqslant y_2, \frac{\pi}{4} < \phi < \frac{3\pi}{4} \tag{3.12}$$

for some $M_3 > 0$ and $y_2 > 0$. Fix the point $iy$, where $0 < y < y_2$. Then from (3.12) we get that the function $F(z)$ is holomorphic in the circle of radius $d = y/\sqrt{2}$ with its centre at the point $iy$, and satisfies the estimate, for some $M_4 > 0$,

$$|F(re^{i\phi})| \leqslant \frac{M_4}{\Delta_\phi^{N_1}} \frac{\rho\left(\frac{1}{y}\right)}{y} \frac{\rho\left(\frac{1}{r}\right)}{\rho\left(\frac{1}{y}\right)} \frac{y}{r},$$

where $\Delta_\phi$ is the angular distance from the point $z = re^{i\phi}$ to the boundary of the sector $\pi/4 < \phi < 3\pi/4$. Now if $z = re^{i\phi}$ lies in the circle $|z - iy| = d = y/\sqrt{2}$ then there exists a positive number $c_i$, such that

$$\Delta_\phi > c_1 \frac{d - |z|}{d}, \quad c_2 r < y < c_3 r$$

and hence, in view of the automodelity of $\rho$ (see Section 3.2),

$$c_4 \rho\left(\frac{1}{y}\right) < \rho\left(\frac{1}{r}\right) < c_5 \rho\left(\frac{1}{y}\right), \quad 0 < y \leqslant y_0.$$

By continuing this inequality for some $M_5 > 0$ we get

$$|F(re^{i\phi})| \leqslant M_5 \left(\frac{d}{d - |z|}\right)^{N_1} \frac{\rho\left(\frac{1}{y}\right)}{y}.$$

Using lemma 1 for $g = F$, $N = N_1$, $d = y/\sqrt{2}$, $M = M_5 d^{N_1}(\rho(1/y)/y)$ we obtain estimate (3.10).

Conversely, let condition (3.10) be satisfied. Again, using lemma 1 we conclude that the estimate of the form (3.12) is fulfilled, and hence estimate (3.10) is also valid. As in the proof of theorem 2 of Section 7.1, we conclude that the estimate of the form (3.9) is valid. □

**Proof of theorem 1.** The statements of this theorem follow from theorem 2 of Section 7.1, in which the corresponding estimates have to be changed by estimate (3.10) of lemma 2. □

Now suppose that $u_0$ is a continuous, tempered function in $\mathbf{R}^1$. Then formula (3.3) for a fixed $\xi = \xi_0$ takes the form

$$\tilde{u}(z, \xi) = \frac{1}{2\sqrt{-iz}} \int e^{-\sqrt{-iz}\,|\eta|} u_0(\xi_0 - \eta)\, d\eta =$$

$$= \frac{1}{\sqrt{-iz}} \int_0^\infty e^{-\eta\sqrt{-iz}} \frac{u_0(\xi_0 - \eta) + u_0(\xi_0 + \eta)}{2}\, d\eta. \tag{3.13}$$

Let $\rho(k) = 1$. In this case theorem 1 takes the following form. *The solution $u(t,\xi)$ of the Cauchy problem* (3.1) *has a quasi-limit* $A \neq 0$ as $t \to +\infty$ at the point $\xi_0$, if and only if the following conditions are fulfilled:*

(i) $$\lim_{r\to+0} r\int e^{-r|\eta|}u_0(\xi_0-\eta)\,d\eta = A, \tag{3.14}$$

(ii) *there exists numbers $M > 0$, $N \geqslant 0$ and $r_0 > 0$, such that*

$$\left|r^{m+1}\int |\eta|^m e^{-r|\eta|}u_0(\xi_0-\eta)\,d\eta\right| \leqslant M\,2^{m/2}m^N m!,\quad 0<r\leqslant r_0,\ m=1,2,\ldots \tag{3.15}$$

Note that condition (3.15) can be interpreted as the boundedness of the Borel moments of the function $u(\xi)$ at the point $\xi_0$.

Condition (3.15) is essential, as the following example shows.

**Example 1.** *There exists a continuous, tempered function $u_0(\xi)$ with support in* $[0,\infty)$ *which satisfies condition* (3.14) *and for which the solution $u(t,\xi)$ of the Cauchy problem* (3.1) *does not have a quasi-limit as $t \to +\infty$ at the point $\xi = 0$.*

At first we note that for such a $u_0$ formula (3.13) has the form (for $\xi_0 = 0$)

$$\begin{aligned}\tilde{u}(z,0) &= \frac{1}{2\sqrt{-iz}}\int_{-\infty}^{\infty} e^{-\sqrt{-iz}\,|\eta|}u_0(-\eta)\,d\eta = \\ &= \frac{1}{2\sqrt{-iz}}\int_0^{\infty} e^{-\sqrt{-iz}\,\xi}u_0(\xi)\,d\xi = \\ &= \frac{i}{2i\sqrt{-iz}}L[u_0](i\sqrt{-iz}) = \\ &= \frac{1}{2}L[u_0^{(-1)}](i\sqrt{-iz}).\end{aligned} \tag{3.16}$$

Consider the function

$$A_\alpha(z) = \exp\left[-\frac{i}{2}\ln^2(e^{i\alpha}z)\right],\quad z = re^{i\phi},\ -\frac{\pi}{2}<\phi<\frac{3\pi}{2} \tag{3.17}$$

which is analytic in the plane $z$ with the cut $(-i\infty, i0)$, and which satisfies the condition

$$\begin{aligned}|A_\alpha(z)| &= \left|\exp\left\{-\frac{i}{2}[\ln^2 r + 2i(\phi+\alpha)\ln r - (\phi+\alpha)^2]\right\}\right| = \\ &= \exp[(\phi+\alpha)\ln r] = r^{\phi+\alpha} = \\ &= |z|^{\phi+\alpha},\quad -\frac{\pi}{2}<\phi<\frac{3\pi}{2}.\end{aligned} \tag{3.18}$$

Now from (3.17) and (3.18) it follows that the function

*See Section 3.1.

$$\Psi_N(\zeta) = (-1)^{N+1}\frac{1 + A_{-(\pi/4)}(-i\zeta^2)}{\zeta^2(\zeta^2 - 1)^N}$$

is analytic in the upper half-plane and satisfies the estimate

$$|\Psi_N(\zeta)| \leqslant 4^{-N}\tau^{-N-2-(3\pi/2)}(1 + \rho^2)^{-N+6}, \quad \zeta = \rho e^{i\theta} = \omega + i\tau. \tag{3.19}$$

As the initial function in (3.1) we take the function $u_{0N}(\xi)$ from $\mathcal{S}'_+$ such that the Laplace transform of its first primitive is equal to $2\psi_N$; that is,

$$\frac{1}{2}L[u_{0N}^{(-1)}](\zeta) = \Psi_N(\zeta). \tag{3.20}$$

In view of estimate (3.19) for a sufficiently large $N$ (e.g. $N > 20$), the function $u_{0N}(\xi)$ is continuously differentiable and tempered. According to formula (3.16) the Laplace transform in $t$ of the solution $u_N(t,\xi)$ of the Cauchy problem (3.1) for $u_0 = u_{0N}$ at point $\xi = 0$ is equal to

$$\tilde{u}_N(z,0) = \Psi_N(i\sqrt{-iz}) = (-1)^{N+1}\frac{1 + A_{-(\pi/4)}(z)}{iz(iz - 1)^N}, \quad z = x + iy, \ y > 0.$$

Condition (3.14) for $u_{0N}$ is fulfilled:

$$\lim_{y\to+0} y\tilde{u}_N(iy,0) = \lim_{y\to+0}(-1)^N\frac{1 + A_{-(\pi/4)}(iy)}{(-1 - y)^N} = 1 \neq 0.$$

However, condition (3.15) is not satisfied. In fact, according to lemma 2 this condition is equivalent to conditon (3.9), which, in turn, is equivalent to estimate (3.11) for $\Phi = \tilde{u}_N(z,0)$ and $\rho(k) \equiv 1$; that is,

$$|r\tilde{u}_N(re^{i\phi},0)| \leqslant \frac{M_1}{\sin^p\phi}, \quad 0 < r \leqslant r_0$$

for some numbers $M_1 > 0$, $p \geqslant 0$ and $r_0 > 0$. In particular, along each ray $\arg z = \phi$, $0 < \phi < \pi$, there must exist the limit

$$\lim_{r\to+0} r\tilde{u}_N(re^{i\phi},0) = 1.$$

But, for $0 < \phi < \pi/4$, e.g. $\phi = \pi/8$, this limit does not exists, because, by virtue of (3.18),

$$\lim_{r\to+0}|r\Psi_N(i\sqrt{-ire^{i\phi}})| = \lim_{r\to+0}|1 + A_{-(\pi/4)}(re^{i\phi})| =$$
$$= \lim_{r\to+0} r^{-(\pi/8)} = \infty.$$

According to what has been said it follows that the function $u_N(t,0)$ cannot have a quasi-limit as $t \to +\infty$ at the point $\xi = 0$, but does not have a usual limit.

**Example 2.** *There exists a continuous, tempered function $u_0(\xi)$ with support in $[0,\infty)$, such that the solution $u(t,\xi)$ of the Cauchy problem (3.1) has a quasi-limit for $t \to +\infty$ at the point $\xi = 0$, but does not have a usual limit.*

As an example of such an initial function $u_0$ we take the function $u_{0N}$ for which the Laplace transform of its first primitive is equal to

$$2\Phi_N(\zeta) = \frac{2(-1)^{N+1}[1 + A_0(-i\zeta^2)\ln(-i\zeta^2)]^2}{\zeta^2(\zeta^2 - 1)^N}$$

where the function $A_0(-i\zeta^2)$ is defined by formula (3.17). Thus, according to formula (3.16), we have

$$\tilde{u}_N(z,0) = \frac{1}{2}L[u_{0N}^{(-1)}](i\sqrt{-iz}) = \Phi_N(i\sqrt{-iz}) =$$

$$= \frac{(-1)^{N+1}[1 + A_0(z)\ln z]^2}{iz(iz-1)^N}.$$

$$z = re^{i\phi} = x + iy,\ y > 0. \tag{3.21}$$

Now we show that the solution $u_N(t,\xi)$ of the Cauchy problem for $u_0 = u_{0N}$ has a quasi-limit for $t \to +\infty$ at the point $\xi = 0$. For this, in view of theorem 1 of Section 7.1, we have for the function $\tilde{u}_N(z,0)$ (when $\rho(k) \equiv 1$)

$$\text{(A)}\ \lim_{y\to+0} y\tilde{u}_N(iy,0) = \lim_{y->+0}[1 + A_0(iy)\ln(iy)] = 1.$$

From the limit relation there immediately follows

$$|A_0(z)||\ln z| = r^{\phi}\sqrt{\phi^2 + \ln^2 r} \to 0,\ r \to +0,\ 0 < \phi < \pi.$$

Now we are going to prove condition (B): there exist numbers $M_2 > 0$, $m \geqslant 0$, and $r_0 > 0$, such that

$$|r\tilde{u}(re^{i\phi},0)] \leqslant \frac{M_2}{\sin^m\phi},\quad 0 < r \leqslant r_0. \tag{3.22}$$

We estimate $A_0(z)\ln z$ for $|z| = r < 1/200$, $0 < \phi < \pi$. We have

$$|A_0(re^{i\phi})\ln(re^{i\phi})| = r^{\phi}\sqrt{\ln^2 r + \phi^2}$$

therefore

$$r^{\phi}|\ln r| \leqslant |A_0(re^{i\phi})\ln(re^{i\phi})| \leqslant 2r^{\phi}|\ln r|. \tag{3.23}$$

The function $\eta(r) = -r^{\phi}\ln r$ for $0 < r < 1/200$, $0 < \phi < \pi$, has the maximum for $r = r^{-1/\phi}$ and

$$\max_{0<r<1/200}\eta(r) = \frac{1}{e\phi},$$

so that

$$|A_0(re^{i\phi})\ln(re^{i\phi})| \leqslant \frac{2}{e\phi},\quad 0 < r < 1/200,\ 0 < \phi < \pi. \tag{3.24}$$

Now from equality(3.21) and from estimate (3.24) we have

$$|r\tilde{u}_N(re^{i\phi},0)| \leqslant \left|\frac{2^{N+3}}{\phi}\right|^2 \leqslant \frac{2^{2N+6}}{\sin^2\phi},\quad 0 < r < 1/200,\ 0 < \phi < \pi. \tag{3.25}$$

In addition, for $r = e^{-(1/\phi)}$ we have the estimate

$$|r\tilde{u}_N(re^{i\phi},0)| \geqslant \frac{1}{100\sin^2\phi},\quad 0 < r = e^{-(1/\phi)} < \frac{1}{200}. \tag{3.26}$$

Thus from estimate (3.25) we obtain estimate (3.22) for $m = 2$. As has been noted, from this there follows the existence of a quasi-limit as $t \to +\infty$ for the function $u_N(t,0)$.

Now we show that the function $u_N(t,0)$ does not have a usual limit as $t \to +\infty$; moreover, it is bounded as $t \to +\infty$. For this we use the following

**Lemma 3.** *Let $u \in \mathcal{S}'_+$, and there exist numbers $M > 0$ and $T_0$ such that $|u(t)| < M$, $t > T_0$. Then the Laplace transform of $u$ satisfies the following estimation for some $c > 0$*

$$|r\tilde{u}(re^{i\phi})| \leqslant \frac{c}{\sin\phi}, \quad 0 < \phi < \pi, \quad 0 < r < 1. \tag{3.27}$$

**Proof.** Let a function $\eta(t)$ from $\mathcal{D}(-(1/2), 2T_0)$ be such that $\eta(t) \equiv 1$ for $0 \leqslant t \leqslant T_0$ and $|1 - \eta(t)| \leqslant 1$, $t \in \mathbf{R}^1$. Then we have (3.27):

$$|r\tilde{u}(re^{i\phi})| = \left| r\left(u(t), e^{itre^{i\phi}}\right)\right| \leqslant$$

$$\leqslant \left| r\left(u(t), \eta(t)e^{itre^{i\phi}}\right)\right| + \left| r\left(u(t), [1-\eta(t)]e^{itre^{i\phi}}\right)\right| \leqslant$$

$$\leqslant rc_1 \left\| \eta(t)e^{itre^{i\phi}} \right\|_{C^m([-(1/2), 2T_0])} + r\left| \int_0^\infty [1-\eta(t)]u(t)e^{itre^{i\phi}}dt \right| \leqslant$$

$$\leqslant c_2 + rM\int_0^\infty e^{-rt\sin\phi}dt \leqslant \frac{c}{\sin\phi}$$

for some positive $c_1, c_2$ and $c$. Here we used estimate (2.1) of Section 1.2. □

Continuing our arguments we show that the function $u_N(t,0)$ has no limit as $t \to +\infty$, otherwise it would be bounded as $t \to +\infty$, and by lemma 3 its Laplace transform $\tilde{u}_N(z,0)$ would satisfy estimate (3.27) which contradicts estimation (3.26) obtained above.

Theorem 1 obviously can be generalized onto the case of many spatial variables and it is also true for more general classes of parabolic (ultra- parabolic) equations with constant coefficients. For a generalized Cauchy problem for the ultra-parabolic equation and the stabilization of its solution see the papers by V.S. Vladimirov and Yu.N. Drozzinov (1967), and by Yu.N. Dorzzinov (1969).

## 12. Quasi-asymptotics of passive operators.

In this section we consider asymptotic properties of solutions of a system of convolution equations which is passive relative to some cone.

**1. The translationally-invariant passive operators.** We consider a physical system obeying the following scheme. Suppose that the original in-disturbance $u(\xi) = (u_1(\xi), \ldots, u_N(\xi))$ is acting on the system, and as its result there arises an out-disturbance, the response of the system, $h(\xi) = (h_1(\xi), \ldots, h_N(\xi))$. Here $\xi = (\xi_1, \xi_2, \ldots, \xi_n)$ are understood as the time, space and other variables. Suppose that the following conditions have been fulfilled:

(a) *Linearity:* if to the original disturbances $u$ and $v$ there correspond disturbances $h$ and $g$, then to their linear combinations $\alpha u + \beta v$ there corresponds the response disturbance $\alpha h + \beta g$.

(b) *Reality:* if the original disturbance $u$ is real then the response disturbance $h$ is also real.

(c) *Continuity:* if all components of the original disturbance $u(\xi)$ tend to zero in $\mathcal{E}'$, then all components of the response disturbance $h(\xi)$ also tend to zero in $\mathcal{D}'$.

(d) *Translational-invariance:* if to the original disturbance $u(\xi)$ there corresponds a disturbance $h(\xi)$, then, for any translation $a \in \mathbf{R}^n$, to the original disturbance $u(\xi + a)$ there corresponds the response disturbance $h(\xi + a)$.

Conditions (a) - (d) are equivalent to the existence of a unique $(N \times N)$-matrix:

$$z(\xi) = (z_{ij}(\xi), i, j = 1, 2, \ldots, N), \quad z_{ij} \in \mathcal{D}'(\mathbf{R}^n)$$

which connects the original $u(\xi)$ disturbance and the response disturbance $h(\xi)$ by the formula (see Section 1.7)

$$Z * u = h. \tag{1.1}$$

Let us impose on the system (1.1) yet another requirement the so-called condition of *passivity relative to the cone* $\Gamma$. Suppose that $\Gamma$ is a closed convex solid cone in $\mathbf{R}^n$, with its vertex at 0.

(e) *Passivity relative to the cone* $\Gamma$*:* for any vector function $\phi(\xi)$ from $\mathcal{D}^{\times N}$ - that is, $\phi = (\phi_1, \ldots, \phi_N)$, $\phi_j \in \mathcal{D}$, $j = 1, \ldots, N$ (see Section 0.4) - the following inequality holds

$$\operatorname{Re} \int_{-\Gamma} < Z * \phi, \ \phi > d\xi \geqslant 0, \tag{1.2}$$

where $<, >$ is the scalar product in $\mathbf{C}^N$ (see Section 0.1).

Note that the function $<Z * \phi, \phi>$ is finite and infinitely differentiable (see Section 1.7), so that the integral in (1.2) always exists. Furthermore, because of the reality of the matrix $Z(\xi)$ the condition of passivity (1.2) is equivalent to the condition

$$\int_{-\Gamma} <Z * \phi, \phi> d\xi \geqslant 0, \quad \phi \in {}_r^{\times N}, \tag{1.3}$$

where $\mathcal{D}_r^{\times N}$ consists of real $N$-vector functions with components in $\mathcal{D}$.

The inequality (1.3) is of the energy type: it reflects the ability of a physical system to absorb and redistribute energy, but not to generate it. Here, the causality relative to the cone $\Gamma$ is taken into account.

The convolution operator $Z*$ is called a *passive translationally-invariant operator relative to the cone* $\Gamma$.

To illustrate the proposed scheme, let us consider a one-dimensional translationally-invariant passive system ($n = N = 1$): an elementary electric circuit consisting of a resistance $L$, a self-inductance $L$, a capacitance $C$, and a source of electromotive force $e(t)$ switched on at time $t = 0$ (Fig. 15).

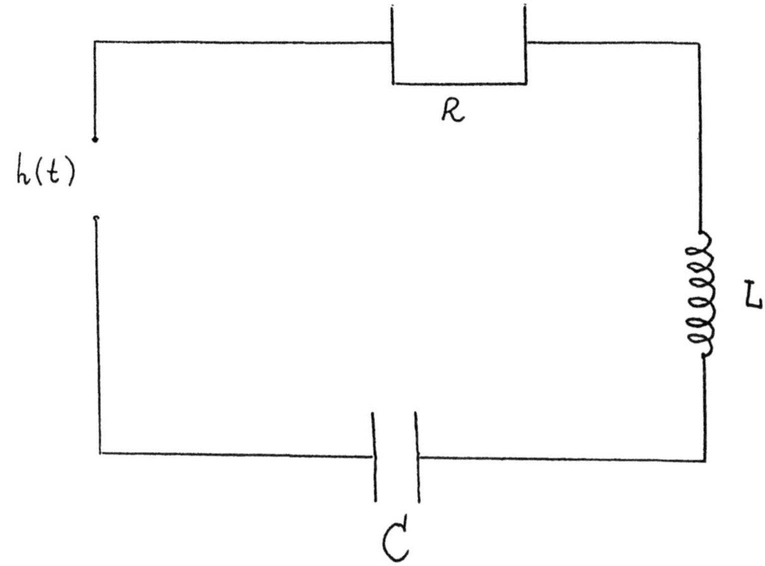


Figure 15.

Then, by the Kirchhoff law, the current $i(t)$ in the circuit satisfies the integro-differential equation

$$L\frac{di}{dt} + Ri + \frac{1}{C}\int_0^t i(\tau)\,d\tau = e(t);$$

that is,

$$Z * i = e,$$

where

$$Z(t) = L\,\delta'(t) + R\,\delta(t) + \frac{1}{C}\theta(t)$$

is the generalized resistance of the circuit. We not verify that the operator $Z*$ satisfies the condition of passivity (1.3) relative to the cone $\Gamma = [0, \infty)$:

$$\int_{-\infty}^{0}(Z * \phi)\phi\,dt = \int_{-\infty}^{0}\left[L\phi'(t) + R\phi(t) + \frac{1}{C}\int_0^t \phi(\tau)\,d\tau\right]\phi(t)\,dt =$$

$$= \frac{L}{2}\phi^2(0) + R\int_{-\infty}^{0}\phi^2(t)\,dt + \frac{1}{2C}\left|\int_{-\infty}^{0}\phi(t)\,dt\right|^2 \geqslant 0, \quad \phi \in \mathfrak{D}_\Gamma.$$

One-dimensional ($n = 1$) linear passive systems describe the relationship between the currents and the voltage in complex electric circuits. They also describe linear thermodynamic systems, the scattering of electromagnetic waves and elementary particles. One-dimensional passive operators have been studied by many authors and the results of their investigations have been summarized in two monographs: A. Zemanian (1965), and E.J. Beltrami and M.R. Wholers (1966).

Many-dimensional ($n \geqslant 2$) linear passive systems are frequently encountered in mathematical physics: they describe physical systems in which their space-time dynamics have been taken into account.

In this subsection we list, without proofs, the main results of the theory of many-dimensional translationally-invariant passive systems. For a detailed exposition see the book by V.S. Vladimirov (1979). From conditions (a) - (e) the following properties follows.

(1) *Dissipativity:*

$$\mathrm{Re}\int \langle Z * \phi, \phi \rangle \, d\xi \geqslant 0, \quad \phi \in \mathcal{D}^{\times N}.$$

(2) *Causality with respect to the cone* $\Gamma$*:*

$$\mathrm{supp}\, Z \subset \Gamma.$$

(3) *Positive definiteness (by Bochner):*

$$\langle Z(\xi)a + \bar{Z}^T(\xi)a, a \rangle \gg 0, \quad a \in C^N.$$

(4) *Growth condition:*

$$Z \in (\mathcal{D}')^{\times N}.$$

(5) *Existence of the Laplace transform:*

$$L[Z] = \tilde{Z}(z), \quad z \in T^C,$$

where the $(N \times N)$-matrix function $\tilde{Z}(z)$ is holomorphic in the tube domain $T^C$, $C = \mathrm{int}\,\Gamma^*$. The matrix function $\tilde{Z}$ is called the *impedance* of the physical system.

(6) *Reality of the impedance:* if the cone $\Gamma$ is acute, then

$$\tilde{Z}(z) = \tilde{Z}(-\bar{z}), \quad z \in T^C,$$

or, equivalently, the matrix function $\tilde{Z}(iy)$ is real for $y \in C$.

(7) *Positivity of the impedance:*

$$\mathrm{Re}\,\tilde{Z}(z) \geqslant 0, \quad z \in T^C.$$

**Definition.** A matrix function $M(z)$ is called *positive real* in a tube domain $T^C$ if it is holomorphic in $T^C$ and satisfies the conditons

$$\mathrm{Re}\, M(z) = \frac{1}{2}[M(z) + \bar{M}^T(z)] \geqslant 0, \quad z \in T^C, \tag{1.4}$$

$$M(iy) \text{ is real for } y \in C. \tag{1.5}$$

**Theorem 1.** (V.S. Vladimirov, 1979). *A matrix* $Z(\xi)$ *defines a passive operator relative to an acute cone* $\Gamma$, *if and only if its impedance* $\tilde{Z}(z)$ *is a positive real matrix function in the tube domain* $T^C$, *where* $C = \mathrm{int}\,\Gamma^*$.

**2. The fundamental solution and the Cauchy problem.** The fundamental solution of the passive operator $Z*$ relative to the cone $\Gamma$ is a matrix

$A = (a_{ij}(\xi), i, j = 1, \ldots, N)$,

$a_{ij} \in \mathcal{D}'$, which satisfies the convolution matrix equation

$$Z * A = I\,\delta(\xi), \tag{2.1}$$

where $I$ is the unit $(N \times N)$-matrix. (By the convolution of matrices we understand the usual product of these matrices in which multiplication of their components is replaced by the convolution.)

The operator $A^*$ is also said to be the *inverse operator* to the operator $Z^*$, and the matrix function $\tilde{A}(z)$, the Laplace transform of the matrix $A(\xi)$, if it exists, is called the *admittance* of the physical system.

The passive operator $Z^*$ relative to the acute cone $\Gamma$ is said to be *nonsingular* if

$$\det \tilde{Z}(z) \neq 0, \quad z \in T^C, \tag{2.2}$$

where $C = \operatorname{int}\Gamma^*$. The passive operator $Z^*$ relative to the acute cone $\Gamma$ is said to be *completely nonsingular* if there exists a point $z^0 \in T^C$ such that

$$\operatorname{Re} \tilde{Z}(z^0) > 0. \tag{2.3}$$

Now we prove the assertion: *if a passive operator $Z^*$ relative to an acute cone $\Gamma$ is completely nonsingular then*

$$\operatorname{Re} \tilde{Z}(z) > 0, \quad z \in T^C. \tag{2.4}$$

**Proof.** Indeed, for any vector $a \in \mathbb{C}^n$, according to theorem 1 of Section 12.1, the function $<\tilde{Z}(z)a, a>$ is holomorphic, and $\operatorname{Re}<\tilde{Z}(z)a, a> \geqslant 0$ in $T^C$. Then by (2.3) the following inequality holds

$$\operatorname{Re} <\tilde{Z}(z^0)a, a>> 0 \quad \text{if } a \neq 0.$$

As is well known (see, for example, V.S. Vladimirov, 1979, Section 16.1), if a real part of some function, which is holomorphic and has a nonnegative real part, vanishes at least at one point then such a function is equal to an imaginary constant. Hence, the following inequality holds

$$\operatorname{Re} <\tilde{Z}(z)a, a>> 0, \quad z \in T^C \quad \text{if } a \neq 0$$

and thus inequality (2.4) holds. □

It is clear that every completely nonsingular passive operator is a nonsingular passive operator relative to the same cone. The converse statement is not true (see example 6, Section 12.5).

**Theorem.** *Let $Z^*$ be a passive operator relative to an acute cone $\Gamma$. If there exists a point $z^0 \in T^C$ such that*

$$\det \tilde{Z}(z^0) \neq 0$$

*then $Z^*$ is a nonsingular passive operator; that is, condition* (2.2) *is fulfilled.*

**Proof.** First we note that if $\operatorname{Re}\tilde{Z}(z^0) > 0$ - that is if the operator $Z^*$ is completely nonsingular - then it is also nonsingular. Therefore, we suppose that the Ker of the operator $\operatorname{Re}\tilde{Z}(z^0)$ is not empty, and let

$$\operatorname{Ker} \operatorname{Re} \tilde{Z}(z^0) = H_2 \subset \mathbf{C}^N, \quad \mathbf{C}^N = H_1 \oplus H_2$$

so that

$$\operatorname{Re} < \tilde{Z}(z^0)a, a > \ > 0, \ a \in H_1, \ a \neq 0.$$

Now we prove that

$$\operatorname{Ker} \tilde{Z}(z) = H_2, \ z \in T^C. \tag{2.5}$$

In fact, if $b \in H_2$, then $\operatorname{Re} < \tilde{Z}(z^0)b, b> = 0$, and hence $\operatorname{Re} <\tilde{Z}(z)b, b> = 0$, $z \in T^C$; since the function $<\tilde{Z}(z)b, b>$ is holomorphic in $T^C$, its real part is non-negative in $T^C$ and it vanishes at the point $z^0 \in T^C$. Now from the polarization formula: if $a_1$ and $a_2$ are in $\mathbf{C}^N$ then

$$<\operatorname{Re} \tilde{Z}(z)a_1, a_2> = \frac{1}{4}\operatorname{Re} <\tilde{Z}(z)(a_1 + a_2), a_1 + a_2> -$$

$$- \frac{1}{4}\operatorname{Re} <\tilde{Z}(z)(a_1 - a_2), a_1 - a_2> +$$

$$+ \frac{i}{4}\operatorname{Re} <\tilde{Z}(z)(a_1 + ia_2), a_1 + ia_2> -$$

$$- \frac{i}{4}\operatorname{Re} <\tilde{Z}(z)(a_1 - ia_2), a_1 - ia_2>,$$

and (2.5) follows.

Without loss of generality we can assume that the first $k$ coordinate vectors of $\mathbf{C}^N$ belong to $H_1$, and the remainder, $N - k$ coordinate vectors of $\mathbf{C}^N$, compose a basis in $H_2$. (If this is not the case, then it is possible to achiev such a situation at the point $z^0$ by using an appropriate unitary transformation; at all other points $z \in T^C$ such a situation will automatically occur, as we have just prove.) In this basis the matrix $\operatorname{Re} \tilde{Z}(z)$ has the form

$$\operatorname{Re} \tilde{Z}(z) = \begin{pmatrix} A_1(z) & A_2(z) \\ \bar{A}_2^T(z) & 0 \end{pmatrix}, \tag{2.6}$$

where $A_1(z)$ is a $(k \times k)$-matrix function and $A_2(z)$ is a $(k \times (N - k))$-matrix function.

Let $a \in H_1$ and $b \in H_2$. Then for $\lambda \in C^1$ we have from (2.6)

$$<\operatorname{Re} \tilde{Z}(a + \lambda b), a + \lambda b> = <A_1(z)a, a>_{H_1} + <A_2(z)\lambda b, a>_{H_1} +$$

$$+ <\bar{A}_2^T(z)a, \lambda b>_{H_2} = <A_1(z)a, a>_{H_1} +$$

$$+ 2\operatorname{Re}[\lambda <A_2(z)b, a>_{H_1}] \geqslant 0. \tag{2.7}$$

As $<A_1(z)a, a>_{H_1}$, then by taking at first the real $\lambda \to \pm\infty$ and then an imaginary $\lambda = i\tau, \tau \to \pm\infty$ we conclude that equality (2.7) is possible only if

$$\operatorname{Re}<A_2(z)b, a>_{H_1} = \operatorname{Im}<A_2(z)b, a>_{H_1} = 0;$$

that is, if $A_2(z) = 0$, $z \in T^C$.

Now, in the basis we have chosen the matrix $\tilde{Z}(z)$ is represented inthe form (see (2.6))

$$\tilde{Z}(z) = \operatorname{Re}\tilde{Z}(z) + i\operatorname{Im}[\tilde{Z}(z) - \tilde{Z}(z^0)] + i\operatorname{Im}\tilde{Z}(z^0) =$$

$$= \begin{pmatrix} A_1(z) & 0 \\ 0 & 0 \end{pmatrix} + i\begin{pmatrix} B_1(z) & B_2(z) \\ \overline{B}_2^T(z) & 0 \end{pmatrix} + i\operatorname{Im}\tilde{Z}(z^0)$$

where $B_1(z)$ is a $(k \times k)$-matrix function and $B_2(z)$ is a $(k \times (N-k))$-matrix function. From this representation it follows that the matrices $B_2(z)$ and $\overline{B}_2^T(z)$ are holomorphic in $T^C$ and therefore $B(z) = \text{const} = B_2(z^0) = 0$. Hence,

$$\tilde{Z}(z) = \begin{pmatrix} A_1(z) & 0 \\ 0 & 0 \end{pmatrix} + i\begin{pmatrix} B_1(z) & 0 \\ 0 & 0 \end{pmatrix} + i\operatorname{Im}\tilde{Z}(z^0).$$

Now we take an arbitrary

$$c = a + b \in \mathbf{C}^N,\ a \in H_1,\ b \in H_2,\ c \neq 0$$

and thus, if $a \neq 0$, we have

$$\operatorname{Re} \langle \tilde{Z}(z)c, c\rangle = \operatorname{Re} \langle \tilde{Z}(z)a, a\rangle > 0;$$

and if $a = 0$, then

$$\tilde{Z}(z)c = i\operatorname{Im}\tilde{Z}(z^0)c = \tilde{Z}(z^0)c \neq 0,$$

as $\det\tilde{Z}(z^0) \neq 0$. Thus $\tilde{Z}(z)c \neq 0$ for any $c \in \mathbf{C}^N$, $c \neq 0$, from which it follows that $\det\tilde{Z}(z)z \neq 0$, $z \in T^C$, and the operator $Z^*$ is nonsingular. □

**Theorem 2.** *Every nonsingular (completely nonsingular) passive operator relative to a cone has a unique fundamental solution which determines a nonsingular (completely nonsingular) passive operator relative to the same cone.*

**Proof.** Let $Z^*$ be a nonsingular passive operator relative to a cone $\Gamma$, such that $\tilde{Z}(z)$ is a positive real matrix function in $T^C$ (by theorem 1, Section 12.1) and $\det\tilde{Z}(z) \neq 0$, $z \in T^C$. Applying the Laplace transform to the matrix convolution equation (2.1), we obtain an equivalent matrix equation

$$\tilde{Z}(z)\tilde{A}(z) = I,\ z \in T^C. \tag{2.8}$$

Equation (2.8) is uniquely solvable for all $z \in T^C$, and its solution, the matrix function $\tilde{A}(z) = \tilde{Z}^{-1}(z)$, is holomorphic, and $\det\tilde{A}(z) \neq 0$ in $T^C$. Furthermore, since $\tilde{Z}(iy)$ is freal from $y \in C$ then $\tilde{A}(iy)$ is also real for $y \in C$. Finally, from the condition $\operatorname{Re}\tilde{Z}(z) \geqslant 0$, $z \in T^C$, and from (2.8), we derive

$$\operatorname{Re}\tilde{A}(z) = \tilde{A}^T(z)\operatorname{Re}\tilde{Z}(z)\tilde{A}(z) \geqslant 0,\ z \in T^C. \tag{2.9}$$

Consequently, the matrix $\tilde{A}(z)$ is positive real in $T^C$ and, in addition, if $\operatorname{Re}\tilde{Z}(z) > 0$ then $\operatorname{Re}\tilde{A}(z) > 0$. By theorem 1 of Section 12.1 the matrix $A(\xi)$ defines a nonsingular (completely nonsingular) passive operator relative to the cone $\Gamma$. The matrix $A(\xi)$ is unique. □

Let $\Gamma$ be a closed convex acute cone, $C = \operatorname{int}\Gamma^*$, let $S$ be a $C$-like surface, and let $S_+$ be the domain situated over $S$ (see Section 0.2 and Fig. 3).

**Definition.** *The generalized Cauchy problem* for a passive operator $Z^*$ relative to a cone $\Gamma$ with a source $h = (h_1, \ldots, h_N)$, $h_j \in \mathcal{D}'_{S_+}$, $j = 1, \ldots, N$ (see Section 1.3) shall be understand to be the problem of finding in $\mathbf{R}^n$ a solution $u(\xi)$ from $(\mathcal{D}'_{S_+})^{\times N}$

of system (1.1) of Section 12.1.

From the assertions of Section 1.8 and Section 1.9 and from theorem 2 there immediately follows

**Theorem 3.** *If a passive operator $Z^*$ relative to a cone $\Gamma$ is nonsingular, then the solution of the generalized Cauchy problem for the operator $Z^*$ exists for any $h$ from $(\mathcal{D}'_{S_+})^{\times N}$, it is also unique and is given by the formula*

$$u = A * h. \tag{2.10}$$

*In addition, if $S$ is a strictly $C$-like surface and $h \in (\mathcal{S}'_{\dot{S}_+})^{\times N}$, then the solution of the generalized Cauchy problem for the operator $Z^*$ exists and is unique in the class* $(\mathcal{S}'_{S_+})^{\times N}$ (and is given by formula (2.10)).

**3. Quasi-asymptotics of passive operators and their fundamental solutions.** Let $\Gamma$ be a regular cone and let $\{U_k, k \in I\}$ be a family of linear nonsingular transformations of the space $\mathbf{R}^n$ which leaves the cone $\Gamma$ invariant (see Section 3.1)

$$U_k\Gamma = \Gamma, \ \det U_k = J_k, \ V_k = U_k^*, \ V_k\Gamma^* = \Gamma^*.$$

A matrix $Z(\xi) = (z_{ij}(\xi), i, j = 1, \ldots, N)$, $z_{ij} \in \mathcal{S}'(\Gamma)$, has a *quasi-asymptotics* $G(\xi) = (g_{ij}(\xi), i, j = 1, \ldots, N)$ in the cone $\Gamma$ over the family $\{U_k, k \in I\}$, with respect to the function $\rho_k$, $k \in I$,

$$\frac{Z(U_k\xi)}{\rho_k} \to G(\xi), \ k \to \infty, \ k \in I \tag{3.1}$$

in $\mathcal{S}'(\Gamma)$, if each its components has a quasi-asymptotics in the cone $\Gamma$ over the family $\{U_k, k \in I\}$ with respect to $\rho_k$, $k \in I$:

$$\frac{z_{ij}(U_k\xi)}{\rho_k} \to g_{ij}(\xi), \ k \to \infty, \ k \in I \tag{3.2}$$

in $\mathcal{S}'(\Gamma)$, $i, j = 1, \ldots, N$, and if at least one distribution $g_{ij}(\xi) \not\equiv 0$, $i, j = 1, \ldots, N$.

If a similar way we define the quasi-asymptotics of a vector distribution (component-wise).

**Theorem 1.** *Let a matrix distribution $Z(\xi)$ define a passive operator $Z^*$ relative to a regular cone $\Gamma$. $Z(\xi)$ has a quasi-asymptotics $G(\xi)$ in the cone $\Gamma$ over a family $\{U_k, k \in I\}$ with respect to a function $\rho_k$, $k \in I$ as in* (3.1) *if and only if the following condition is fulfilled:*

(A) *There exists an open set $\Omega \subset C$ such that*

$$\frac{1}{J_k\rho_k}\tilde{Z}(iV_k y) \to H(iy) \not\equiv 0, \ k \to \infty, \ k \in I, \ y \in \Omega. \tag{3.3}$$

*In addition*

$$H(z) = \tilde{G}(z) = L[G], \ z \in T^C, \tag{3.4}$$

*and the matrix function $\tilde{G}(z)$ is positive real in $T^C$, so that the matrix distribution $G(\xi)$ defines a passive operator relative to the cone $\Gamma$.*

**Proof.** The necessity of the condition immediately follows from the general Tauberian theorem (theorem 1, Section 4.1).

Let us now prove sufficiency. By theorem 1 of Section 12.1, for any $a \in C^N$ the function $<\tilde{Z}(z)a, a>$ has a nonnegative real part, and thus, it is of bounded argument in $T^C$. From condition (3.3) it follows that

$$\frac{1}{J_k \rho_k} <\tilde{Z}(iV_k y)a, a> \to <H(iy)a, a>, \quad k \to \infty, \quad k \in I, \quad y \in \Omega$$

and therefore by theorem 1 of Section 4.1 the distribution $<Z(\xi)a, a>$ has a quasi-asymptotics $<G(\xi)a, a>$ in the cone $\Gamma$ over the family $\{U_k, k \in I\}$ with respect to the function $\rho_k$, $k \in I$, and $L[G] = H(z)$. Now from the polarization formula

$$<Z(\xi)a, b> = \frac{1}{4}<Z(\xi)(a+b), a+b> -$$
$$-\frac{1}{4}<Z(\xi)(a-b), a-b> +$$
$$+\frac{i}{4}<Z(\xi)(a+ib), a+ib> -$$
$$-\frac{i}{4}<Z(\xi)(a-ib), a-ib>, \quad a, b \in C^N$$

it follows that the distribution $<Z(\xi)a, b>$ has a quasi-asymptotics $<G(\xi)a, b>$, from which (3.1) follows.

Further, we have

$$\frac{1}{J_k \rho_k} \mathrm{Re} <\tilde{Z}(V_k z)a, a> \geqslant 0, \quad k \in I, \quad z \in T^C, \quad a \in C^N. \tag{3.5}$$

Taking the limit in this inequality as $k \to \infty$, $k \in I$, we get

$$\mathrm{Re} <H(z)a, a> \geqslant 0, \quad z \in T^C, \quad a \in C^N;$$

that is, $\mathrm{Re}\, H(z) = \mathrm{Re}\, L[G] \geqslant 0$. Because the matrix $\tilde{Z}(iV_k y)$ is real, the matrix $H(iy)$ is also real, and thus the matrix $H(z)$ is positive real. Owing to theorem 1 of Section 12.1 the matrix distribution $G(\xi)$ defines a passive operator $G^*$ relative to the cone $\Gamma$.

Hence, it is possible to make the following statement on quasi-asymptotics of passive operators. The quasi-asymptotics of a passive operator relvative to a cone $\Gamma$, if it exists, defines in turn a passive operator relative to the same cone $\Gamma$. □

**Corollary 1.** *If under the hypothese of theorem 1 there exists a point* $y^0 \in \Omega \subset C$ *such that*

$$\det H(iy^0) \neq 0 \tag{3.6}$$

*then there exists a fundamental solution* $A(\xi)$ *of the operator* $Z^*$, *and it has a quasi-asymptotics* $Q(\xi)$ *in the cone* $\Gamma$ *over the family* $\{U_k, k \in I\}$ *with respect to the function* $(J_k^2 \rho_k)^{-1}$, $k \in I$, *and also*

$$\tilde{Q}(z) = L[Q] = H^{-1}(z), \quad z \in T^C. \tag{3.7}$$

**Proof.** Since $H(z)$ is a positive real matrix function in $T^C$ then from (3.6) by vritue of theorem 1 of Section 12.2 it follows that $\det H(z) \neq 0$ in $T^C$. Therefore, from (3.3) it follows for sufficiently large $k$ that $\det \tilde{Z}(iV_k y) \neq 0$ and hence $\det \tilde{Z}(z) \neq 0$, $z \in T^C$. Thus, the passive operator $Z^*$ relative to the cone $\Gamma$ is nonsingular. Owing to theorem 2 of Section 12.2 there exists a unique fundamental solution $A(\xi)$, such that $\tilde{Z}(z)\tilde{A}(z) = I$. Now if we pass to the limit as $k \to \infty$, $k \in I$ in the equality

$$\frac{1}{J_k \rho_k}\tilde{Z}(iV_k y)J_k \rho_k \tilde{A}(iV_k y) = I, \; y \in \Omega \subset C$$

and if we take account of (3.3), we get

$$\frac{J_k^2 \rho_k}{J_k}\tilde{A}(iV_k y) \to H^{-1}(iy), \; k \to \infty, \; k \in I, \; y \in \Omega \tag{3.8}$$

from which equality (3.7) follows. □

**Corollary 2.** *If a matrix distribution* $Z(\xi)$ *defines a nonsingular passive operator* $Z^*$ *relative to a regular cone* $\Gamma$, *and for some domain* $\Omega \subset C$ *there exists a function* $\rho_k$, $k \in I$, *such that*

$$\rho_k \tilde{Z}^{-1}(iV_k y) \to \tilde{Q}(iy), \; k \to \infty, \; k \in I, \; y \in \Omega \tag{3.9}$$

*then the fundamental solution* $A(\xi)$ *has a quasi-asymptotics* $Q(\xi)$ *in the cone* $\Gamma$ *over the family* $\{U_k, k \in I\}$ *with respect to the function* $(J_k \rho_k)^{-1}$, $k \in I$.

**Proof.** This statement immediately follows from theorem 2 of Section 12.2 and theorem 1. □

We consider the case $n = 1$, $\Gamma = [0, \infty)$ in more detail.

*Let a passive nonsingular operator* $Z^*$ *have a quasi-asymptotics at infinity with respect to an automodel function* $\rho$ *of order* $\beta$,

$$\frac{Z(k\xi)}{\rho(k)} \to f_{\beta+1}(\xi)C, \; k \to \infty \tag{3.10}$$

*in* $\mathcal{S}'_+$ ($\det C \neq 0$). *Then the fundamental solution* $A(\xi)$ *has a quasi-asymptotics at infinity with respect to the automodel function* $\rho_1 = [\xi^2\rho(\xi)]^{-1}$ *of order* $-\beta - 2$ (by Corollary 1).

From the results of the previous chapter (see Corollary 1, Section 9.2), and from the positive reality of $\tilde{A}(z)$, it follows that *if* $\gamma \geqslant 2$ *and* $-2 < \beta \leqslant 0$, *or* $\gamma > 2$ *and* $\beta = -2$, *then the matrix function* $A^{(-\gamma)}(\xi)$ *(the primitive of order* $\gamma$ *is defined component-wise) is continuous and there exists the limit*

$$\lim_{\xi \to \infty} \frac{1}{\xi^\gamma \rho_1(\xi)} A^{(-\gamma)}(\xi) = \frac{1}{\Gamma(\gamma - \beta - 1)} C^{-1}; \tag{3.11}$$

where $\Gamma(t)$ is the gamma function.

**Theorem 2.** *Let a passive nonsingular operator* $Z^*$ *have a nonsingular quasi-asymptotics* (3.10) *at infinity with respect to an automodel function* $\rho$ *of order* $\beta$ *and let a vector function* $h$ *have a continuous derivative of order* $\gamma$ ($\gamma = 2$ *for* $-2 < \beta \leqslant 0$; $\gamma > 2$ *for* $\beta = -2$) *which has an asymptote at infinity with respect to an automodel function* $\rho_1$ *of order* $\omega > -1$,

$$\frac{1}{\rho_1(\xi)}h^{(\gamma)}(\xi) \to \frac{1}{\Gamma(\omega+1)}H,\ H \neq 0. \tag{3.12}$$

*Then the solution $u(\xi)$ of the generalized Cauchy problem*

$$Z * u = h,\ h \in (\mathcal{S}'_+)^{\times N} \tag{3.13}$$

*is continuous for $\xi > 0$ and has an asymptotics at infinity with respect to the automodel function $\xi^{\gamma-1}\rho_1(\xi)/\rho(\xi)$,*

$$\lim_{\xi\to\infty}\frac{\rho(\xi)}{\xi^{(\gamma-1)}\rho_1(\xi)}u(\xi) = \frac{1}{\Gamma(\gamma-\beta+\omega)}C^{-1}H. \tag{3.14}$$

**Proof.** Assuming $\gamma = 2$ for $-2 < \beta \leqslant 0$, or $\gamma > 0$, for $\beta = -2$ we have

$$u(\xi) = A * h = A^{(-\gamma)} * h^{(\gamma)}.$$

Hence, $u(\xi)$ is the convolution of the continuous matrix function on $A^{(-\gamma)}(\xi)$ and the vector function $h^{(\gamma)}(\xi)$, which have asymptotics (3.11) and (3.12), respectively. From this there follows the limit relation (3.14). □

**4. Differential operators of the passive type.** A system of $N$ linear differential equations with constant coefficients of order $\leqslant m$ is defined by the matrix (see Section 0.4)

$$Z(\xi) = \sum_{0\leqslant|\alpha|\leqslant m} Z_\alpha \partial^\alpha \delta(\xi), \tag{4.1}$$

where $Z_\alpha$ are constant $(N \times N)$-matrices.

**Theorem 1.** *A system of $N$ linear differential equations with constant coefficients is passive relative to an acute cone $\Gamma$, if and only if*

$$Z(\xi) = \sum_{1\leqslant j\leqslant n} Z_j\,\partial_j\,\delta(\xi) + Z_0\delta(\xi), \tag{4.2}$$

*where $Z_1, \ldots, Z_n$ are real symmetric $(N \times N)$- matrices such that for all $y \in C = \operatorname{int}\Gamma^*$*

$$\sum_{1\leqslant j\leqslant n} y_j Z_j \geqslant 0;$$

*the matrix $Z_0$ is real and* $\operatorname{Re} Z_0 = 1/2(Z_0 + Z_0^T) \geqslant 0$.

**Proof. Necessity.** Let the differential operator $Z*$, which is defined by formula (4.1), be passive relative to the cone $\Gamma$. By theorem 1 of Section 12.1 the matrix function

$$\tilde{Z}(z) = \sum_{0\leqslant|\alpha|\leqslant m} (-iz)^\alpha Z_\alpha \tag{4.3}$$

is positive real in $T^C$. Therefore, for each $a \in \mathbb{C}^N$ the function $<\tilde{Z}(z)a, a>$ is holomorphic and $\operatorname{Re} <\tilde{Z}(z)a, a> \geqslant 0$ in $T^C$. For such functions the following estimate is valid: for any cone $C' \subset\subset C$ there exists a number $M(C')$ such that

$$|<\tilde{Z}(z)a, a>| \leqslant M(C')\frac{1+|z|^2}{|y|},\ z \in T^{C'}$$

(see V.S. Vladimirov, 1979, Section 16.1), from which it follows that all matrix elements $\tilde{z}_{ij}(z)$, $i, j = 1, \dots, N$, satisfy this estimate; that is,

$$\left| \sum_{0 \leq |\alpha| \leq m} (-iz)^\alpha z_{\alpha, ij}(z) \right| \leq M(C')\frac{1+|z|^2}{|y|}, \quad z \in T^{C'}.$$

But the last estimate is possible only when $z_{\alpha, ij}(z) = 0$, $1 \leq 1, j \leq N$, $|\alpha| \geq 2$. Therefore matrix (4.3) takes the form

$$\tilde{Z}(z) = -i \sum_{1 \leq j \leq n} z_j Z_j + Z_0 \tag{4.4}$$

and thus the representation (4.2) has been proved.

Now the conditions of positive reality give:

$$\tilde{Z}(iy) = \sum_{1 \leq j \leq n} y_j Z_j + Z_0$$

is real for $y \in C$,

$$-i \sum_{1 \leq j \leq n} z_j Z_j + Z_0 + i \sum_{1 \leq j \leq n} \bar{z}_j \bar{Z}^T + \bar{Z}_0^T \geq 0, \quad z \in T^C.$$

From this we conclude that the matrices $Z_j$, $j = 0, 1, \dots, n$, are real, the matrices $Z_j$, $j = 1, \dots, n$, are symmetric and

$$\sum_{1 \leq j \leq n} y_j Z_j \geq 0, \quad y \in C; \quad \operatorname{Re} Z_0 \geq 0.$$

**Sufficiency.** Let $Z(\xi)$ satisfy the conditions of the theorem. Then its Laplace transform $\tilde{Z}(z)$ has the form (4.4) and

$$\operatorname{Re} \tilde{Z}(z) = \sum_{1 \leq j \leq n} y_j Z_j + \operatorname{Re} Z_0 \geq 0, \quad z \in T^C,$$

$\tilde{Z}(iy)$ is real for $y \in C$. Therefore the matrix function is positive real in $T^C$ and by theorem 1 of Section 12.1 the operator $Z^*$ is passive relative to the cone $\Gamma$. □

**Corollary 1.** *Under the hypothesis of theorem 1, if there exists a point $y^0 \in C$ such that*

$$\det\left( \sum_{1 \leq j \leq n} y_j^0 Z_j + Z_0 \right) \neq 0 \quad \left( \sum_{1 \leq j \leq n} y_j^0 Z_j + \operatorname{Re} Z_0 > 0 \right) \tag{4.5}$$

*then the differential operator $Z^*$ is nonsingular (completely nonsingular).*

**Proof.** This immediately follows from theorem 1 of Section 12.2 (see formula (2.3), Section 12.2).

Let real symmetric $(N \times N)$-matrices $Z_1, \dots, Z_n$ be given, which possess the property that for some point $l \in \mathbf{R}^n$, $\sum_{1 \leq j \leq n} l_j Z_j > 0$. We denote

$$\Gamma_c = [\xi : \xi_1 = \langle Z_1 a, a \rangle, \dots, \xi_n = \langle Z_n a, a \rangle, a \in C^n],$$

$$\Gamma_r = [\xi : \xi_1 = \langle Z_1 a, a \rangle, \dots, \xi_n = \langle Z_n a, a \rangle, a \in \mathbf{R}^n].$$

For the mapping

$$a \to \xi = (\langle Z_1 a, a \rangle, \dots, \langle Z_n a, a \rangle) \tag{4.6}$$

the preimage of zero is zero because of the inequality

$$(l,\xi) = < \sum_{1\leqslant j\leqslant n} l_j Z_j a, a > \geqslant \kappa |a|^2 \tag{4.7}$$

for some $\kappa > 0$. It is clear that $\Gamma_c$ and $\Gamma_r$ are cones with their vertices at zero, and $\Gamma_r \subset \Gamma_c$. □

**Lemma 1.** *The cone $\Gamma_c$ and $\Gamma_r$ are closed and acute, and $\Gamma_c = \Gamma_r + \Gamma_r$, $l \in \operatorname{int}\Gamma_c^*$.*

**Proof.** The mapping (4.6) is continuous from $C^n$ into $\mathbf{R}^n$, and owing to inequality (4.7) the preimage of any compact set is compact. Therefore cones $\Gamma_c$ and $\Gamma_r$ are both closed. Further, from the equalities

$$<Z_j a, a> = <Z_j b, b> + <Z_j c, c>, \quad a = b + ic, \quad j = 1, \dots, n$$

we conclude that $\Gamma_c = \Gamma_r + \Gamma_r$. Finally, by virtue of inequality (4.7) the plane $(l,\xi) = 0$ has a single common point with the cone $\Gamma_c$, its vertex. Therefore, the cones $\Gamma_c$ and $\Gamma_r$ are acute and $l \in \operatorname{int}\Gamma_c^*$. □

Note that the cones $\Gamma_c$ and $\Gamma_r$ can be nonsolid as the following example shows: if $n = 2$ and $Z_1 > 0$, $Z_2 = 0$ then $\Gamma_c$ and $\Gamma_r$ lie in the plane $\xi_2 = 0$.

**Theorem 2.** *The matrix (4.2) defines a passive nonsingular operator if and only if the matrices $Z_1, \dots, Z_n$ are real symmetric, the matrix $Z_0$ is real, and*

$$\operatorname{Re} Z_0 = \frac{1}{2}(Z_0 + Z_0^T) \geqslant 0.$$

*and there exists a point $l = (l_1, \dots, l_n)$ such that*

$$\sum_{1\leqslant j\leqslant n} l_j Z_j + \operatorname{Re} Z_0 > 0. \tag{4.8}$$

*In addition, the operator $Z^*$ is passive and completely nonsingular relative to any acute cone $\Gamma$ which contains the cone $\Gamma_c$, and $l \in \operatorname{int}\Gamma^*$.*

**Proof.** Necessity immediately follows from corollary 1 of theorem 1 (see (4.5)). Let us prove the sufficiency. Let the matrices $Z_0, Z_1, \dots, Z_n$ in (4.2) satisfy the hypothesis of theorem 2. Let $\Gamma$ be an acute cone which contains the cone $\Gamma_c$ and such that $l \in C = \operatorname{int}\Gamma^*$. From this it follows that $(y,\xi) \geqslant 0$ for all $y \in C$, $\xi \in \Gamma_c \subset \Gamma$; that is,

$$(y,\xi) = \sum_{1\leqslant j\leqslant n} y_j <Z_j a, a> \geqslant 0, \quad y \in C, \quad a \in C^N,$$

which means that the matrix

$$\sum_{1\leqslant j\leqslant n} y_j Z_j \geqslant 0, \quad y \in C.$$

By theorem 1 the matrix $Z$ defines a passive operator relative to the cone $\Gamma$. Further, as $l \in C$, and owing to (4.8)

$$\operatorname{Re} Z(il) = \sum_{1\leqslant j\leqslant n} l_j Z_j + \operatorname{Re} Z_0 > 0$$

we conclude, according to corollary 1 of theorem 1 (see (4.5)), that the operator $Z^*$ is completely nonsingular. □

Now we calculate the quasi-asymptotics of differential passive operators relative to a regular cone $\Gamma$ over the group of dilatations of the cone; $\xi \to k\xi$, $k > 0$.

By taking into account that $J_k = k^n$, $V_k = (1/k)I$ and, using formula (4.4), we have

$$\frac{1}{J_k \rho(k)} \tilde{Z}(iV_k y) = \frac{1}{k^{n+1}\rho(k)} \sum_{1 \leqslant j \leqslant n} y_j Z_j + \frac{1}{k^n \rho(k)} Z_0; \tag{4.9}$$

where $\rho(k)$ is an automodel function. This follows from (4.9) and from theorem 1 of Section 12.3.

**Theorem 3.** *Let the matrix* (4.2) *define a passive differential operator relative to a regular cone* $\Gamma$. (i) *If* $Z_0 \neq 0$,*then* $Z(\xi)$ *has a quasi-asymptotics* $Z_0\delta(\xi)$ *in the cone* $\Gamma$ *with respect to the automodel function* $\rho(k) = k^{-n}$ *of order* $-n$.

(ii) *If* $Z_0 = 0$, *then* $Z(\xi)$ *has a quasi-asymptotics*

$$\sum_{1 \leqslant j \leqslant n} Z_j \partial_j \delta(\xi)$$

*in the cone* $\Gamma$ *with respect to the automodel function* $\rho(k) = k^{-n-1}$ *of order* $-n-1$.

**Corollary 2.** *In case* (i) *of theorem 3: if* $\det Z_0 \neq 0$ *then the fundamental solution of the operator* $Z^*$, *which is passive relative to the cone* $\Gamma$, *has a quasi-asymptotics*

$$Q(\xi) = Z_0^{-1}\delta(\xi) \tag{4.10}$$

*in* $\Gamma$ *with respect to the automodel function* $k^{-n}$ *of order* $-n$.

**Proof.** This follows immediately from Corollary 1 of Section 12.3 and from theorem 3. □

**5. Examples.** In this subsection $t$ is the time variable and $\xi \in \mathbf{R}^n$ are the space variables. By

$$V_n^+(a) = [(t, \xi) \in \mathbf{R}^{n+1}: at > |\xi|]$$

we denote the future cone in $\mathbf{R}^{n+1}$ which corresponds to the propagation velocity $a$; $V_n^+ = V_n^+(1)$ (see Section 0.2).

(1) *Elementary electric circuit*, $N = n = 1$.The distribution

$$Z(t) = L\delta'(t) + R\delta(t) + \frac{1}{C}\theta(t)$$

defines a passive nonsingular operator $Z^*$ relative to the cone $[0, \infty)$ (see Section 12.1). It has a quasi-asymptotics with respect to the function $\rho \equiv 1$ which is equal to $\theta(t)/C$, but its fundamental solution has a quasi-asymptotics $Q(t)$ with respect to the function $\rho = t^{-2}$, which is equal to

$$Q(t) = C\delta'(t). \tag{5.1}$$

(2) *The Dirac equation*, $N = n = 4$. The corresponding differential operator has the form

$$i\gamma^0\partial_t + i \sum_{1 \leqslant \mu \leqslant 3} \gamma^\mu \partial_\mu - mI, \tag{5.2}$$

where $\gamma^\mu$, $\mu = 0, 1, 2, 3$, are $(4 \times 4)$-Dirac matrices. In the Majorana basis (see, for example, N.N. Bogolyubov, A.A. Logunov and I.T. Todorov, 1974, ch. 2) they have the form

$$\gamma^0 = \begin{pmatrix} 0 & i\sigma_1 \\ -i\sigma_1 & 0 \end{pmatrix}, \quad \gamma^1 = \begin{pmatrix} iI & 0 \\ 0 & -iI \end{pmatrix}, \quad \gamma^2 = \begin{pmatrix} 0 & \sigma_2 \\ -\sigma_2 & 0 \end{pmatrix}, \quad \gamma^3 = \begin{pmatrix} 0 & iI \\ iI & 0 \end{pmatrix}$$

where $\sigma_k$ are$(2 \times 2)$-Pauli matrices

$$\sigma_1 = \begin{pmatrix} 0 & 1 \\ 1 & 0 \end{pmatrix}, \quad \sigma_2 = \begin{pmatrix} 0 & -i \\ i & 0 \end{pmatrix}, \quad \sigma_3 = \begin{pmatrix} 1 & 0 \\ 0 & -1 \end{pmatrix}.$$

Multiplying operator (5.2) by the matrix $-\gamma^0$ we obtain the operator

$$I\,\partial_t + \sum_{1 \leqslant \mu \leqslant 3} \gamma^0\gamma^\mu\partial_\mu + im\,\gamma^0 \tag{5.3}$$

which is a passive and completely nonsingular relative to the cone $\overline{V}_3^+$.

In fact, the matrices $\gamma^0\gamma^\mu$ are real and symmetric and the matrix $\mathrm{Im}\gamma^e$ is real and skew-symmetric. Further, the cone $\Gamma_r$ is the boundary of the cone $V_3^+$ because mapping (4.6) of Section 12.4 for $a \in \mathbf{R}^4$ has the form

$$t = \langle Ia, a \rangle = a_1^2 + \cdots + a_4^2,$$

$$\xi_1 = \langle \gamma^0\gamma^1 a, a \rangle = 2a_1a_4 + 2a_2a_3,$$

$$\xi_2 = a_1^2 - a_2^2 + a_3^2 - a_4^2,$$

$$\xi_3 = -2a_1a_2 + 2a_3a_4$$

so that $t \geqslant 0$ and

$$t^2 - |\xi|^2 = (a_1^2 + \cdots + a_4^2)^2 - 4(a_1a_4 + a_2a_3)^2 - (a_1^2 - a_2^2 + a_3^2 - a_4^2)^2 -$$
$$- 4(a_1a_2 - a_3a_4)^2 = 0.$$

By theorem 2 of Section 12.4 the operator (5.3) is passive and completely nonsingular relative to the cone $\Gamma_c = \overline{V}_3^+$ which is the convex hull of the cone $\Gamma_r$ (cf. lemma 1, Section 12.4).

According to corollary 2 of Section 12.4 the fundamental solution of the operator (5.3) has a quasi-asymptotics $Q(t, \xi)$ in $\overline{V}_3^+$ with respect to the function $\rho(k) = k^{-4}$, and

$$Q(t, \xi) = mi\,\gamma^0\delta(t, \xi). \tag{5.4}$$

Now in order to find a quasi-asymptotics of the fundamental solution of the Dirac equation it is necessary to multiply the matrix (5.4) from the right by the matrix $(-i\,\gamma^0)^{-1} = i\,\gamma^0$.

(3) *The equation of a rotation fluid and acoustics*, $N = n = 4$. The corresponding differential operator has the form

$$\alpha\partial_t p + \rho\,\mathrm{div}\,\mathbf{v}, \quad \rho\partial_t \mathbf{v} + \partial p + \mathbf{v} \times \mathbf{w}. \tag{5.5}$$

Here $\rho = \mathrm{const}$ denotes the density, $p$ the pressure, $\mathbf{v}$ the velocity, $\alpha$ the coefficient of compressibility of a fluid (gas), and $\mathbf{w}$ is a vector of rotation, $|\mathbf{w}| = \omega$. In the system of space coordinates $\xi = (\xi_1, \xi_2, \xi_3)$, where the axis $\xi_3$ is directed along the axis of the rotation $\mathbf{w}$, the corresponding operator $Z^*$ has the form

$$Z^* = \begin{vmatrix} \alpha\partial_t & \partial_1 & \partial_2 & \partial_3 \\ \partial_1 & \partial_t & \frac{\omega}{\rho} & 0 \\ \partial_2 & -\frac{\omega}{\rho} & \partial_t & 0 \\ \partial_3 & 0 & 0 & \partial_t \end{vmatrix}. \tag{5.6}$$

For all $\alpha > 0$ the operator (5.6) is passive and completely nonsingular relative to the cone $\bar{V}_3^+(1/\sqrt{\alpha})$. In fact, the mapping (4.6) of Section 12.4), for $\alpha \in \mathbf{R}^4$ has the form

$$t = \alpha a_1^2 + a_2^2 + a_3^2 + a_4^2, \quad \xi_1 = 2a_1a_2,$$

$$\xi_2 = 2a_1a_2, \quad \xi_3 = 2a_1a_4$$

so that $t \geqslant 0$ and

$$t^2 - \alpha|\xi|^2 = (\alpha a_1^2 + a_2^2 + a_3^2 + a_4^2)^2 - 4\alpha a_1^2(a_2^2 + a_3^2 + a_4^2) =$$

$$= (\alpha a_1^2 - a_2^2 - a_3^2 - a_4^2)^2 \geqslant 0.$$

Therefore $\Gamma_c = \Gamma_r = \bar{V}_3^+(1/\sqrt{\alpha})$.

It is not difficult to see that $\det Z_0 = 0$. Therefore, for the calculation of a quasi-asymptotics for the fundamental solution we can use corollary 2 of Section 12.4. For that we use corollary 2 of Section 12.3. Performing the Laplace transform and inverting the matrix $\tilde{Z}(z)$ we conclude that in(3.9) of Section 12.3 there exists, for $\rho_k = k^{-1}$, the limit

$$\lim_{k\to\infty}\frac{1}{k}\tilde{Z}^{-1}\left(\frac{iy}{k}\right) = \tilde{Q}(iy) = \frac{1}{y_0^2 - \frac{1}{\alpha}y_3^2}\begin{vmatrix} y_0 & 0 & 0 & -y_3 \\ 0 & 0 & 0 & 0 \\ 0 & 0 & 0 & 0 \\ -y_3 & 0 & 0 & \alpha y_0 \end{vmatrix}. \tag{5.7}$$

Inverting the Laplace transform and taking into account the equality

$$L\left[\frac{\theta(t - \sqrt{\alpha}|\xi_3|)}{2} \times \delta(\xi_1, \xi_2)\right] = \frac{1}{\frac{1}{\alpha}z_3^2 - z_0^2},$$

we obtain from (5.7) the quasi-asymptotics $Q(t, \xi)$ of the fundamental solution of the operator $Z^*$ (see (5.6)) in the following form

$$Q(t, \xi) = \begin{vmatrix} \partial_t & 0 & 0 & -\partial_3 \\ 0 & 0 & 0 & 0 \\ 0 & 0 & 0 & 0 \\ -\partial_3 & 0 & 0 & \alpha\partial_t \end{vmatrix} \frac{\theta(t - \sqrt{\alpha}|\xi_3|)}{2} \times \delta(\xi_1, \xi_2). \tag{5.8}$$

Formula (5.8) shows that the fundamental matrix of solutions of the system of equations of a rotating fluid and acoustics possesses a quasi- asymptotics in the cone $\bar{V}_3^+(1/\sqrt{\alpha})$ with respect to the automodel function $\rho(k) = k^{-3}$ (see corollary 2, Section 12.3) which differs from zero only on two rays on the bondary of the cone $V_3^+(1/\sqrt{\alpha})$ whose spatial projections coincide with the axis of rotation **w**.

(4) *The telegraph equation*, $N = n = 2$. The corresponding differential operator has the form

$$\left[\begin{pmatrix} L & 0 \\ 0 & C \end{pmatrix}\partial_t + \begin{pmatrix} 0 & 1 \\ 1 & 0 \end{pmatrix}\partial_\xi\right]u, \tag{5.9}$$

where $u = (u_1, u_2)$, $u_1$ is the current, $u_2$ is the voltage in the circuit, $L$ is the self-inductance, and $C$ is the capacitance. The operator (5.9) for $L > 0$ and $C > 0$ is passive and completely nonsingular relative to the cone $\bar{V}_1^+(1/\sqrt{LC})$. The fundamental solution $\mathcal{E}(t, \xi)$ has a quasi-asymptotics $A(t, \xi)$ with respect to the automodel function $\rho(k) = k^{-1}$ which coincides with its fundamental solution

$$A(t, \xi) = \mathcal{E}(t, \xi) = \begin{pmatrix} C\partial_t & -\partial_\xi \\ -\partial_\xi & L\partial_t \end{pmatrix}\frac{\theta(t - \sqrt{LC}\,|\xi|)}{2LC}.$$

(5) Consider the following differential operator ($N = 4, n = 3$):

$$\partial_t u + B_1\partial_1 u + B_2\partial_2 u + B_0 u, \tag{5.10}$$

where

$$B_1 = \begin{pmatrix} \frac{1}{3}I & 0 \\ 0 & \sigma_3 \end{pmatrix}, \quad B_2 = \begin{pmatrix} \frac{1}{3}\sigma_1 & 0 \\ 0 & \sigma_3 \end{pmatrix}, \quad B_0 = \frac{1}{3}\begin{pmatrix} 0 & i\sigma_2 \\ i\sigma & 0 \end{pmatrix}$$

$\sigma_k$, $k = 1, 2, 3$, are the Pauli matrices (see example 2). The operator (5.10) is passive and completely nonsingular. Since $\det B_0 \neq 0$, then according to corollary 2 of Section 12.4 the fundamental solution of the operator (5.10) has the quasi- asymptotics $Q(t, \xi)$ with respect to the automodel function $\rho(k) = k^{-3}$ which has the form

$$Q(t, \xi) = 3\begin{pmatrix} 0 & -i\sigma_2 \\ -i\sigma_2 & 0 \end{pmatrix}\delta(t, \xi). \tag{5.11}$$

The system (5.10) arises in crystal optics. This was investigated by D. Ludwig (1963) who showed, by use of the Jost method of expanding the $\delta$-function into plane waves and the method of the stationary phase, that in the fundamental matrix of solutions there are components which have asymptotics of the form

$$\frac{c_1(\xi)}{\sqrt{t}}e^{i(t/2)} + \frac{c_2(\xi)}{\sqrt{t}}e^{-i(t/2)} + O\left(\frac{1}{\sqrt{t}}\right), \quad t \to \infty, \tag{5.12}$$

where $c_1(\xi)$ and $c_2(\xi)$ are some functions of the space variables $\xi = (\xi_1, \xi_2)$.

Asymptotics along other directions in the causal cone are difficult to study with standard methods. We note that formula (5.11) does not contradict formula (5.12), but gives a new characterization (in terms of quasi-asymptotics) of the asymtotic behaviour of the fundamental solution in the whole causual cone.

(6) Consider the following system of differential equations, $N = n = 2$:

$$Z * u \equiv \begin{pmatrix} 1 & -1 \\ -1 & 1 \end{pmatrix} \partial_1 u + \begin{pmatrix} 1 & -1 \\ -1 & 1 \end{pmatrix} \partial_2 u + \begin{pmatrix} 0 & -1 \\ 1 & 0 \end{pmatrix} u = h. \tag{5.13}$$

where $u = (u_1, u_2)$ and $h = (h_1, h_2)$. This is a passive nonsingular system relative to the cone $\mathbf{R}_+^2$ ($\det \tilde{Z}(z) \equiv 1$). But it is not a completely nonsingular system, because

$$\operatorname{Re} \tilde{Z}(z) = \begin{pmatrix} 1 & -1 \\ -1 & 1 \end{pmatrix} (y_1 + y_2) \geqslant 0, \; y \in \mathbf{R}_+^2.$$

The fundamental solution has a quasi-asymptotics $Q(\xi)$ with respect to the function $\rho(k) = k^{-2}$ which is equal to

$$Q(\xi) = \begin{pmatrix} 0 & 1 \\ -1 & 0 \end{pmatrix} \delta(\xi).$$

If we pose the generalized Cauchy problem for some $\mathbf{R}_+^2$- like surface $S$, $\operatorname{supp} h \subset \bar{S}_+$ (see Seciton 0.2), then its solution is expressed by

$$u = \begin{pmatrix} 1 & 1 \\ 1 & 1 \end{pmatrix} \partial_1 h + \begin{pmatrix} 1 & 1 \\ 1 & 1 \end{pmatrix} \partial_2 h + \begin{pmatrix} 0 & 1 \\ -1 & 0 \end{pmatrix} h. \tag{5.14}$$

Using formula (5.14) we note that the smoothness of the solution $u(\xi)$ can be even worse than the smoothness of the source $h(\xi)$ (if, of course, $h$ is not infinitely differentiable). In particular, if $\operatorname{supp} h \subset S$ then $\operatorname{supp} u \subset S$.

Linear passive operators are frequently encountered in mathematical physics. Among many-dimensional linear passive systems are: the Maxwell equations, the Dirac equations, the equations of a rotating fluid and acoustics, the linearized system of equations of magnetic hydrodynamics, the linearized system of equations of the theory of elasticity, transport equations, and others (for details see V.S. Vladimirov, 1979, Section 19).

## Comments on Chapter 4

The asymptotics behaviour of solutions of convolution equations in the one-dimensional case was studied by N. Wiener (1932) and T.H. Ganelius (1971). The quasi-asymptotics of solutions of many-dimensional (and one-dimensional) convolution equations in the algebra $\mathcal{S}'_\Gamma$ has been studied by Yu.N. Drozzinov and B.I. Zavialov (1977, 1984, 1985b), and Yu.N. Drozzinov (1981, 1983a).

The theory of many-dimensional translationally-invariant passive systems has been constructed by V.S. Vladimirov on the basis of his theory of holomorphic matrix functions with nonnegative real parts in domains of tubes over cones. The detailed exposition and bibliography of the problem can be found in the book by V.S. Vladimirov (1979).

It should be mentioned that the theory of translationally-invariant passive systems with value in a Hilbert space in the case of one variable was presented in the book A. Zemanian (1972), and in the case of many variables in the papers by R.H. Galeev (1981, 1982). Passive systems of a more general form (nontranslationally-invariant) have been considered in the paper by Yu.N.

Drozzinov (1983b).

Investigations of the asymptotics of solutions of the Cauchy problem for one-dimensional passive operators began in the late fifties. One of the first papers in this direction was by H. König and T.Meixner (1958). Quasi-asymptotics of the passive operators and their fundamental solutions has been studied by Yu.N. Drozzinov (1983a), and by Yu.N.Drozzinov and B.I.Zavialov (1985b).

## Chapter 5

# TAUBERIAN THEOREMS FOR CAUSAL FUNCTIONS

In this chapter some results concerning the Tauberian theory for causal distributions are discussed. A distribution from $\mathcal{S}'(\mathbf{R}^4)$ is called causal if the support of its Fourier transform belongs to the closed light cone

$$\bar{V} = \bar{V}_3^+ \bigcup \bar{V}_3^- = [x = (x_0, \vec{x}) \in \mathbf{R}^4 : x_0^2 \geqslant |\vec{x}|^2],$$

where $V_3^+$ and $V_3^- = -V_3^+$ denote the future light cone and past light cone, respectively (see Section 0.2). These results were initiated by some problems of quantum field theory, namely by the problems of deep inelastic lepton-hadron scattering (see I.D. Björken, 1967; N.N. Bogolyubov, V.S. Vladimirov and A.N. Tavkhelidze, 1972; B.I. Zavialov, 1973 a,b, 1974; V.S. Vladimirov and B.I. Zavialov, 1979, 1980, 1982 and the bibliography there). We shall start with the Jost-Lehmann-Dyson representation which is a very useful tool in such problems.

## 13. The Jost-Lehmann-Dyson representation.

The Jost-Lehmann-Dyson representation in the symmetric case was introduced by R. Jost and H. Lehmann (1957) and in the general case by F.J. Dyson (1958). The proof of this representation is contained in the book V.S. Vladimirov (1966). Here we give a construction which in some way differs from the original and which is more suitable for our purposes.

**1. The Jost-Lehmann-Dyson representation in the symmetric case.** Further, we shall consider distributions from $\mathcal{S}'(\mathbf{R}^4)$. By $x, q, \cdots$, and so on, we denote vectors from $\mathbf{R}^4$, so that, for instance, $x = (x_0, x_1, x_2, x_3) = (x_0, \vec{x})$. The form

$$xq = x_0 q_0 - x_1 q_1 - x_2 q_2 - x_3 q_3 = x_0 q_0 - (\vec{x}, \vec{q}) \tag{1.1}$$

endows $\mathbf{R}^4$ with a Lorentz scalar product, so that $x^2 = x_0^2 - |\vec{x}|^2$ and $q^2 = q_0^2 - |\vec{q}|^2$. As before $\mathcal{S}'(\bar{V})$ denotes the subspace of the distributions from $\mathcal{S}'(\mathbf{R}^4)$ with supports in the closed light cone $\bar{V}$. It is simultaneously the space of linear continuous functionals over the test space $\mathcal{S}(\bar{V})$ (see Section 1.3). Let $\mathcal{S}_a(\bar{V})$ and $\mathcal{S}_s(\bar{V})$ denote the subspaces of $\mathcal{S}(\bar{V})$ consisting of odd (antisymmetric) and even (symmetric) functions with rspect to $x_0$, respectively (see Section 1.6). The corresponding dual spaces $\mathcal{S}_a'(\bar{V})$ and $\mathcal{S}_s'(\bar{V})$ are obviously the spaces of odd and even distributions over $x_0$ from $\mathcal{S}'(\bar{V})$.

Let us introduce the mappings $\phi(x_0, \vec{x}) \to \hat{\phi}(\kappa, \vec{x})$ and $\phi(x_0, \vec{x}) \to \phi(\kappa, \vec{x})$, where $\phi \in \mathcal{S}(\mathbf{R}^4)$ and $\hat{\phi}, \phi \in \mathcal{S}_{+}(3)$ are defined as follows. Every function from $\mathcal{S}(\mathbf{R}^4)$ has a unique representation in the form

$$\phi(x_0, \vec{x}) = \phi_a(x_0, \vec{x}) + \phi_s(x_0, \vec{x}),$$

where $\phi_s$ is even, and $\phi_a$ is odd with respect to $x_0$ (see Section 1.6). There exists a unique function $\hat{\phi} \in \mathcal{S}_{+}(3)$ such that

$$\phi_a(x_0, \vec{x}) = x_0 \hat{\phi}(x_0^2, \vec{x}). \tag{1.2}$$

Similarly there exists a unique fucntion $\phi \in \mathcal{S}_{+}(3)$ such that

$$\phi_s(x_0, \vec{x}) = \check{\phi}(x_0^2, \vec{x}). \tag{1.3}$$

The mapping

$$\phi \to (\hat{\phi}, \check{\phi})$$

is an isomorphism from the space $\mathcal{S}(\mathbf{R}^4)$ onto the direct sum $\mathcal{S}_{+}(3) \oplus \mathcal{S}_{+}(3)$.

Now let $\phi \in \mathcal{S}(\bar{V})$. Then the function $\phi$ can be continued outside the cone $\bar{V}$ as a function from $\mathcal{S}(\mathbf{R}^4)$ (of course, in many ways) (see Section 1.1). Then one can apply the operators $\hat{\phi}$ and $\phi$ to this continuation. For the functions from $\mathcal{S}(\bar{V})$ let us introduce the mappings

$$\phi(x) \to \phi_1(\kappa, \vec{x}) = \hat{\phi}(\kappa + |\vec{x}|^2, \vec{x}), \tag{1.4}$$

$$\phi(x) \to \phi_2(\kappa, \vec{x}) = \phi(\kappa + |\vec{x}|^2, \vec{x}),$$

where $\phi_1, \phi_2 \in {}_{+}(3)$. It is clear that the functions $\phi_1$ and $\phi_2$ do not depend on a particular continuation of $\phi$.

The mapping $\phi \to \phi_1$ is an isomorphism from the space $\mathcal{S}_a(\bar{V})$ onto the space $\mathcal{S}_{+}(3)$. Similarly the mapping $\phi \to \phi_2$ is an isomorphism from the space $\mathcal{S}_s(\bar{V})$ onto the space $\mathcal{S}_{+}(3)$. So the mapping $\phi \to (\phi_1, \phi_2)$ is an isomorphism from the space $\mathcal{S}(\bar{V})$ onto the direct sum $\mathcal{S}_{+}(3) \oplus \mathcal{S}_{+}(3)$. This fact make it possible to establish the isomorphisms from the spaces $\mathcal{S}_a'(\bar{V})$ and $\mathcal{S}_s'(\bar{V})$ onto $\mathcal{S}_{+}'(3) = \mathcal{S}'(\mathbf{R}_{+}^{1} \times \mathbf{R}^3)$ (see Section 1.3) and therefore the isomorphism from the space $\mathcal{S}'(\bar{V})$ onto the direct sum $\mathcal{S}_{+}'(3) \oplus \mathcal{S}_{+}'(3)$. Let us describe this isomorphism in a more detailed way.

If $\tilde{F} \in \mathcal{S}'(\bar{V})$ then it has a unique decomposition into the sum of its odd $\tilde{F}_a$ and even $\tilde{F}_s$ parts in $x_0$:

$$\tilde{F} = \tilde{F}_a + \tilde{F}_s;$$

$$\tilde{F}_a = \frac{1}{2}[\tilde{F}(x_0, \vec{x}) - \tilde{F}(-x_0, \vec{x})],$$

$$\tilde{F}_s = \frac{1}{2}[\tilde{F}(x_0, \vec{x}) + \tilde{F}(-x_0, \vec{x})].$$

If $\tilde{F}$ is given, then one can construct two distributions $\Phi_a(\kappa, \vec{x})$ and $\Phi_2(\kappa, \vec{x})$ from $\mathcal{S}_{+}'(3)$ by the following formulae: for any $\phi(\kappa, \vec{x})$ from $\mathcal{S}_{+}(3)$

$$(\Phi_a, \phi) = (\tilde{F}(x), x_0 \phi(x^2, \vec{x})) = (\tilde{F}_a(x), x_0 \phi(x^2, \vec{x})), \tag{1.6'}$$

$$(\Phi_s, \phi) = (\tilde{F}(x), \phi(x^2, \vec{x})) = (\tilde{F}_s(x), \phi(x^2, \vec{x})). \tag{1.6''}$$

Conversely, if two distributions $\Phi_a$ and $\Phi_s$ from $\mathcal{S}'_+(3)$ are given, then the distribution $\tilde{F}$ from $\mathcal{S}'(\bar{V})$ is reconstructed in the following way

$$\tilde{F} = \tilde{F}_a + \tilde{F}_s, \tag{1.7}$$

where $\tilde{F}_a$ and $\tilde{F}_s$ are defined by the formulae

$$(\tilde{F}_a, \phi) = (\Phi_a(\kappa, \vec{x}), \hat{\phi}(\kappa + |\vec{x}|^2, \vec{x})), \tag{1.8'}$$

$$(\tilde{F}_s, \phi) = (\Phi_s(\kappa, \vec{x}), \check{\phi}(\kappa + |\vec{x}|^2, \vec{x})), \tag{1.8''}$$

where $\phi \in \mathcal{S}(\mathbf{R}^4)$, and operations $\hat{\phi}$ and $\check{\phi}$ have been introduced by (1.2) and (1.3), so that

$$\phi(x_0, \vec{x}) = x_0\hat{\phi}(x_0^2, \vec{x}) + \check{\phi}(x_0^2, \vec{x}). \tag{1.9}$$

The relations (1.7) can formally be written as

$$\tilde{F}(x) = \epsilon(x_0)\Phi_a(x^2, \vec{x}) + |x_0|\Phi_s(x^2, \vec{x}). \tag{1.10}$$

Let us introduce the distributions $\Psi_a(\lambda, \vec{u})$ and $\Psi_s(\lambda, \vec{u})$ from $\mathcal{S}'_+(3)$ by

$$\Psi_a(\lambda, \vec{u}) = \frac{1}{4\pi^2 i} F_{\vec{x}}^{-1}[B_\kappa[\Phi_a(\kappa, \vec{x})]], \tag{1.11'}$$

$$\Psi_s(\lambda, \vec{u}) = \frac{1}{4\pi^2} F_{\vec{x}}^{-1}[B_\kappa[\Phi_s(\kappa, \vec{x})]], \tag{1.11''}$$

where $B_\kappa$ is the $B$-transform with respect to $\kappa$ and $F_{\vec{x}}^{-1}$ is the inverse Fourier transform with respect to x (see Section 2.1 and Section 2.3).

Let us note also that the mappings

$$\phi(q) \to \int \hat{\phi}(|\vec{q} - \vec{u}|^2 + \lambda, \vec{q})\, d\vec{q} = \phi_1(\lambda, \vec{u})$$

$$\phi(q) \to \int \check{\phi}(|\vec{q} - \vec{u}|^2 + \lambda, \vec{q})\, d\vec{q} = \phi_2(\lambda, \vec{u})$$

are linear and continuous from $\mathcal{S}(\mathbf{R}^4)$ into $\mathcal{S}_+(3)$.

**Theorem 1.** *Let $F$ be a causal distribution; that is, Fourier transform $\tilde{F} \in \mathcal{S}'(\bar{V})$. Then for any $\phi \in \mathcal{S}(\mathbf{R}^4)$ the following formulae hold*

$$(F_a, \phi) = (\Psi_a(\lambda, \vec{u}), \int \hat{\phi}(|\vec{q} - \vec{u}|^2 + \lambda, \vec{q})\, d\vec{q}), \tag{1.12'}$$

$$(F_s, \phi) = 2(\partial_\lambda \Psi_s(\lambda, \vec{u}), \int \check{\phi}(|\vec{q} - \vec{u}|^2 + \lambda, \vec{q})\, d\vec{q}). \tag{1.12''}$$

*Here*

$$F = F_a + F_s \tag{1.13}$$

*and*

$$\phi(q) = q_0\hat{\phi}(q_0^2, \vec{q}) + \phi(q_0^2, \vec{q}) \tag{1.14}$$

*are decompositions of $F$ and $\phi$ into the sums of odd and even parts, respectively. The distributions $\Psi_a$ and $\Psi_s$ are given by formulae (1.11) and (1.10).*

**Proof.** First, let us prove (1.12'). It is sufficient to prove (1.12') only for the odd test functions from $\mathcal{S}'(\mathbf{R}^4)$ with respect to $q_0$; that is, for functions of the form

$$\phi(q) = q_0\hat{\phi}(q_0^2, \vec{q}). \tag{1.15}$$

Let us note the following formula (see N.N. Bogolyubov and D.V. Shirkov, p. 464):

$$F\left[\epsilon(x_0)\delta(x^2) - \frac{1}{2}\sqrt{\frac{\lambda}{x^2}}\ \epsilon(x_0)\theta(x^2)J_1(\sqrt{\lambda x^2})\right] = 4\pi^2 i\ \epsilon(q_0)\delta(q^2-\lambda).$$

Let us apply both sides of this equality to the test function $q_0$, $\hat{\phi}(q_0^2, \vec{q})$, $\hat{\phi} \in \mathcal{S}_+(3)$. Noting that this function is the Fourier transform of some odd function $\phi(x_0, \vec{x}) = x_0\phi_1(x_0^2, \vec{x})$; that is,

$$q_0\hat{\phi}(q_0^2, \vec{q}) = F[x_0\phi_1(x_0^2, \vec{x})],$$

we have for any $\vec{u} \in \mathbf{R}^2$ and $\lambda \geqslant 0$

$$\int e^{i(\vec{u},\vec{x})}\int_0^\infty J_0(\sqrt{\lambda\kappa})\partial_\kappa\phi_1(\kappa + |\vec{x}|^2, \vec{x})\, d\kappa\, d\vec{x} =$$

$$= \frac{1}{4\pi^2}\int\hat{\phi}(|\vec{q} - \vec{u}|^2 + \lambda, \vec{q})\, d\vec{q}. \tag{1.16}$$

Taking into account formulae (1.8), (1.11'') and (3.6), (3.1) of Section 2 we get the equations

$$(F_a, \phi) = (\tilde{F}_a, F^{-1}[\phi]) =$$
$$= (\Phi_a(\kappa, \vec{x}), \phi_1(\kappa + |\vec{x}|^2, \vec{x})) =$$
$$= -4\pi^2 i(\Psi_a(\lambda, \vec{u}), \int e^{i(\vec{u},\vec{x})}\int_0^\infty J_0(\sqrt{\lambda\kappa})\partial_\kappa\phi_1(\kappa + |\vec{x}|^2, \vec{x})\, d\kappa\, d\vec{x}),$$

where $\phi$ is a function of the form (1.15).

Now (1.12') follows from (1.16).

To prove formula (1.12'') let us consider the odd distribution $F_1(q) = (1/2q_0)F_s(q)$, which is defined by the equality

$$(F_1, \phi) = \frac{1}{2}(F_s(q), \hat{\phi}(q_0^2, \vec{q})). \tag{1.17}$$

Its Fourier transform $\tilde{F}_1(x)$ is (up to some factor) one of the primitives of $\tilde{F}_s(x)$ with respect to $x_0$, namely the one which belongs to $\mathcal{S}_a'(\bar{V})$, because $F_1$ is odd. On the other hand, it follows from (1.17), (1.6) and (1.8) that the corresponding distribution $\Phi_1(\kappa, \vec{x})$ is related to $\Phi_s(\kappa, \vec{x})$ by means of the condition

$$\Phi_s(\kappa, \vec{x}) = \frac{4}{i}\partial_\kappa\Phi_1(\kappa, \vec{x}). \tag{1.18}$$

Indeed, for any $\phi \in \mathcal{S}_+(3)$ we have

$$(\Phi_s, \phi) = (F_s(x), \phi(x^2, \vec{x})) = \frac{2}{i}(\partial_0\tilde{F}_1(x), \phi(x^2, \vec{x})) =$$
$$= -\frac{4}{i}(\tilde{F}_1(x), x_0\partial_{x^2}\phi(x^2, \vec{x})) = \frac{4}{i}(\partial_\kappa\Phi_1, \phi).$$

Taking into account (1.18) and (1.11) and the fact that the $B$-transform of a primitive of a distribution is a derivative of the $B$-transform (see (3.14), Section 2.3) we get for the weight function $\Psi_1(\lambda, \vec{u})$

$$\Psi_1(\lambda, \vec{u}) = \frac{1}{4\pi^2 i} F_{\vec{x}}^{-1}\left[B_\kappa\left[\frac{i}{4}\Phi_s^{(-1)}(\kappa, \vec{x})\right]\right] =$$

$$= \frac{1}{4\pi^2} F_{\vec{x}}^{-1}[\partial_\lambda B_\kappa[\Phi_s(\kappa, \vec{x})]].$$

so by (1.11")

$$\Psi_1(\lambda, \vec{u}) = \partial_\lambda \Psi_s(\lambda, \vec{u}) \tag{1.19}$$

and (1.12") follows.

Formula (1.12)-(1.14) can be expressed by the formal relationship

$$F(q) = \int \epsilon(q_0)\delta[(q-u)^2 - \lambda]\Psi_0(\lambda, u)\, d\lambda\, du, \tag{1.20}$$

where

$$\Psi_0(\lambda, u) = \Psi_a(\lambda, \vec{u}) \times \delta(u_0) + \Psi_s(\lambda, \vec{u}) \times \delta'(u_0). \tag{1.21}$$

Formula (1.20) is usually called the symmetric Jost-Lehmann-Dyson representation for $F(q)$, the distributions $\Psi_a$ and $\Psi_s$ being the odd and even weight functions, respectively. Of course, later we shall use only the strict formulae (1.12).

The following theorem holds.

**Theorem 2.** *(Jost-Lehmann). Let a causal distribution $F(q)$ vanish in some neighbourhood of the segment $I(a) = [q \in \mathbf{R}^4\colon |q_0| \leqslant a_0, \vec{q} = \vec{a}]$, where $\vec{a}$ is some vector from $\mathbf{R}^3$. Then its weight function $\Psi_0(\lambda, u)$ vanishes in the set $O_{I(a)}$ which consists of all the points $(\lambda \geqslant 0, u_0 = 0, \vec{u} \in \mathbf{R}^3)$ for which the hyperboloid, defined by the equation $q_0^2 - |\vec{q} - \vec{u}|^2 = \lambda$, intersects the segment $I(a)$* (see Fig. 16). *In other words* (see Fig. 17).

$$O_{I(a)} = [(\lambda, u) \in \mathbf{R}_+^1 \times \mathbf{R}^4\colon u_0 = 0,\ |\vec{u} - \vec{a}|^2 + \lambda < a_0^2].$$

Later we shall give the proof of this theorem for a more general case (in Section 13.3). Now let us point out the following corollary.

**Corollary.** *Let $F$ be a causal distribution. Then the support of $F$ is in the set* (see Fig. 18)

$$T = [q \in \mathbf{R}^4\colon |\vec{q}| \leqslant |q_0| + 1] \tag{1.22}$$

*if and only if the supports of the weight functions $\Psi_a$ and $\Psi_s$ are in the set* (see Fig. 19)

$$T_0 = [(\lambda, \vec{u}) \in \mathbf{R}_+^1 \times \mathbf{R}^3\colon |\vec{u}| \leqslant 1]. \tag{1.23}$$

## 2. Inversion of the Jost-Lehmann-Dyson representation in the symmetric case.

**Theorem 1.** *Let $F$ be a causal distribution and let $(\Psi_a, \Psi_s)$ be its weight functions in the Jost-Lehmann-Dyson representation. Then there is a number $N \geqslant 0$ such that for any $\phi \in \mathcal{S}(\mathbf{R}^4)$ satisfying the conditions*

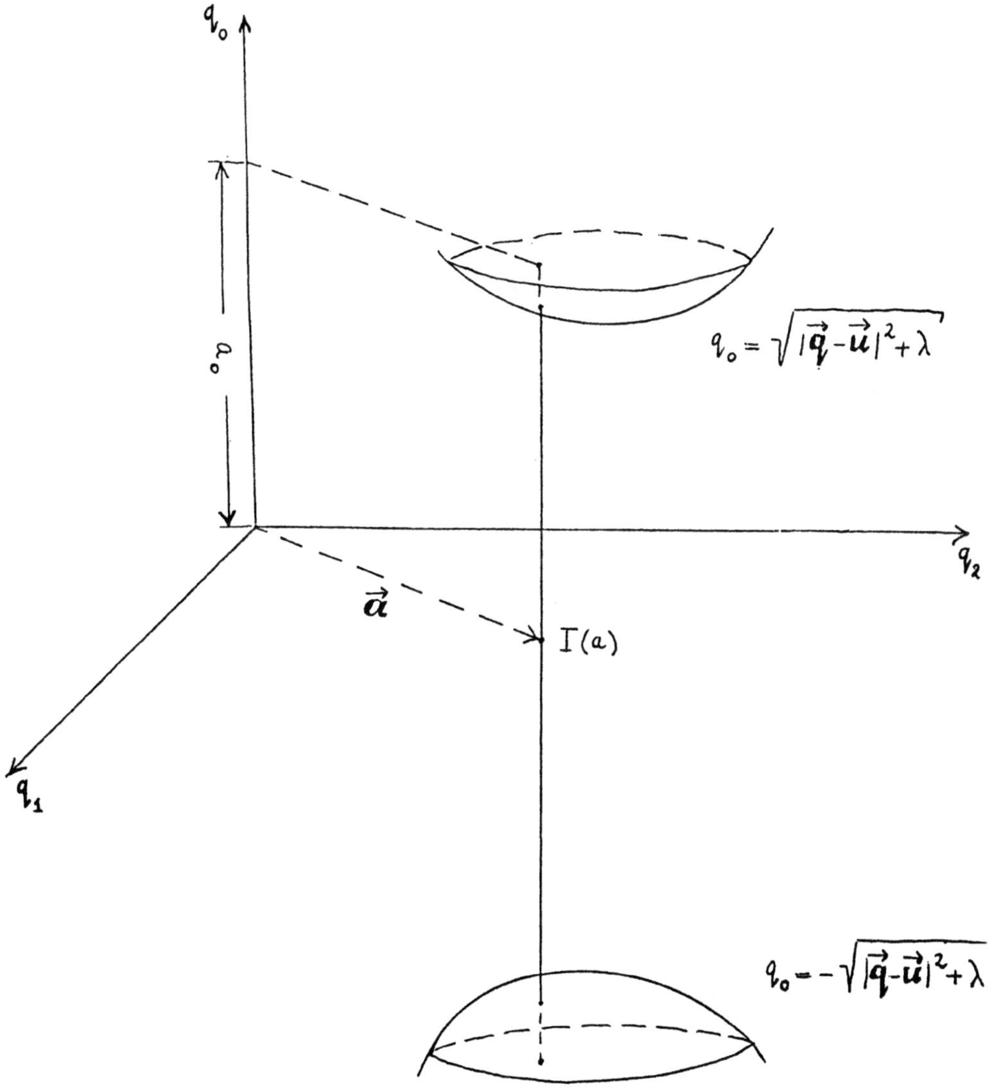
$q_0$
$a_0$
$q_0 = \sqrt{|\vec{q}-\vec{u}|^2+\lambda}$
$q_2$
$\vec{a}$
$\Gamma(a)$
$q_1$
$q_0 = -\sqrt{|\vec{q}-\vec{u}|^2+\lambda}$

Figure 16.

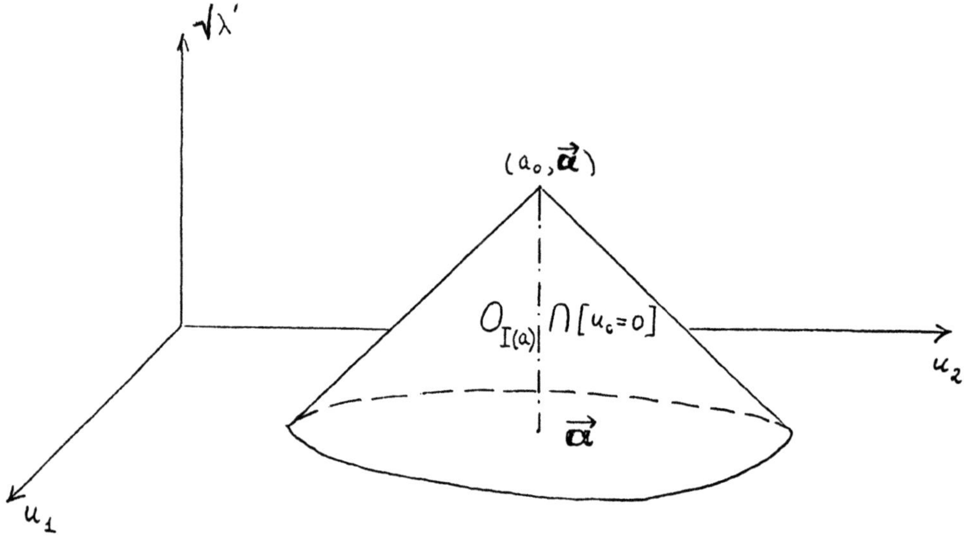


Figure 17.

$$\int_{-\infty}^{+\infty} \phi(\lambda, \vec{u})\lambda^k \, d\lambda = 0, \ k = 0, 1, \ldots, N, \tag{2.1}$$

*the following formulae are valid*

$$(\Psi_a, \phi) = \frac{1}{\pi^3}\left\{F_a(q), q_0 \int {}^{(3/2)}\phi^{(3/2)}(q_0^2 - |\vec{y}|^2, \vec{q} - \vec{u})\, d\vec{y}\right|. \tag{2.2'}$$

$$(\Psi_s, \phi) = \frac{1}{2\pi^3}\left(F_s(q), \int {}^{(5/2)}\phi^{(3/2)}(q_0^2 - |\vec{y}|^2, \vec{q} - \vec{y})\, d\vec{y}\right|. \tag{2.2''}$$

*where the fractional derivatives of $\phi(\lambda, \vec{u})$ act on the first argument.**

**Proof.** Formula (2.2'') follows from (2.2') for the distribution $F_1(q) = (1/2q_0)F_s(q)$ and its weight function $\Psi_1 = \partial_\lambda \Psi_s$ (see (1.17) and (1.19)) by virtue of the equality

$${}^{(3/2)}(\partial_\lambda \phi)^{(3/2)} = -\,{}^{(5/2)}\phi^{(3/2)}.$$

Let us prove the formula (2.2'). By (1.11') we have

$$(\Psi_a, \phi) = \frac{1}{4\pi^2 i}(\Phi_a(\kappa, \vec{x}), b_\lambda[F_{\vec{u}}^{-1}[\phi(\lambda, \vec{u})]]). \tag{2.3}$$

But

$$F_{\vec{u}}^{-1}[\phi(\lambda, \vec{u})] = \int \underset{\sim}{\phi}(x_0, \vec{x}) e^{i\lambda x_0} dx_0,$$

where $\underset{\sim}{\phi}(x_0, \vec{x})$ is the inverse Fourier transform of $\phi(\lambda, \vec{u})$ with respect to all

*For the right and left derivatives see Section 2.4. Note also that due to the lemmas of Section 2.4 the right-hand sides of formulae (2.2) are well defined if $N$ is sufficiently large.

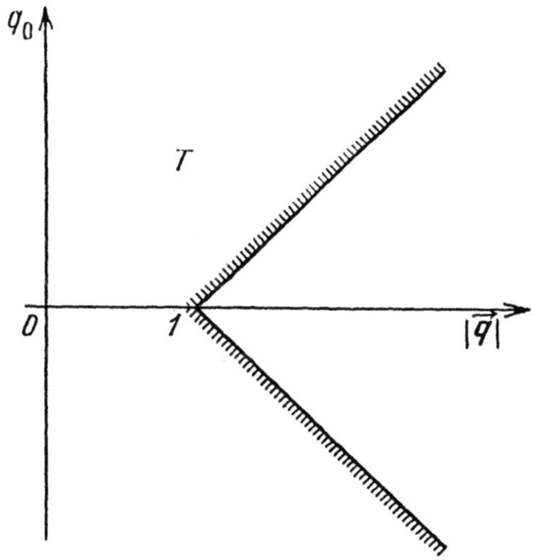


Figure 18

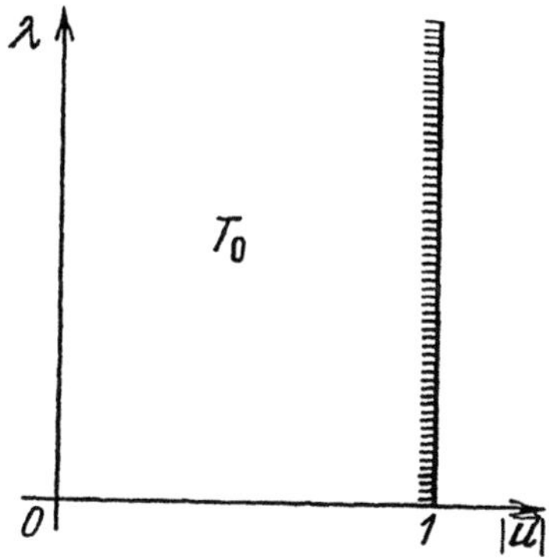


Figure 19

variables. Further, we have

$$(\Psi_a,\phi) = \lim_{\epsilon\to+0}\frac{1}{4\pi^2 i}\left(\Phi_a(\kappa,\vec{x}),\, b_\lambda\left[\int\phi(x_0,\vec{x})e^{i\lambda(x_0+i\epsilon)}dx_0\right]\right) =$$

$$= \lim_{\epsilon\to+0}\frac{1}{4\pi^2 i}\left(\Phi_a(\kappa,\vec{x}),\, \int\phi(x_0,\vec{x})b_\lambda\left[e^{i\lambda(x_0+i\epsilon)}\right]dx_0\right) =$$

$$= \lim_{\epsilon\to+0}\frac{1}{4\pi^2 i}\left(\Phi_a(\kappa,\vec{x}),\, \int\phi(x_0,\vec{x})e^{-i(\kappa/4(x_0+i\epsilon))}dx_0\right). \tag{2.4}$$

Here we used formula (3.4) of Section 2.3.

Since the distribution $\Phi_a(\kappa,\vec{x})$ has a finite order (see Section 1.3) then $\Phi_a \in \mathcal{S}_p'(\mathbf{R}_+^1\times\mathbf{R}^3)$ for some $p \geqslant 0$. On the other hand, due to (2.1)

$$\underset{\sim}{\phi}(x_0,\vec{x}) = x_0^N\phi_1(x_0,\vec{x}),$$

where $\phi_1 \in \mathcal{S}(\mathbf{R}^4)$. So it follows that

$$\int\underset{\sim}{\phi}(x_0,\vec{x})e^{-i(\kappa/4(x_0+i\epsilon))}dx_0 \to \int\underset{\sim}{\phi}(x_0,\vec{x})e^{i(\kappa/4x_0)}dx_0,\quad \epsilon\to+0$$

in $\mathcal{S}_p(\mathbf{R}_+^1\times\mathbf{R}^3)$ if $N$ is sufficiently large. Therefore one can take a limit in (2.4) before the functional action. As a result we get

$$(\Psi_a, \underset{\sim}{\phi}) = \frac{1}{4\pi^2 i}\left(\Phi_a(\kappa, \vec{x}), \underset{\sim}{\int}\phi(x_0, \vec{x})e^{-i(\kappa/4x_0)}dx_0\right). \tag{2.5}$$

Now using (1.6') we obtain

$$(\Psi_a, \phi) = \frac{1}{4\pi^2 i}\left(\tilde{F}_a(x), x_0\int\phi(x_0', \vec{x})e^{-i(x^2/4x_0')}dx_0'\right) =$$

$$= \frac{1}{4\pi^2 i}\left(F_a(q), F_{\vec{x}}\left[\int\phi(x_0', \vec{x})e^{i(|\vec{x}|^2/4x_0')}F_{x_0}\left[x_0 e^{-i(x_0^2/4x_0')}\right]dx_0'\right]\right).$$

Taking into account formula (1.3) of Section 2.1 we have

$$F_{x_0}\left[x_0 e^{-i(x_0^2/4x_0')}\right] = \frac{1}{i}\partial_{q_0}F_{x_0}\left[e^{-i(x_0^2/4x_0')}\right] =$$

$$= \frac{1}{i}\partial_{q_0}\sqrt{4\pi|x_0'|}e^{-i(\pi/4)\epsilon(x_0')+iq_0^2x_0'} =$$

$$= 4\sqrt{\pi}x_0'\sqrt{|x_0'|}q_0 e^{-i(\pi/4)\epsilon(x_0')+iq_0^2x_0'}.$$

Therefore,

$$(\Psi_a, \phi) = -\frac{i}{\pi\sqrt{\pi}}\left(F_a(q), q_0F_{\vec{x}}\left[\int\underset{\sim}{\phi}(x_0', \vec{x})x_0'\sqrt{|x_0'|}e^{-i(\pi/4)\epsilon(x_0')+iq_0^2x_0'+i(|\vec{x}|^2/4x_0')}dx_0'\right]\right) =$$

$$= -\frac{i}{\pi\sqrt{\pi}}\left(F_a(q), q_0\int x_0'\sqrt{|x_0'|}e^{-i(\pi/4)\epsilon(x_0'+iq_0^2x_0')}\right.$$

$$\left. F_{\vec{x}}\left[\phi(x_0', \vec{x})e^{i(|\vec{x}|^2/4x_0')}\right]dx_0'\right). \tag{2.6}$$

Using formula (1.3) of Section 2.1 again, we get

$$F_{\vec{x}}\left[\phi(x_0', \vec{x})e^{i(|\vec{x}|^2/4x_0')}\right] = \frac{1}{9\pi^3}F_{\vec{x}}\left[\phi(x_0', \vec{x})\right] * F_{\vec{x}}\left[e^{i(|\vec{x}|^2/4x_0')}\right] =$$

$$= \frac{|x_0'|^{3/2}}{\pi\sqrt{\pi}}e^{i(3\pi/4)\epsilon(x_0')}\left(F_{q_0}^{-1}[\phi](x_0', \vec{q}) * e^{-i|\vec{q}|^2x_0'}\right) =$$

$$= \frac{|x_0'|^{3/2}}{\pi\sqrt{\pi}}e^{i(3/4)\pi\epsilon(x_0')}\int e^{-i|\vec{y}|^2x_0'}F_{q_0}^{-1}[\phi](x_0', \vec{q}-\vec{y})\,d\vec{y}. \tag{2.7}$$

It is clear that all those manipulations are correct and all the expressions make sense if $N$ is sufficiently large. Substituting (2.6) into (2.7) we get

$$(\Psi_a, \phi) = \frac{1}{\pi^3}\left(F_a(q), q_0\int F_{q_0}^{-1}[\phi](x_0, \vec{q}-\vec{y})|x_0|^3 e^{ix_0(q_0^2-|\vec{y}|^2)}dx_0\,d\vec{y}\right) =$$

$$= \frac{1}{\pi^3}\left(F_a(q), q_0\int F_{x_0}\left[F_{q_0}^{-1}[\phi](x_0, \vec{q}-\vec{y})|x_0|^3\right](q_0^2-|\vec{y}|^2)\,d\vec{y}\right).$$

So formula (2.2) follows if one takes into account that for $\phi \in \mathcal{S}(\mathbf{R}^1)$ the following equalities are valid

$$
\begin{aligned}
\underset{\sim}{F}[\phi(x_0)|x_0|^3] &= \underset{\sim}{F}[\phi(x_0)(x_0+i0)^{3/2}(x_0-i0)^{3/2}] = \\
&= \underset{\sim}{F}[\phi(x_0)F[f_{-(3/2)}(\xi)]\,F[f_{-(3/2)}(-\xi)]] = \\
&= \frac{1}{4\pi^2}\phi(q_0) * F^2[f_{-(3/2)}(\xi)] * F^2[f_{-(3/2)}(-\xi)] = \\
&= \phi(q_0) * f_{-(3/2)}(q_0) * f_{-(3/2)}(-q_0) = {}^{(3/2)}\phi^{(3/2)}(q_0).
\end{aligned}
$$

The following lemma holds.

**Lemma 1.** *Let $\phi(\lambda, \vec{u})$ belong to the space $\mathcal{S}_+(3)$. Then for any natural number $N$ there exists $\phi_1 \in \mathcal{S}(\mathbf{R}^3)$ such that*

$$\phi_1(\lambda, \vec{u}) = \phi(\lambda, \vec{u}), \quad \lambda \geqslant 0$$

*and $\phi_1$ satisfies conditions* (2.1).

**Proof.** The function $\phi$ can be continued into the domain $[\lambda < 0, \vec{u} \in \mathbf{R}^3]$ as a function from $\mathcal{S}(\mathbf{R}^4)$ (see Section 1.1). Preserving the same notation for this continuation let us introduce the functions

$$\eta_k(\vec{u}) = \int \lambda^k \phi(\lambda, \vec{u})\, d\lambda, \quad k = 0, 1, \cdots$$

It is clear that $\eta_k \in \mathcal{S}(\mathbf{R}^3)$, $k = 0, 1, \cdots$. For any natural number $N$ there exist functions $\psi_j \in \mathcal{S}(\mathbf{R}^1)$, $j = 0, 1, \ldots, N$, which have the properties

$$\operatorname{supp}\psi_j \subset (-\infty, 0], \quad \int_{-\infty}^{0} \lambda^k \psi_j(\lambda)\, d\lambda = \delta_{jk}, \quad kj = 0, 1, \ldots, N.$$

Now one can take

$$\phi_1 = \phi(\lambda, \vec{u}) - \sum_{j=0}^{N} \psi_j(\lambda)\eta_j(\vec{u}).$$

According to theorem 1 the following formulae hold:

$$
\begin{aligned}
(\Psi_a, \phi) &= \frac{1}{\pi^3}\left(F_a(q), q_0 \int {}^{(3/2)}\phi_1^{(3/2)}(q_0^2 - |\vec{y}|^2, \vec{q} - \vec{y})\, d\vec{y}\right), \\
(\Psi_s, \phi) &= \frac{1}{2\pi^3}\left(F_s(q), \int {}^{(5/2)}\phi_1^{(3/2)}(q_0^2 - |\vec{y}|^2, \vec{q} - \vec{y})\, d\vec{y}\right).
\end{aligned} \tag{2.8}
$$

Thus, these formulae are in fact the inversion formulae for the Jost-Lehmann-Dyson representation.

Now let us suppose that the distribution $F(q)$ satisfies condition (1.22), that is, $\operatorname{supp} F \subset T$. As previously mentioned, this is equivalent to the condition $\operatorname{supp}\Psi_{a,s} \subset T_0$ (see (1.23)). If $\phi \in \mathcal{S}(\mathbf{R}^1)$ and $\vec{a}$ is some vector from $\mathbf{R}^3$ then the following expressions make sense

$$(\Psi_{a,s}(\lambda, \vec{u}), \phi(-\lambda + (\vec{a}, \vec{u}))).$$

Later, we shall require the following theorem:

**Theorem 2.** *Let $F$ be a causal distribution with support in $T$, and let $(\Psi_a, \Psi_s)$ be its weight functions. Then there exists a natural number $N$ (which depends on the order $F$ in $\mathcal{S}'(T)$) such that for any $\phi \in \mathcal{S}(\mathbf{R}^1)$ satisfying the conditions*

$$\int_{-\infty}^{+\infty} \lambda^k \phi(\lambda)\, d\lambda = 0,\ k = 0, 1, \ldots, N \tag{2.1'}$$

*the following formulae hold:*

$$(\Psi_a(\lambda, \vec{u}), \phi(-\lambda + (\vec{a}, \vec{u}))) = \frac{1}{\pi\sqrt{\pi}} \left( F_a(q), q_0 \phi^{(3/2)} \left( -\frac{|\vec{a}|^2}{4} - q_0^2 + (\vec{a}, \vec{q}) \right) \right), \tag{2.9'}$$

$$(\Psi_s(\lambda, \vec{u}), \phi(-\lambda + (\vec{a}, \vec{u}))) = \frac{1}{2\pi\sqrt{\pi}} \left( F_s(q), \phi^{(5/2)} \left( -\frac{|\vec{a}|^2}{4} - q_0^2 + (\vec{a}, \vec{q}) \right) \right), \tag{2.9''}$$

*where* $\vec{a}$ *is any vector from* $\mathbf{R}^3$. *In this case the function*

$$\phi^{(\alpha)} \left( -\frac{|\vec{a}|^2}{4} - q_0^2 + (\vec{a}, \vec{q}) \right),\quad \alpha = \frac{3}{2}, \frac{5}{2}$$

*belongs to* $\mathcal{S}_p(T)$ *if* $N$ *is sufficiently large,* $p$ *is the order of* $F$, *so that the right hand sides of* (2.9) *make sense.*

**Proof.** Let us prove (2.9') for instance. Assuming $\vec{a} = (2a, 0, 0)$ let us put in (2.2')

$$\phi(\lambda, \vec{u}) = \phi_0(-\lambda + 2au_1) g_1 \left( \frac{u_1}{R} \right) g_2(\tilde{u}),$$

where $\phi_0 \in \mathcal{S}(\mathbf{R}^1)$ satisfies the conditions (2.1'), $\tilde{u} = (u_2, u_2)$, $g_2 \in \mathcal{D}(\mathbf{R}^2)$, $g_1 \in \mathcal{D}(\mathbf{R}^1)$, $g_1(0) = 1$, $R > 0$. We have

$$\left( \Psi_a(\lambda, \vec{u}), \phi_0(-\lambda + 2au_1) g_1 \left( \frac{u_1}{R} \right) g_2(\tilde{u}) \right) = \frac{1}{\pi^3} (F_a, \eta_R). \tag{2.10}$$

where

$$\eta_R(q) = q_0 \int^{(3/2)} \phi_0^{(3/2)} (|\vec{y}|^2 - q_0^2 + 2a(q_1 - y_1)) g_1 \left( \frac{q_1 - y_1}{R} \right) g_2(\tilde{q} - \tilde{y})\, d\tilde{y}.$$

Introducing a new variable of integration $x = (y_1 - a^2)$ we get

$$\eta_R(q) = \frac{q_0}{2} \int g_2(\tilde{q} - \tilde{y}) \int_0^{\infty} {}^{(3/2)} \phi_0^{(3/2)} (x - a^2 - q_0^2 + 2aq_1 + |\tilde{y}|^2) \times$$

$$\times \left[ g_1 \left( \frac{q_1 - a - \sqrt{x}}{R} \right) + g_1 \left( \frac{q_1 - a + \sqrt{x}}{R} \right) \frac{dx}{\sqrt{x}} \right] d\tilde{y}. \tag{2.11}$$

It is clear that if $N$ is sufficiently large, then for $R \to \infty$

$$\eta_R(q) \to q_0 \int g_2(\tilde{q} - \tilde{y}) \int_0^{\infty} {}^{(3/2)} \phi_0^{(3/2)} (x - a^2 - q_0^2 + 2aq_1 + |\tilde{y}|^2) \frac{dx}{\sqrt{x}}\, d\tilde{y} =$$

$$= \sqrt{\pi} q_0 \int^{(1)} \phi_0^{(3/2)} (-a^2 - q_0^2 + 2aq_1 + |\tilde{y}|^2) g_2(\tilde{q} - \tilde{y})\, d\tilde{y} \equiv \eta(q). \tag{2.12}$$

where convergence is uniform over $q$ on every compact set and the same is true for all derivatives of $\eta_R(q)$. Let us now show that convergence in (2.12) is in fact convergence in $\mathcal{S}(T)$. To show this let us prove the following lemma.

**Lemma 2.** *Let* $g \in C^n([0,\infty))$ *and let* $\phi \in C^\infty(\mathbf{R}^1)$ *satisfy conditions* (2.1'). *Then for any* $N \geqslant 2n+3$ *and* $R > 0$ *the following inequality holds:*

$$\left|\int_0^\infty \phi(\lambda+\xi) g\left(\frac{\lambda}{R}\right)\frac{dx}{\sqrt{\lambda}}\right| \leqslant c(\phi) P_n[g](1+\xi^2)^{-[n/2]}, \tag{2.13}$$

*where*

$$P_n[g] = \max_{0\leqslant k\leqslant n}\ \sup_{\lambda\geqslant 0}\lambda^k\,|g^{(k)}(\lambda)|.$$

**Proof.** Denoting $m = [n/2]$ we have

$$(1+\xi^2)^m \int_0^\infty \phi(\lambda+\xi) g\left(\frac{\lambda}{R}\right)\frac{dx}{\sqrt{\lambda}} =$$

$$\int_0^\infty [1+(\lambda+\xi)^2+\lambda^2-2\lambda(\lambda+\xi)]^m \phi(\lambda+\xi) g\left(\frac{\lambda}{R}\right)\frac{dx}{\sqrt{\lambda}} =$$

$$= \sum_{l,k=0}^{2m} C_{l,k}\int_0^\infty (\lambda+\xi)^l \lambda^{k-1/2}\phi(\lambda+\xi) g\left(\frac{\lambda}{R}\right) d\lambda. \tag{2.14}$$

From condition s(2.1') it follows that there are functions $\phi_l \in C^\infty(\mathbf{R}^1)$ such that

$$\lambda^l \phi(\lambda) = \phi_l^{(N+1-l)}(\lambda). \tag{2.15}$$

By the lemma of Section 2.4 the primitive of order $k$, $\phi_l^{(N+1-l-k)}$ of the function $\lambda^l\phi(\lambda)$ satisfies the conditions

$$\int_{-\infty}^{+\infty} \left|\phi_l^{(N+1-l-k+j)}|\lambda^s|\, d\lambda\right| < \infty;\ s = 0,1,\ldots,N-2-l-k,\ j = 0,1,\cdots \tag{2.16}$$

where $l,k = 0,1,\ldots,2m \leqslant n$.

From (2.15) we derive

$$\left|\int_0^\infty (\lambda+\xi)^l\phi(\lambda+\xi)g\left(\frac{\lambda}{R}\right)\lambda^{k-1/2}d\lambda\right| = \left|\int_0^\infty \phi_l^{(N+1-l)}(\lambda+\xi)g\left(\frac{\lambda}{R}\right)\lambda^{k-1/2}d\lambda\right| =$$

$$= \left|\int_0^\infty \phi_l^{(N+1-l-k)}(\lambda+\xi)\left(\lambda^{k-1/2}g\left(\frac{\lambda}{R}\right)\right)^{(k)} d\lambda\right| \leqslant$$

$$\leqslant \sup_{\lambda\geqslant 0}\sqrt{\lambda}\left|\left(\lambda^{k-1/2}g\left(\frac{\lambda}{R}\right)\right)^{(k)}\right|\int_0^\infty |\phi_l^{(N+1-l-k)}(\lambda+\xi)|\frac{d\lambda}{\sqrt{\lambda}} \leqslant$$

$$\leqslant \sup_{\lambda\geqslant 0}\sqrt{\lambda}\sum_{j=0}^k\left|\binom{j}{k}(\lambda^{k-1/2})^{(k-j)}g^{(j)}\left(\frac{\lambda}{R}\right)\frac{1}{R^j}\right|\int_\xi^\infty |\phi_l^{(N+1-l-k)}(\lambda)|\frac{d\lambda}{\sqrt{\lambda-\xi}} \leqslant$$

$$\leqslant \int_\xi^\infty |\phi_l^{(N+1-l-k)}(\lambda)|\frac{d\lambda}{\sqrt{\lambda-\xi}}\sum_{j=0}^k C'_{k,j}\sup_{\lambda\geqslant 0}\left|\left(\frac{\lambda}{R}\right)^j g^{(j)}\left(\frac{\lambda}{R}\right)\right|.$$

From (2.16) it is easy to see that

$$\int_{\xi}^{\infty} \left|\phi_l^{(N+1-l-k)}(\lambda)\right| \frac{d\lambda}{\sqrt{\lambda-\xi}} \leqslant C''_{l,k}.$$

Therefore,

$$\left|\int_0^{\infty}(\lambda+\xi)^l \phi(\lambda+\xi) g\left(\frac{\lambda}{R}\right)\lambda^{k-1/2} d\lambda\right| \leqslant C'''_{l,k} P_k[g].$$

From here and from (2.14) we get estimate (2.13).

Let us continue the proof of the theorem, returning to formulae (2.11) and (2.12). We have to prove that for any natural $p$ there is a natural $N$ such that the convergence in (2.12) takes place in $\mathcal{S}_p(T)$, where $T = [q \in \mathbf{R}^4 \colon |\tilde{q}| \leqslant |q_0| + 1]$ as before, and that $\phi$ satisfies condition (2.1'). To show this it is sufficient to prove that the sequence $\{\eta_R, R \to \infty\}$ is bounded in $\mathcal{S}_p(T)$. This will be proved if we show that the expressions of the form

$$(1+|q|^{2p+1})\int G(\tilde{g}-\tilde{y})\int_0^{\infty}\phi_0(x-a^2-q_0^2+2aq_1+|\tilde{y}|^2)\times$$

$$\times\left[g\left(\frac{q_1-a+\sqrt{x}}{R}\right)+g\left(\frac{q_1-a-\sqrt{x}}{R}\right)\right]\frac{dx}{\sqrt{x}}\,d\tilde{y} \tag{2.17}$$

are uniformly bounded for $q \in T$ if $G \in D(\mathbf{R}^2)$, $g \in D(\mathbf{R}^1)$ and $\phi_0 \in C^{\infty}(\mathbf{R}^1)$ satisfies conditon (2.1').

Let us first show that for $2s \leqslant N-5$ the following representation takes place

$$\int\phi(\nu+|\tilde{y}|^2)G(\tilde{q}-\tilde{y})\,d\tilde{y} =$$

$$= \frac{1}{|\tilde{q}|^{2s}}\sum_{l=0}^{2s}\int\phi^{(-l)}(\nu+|\tilde{y}|^2)G_l(\tilde{q}-\tilde{y})\,d\tilde{y}, \tag{2.18}$$

where $G_l$ are functions from $D(\mathbf{R}^2)$. $\phi^{(-l)}$ is the primitive of $\phi$ of order $l$ and therefore, by lemma of Section 2.4, it satisfies conditons (4.2)–(4.3) of Section 2.4, where $N$ should be replaced by $N-l-2$.

Indeed, after the change of variables $\tilde{g}-\tilde{y}=\xi$, we have

$$\int\phi(\nu+|\tilde{y}|^2)G(\tilde{q}-\tilde{y})\,d\tilde{y} = \int\phi(\nu+|\tilde{q}|^2+|\tilde{y}|^2-2(\tilde{q},\xi))G(\xi)\,d\xi =$$

$$= \left(\frac{1}{2|\tilde{q}|}\right)^{2s}\int\Delta_{\tilde{\sigma}}^{s}\phi^{(-2s)}(\nu+|\tilde{q}|^2+|\xi|^2-2(\tilde{q},\tilde{\sigma}))\Big|_{\tilde{\sigma}=\xi}G(\xi)\,d\xi.$$

Using, repeatedly, the formula of integration by parts of the type

$$\int_{-\infty}^{+\infty}\partial_\sigma\psi(\xi,\sigma)\Big|_{\sigma=\xi}d\xi+\int_{-\infty}^{+\infty}\partial_\sigma\psi(\sigma,\xi)\Big|_{\sigma=\xi}d\xi=\int_{-\infty}^{+\infty}\frac{d}{d\xi}\psi(\xi,\xi)\,d\xi=0$$

and returning to the variables $\tilde{y}=\tilde{q}-\xi$, we get representation (2.18).

Substituting expansion (2.18) into (2.17) for $\nu = x - a^2 - q_0^2 + 2aq_1$ we get an expression for a typical summand

$$\frac{(1+|q|)^{2p+1}}{|\tilde{q}|^{2s}} \int G_l(\tilde{q}-\tilde{y}) \int_0^\infty \phi_0^{(-l)}(x - a^2 - q_0^2 + 2aq_1 + |\tilde{y}|^2) \times$$

$$\times \left[ g\left(\frac{q_1 - a + \sqrt{x}}{R}\right) + g\left(\frac{q_1 - a - \sqrt{x}}{R}\right)\right] \frac{dx}{\sqrt{x}} d\tilde{y}. \tag{2.19}$$

Let us denote

$$g_{R,q_1}(x) = g\left(\frac{q_1 - a}{R} - \sqrt{x}\right) + g\left(\frac{q_1 - a}{R} + \sqrt{x}\right).$$

Since

$$x\frac{d}{dx} g_{R,q_1}(x) = \frac{1}{2} s_- g'(s_-) + \frac{1}{2} s_+ g'(s_+) - \frac{q_1 - a}{2R}\left[g'(s_-) - g'(s_+)\right],$$

where $s_\pm = (q_1 - a/R) \pm \sqrt{x}$, then applying the operator $x(d/dx)$ $k$ times we get that

$$\left(x\frac{d}{dx}\right)^k g_{R,q_1}(x) = \sum_{i=0}^k a_i(R) q_1^i g_i(s_-) + \sum_{i=0}^k b_i(R) q_1^i g_i(s_+),$$

where $a_i(R)$ and $b_i(R)$ do not depend on $q_1$ and they are bounded for $R \geqslant 1$. The functions $g_i(s)$ are of the form $s^{n_i} g^{(m_i)}(s)$ for some nonnegative integers $n_i$ and $m_i$. So, it follows that for $R \geqslant 1$ the following estiamte holds

$$\left|\left(x\frac{d}{dx}\right)^k g_{R,q_1}(x)\right| \leqslant A_k(1 + |q_1|^k),$$

where the $A_k$ do not depend on $(x, q_1, R)$.

Since the operator $x^k(d^k/dx^k)$ is the linear combination of the operators $(x(d/dx))^j$, $j = 0, 1, \dots, k$, we get the following estimate for the norm $P_{2n}[g_{R,q_1}(x)]$ in (2.13)

$$P_{2n}[g_{R,q_1}(x)] \leqslant A(1 + |q_1|^{2n}), \tag{2.20}$$

where $A$ does not depend on $(q_1, R)$.

Let us now apply lemma 2 to the internal integral in (2.19). Using estimate (2.20) and assuming

$$N - 2s - 2 \geqslant 4n + 2 \tag{2.21}$$

we obtain

$$\left| \int_0^\infty \phi_0^{(-l)}(x - a^2 - q_0^2 + 2aq_1 + |\tilde{y}|^2) \left[ g\left(\frac{q_1 - a - \sqrt{x}}{R}\right) + g\left(\frac{q_1 - a + \sqrt{x}}{R}\right)\right] \frac{dx}{\sqrt{x}} \right| =$$

$$= \left| \int_0^\infty \phi_0^{(-l)}(x - a^2 - q_0^2 + 2aq_1 + |\tilde{y}|^2) g_{R,q_1}\left(\frac{x}{R^2}\right) \frac{dx}{\sqrt{x}} \right| \leqslant$$

$$\leqslant C_l(1+|q_1|^{2n})\left[1+(|\tilde{y}|^2+2aq_1-q_0^2-a^2)^2\right]^{-n}. \qquad (2.22)$$

where $C_l$ does not depend on $(q, \tilde{y})$. Taking into account expansion (2.18) and formulae (2.19) and (2.22), and nothing that $|\vec{q}| \leqslant 1+|q_0|$ in $T$, we get that (2.17) is estimated by the expression

$$C\left(1+|q_0|^{2p+2n+1}\right)\int\frac{|G(\tilde{q}-\tilde{y})|\,d\tilde{y}}{[1+(|\tilde{y}|^2+2aq_1-q_0^2-a^2)^2]^n},\quad |\tilde{q}|\leqslant 1, \qquad (2.23)$$

$$C\frac{1+|q_0|^{2p+2n+1}}{|\tilde{q}|^{2s}}\int\frac{\sum_l|G_l(\tilde{q}-\tilde{y})|\,d\tilde{y}}{[1+(|\tilde{y}|^2+2aq_1-q_0^2-a^2)^2]^n},\quad |\tilde{q}|\geqslant 1. \qquad (2.24)$$

It is clear that (2.23) is uniformly bounded if $|\tilde{q}| \leqslant 1$ and

$$2p+2n+1\leqslant 4n. \qquad (2.25)$$

To prove that expression (2.24) is bounded, note that if (2.25) holds then

$$\sup_{q\in T}\frac{1+|q_0|^{2p+2n+1}}{[1+(|\tilde{y}|^2+2aq_1-q_0^2-a^2)^2]^n}\leqslant B(1+|\tilde{y}|^{2p+2n+1}), \qquad (2.26)$$

where $B$ does not depend on $\tilde{y}$. Inequality (2.26) implies that (2.24) is estimated by the expression

$$C_1|\tilde{q}|^{-2s}\int(1+|\tilde{y}|^{2p+2n+1})\sum_l|G_l(\tilde{q}-\tilde{y})|\,d\tilde{y}\leqslant C_2(1+|\tilde{q}|^{2p+2n+1})|\tilde{q}|^{-2s},$$

which is uniformly bounded for $|\tilde{q}| \geqslant 1$ if

$$2p+2n+1\leqslant 2s. \qquad (2.27)$$

Note that it is possible to choose the numbers $n$ and $s$ which satisfy conditions (2.21), (2.25) and (2.27) if $N \geqslant 8(p+1)$.

Thus we have proved that for any natural $p$ there exists an $N$ such that the sequence $\{\eta_R(q), R\to\infty\}$ in (2.10)-(2.12) converges in $\mathcal{S}_p(T)$ if $\phi_0$ satisfies conditions (2.1'). Assuming $p$ to be equal to the order of $F_a$ in $\mathcal{S}'(T)$, we conclude that the right-hand side of (2.10) converges, when $R\to\infty$, to the limit equal to

$$\frac{1}{\pi^2\sqrt{\pi}}(F_a(q), q_0\int^{(1)}\phi_0^{(3/2)}(-a^2-q_0^2+2aq_1+|\tilde{y}|^2)g_2(\tilde{q}-\tilde{y})\,d\tilde{y}).$$

On the other hand, it is clear that

$$\phi_0(-\lambda+2au_1)g_1\left(\frac{u_1}{R}\right)g_2(\tilde{u})\to\phi_0(-\lambda+2au_1)g_2(\tilde{u}),\quad R\to\infty$$

in $\mathcal{S}(T_0)$, where $T_0$ is defined by (1.23). Thus the left-hand side of (2.10) tends to

$$(\psi_a(\lambda,\vec{u}), \phi_0(-\lambda+2au_1)g_2(\tilde{u})),$$

when $R\to\infty$. Therefore, the following formula holds:

$$(\psi_a(\lambda,\vec{u}), \phi_0(-\lambda+2au_1)g_2(\tilde{u}))=$$

$$=\frac{1}{\pi^2\sqrt{\pi}}(F_a(q), q_0\int^{(1)}\phi_0^{(3/2)}(-a^2-q_0^2+2aq_1+|\tilde{y}|^2)g_2(\tilde{q}-\tilde{y})\,d\tilde{y}). \qquad (2.28)$$

which does not include integration over $y_1$.

Similarly, by putting in (2.28)

$$g_2(\tilde{q}-\tilde{y}) = g_3(q_3-y_3)g_4\left(\frac{q_2-y_2}{R}\right). \tag{2.29}$$

where $g_3, g_4 \in D(\mathbf{R}^1)$, $g_4(0)=1$ and repeating almost literally all the previous calculations we get that for a sufficiently large $N$

$$q_0\int \phi_0^{(5/2)}(-a^2-q_0^2+2aq_1+y_2^2+y_3^2)g_3(q_3-y_3)g_4\left(\frac{q_2-y_2}{R}\right)dy_2\,dy_3 =$$

$$= \frac{q_0}{2}\int g_3(q_3-y_3)\int_0^\infty{}^{(1)}\phi_0^{(3/2)}(x-a^2-q_0^2+2aq_1+y_3^2)$$

$$\left[g_4\left(\frac{q_3-\sqrt{x}}{R}\right)+g_4\left(\frac{q_3+\sqrt{x}}{R}\right)\right]\frac{dx}{\sqrt{x}}\,dy_3 \to$$

$$\to q_0\int g_3(q_3-y_3)\int_0^\infty{}^{(1)}\phi_0^{(3/2)}(x-a^2-q_0^2+2aq_1+y_3^2)\frac{dx}{\sqrt{x}}\,dy_3 =$$

$$= \sqrt{\pi}q_0\int g_3(q_3-y_3)^{(1/2)}\phi_0^{(3/2)}(-a^2-q_0^2+2aq_1+y_3^2)\,dy_3$$

the convergence being in $\mathcal{S}(T)$. From this and from (2.28)-(2.29) we get the formula

$$(\psi_a(\lambda,\vec{u}),\phi_0(-\lambda+2au_1)g_3(u_3)) =$$

$$= \frac{1}{\pi^2}(F_a(q),q_0\int{}^{(1/2)}\phi_0^{(3/2)}(-a^2-q_0^2+2aq_1+y_3^2)g_3(q_3-y_3)\,dy_3). \tag{2.30}$$

Finally, if we put into (2.30) $g_3(u_3)=g_5(u_3/R)$, $g_5 \in D(\mathbf{R}^1)$, $g_5(0)=1$ and let $R\to\infty$ we get (2.9') for $\vec{a}=(2a,0,0)$. Since the last condition is not essential, formula (2.9') is proved for all $\vec{a}\in\mathbf{R}^3$. Formula (2.9'') follows from (2.9') for the distribution $F_1(q)$ (see (1.17) and (1.19)).

Now let $\phi$ be an arbitrary function from (R¹), and let $\vec{n}$ be a unit vector from $S^2$. Then for any natural $N$ one can take a function $\phi_1\in\mathcal{S}(\mathbf{R}^1)$, such that $\phi_1(\lambda)=\phi(\lambda)$ for $\lambda\leqslant 1$, and $\phi_1$ satisfies conditons (2.1'). Then we have

$$(\psi_j(\lambda,\vec{u}),\phi(-\lambda+(\vec{n},\vec{u}))) = (\psi_j(\lambda,\vec{u}),\phi_1(-\lambda+(\vec{n},\vec{u}))),\ j=a,s.$$

So, if $N$ is sufficiently large, then by theorem 2 the following formulae are valid

$$(\psi_a(\lambda,\vec{u}),\phi(-\lambda+(\vec{n},\vec{u}))) = \frac{1}{\pi\sqrt{\pi}}(F_a(q),q_0\phi_1^{(3/2)}(-\frac{1}{4}-q_0^2+(\vec{n},\vec{u}))). \tag{2.31'}$$

$$(\psi_s(\lambda,\vec{u}),\phi(-\lambda+(\vec{n},\vec{u}))) = \frac{1}{2\pi\sqrt{\pi}}(F_s(q),\phi_1^{(5/2)}(-\frac{1}{4}-q_0^2+(\vec{n},\vec{u}))). \tag{2.31''}$$

In particular the test functions in the right-hand sides of (2.31') and (2.31'') belong to $\mathcal{S}_p(T)$, where $p$ is of the order of $F$.

**3. The Jost-Lehmann-Dyson representation in the general case.** First, let us make some remarks on the notations that we shall use in this subsection. Latin letters $x, q, u, \cdots$ will denote vectors from $\mathbf{R}^4$ so that, for instance, $q = (q_0, \vec{q}) = (q_0, q_1, q_2, q_3)$. As before $q^2 = q_0^2 - |\vec{q}|^2$. Greek characters in roman "bold" type $\vec{\lambda}, \vec{\kappa}, \cdots$ will denote vectors from $\mathbf{R}^2$, so that, for instance, $\vec{\lambda} = (\lambda_1, \lambda_2)$. As usual, $|\vec{\lambda}|^2 = \lambda_1^2 + \lambda_2^2$.

For any $a > 1$ let us put (see Fig. 20)

$$\Gamma_a = \{(\beta, u) \in \mathbf{R}_+^1 \times \mathbf{R}^4 : u_0^2 - (1 - \frac{1}{a})(|\vec{u}|^2 + \beta) \leqslant a\}. \tag{3.1}$$

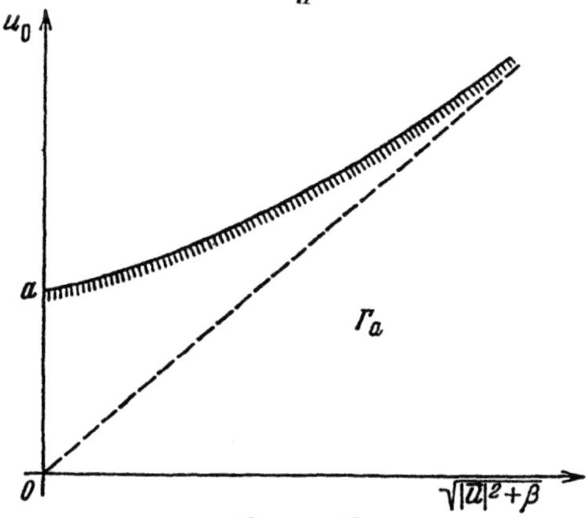


Figure 20.

$\mathcal{S}(\Gamma_a)$ as usual denotes the space of infinitely differentiable functions on $\Gamma_a$ for which see Section 1.1)

$$\|\phi\|_{p,\Gamma_a} = \sup_{(\beta,u)\in\Gamma_a,\, |\alpha| \leqslant p} (1 + |u|^2 + \beta^2)^{p/2} \left|\partial_0^{\alpha_0}\partial_1^{\alpha_1}\partial_2^{\alpha_2}\partial_3^{\alpha_3}\partial_\beta^{\alpha_4}\phi(\beta, u)\right| < \infty, \tag{3.2}$$

$p = 0, 1, \cdots$

The system of norms (3.2) endows $\mathcal{S}(\Gamma_a)$ with a natural topology. Since $\Gamma_a$ is a regular set, the space $\mathcal{S}'(\Gamma_a)$ is identified with the subspace of $\mathcal{S}'(\mathbf{R}^5)$ consisting of tempered distributions with supports in $\Gamma_a$ (see Section 1.3).

For $a_1 > a_2$ there is the restriction map: $\mathcal{S}(\Gamma_{a_1}) \to \mathcal{S}(\Gamma_{a_2})$. This facts makes it possible to construct a projective limit on a decreasing sequence of spaces $\mathcal{S}(\Gamma_a)$, $a \to \infty$, which we denote by $\mathcal{S}$. The elements of $\mathcal{S}$ are differentiable functions on $\mathbf{R}_+^1 \times \mathbf{R}^4$ whose restrictions to any $\Gamma_a$ belong to $\mathcal{S}(\Gamma_a)$. The sequence of functions $\{\phi_k, k \to \infty\}$ converges to zero in $\mathcal{S}$ if and only if for every $p = 0, 1, \cdots$ and $a > 1$ $\|\phi_k\|_{p,\Gamma_a} \to 0$, $k \to \infty$. The space $\mathcal{S}'$ is just the same as the subspace of $\mathcal{S}'(\mathbf{R}^5)$ consisting of distributions with supports in some $\Gamma_a$ (which depends on the particular distribution).

For $\phi \in \mathcal{S}(\mathbf{R}^4)$ let us introduce a mapping $d$:

$$\phi \to d[\phi](\beta, u) = [\phi(q) * \epsilon(q_0)\delta'(q^2 - \beta)](u). \tag{3.3}$$

Here the convolution is taken with respect to the variables $q$ while $\beta \geqslant 0$ is a parameter. The distribution $\epsilon(q_0)\delta'(q^2 - \beta)$ from $\mathcal{S}'(\mathbf{R}^4)$ is defined in a standard way:

if $\phi \in \mathcal{S}(\mathbf{R}^4)$ then

$$(\epsilon(q_0)\delta^{(n)}(q^2-\beta),\phi(q)) = (-1)^n \partial_\beta^n \int \tilde{\phi}(\beta+|\vec{q}\,|^2,\vec{q}\,)\,d\vec{q}\,,\quad n=0,1,\cdots \tag{3.4}$$

where, as before,

$$\phi(q_0,\vec{q}\,) = \phi(q_0^2,\vec{q}\,) + q_0\tilde{\phi}(q_0^2,\vec{q}\,) \tag{3.5}$$

is the decomposition of $\phi$ into the sum of its even and odd parts (see Section 13.1 and Section 1.5).

**Theorem 1.** *The mapping $d$ defined by* (3.3) *is linear and continuous from $\mathcal{S}(\mathbf{R}^4)$ into $\mathcal{S}$.*

**Proof.** For any $\phi \in \mathcal{S}(\mathbf{R}^4)$ we have

$$\left|\partial_0^{\alpha_0}\partial_1^{\alpha_1}\partial_2^{\alpha_2}\partial_3^{\alpha_3}\partial_\beta^n d[\phi](\beta,u)\right| = \left|(\epsilon(q_0)\delta^{(n+1)}(q^2-\beta),\partial^\alpha\phi(q+u))\right|,$$

$$\alpha=(\alpha_0,\alpha_1,\alpha_2,\alpha_3)\in \mathbf{Z}_+^4,\quad n\in \mathbf{Z}_+^1.$$

On the other hand, from (3.4) and (3.5) it follows that for any $\phi \in \mathcal{S}(\mathbf{R}^4)$, $\beta \geqslant 0$ and $n=0,1,\cdots$ the following estimate holds:

$$|(\epsilon(q_0)\delta^{(n+1)}(q^2-\beta),\phi)| \leqslant CP_{\beta,n}[\phi],$$

where $C$ does not depend on $\phi$ and $\beta$, and

$$P_{\beta,n}[\phi] = \sup_{q^2=\beta,\,0\leqslant i\leqslant n+2} (1+|\vec{q}\,|^4)|\partial_0^i\phi(q_0,\vec{q}\,)|.$$

But for any $a>1$ and $N=0,1,\cdots$ there exists a norm $\|\cdot\|_m$ in $\mathcal{S}(\mathbf{R}^4)$ such that

$$P_{\beta,n}[\phi(q+u)] \leqslant C'_{a,N}(1+|u|+\beta)^{-N}\|\phi\|_m,\quad (\beta,u)\in\Gamma_a.$$

Therefore,

$$\left|\partial_0^{\alpha_0}\partial_1^{\alpha_1}\partial_2^{\alpha_2}\partial_3^{\alpha_3}\partial_\beta^n d[\phi](\beta,u)\right| \leqslant C_{a,N}(1+|u|+\beta)^{-N}\|\partial^\alpha\phi\|_m,$$

$$(\beta,u)\in\Gamma_a.$$

The theorem has been proved.

One can easily verify that the images of the mapping satisfy the equation

$$(\Box_3 - 4\partial_\beta - 4\beta\partial_\beta^2)\phi(\beta,u) = 0 \tag{3.6}$$

or, after replacing $\beta$ by $|\vec{\lambda}|^2$, $\vec{\lambda}\in\mathbf{R}^2$,

$$\Box_5\phi(|\vec{\lambda}|^2,u) = 0, \tag{3.7}$$

where $\Box_3$ and $\Box_5$ are wave operators in four-dimensional and six-dimensional spaces, respectively (see Seciton 0.5).

Let $\mathcal{S}_0$ denote the subspace of $\mathcal{S}$ consisting of the functions which satisfy equation (3.7) or (3.6). It is closed in $\mathcal{S}$. So, the mapping $d$ is linear and continuous from $\mathcal{S}(\mathbf{R}^4)$ into $\mathcal{S}_0$.

Further, let $\mathcal{S}_{\mathrm{inv}}(\mathbf{R}^5)$ denote the subspace of $\mathcal{S}(\mathbf{R}^5)$ consisting of all the functions $\phi(\vec{\kappa},\vec{x}\,)$ which are invariant under the rotations in the plane of two variables $\vec{\kappa}=(\kappa_1,\kappa_2)$ (see Section 1.5). It is closed in $\mathcal{S}(\mathbf{R}^5)$. Now we introduce the mapping $l$ from the space $\mathcal{S}(\mathbf{R}^4)$ into the direct sum $\mathcal{S}_{\mathrm{inv}}(\mathbf{R}^5)\oplus\mathcal{S}_{\mathrm{inv}}(\mathbf{R}^5)$ by the formula

$$l[\phi] = (\tilde{\phi}(|\vec{\kappa}|^2 + |\vec{x}|^2, \vec{x}), \phi(|\vec{\kappa}|^2 + |\vec{x}|^2, \vec{x})), \tag{3.8}$$

where similarly to (3.5)

$$\phi(x_0, \vec{x}) = \phi(x_0^2, \vec{x}) + x_0\tilde{\phi}(x_0^2, \vec{x}). \tag{3.9}$$

**Theorem 2.** *The mapping $l$ is linear and continuous from $\mathcal{S}(\mathbf{R}^4)$ onto $\mathcal{S}_{inv}(\mathbf{R}^5) \oplus \mathcal{S}_{inv}(\mathbf{R}^5)$, its null space is the space $\mathcal{S}_0(\mathbf{R}^4 \backslash V)$, which is the space of all functions from $\mathcal{S}(\mathbf{R}^4)$ with supports outside the light cone $V = [x : x^2 > 0]$.*

**Proof.** The statements about continuity, linearity and null space directly follow from the definition of the mapping $l$ by formulae (3.8) and (3.9). Let us verify that the image of $l$ is the whole space $\mathcal{S}_{inv}(\mathbf{R}^5) \oplus \mathcal{S}_{inv}(\mathbf{R}^5)$. Let $(\omega_1, \omega_2) \in \mathcal{S}_{inv}(\mathbf{R}^5) \oplus \mathcal{S}_{inv}(\mathbf{R}^5)$. Because of the radial symmetry of $\omega_1$ and $\omega_2$ with respect to $\vec{\kappa}$, there are functions $\psi_1(\beta, \vec{x})$ and $\psi_2(\beta, \vec{x})$ from $\mathcal{S}_+(3)$ such that (see Section 1.6)

$$\omega_i(\vec{\kappa}, \vec{x}) = \psi_i(|\vec{\kappa}|^2, \vec{x}), \; i = 1, 2.$$

If one now puts

$$\phi(x) = x_0\psi(x^2, \vec{x}) + \psi_2(x^2, \vec{x}),$$

then by (3.8) and (3.9) we get $l[\phi] = (\omega_1, \omega_2)$.

So, due to theorem 2, the mapping $l$ defines an isomorphism for the space $\mathcal{S}(\vec{V}) = \mathcal{S}(\mathbf{R}^4)/\mathcal{S}_0(\mathbf{R}^4 \backslash V)$ onto $\mathcal{S}_{inv}(\mathbf{R}^5) \oplus \mathcal{S}_{inv}(\mathbf{R}^5)$ (see Section 1.1). From the same formulae (3.8) and (3.9) it follows that this isomorphism is topological.

Let us also introduce a mapping $h$ from the space $\mathcal{S}_{inv}(\mathbf{R}^5) \oplus \mathcal{S}_{inv}(\mathbf{R}^5)$ into the space $\hat{\mathcal{S}}$ in the following way. If the functions $(\phi_1, \phi_2)$ are given we take their Fourier transform and the resulting functions

$$\left(\frac{1}{4\pi i}\tilde{\phi}_1, \frac{1}{4\pi}\tilde{\phi}_2\right)$$

will be the data for the Cauchy problem on the hyperplane $u_0 = 0$ for the wave equation. Then in the solution of this Cauchy problem (which is obviously invariant under the rotations we are considering) we should put $|\vec{\lambda}|^2 = \beta$. More precisely, if $(\phi_1, \phi_2) \in \mathcal{S}_{inv}(\mathbf{R}^5) \oplus \mathcal{S}_{inv}(\mathbf{R}^5)$ then we put

$$\tilde{\phi}_j(\vec{\lambda}, \vec{u}) = F[\phi_j], \; j = 1, 2,$$

where the Fourier transform is taken with respect to all the variables $(\vec{\kappa}, \vec{x})$. Then we construct a function $\phi(\vec{\lambda}, u)$ (unique), which satisfies the wave equation (3.7), and, moreover,

$$\phi(\vec{\lambda}, 0, \vec{u}) = \frac{1}{4\pi i}\tilde{\phi}_1(\vec{\lambda}, \vec{u});$$

$$\partial_0\phi(\vec{\lambda}, 0, \vec{u}) = \frac{1}{4\pi}\tilde{\phi}_2(\vec{\lambda}, \vec{u}). \tag{3.10}$$

Since the data $\tilde{\phi}_1$ and $\tilde{\phi}_2$ are infinitely differentiable and invariant under the rotation in the plane $\vec{\lambda}$, then the solution $\phi(\vec{\lambda}, u)$ has the same properties. Thus $\phi(\vec{\lambda}, u)$ depends on $\vec{\lambda}$ by a factor $|\vec{\lambda}|^2$. Putting

$$\phi_0(\beta, u) = \phi(\vec{\lambda}, u)|_{|\vec{\lambda}|^2=\beta} \tag{3.11}$$

we get a function which is the image of $(\phi_1, \phi_2)$ under the mapping $h$.

**Theorem 3.** *The mapping $h$ is an isomorphism from $S_{\mathrm{inv}}(\mathbf{R}^5) \oplus S_{\mathrm{inv}}(\mathbf{R}^5)$ onto $S_0$.*

**Proof.** Since the Fourier transformation is the isomorphism $S(\mathbf{R}^5)$ onto $S(\mathbf{R}^5)$ preserving radial symmetry, the only thing we have to prove is the fact that the solution $\phi$ of the Cauchy problem belongs to $S$, if the data $\tilde{\phi}_1$ and $\tilde{\phi}_2$ belong to $S(\mathbf{R}^5)$, and continuously depends on them. By the Kirchhoff formula (see, for instance, V.S. Vladimirov, 1976) we have

$$\phi(\vec{\lambda}, u) = \frac{1}{4\pi} D_5(\vec{\lambda}, u_0, \vec{u}) * \tilde{\phi}_2(\vec{\lambda}, \vec{u}) +$$

$$+ \frac{1}{4\pi i} \partial_0 D_5(\vec{\lambda}, u_0, \vec{u}) * \tilde{\phi}_1(\vec{\lambda}, \vec{u}), \tag{3.12}$$

where the convolution is taken only with respect to $(\vec{\lambda}, \vec{u})$ and

$$D_5(\vec{\lambda}, u) = \frac{1}{2\pi^2} \epsilon(u_0) \delta'(u^2 - |\vec{\lambda}|^2). \tag{3.13}$$

Let us write, say, the first summand in (3.12) in a more detailed way. Assuming $u_0 > 0$, we get (up to a constant)

$$D_5 * \tilde{\phi}_2 = \frac{1}{2\pi^2} (\delta'(u_0^2 - |\vec{u}'|^2 - |\vec{\lambda}'|^2), \tilde{\phi}_2(\vec{\lambda} - \vec{\lambda}', \vec{u} - \vec{u}')), \tag{3.14}$$

where $u_0$ is a parameter. Taking into account the formulae

$$(\delta(f(x)), \phi) = \int_S \frac{\phi(x)}{|\partial f(x)|} dS, \quad \phi \in S(\mathbf{R}^4),$$

$$\delta^{(k)}(f(x)) = \left[ \frac{1}{|\partial f(x)|^2} \sum_{i=1}^{n} \partial_i f(x) \partial_i \right]^k \delta(f(x)),$$

where $S$ is the surface given by the equation $f(x) = 0$ and $dS$ is the Lebesgue measure on it, let us write (3.14) in the form

$$\frac{1}{2u_0} \int_{u_0 = \sqrt{|\vec{u}'|^2 + |\vec{\lambda}'|^2}} \Phi_{(\vec{\lambda}, u)}(\vec{\lambda}', \vec{u}')\, dS' \equiv \phi_2(\vec{\lambda}, u) \tag{3.15}$$

where

$$\Phi_{(\vec{\lambda}, u)}(\vec{\lambda}', \vec{u}') = -\left( \sum_{j=1}^{2} \partial_{\lambda_j} \lambda_j + \sum_{j=1}^{3} \partial_{u_j} u_j \right) \frac{\tilde{\phi}_2(\vec{\lambda} - \vec{\lambda}', \vec{u} - \vec{u}')}{2(|\vec{\lambda}'|^2 + |\vec{u}'|^2)}. \tag{3.16}$$

From (3.15) and (3.16) one can easily see that for any $a > 1$ the function $\phi_2(\vec{\lambda}, u)$ belongs to the space $(\Gamma_a' \bigcap [u_0 \geqslant 1])$, and the mapping (3.15) is continuous from $(\mathbf{R}^5)$ into $(\Gamma_a' \bigcap [u_0 \geqslant 1])$. Here

$$\Gamma_a' = [(\vec{\lambda}, u) \in \mathbf{R}^6 : (|\vec{\lambda}|^2, u) \in \Gamma_a].$$

The same argument can be applied to the second summand in (3.12). So the mapping (3.12) is continuous from $S(\mathbf{R}^5) \oplus S(\mathbf{R}^5)$ into $(\Gamma_a' \bigcap [u_0 \geqslant 1])$. Writing the Kirchhoff formula for the plane $u_0 = 3$, and considering the domain $u_0 \leqslant 2$, we conclude by similar arguments that the mapping (3.12) is continuous from $\mathcal{S}(\mathbf{R}^5) + \mathcal{S}(\mathbf{R}^5)$ into $\mathcal{S}(\Gamma_a')$.

The following theorem holds.

**Theorem 4.** *The following diagram of mappings is commutative:*

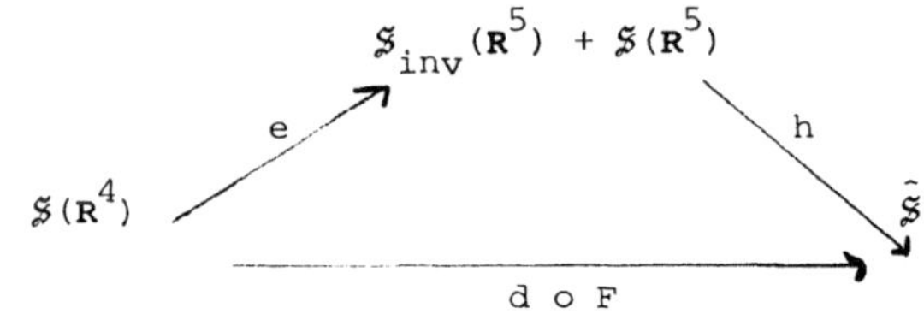


*Here $d \circ F$ is the composition of the Fourier transformation and the mapping $d$.*

**Proof.** We have to verify that

$$d \circ F = h \circ l. \tag{3.18}$$

Taking into account that the Fourier transform of the distribution $\epsilon(q_0)\delta'(q^2 - \beta)$ is equal to

$$\frac{1}{4\pi^2 i}\epsilon(x_0)\theta(x^2)J_0(\sqrt{x^2\beta}).$$

we get for any $\phi \in \mathcal{S}(\mathbf{R}^4)$

$$\begin{aligned}[d \circ F][\phi] &= d \circ F[\phi] = \epsilon(q_0)\delta'(q^2 - \beta) * F[\phi] = \\ &= (\epsilon(q_0)\delta'(q^2 - \beta), F[\phi](u - q)) = \\ &= (F[\epsilon(q_0)\delta'(q^2 - \beta)], \phi(x)e^{iux}) = \\ &= \frac{1}{4\pi^2 i}\int e^{iux}\epsilon(x_0)\theta(x^2)J_0(\sqrt{x^2\beta})\phi(x)\,dx = \phi_1(\beta, u).\end{aligned}$$

Now using the following property of the Bessel transform (see Section 2.3)

$$\int_0^\infty J_0(\sqrt{\gamma\beta})\phi(\gamma)\,d\gamma = \pi F[\phi(|\vec{\kappa}|^2](\vec{\lambda}),$$

where $\vec{\kappa}, \vec{\lambda} \in \mathbf{R}^2$, we obtain

$$\phi_1(\beta, o, \vec{u}) = \frac{1}{4\pi i}F[\hat{\phi}(|\vec{\kappa}|^2 + |\vec{x}|^2, \vec{x})](\vec{\lambda}, \vec{u})|_{|\vec{\lambda}|^2 = \beta},$$

$$\partial\phi_1(\beta, 0, \vec{u}) = \frac{1}{4\pi}F[\phi(|\vec{\kappa}|^2 + |\vec{x}|^2, \vec{x})](\vec{\lambda}, \vec{u})|_{|\vec{\lambda}|^2 = \beta}, \tag{3.19}$$

where the Fourier transform is taken with respect to all the variables $(\vec{\kappa}, \vec{x})$ and the functions $\hat{\phi}$ and $\phi$ are defined by (3.9).

Comparing now (3.19) and (3.10)-(3.11) and using the uniqueness of the solution of the Cauchy problem of the wave equation, we obtain relation (3.18).

**Corollary.** *The transformation $d \circ F$ is a (topological) isomorphism from $\mathcal{S}(\bar{V})$ onto $\hat{\mathcal{S}}_0$.*

**Proof.** According to theorems 2 and 3, the mapping $h \circ l$ is just of this type. But it is equal to $d \circ F$ by the previous theorem.

Now let $F$ be a causal distribution; that is, its Fourier transform $\tilde{F}$ vanishes outside the light cone $\bar{V}$ (and therefore it is functional on the space $\mathcal{S}(\bar{V})$. Due to the corollary $F$ defines some linear, continuous functional $\eta_0$, on $\mathcal{S}_0$. By the Hahn-Banach theorem it can be extended (of course in a non unique way) to the whole $\hat{\mathcal{S}}$. Let us denote this distribution (which belongs to $\hat{\mathcal{S}}'$) by $\eta(\beta, u)$. For any $\phi \in \mathcal{S}(\mathbf{R}^4)$ we have

$$(F, \phi) = (\eta, d[\phi]). \tag{3.20}$$

From this by (3.3) we get the formula

$$F(q) = \int \epsilon(q_0 - u_0)\delta'((q - u)^2 - \beta)\eta(\beta, u)\, du\, d\beta, \tag{3.21}$$

which is, in fact, a formal form of the formula (3.20).

Putting

$$\psi(\beta, u) = -\partial_\beta \eta(\beta, u) \tag{3.22}$$

we get by (3.20) and (3.3) that for any $\phi \in \mathcal{S}(\mathbf{R}^4)$

$$(F, \phi) = (\psi(\beta, u), [\epsilon(q_0)\delta(q^2 - \beta) * \phi(q)](u)) \tag{3.23}$$

or formally

$$F(q) = \int \epsilon(q_0 - u_0)\delta((q - u)^2 - \beta)\psi(\beta, u)\, du\, d\beta. \tag{3.24}$$

The last formula is precisely the well-known Jost-Lehmann-Dyson representation for the causal distribution $F$. It remains only to prove that the support of $\psi$ can be situated on an arbitrary strictly space- like hypersurface.

Note that theorem 4 and its corollary imply that the set of weight functions $\eta$ in representation (3.20) for a given causal distribution is exactly the set of all possible extensions of a given linear, continuous functional $\eta_0$ from the space $\hat{\mathcal{S}}_0$ up to a linear, continuous functional on $\hat{\mathcal{S}}$. The following construction gives a series of such extensions.

Let $\Sigma$ be a $C^\infty$-hypersurface in $\mathbf{R}^4$. We say that $\Sigma$ is strictly space-like if at any point of $\Sigma$ the normal vector belongs to some cone $\Gamma \subset\subset V_3^+$, which does not depend on this point. Note that every strictly space-like hypersurface is strictly $V_3^+$-like (see Section 1.9). The converse statement is not true.

Any strictly space-like hypersurface can be defined by the equation $u_0 = \omega(\vec{u})$, where $\omega \in C^1(\mathbf{R}^3)$. We say that a strictly space-like surface is of class $C^\infty$ if $\omega \in \theta_M(\mathbf{R}^3)$ (see Section 1.4).

If a strictly space-like hypersurface $\Sigma$ is given we denote by $\hat{\mathcal{S}}_\Sigma$ the subspace of all functions $\phi(\beta, u)$ from $\hat{\mathcal{S}}$ which vanish on $\mathbf{R}_+^1 \times \Sigma$ together with its first normal derivatives. Obviously is $\hat{\mathcal{S}}_\Sigma$ closed in $\hat{\mathcal{S}}$.

Let $\phi \in \hat{\mathcal{S}}$, and let $\phi_0$ be a solution of the Cauchy problem for equation (3.7) with the data obtained by the restriction of $\phi(|\vec{\lambda}|^2, u)$ and its first normal derivative to the surface $\mathbf{R}^2 \times \Sigma$. It always exists and is unique. Moreover $\phi_0 \in \hat{\mathcal{S}}_0$. The mapping from $\hat{\mathcal{S}}$ into $\hat{\mathcal{S}}_0$ which we have just defined is continuous. That is, if the sequence $\{\phi_n, n \to \infty\}$ tends to zero in $\hat{\mathcal{S}}$, then the corresponding sequence $\{\phi_{0,n}, n \to \infty\}$ tends to zero in $\hat{\mathcal{S}}_0$. So, any function $\phi \in \hat{\mathcal{S}}$ has a unique representation as the $\Sigma$:

$$\phi = \phi_0 + \phi_\Sigma$$

where $\phi_0 \in \hat{\mathcal{S}}_0$, $\phi_\Sigma = \phi - \phi_0 \in \hat{\mathcal{S}}_\Sigma$. Thus, the space $\hat{\mathcal{S}}$ is the topological direct SUM of subspaces $\hat{\mathcal{S}}_\Sigma$ and $\hat{\mathcal{S}}_0$:

$$\hat{\mathcal{S}} = \hat{\mathcal{S}}_0 + \hat{\mathcal{S}}_\Sigma. \tag{3.25}$$

The representation (3.25) gives a natural continuation of the functional $\eta_0$, which is denoted by $\eta_\Sigma$. Namely, for any $\phi \in \hat{\mathcal{S}}$ we put

$$(\eta, \phi) = \eta_0[\phi_0],$$

where $\phi = \phi_0 + \phi_1$, $\phi_0 \in \hat{\mathcal{S}}_0$, $\phi_1 \in \hat{\mathcal{S}}_\Sigma$. The corresponding distribution, defined by $\eta_\Sigma$ according to (3.22) is denoted by $\psi_\Sigma$. It is clear that

$$\operatorname{supp} \eta_\Sigma \subset \mathbf{R}_+^1 \times \Sigma, \quad \operatorname{supp} \psi_\Sigma \subset \mathbf{R}_+^1 \times \Sigma.$$

Moreover, $\eta_\Sigma$ (and therefore $\psi_\Sigma$) contains no more than the first derivatives of the $\delta$-function on $\Sigma$. These conditions uniquely define $\eta_\Sigma$ and $\psi_\Sigma$. Thus, we have proved the following theorem.

**Theorem 5.** *Let $F$ be a causal distribution. Then for any strictly space-like hypersurface $\Sigma$ in $\mathbf{R}^4$ of class $C^\infty$ there is a unique distribution $\psi_\Sigma$ such that*

$$F(q) = \int \epsilon(q_0 - u_0)\delta[(q - u)^2 - \beta]\psi_\Sigma(\beta, u)\, du\, d\beta, \tag{3.26}$$

$$\operatorname{supp} \psi_\Sigma \subset \mathbf{R}_+^1 \times \Sigma, \tag{3.27}$$

$$(\psi_\Sigma, \phi) = 0, \quad \phi \in \hat{\ }_\Sigma. \tag{3.28}$$

*If a surface $\Sigma$ is given by the equation $u_0 = \omega(\vec{u})$ where $\omega \in \theta_M(\mathbf{R}^3)$ then*

$$\psi_\Sigma(\beta, u) = \delta(u_0 - \omega(\vec{u}))\psi_a(\beta, \vec{u}) + \delta'(u_0 - \omega(\vec{u}))\psi_s(\beta, \vec{u}) \tag{3.29}$$

*where $\psi_j \in \mathcal{S}_+'(3)$, $j = a, s$. In particular, if $\omega \equiv 0$, we have the formula* (1.21)

The following theorem holds.

**Theorem 6.** *Let $F$ be a causal distribution and let $\Sigma_0$ be some strictly space-like hypersurface in $\mathbf{R}^n$ of class $C^\infty$. Then if the weight function $\psi_{\Sigma_0}(\beta, u)$ in* (3.26) *has a quasi-asymptotics in $\beta$ at infinity with respect to an automodel function $\rho$, then any weight function $\psi_\Sigma(\beta, u)$ has a quasi-asymptotics with respect to $\rho$. Here $\Sigma$ is an arbitrary strictly space-like hypersurface in $\mathbf{R}^4$ of class $C^\infty$.*

**Proof.** Due to (3.22) the distribution $\eta_{\Sigma_0}(\beta, u)$ has a quasi-asymptotics over $\beta$ at infinity with respect to $k\rho(k)$ (see Section 3.4) and it is sufficient to prove that the same is true for the distribution $\eta_\Sigma(\beta, u)$. Let us denote by $\hat{\mathcal{S}}_{0k}$, $k > 0$ the subspace of $\hat{\mathcal{S}}$ consisting of all the functions $\phi(\beta, u)$ satisfying the equation

$$\left[\Box_3 - \frac{1}{k}\left(\partial_{X_1}^2 + \partial_{X_2}^2\right)\right]\phi(|\vec{X}|^2, u) = 0. \tag{3.30}$$

It is clear that if $\phi_k(\beta, u)$ belongs to $\hat{\mathcal{S}}_{0,k}$ then $\phi_k(\beta/k, u)$ belongs to $\hat{\mathcal{S}}_0$. Taking into account the relation

$$(\eta(k\beta, u), \phi(\beta, u)) = \frac{1}{k}\left(\eta(\beta, u), \phi\left(\frac{\beta}{k}, u\right)\right)$$

and the fact that $\eta_{\Sigma_0}$ and $\eta_\Sigma$ give the same functional on $\hat{\mathcal{S}}_0$ (corollary to theorem 4), we conclude that

$$(\eta_{\Sigma_0}(k\beta,u),\phi_k(\beta,u)) = (\eta_\Sigma(k\beta,u),\phi_k(\beta,u)). \tag{3.31}$$

Let $\phi$ be a function from $\mathcal{S}$. Let us denote by $\phi_{1,k}(\beta,u)$ the function which is the solution of the Cauchy problem for equation (3.30), with data that are equal to the restriction of $\phi(|\vec{\lambda}|^2,u)$ and its co-normal derivatives to the surface $\mathbf{R}^2\times\Sigma$, followed by the substitution $|\vec{\lambda}|^2=\beta$. Since

$$(\eta_\Sigma(k\beta,u),\phi(\beta,u)-\phi_{1,k}(\beta,u)) = \left(\eta_\Sigma(\beta,u),\frac{1}{k}\left[\phi\left(\frac{\beta}{k},u\right)-\phi_{1,k}\left(\frac{b}{k},u\right)\right]\right),$$

$$\frac{1}{k}\left[\phi\left(\frac{\beta}{k},u\right)-\phi_{1,k}\left(\frac{\beta}{k},u\right)\right]\in\mathcal{S}_\Sigma,$$

then by theorem 5

$$(\eta_\Sigma(k\beta,u),\phi(\beta,u)) = (\eta_{\Sigma_0}(k\beta,u),\phi_{1,k}(\beta,u)). \tag{3.32}$$

From the identities (3.31) and (3.32) we get

$$(\eta_\Sigma(k\beta,u),\phi(\beta,u)) = (\eta_{\Sigma_0}(k\beta,u),\phi_{1,k}(\beta,u)). \tag{3.33}$$

Since the sequence $\{(1/k\rho(k))\eta_{\Sigma_0}(k\beta,u),k\to\infty\}$ converges in $\mathcal{S}'(\mathbf{R}^5)$ (and therefore in $\mathcal{S}'$) then, due to (3.33), we can complete the proof of the theorem if we show that

$$\phi_{1,k}(\beta,u)\to\phi_\infty(\beta,u),\ k\to\infty \text{ in } \mathcal{S}, \tag{3.34}$$

where $\phi_\infty$ is some function from $\mathcal{S}$.

To prove (3.34) let us use the Kirchhoff formula. We have

$$\phi_{1,k}(\beta,u) = \left\{\left(\delta(\Sigma')\partial_\gamma\phi\left(\frac{|\vec{\lambda}|^2}{k},u\right)+\partial_\gamma\left[\delta(\Sigma')\phi\left(\frac{|\vec{\lambda}|^2}{k},u\right)\right]\right)\right. *$$

$$\left. * \frac{1}{2\pi^2}\epsilon(u_0)\delta'(u^2-|\vec{\lambda}|^2)\right\}(\lambda,u)\Big|_{|\vec{\lambda}|^2=k\beta}. \tag{3.35}$$

Here $\partial_\gamma=\gamma_0\partial_0+\cdots+\gamma_3\partial_3+\gamma_4\partial_{\lambda_1}+\gamma_5\partial_{\lambda_2}$ is the co-normal derivative, $\gamma=(\gamma_0,\ldots,\gamma_5)$, is the co-normal to the surface $\Sigma'=\mathbf{R}^2\times\Sigma$. In particular, if $\Sigma$ is given by the equation $u_0=\omega(\vec{u})$, then $\gamma_0=(1+|\partial\omega|^2)^{-1/2}$, $\gamma_i=\gamma_0\partial\omega$, $i=1,2,3$, $\gamma_4=\gamma_5=0$. The distributions $\partial(\Sigma')f$ and $\partial_\gamma[\delta(\Sigma')f]$ in (3.35), where $f$ is a smooth function, are defined in the following way: for any $\phi \in \mathcal{S}(\mathbf{R}^5)$

$$(\delta(\Sigma')f\,,\phi) = \int_{\Sigma'} f\,\phi\,dS\,;$$

$$(\partial_\gamma[\delta(\Sigma')f\,],\phi) = -\int_{\Sigma'} f\,\partial_\gamma\phi\,dS\,,$$

where $dS = \gamma_0^{-1}\,d\vec{u}\,d\vec{\lambda}$ is the Lebesgue measure on $\Sigma'$.

Let us rewrite (3.35) in the following way

$$\phi_{1,}k\,(|\vec{\lambda}|^2,u) = [h_\Sigma(\vec{\lambda},u) * k\,\epsilon(u_0)\delta'(u^2 - k\,|\vec{\lambda}|^2)](\vec{\lambda},u), \tag{3.36}$$

where

$$h_\Sigma(\vec{\lambda},u) = \delta(\Sigma')\partial_\gamma\phi(\vec{\lambda},u) + \partial_\gamma[\delta(\Sigma')\phi(\vec{\lambda},u)]. \tag{3.36'}$$

Note (see Section 1.6) that

$$F[k\,\epsilon(u_0)\delta'(u^2 - k\,|\vec{\lambda}|^2)] = C_1\epsilon(q_0)\delta\left(q^2 - \frac{|\vec{\kappa}|^2}{k}\right) \rightarrow$$

$$\rightarrow C_1\epsilon(q_0)\delta(q^2),\ k \rightarrow \infty$$

in $\mathcal{S}'(\mathbf{R}^6)$. Hence

$$k\,\epsilon(u_0)\delta'(u^2 - k\,|\vec{\lambda}|^2) \rightarrow C_2\epsilon(u_0)\delta(u^2) \times \delta(\vec{\lambda}),\ k \rightarrow \infty \tag{3.37}$$

in $\mathcal{S}'(\mathbf{R}^6)$ and

$$\operatorname{supp}[k\,\epsilon(u_0)\delta'(u^2 - k\,|\vec{\lambda}|^2)] \subset \bar{V}_5',\ k \geqslant 1, \tag{3.38}$$

where $V_5' = [(\vec{\lambda},u) \in \mathbf{R}^6\colon u^2 - |\vec{\lambda}|^2 > 0]$ is the light cone in $\mathbf{R}^6$ (see Section 0.2) with permuted coordinates.

On the other hand, it follows from (3.36') that

$$F[h_\Sigma](\vec{\kappa},q) = \int e^{i[q_0\omega(\vec{u}) - (\vec{q},\vec{u}) - (\vec{\kappa},\vec{\lambda})]}g\,(q\,,\vec{\kappa},\vec{u}\,,\vec{\lambda})\,d\vec{u}\,d\vec{\lambda},$$

where $\omega(\vec{u})$ is the function which defines the surface $\Sigma$ in $\mathbf{R}^5$, and $g\,(q\,,\vec{\kappa},\vec{u}\,,\vec{\lambda})$ is a polynomial of the first order in $(q\,,\vec{\kappa})$ with coefficients from $\mathcal{S}(\mathbf{R}^5(\vec{u}\,,\vec{\lambda}))$. Since $\Sigma$ is strictly space-like we have $|\partial\omega| \leqslant 1 - 2\epsilon$ for some $\epsilon > 0$.

Let us now prove the following lemma.

**Lemma 1.** *Let* $\omega \in \theta_M(\mathbf{R}^n)$ *and* $|\partial w| < 1 - 2\epsilon$, $\epsilon > 0$. *Moreover let* $g \in \mathcal{S}(\mathbf{R}^n)$. *Then the function*

$$\psi(q) = \int e^{i[\omega(\vec{u})q_0 - (\vec{u},\vec{q})]}g\,(\vec{u})\,d\vec{u},\ q \in \mathbf{R}^{n+1},$$

*together with all its derivatives decreases outside the cone* $|\vec{q}| < (1-\epsilon)|q_0|$ *faster then any negative power of its arguments.*

**Proof.** It is sufficient to verify this assertion only for $\psi$ because all its derivatives have the same form. To do this it is sufficient to show that for any $k = 0, 1, \cdots$

$$|\sigma^k\psi(\sigma q)| \leqslant C_k,\ (1-\epsilon)|q_0| < 1,\ |\vec{q}| = 1.$$

Taking into account that for $|\vec{q}\,| = 1$

$$(\vec{q}, \partial)e^{i\sigma[\omega(\vec{u})q_0-(\vec{u},\vec{q})]} = i\sigma[q^0(\partial\omega(\vec{u}), \vec{q}) - 1]e^{i\sigma[\omega(\vec{u})q_0-(\vec{u},\vec{q})]}$$

$$\partial = (\partial_1, \partial_2, \ldots, \partial_n),$$

we get

$$\sigma^k\psi(\sigma q) = \int\left\{\frac{1}{i[q_0(\partial\omega(\vec{u}), \vec{q}) - 1]}(\vec{q}, \partial)\right\}^k e^{i\sigma[\omega(\vec{u})q_0-(\vec{u},\vec{q})]}g(\vec{u})\,d\vec{u}.$$

Integrating by parts and noting that for $|\vec{q}\,| = 1$, $|q_0| < (1/1-\epsilon)$

$$|q_0(\partial\omega(\vec{u}), \vec{q}) - 1| > \frac{\epsilon}{1-\epsilon}$$

we complete the proof of the lemma.

Let us continue the proof of the theorem. Due to lemma 1 the function $F[h_\Sigma](\vec{\kappa}, q)$ is infinitely differentiable and decreases outside the cone $|\vec{q}\,|^2 + |\vec{\kappa}|^2 < (1-\epsilon)^2 q_0^2$ faster than any negative power of its variables. Let $\tilde{\alpha}(\vec{\kappa}, q)$ be the function from $\mathcal{S}(\mathbf{R}^6)$ which is equal to $F[h_\Sigma]$ outside the cone $|\vec{q}\,|^2 + |^{\rightarrow}|^2 \leqslant (1-(\epsilon/2))q_0^2$, and let $\tilde{\alpha} = F[\alpha]$. Now we get from (3.36)

$$F[\phi_{1,k}(|\vec{\lambda}|^2, u)] = C_3 F[h_\Sigma](\vec{\kappa}, q)\epsilon(q_0)\delta\left(q^2 - \frac{|\vec{\kappa}|^2}{k}\right) =$$

$$= C_3\tilde{\alpha}(\vec{\kappa}, q)\epsilon(q_0)\delta\left(q^2 - \frac{|\kappa|^2}{k}\right), \quad k > 1.$$

Thus taking the inverse Fourier transforms we get

$$\phi_{1,k}(|\vec{\lambda}|^2, u) = [\alpha(\vec{\lambda}, u) * k\,\epsilon(u_0)\delta'(u^2 - k\,|\vec{\lambda}|^2)](\vec{\lambda}, u).$$

where $\alpha$ belongs to $\mathcal{S}(\mathbf{R}^6)$. From this and from (3.37) and (3.38) the limit relation (3.34) follows.

In conclusion, let us prove the theorem about the support of a weight function $\psi_\Sigma$.

**Theorem 7.** *Let $F$ be a causal distribution and let $\psi_\Sigma$ be its weight function where $\Sigma$ is some strictly space-like hypersurface in $\mathbf{R}^4$ of class $C^\infty$. Further, let $F(q)$ vanish in some neighbourhood of the time-like segment $[q_1, q_2]$ connecting points $q_1$ and $q_2$ ($(q_2 - q_1)^2 > 0$, $q_{2,0} > q_{1,0}$). Then $\psi_\Sigma(\beta, u)$ vanishes in the domain* (Fig. 21)

$$O_{[q_1,q_2]} =$$

$$= \left\{(\beta, u) \in \mathbf{R}_+^1 \times \mathbf{R}^4 : \beta = |\vec{\lambda}|^2, (u, \vec{\lambda}) \in \left[(q_1, \vec{o}) + V_5^+\right] \cap \left[(q_2, \vec{o}) + V_5^-\right]\right\}.$$

*This domain is precisely the set of all points $(\beta, u) \in \mathbf{R}_+^1 \times \mathbf{R}^4$ for which both wings of the hyperboloid $[q \in \mathbf{R}^4 : (q-u)^2 = \beta]$ intersect the segment $[q_1, q_2]$.*

To prove the theorem let us first prove the following lemma.

**Lemma 2.** *Let $[q_1, q_2]$ be some time-like segment in $\mathbf{R}^4$ and let $\Sigma$ be some strictly space-like hypersurface of class $C^\infty$. If $\phi(\beta, u)$ is a function from $\mathcal{S}_0$ such that $\phi(|\vec{\lambda}|^2, u)$ vanishes together with its first derivatives in $\mathbf{R}^2 \times \Sigma$ with a possible*

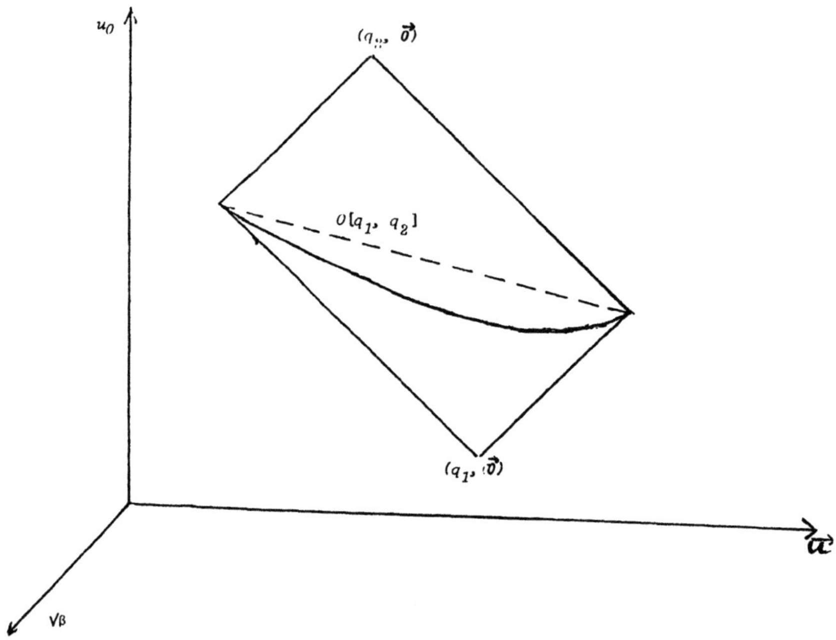


Figure 21.

*exception of the set*

$$Q = \left\{ (\vec{\lambda}, u): (u, \vec{\lambda}) \in (\Sigma \times \mathbf{R}^{2)} \bigcap \left[(q_1, \vec{0}) + \bar{V}_5^+\right] \bigcap \left[(q_2, \vec{0}) + \bar{V}_5^-\right] \right\}$$

*then there exists a sequence of functions from* $\mathcal{S}(\mathbf{R}^4)$, $\{\phi_n, n \to \infty\}$, *such that the support of* $\phi_n$ *is in a given neighbourhood of the segment* $[q_1, q_2]$ *and*

$$d[\phi_n] \to \phi(\beta, u), \; n \to \infty$$

*in* $\mathcal{S}$.

**Proof.** One can assume, without loss of generality, that $q_2 = -q_1 = (a, \vec{0})$, $a > 0$. Let us put

$$\phi_1(\beta, \vec{u}) = \phi(\beta, 0, \vec{u}), \quad \phi_2(\beta, \vec{u}) = \partial_0 \phi(\beta, 0, \vec{u}).$$

Since $\phi(|\vec{\lambda}|^2, u)$ satisfies equation (3.7) it is sufficient to prove that there is a sequence of functions from $\mathcal{S}(\mathbf{R}^4)$ $\{\phi_n, n \to \infty\}$ with supports in an arbitrarily small neighbourhood of the segment $[q_1, q_2]$ such that

$$d[\phi_n](\beta, 0, \vec{u}) \to \phi_1(\beta, \vec{u});$$

$$\partial_0 d[\phi_n](\beta, 0, \vec{u}) \to \phi_2(\beta, \vec{u}).$$

as $n \to \infty$, in the topology of $\mathcal{S}_+(3)$.

Note that since the domain of influence of a given point for the wave equation (3.7) is the boundary of the light cone $V_5$, with its vertex at this point, and since

$$\bigcup_{(q,\vec{\lambda}')\in Q} [(q-u)^2 - |\vec{\lambda}' - \vec{\lambda}|^2 = 0]\Big|_{|\vec{\lambda}|^2=\beta, u_0=0} \subset [(\beta,\vec{u}) \in \mathbf{R}^1_+ \times \mathbf{R}^3 : |\vec{u}|^2 + \beta \leqslant a^2]$$

the following embedding exists:

$$\operatorname{supp}\phi_j \subset [(\beta,\vec{u}) \in \mathbf{R}^1_+ \times \mathbf{R}^3 : |\vec{u}|^2 + \beta \leqslant a^2],\ j = 1,2. \tag{3.39}$$

Now let $\epsilon > 0$ be an arbitrarily small number. Let us consider a family of functions of the form $d[q_0 h(q_0^2) g(\vec{q})]$, where

$$h \in \mathcal{S}_+,\ \operatorname{supp} h \subset [t : 0 \leqslant t \leqslant (a+\epsilon)^2],$$
$$g \in \mathcal{S}(\mathbf{R}^3),\ \operatorname{supp} g \subset [\vec{q} : \vec{q} \in \mathbf{R}^3,\ |\vec{q}| \leqslant \epsilon]. \tag{3.40}$$

According to (3.3) and (3.4) we have

$$d[q_0 h(q_0^2) g(\vec{q})] = \int h'(|\vec{q} - \vec{u}|^2 + \beta) g(\vec{q})\, d\vec{q}. \tag{3.41}$$

As we already know (see Section 1.1), the mapping $h \to h'$ is an automorphism of $\mathcal{S}_+$. So the function of the form $h'$, $h \in \mathcal{S}_+$ span the whole space $\mathcal{S}_+$. Letting $g(\vec{q})$ in (3.41) tend to various derivatives of $\delta(\vec{q})$ in such a way that the second condition in (3.40) holds, and noting that the right- hand side of (3.41) has a limit in $\mathcal{S}_+(3)$, we conclude that the closure of the linear hull of functions (3.41) contains all the functions of the form $P(\beta,\vec{u}) h(|\vec{u}|^2 + \beta)$, where $P$ is an arbitrary polynomial and $h \in \mathcal{S}_+$ satisfies the first condition in (3.40). But every function satisfying condition (3.39) can be approximated (in the topology of $\mathcal{S}_+(3)$) by the functions of this type in such a way that

$$\partial_0 d[q_0 h(q_0^2) g(\vec{q})]\Big|_{u_0=0} = 0.$$

On the other hand, any function satisfying condition (3.39) can be approximated by functions of the form $\partial_0 d[h(q_0^2) g(\vec{q})]$ and $d[h(q_0^2) g(\vec{q})](\beta, 0, \vec{u}) = 0$. This means that any pair of functions $(\phi_1, \phi_2)$ from $\mathcal{S}_+(3)$ satisfying conditions (3.39) can be simulatenously approximated by functions of the form

$$d[q_0 h_1(q_0^2) g_1(\vec{q}) + h_2(q_0^2) g_2(\vec{q})](\beta, 0, \vec{u}),$$
$$\partial_0 d[q_0 h_1(q_0^2) g_1(\vec{q}) + h_2(q_0^2) g_2(\vec{q})](\beta, 0, \vec{u}),$$

where $h_1, h_2, g_1, g_2$ satisfy (3.40).

Let us prove now theorem 7. Let $\phi \in \mathcal{S}$ and $\operatorname{supp}\phi \subset O_{[q_1,q_2]}$. Let $\phi_0(\beta, u)$ denote the function which is the solution of the Cauchy problem for equation (3.7) with the data given by the restriction $\phi(|\vec{\lambda}|^2, u)$ on $\mathbf{R}^2 \times \Sigma$, where we put $|\vec{\lambda}|^2 = \beta$.

Then by the corollary to theorem 4 we have

$$(\eta_\Sigma, \phi) \equiv (\eta_\Sigma, \phi_0). \tag{3.42}$$

On the other hand, $\phi_0$ satisfies the hypothesis of the lemma. Thus, there exists a sequence of functions from $\mathcal{S}(\mathbf{R}^4)$ $\{\phi_n, n \to \infty\}$ with supports in a neighbourhood of the

segment $[q_1, q_2]$ (where $F = 0$) such that

$$d[\phi_n] \to \phi_0, \; n \to \infty \tag{3.43}$$

in $\mathfrak{S}'$. Therefore, by (3.20), (3.42), and (3.43) we get

$$(\eta_\Sigma, \phi) = (\eta_\Sigma, \phi_0) = \lim_{n \to \infty} (\eta_\Sigma, d[\phi_n]) =$$

$$= \lim_{n \to \infty} (F, \phi_n) = 0.$$

Hence $\eta_\Sigma = 0$ in $O_{[q_1, q_2]}$ and by (3.22) the same true for $\psi_\Sigma$.

## 14. Automodel asymptotics for the causal functions and singularities of their Fourier transforms on the light cone.

In this section we study the connection between the asymptotic properties of causal functions in the so-called Bjŏrken domain, singularities of their Fourier transforms on the light cone, and the quasi-asymptotics properties of their weight functions in the Jost-Lehmann-Dyson representation. The support of a causal function under consideration will satisfy an additional condition, the so-called spectral condition. This condition holds for the commutators of local currents in quantum field theory.

**1. Some preliminary results and definitions.** Let a causal function $F(q)$ satisfy the spectral condition (see Fig. 18)

$$(S) \operatorname{supp} F \subset T = [q \in \mathbf{R}^4 \colon |\vec{q}| \leqslant |q_0| + 1].$$

Together with the variables $q \in \mathbf{R}^4$ let us introduce the Bjŏrken variables $(\xi, \nu, \vec{n})$, $\vec{n} \in S^2$ by the formulae

$$\xi = -\frac{q^2}{2|q_0|}, \quad \nu = 2q_0, \quad \vec{n} = \frac{\vec{q}}{|\vec{q}|}. \tag{1.1}$$

Conversely

$$q_0 = \frac{\nu}{2}, \quad \vec{q} = \sqrt{\frac{\nu^2}{4} + \xi|\nu|}\,\vec{n}, \quad \left|\frac{\partial(\xi, \nu, \vec{n})}{\partial(q_0, \vec{q})}\right| = \frac{2}{|q_0||\vec{q}|}. \tag{1.2}$$

The mapping (1.1) transforms the domain $[q \in \mathbf{R}^4 \colon |\vec{q}| \geqslant a]$, $a > 0$ into the domain

$$\left[(\xi, \nu, \vec{n}) \colon \xi \geqslant -\frac{|\nu|}{4} + \frac{a^2}{|\nu|}\right]$$

and the set $T = [q \in \mathbf{R}^4 \colon |\vec{q}| \leqslant |q_0| + 1]$ into the set $\tilde{T} \times S^2$, where $S^2$ is the unit sphere in $\mathbf{R}^3$ and

$$\tilde{T} = \left[(\xi, \nu) \in \mathbf{R}^2 \colon \xi \leqslant 1 + \frac{1}{|\nu|}\right]$$

(see Fig. 22). Now let $\chi \in \theta_M(\mathbf{R}^2)$ (see Section 1.4) and

$$\operatorname{supp}\chi \subset \left[(\xi,\nu) \in \mathbf{R}^2 : \xi \geqslant -\frac{|\nu|}{4} + \frac{a^2}{|\nu|}\right], \quad a > 0. \tag{1.3}$$

The following inequality holds in the intersection of $\tilde{T}$ and the set (1.3)

$$1 + \frac{1}{|\nu|} \geqslant -\frac{|\nu|}{4} + \frac{a^2}{|\nu|}.$$

Therefore, if $a > 1$, then we have $|\nu| \geqslant 2(a-1)$. Hence,

$$\tilde{T} \bigcap \operatorname{supp}\chi \subset \left[(\xi,\nu)\colon |\nu| \geqslant 2(a-1), -\frac{|\nu|}{4} + \frac{a^2}{|\nu|} \leqslant \xi \leqslant 1 + \frac{1}{|\nu|}\right]. \tag{1.4}$$

Note that

$$\tilde{T} \bigcap \operatorname{supp}\chi \subset \left[(\xi,\nu)\colon \xi \leqslant 1 + \frac{1}{2(a-1)}\right]. \tag{1.5}$$

The change of variables (1.1)- (1.2) is not singular in the neighbourhood of the set (1.4). Hence, for any multiplier $\chi(\xi,\nu)$ with properties (1.3) for $a > 1$, one can introduce a distribution $F_\chi(\xi,\nu,\vec{n})$ from $\mathcal{S}'(\mathbf{R}^2)\otimes\mathcal{E}'(S^2)$ by the formula (see Section 1.5)

$$_\chi(\xi,\nu,\vec{n}) = \chi\left(-\frac{q^2}{2|q_0|}, 2q_0\right)F(q), \tag{1.6}$$

or, more precisely, for any $\phi \in \mathcal{S}(\mathbf{R}^2)\times\mathcal{E}(S^2)$

$$(F_\chi,\phi) = \left(F(q), \chi\left(-\frac{q^2}{2|q_0|}, 2q_0\right)\frac{2}{|q_0||\vec{q}\,|}\phi\left(-\frac{q^2}{2|q_0|}, 2q_0, \frac{\vec{q}}{|\vec{q}\,|}\right)\right). \tag{1.7}$$

It is clear that by (1.4) we have

$$\operatorname{supp} F_\chi \subset \left[(\xi,\nu)\colon |\nu| \geqslant 2(a-1), -\frac{|\nu|}{4} + \frac{a^2}{|\nu|} \leqslant \xi \leqslant 1 + \frac{1}{|\nu|}\right] \times S^2. \tag{1.8}$$

The following theorem holds.

**Theorem 1.** *Let $F(q)$ be a causal distribution satisfying the spectral condition $(S)$ and let $\chi(\xi,\nu)$ be a multiplier in $\mathcal{S}(\mathbf{R}^2)$ with property (1.3) with $a > 2$. Then the distribution $F_\chi(\xi,\nu,\vec{n}) \in \mathcal{S}'(\mathbf{R}^2)\otimes\mathcal{E}'(S^2)$ is in fact infinitely differentiable with respect to $(\nu,\vec{n}) \in \mathbf{R}^1\otimes S^2$. In other words, for any $\phi \in \mathcal{S}(\mathbf{R}^1)$ the function $(F_\chi(\xi,\nu,\vec{n}),\phi(\xi))$ is infinitely differentiable with respect to $(\nu,\vec{n}) \in \mathbf{R}^1\otimes S^2$. Moreover, there are numbers $c > 0$, $m \geqslant 0$, and an integer $p \geqslant 0$, such that*

$$|(F_\chi(\xi,\nu,\vec{n}),\phi(\xi))| \leqslant c(1+|\nu|)^m\|\phi\|_p, \tag{1.9}$$

*where $\|\cdot\|_p$ is the norm in $\mathcal{S}_p(\mathbf{R}^1)$.*

**Proof.** Due to (1.17) of Section 13.1, (1.6) and (1.4) it is sufficient to consider only the case of a distribution $F(q_0,\vec{q})$ antisymmetric with respect to $q_0$, and to consider $F_\chi(\xi,\nu,\vec{n})$ for $\nu > 0$. Let $\Psi = \psi_a$ be a weight function for $F$ in the Jost-Lehmann-Dyson representation (1.12') of Section 13.1 for $F$. According to (1.7),

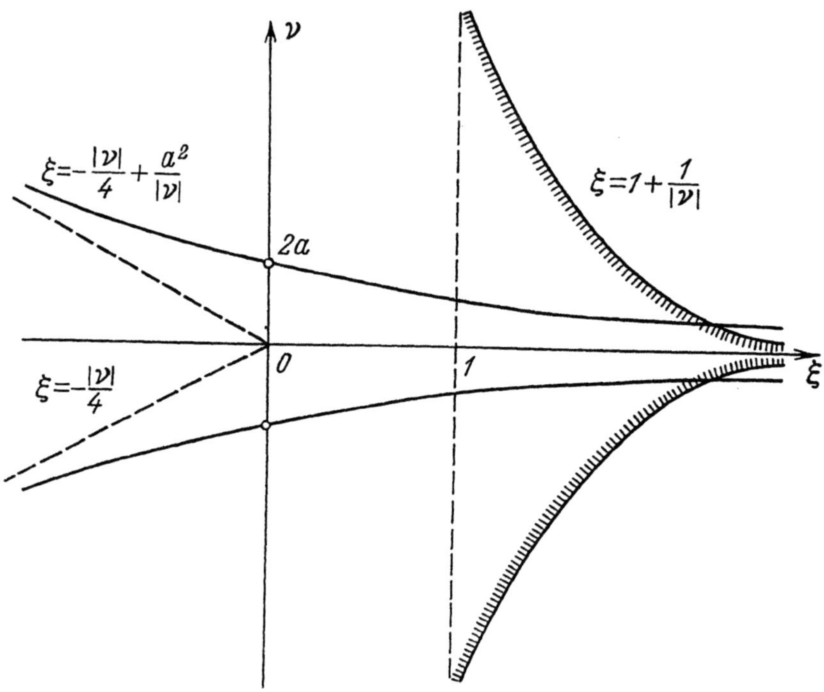


Figure 22.

for any

$$\phi \in \mathcal{S}(\mathbf{R}^2) \otimes \mathcal{E}(S^2), \quad \operatorname{supp} \phi \subset [(\xi, \nu, \vec{n}): \nu > 0]$$

we have

$$(\mathcal{F}_\chi, \phi) =$$

$$= \left( \psi(\lambda, \vec{u}), \int \frac{\phi_1\left( \dfrac{|\vec{q}\,|^2}{2\sqrt{|\vec{q} - \vec{u}\,|^2 + \lambda}} - \dfrac{1}{2}\sqrt{|\vec{q} - \vec{u}\,|^2 + \lambda}, 2\sqrt{|\vec{q} - \vec{u}\,|^2 + \lambda}, \dfrac{\vec{q}}{|\vec{q}\,|} \right)}{|\vec{q}\,|(|\vec{q} - \vec{u}\,|^2 + \lambda)} \, d\vec{q} \right) =$$

$$= \left( \psi(\lambda, \vec{u}), \int_{S^2} \int_0^\infty \phi_1 \left( \frac{2r(\vec{n}, \vec{u}) - |\vec{u}\,|^2 - \lambda}{2\sqrt{r^2 + |\vec{u}\,|^2 + \lambda - 2r(\vec{n}, \vec{u})}}, 2\sqrt{r^2 + |\vec{u}\,|^2 + \lambda - 2r(\vec{n}, \vec{u})}, \vec{n} \right) \times \right.$$

$$\left. \times \frac{r \, dr \, d\vec{n}}{r^2 + |\vec{u}\,|^2 + \lambda - 2r(\vec{n}, \vec{u})} \right). \tag{1.10}$$

where

$$\phi_1(\xi, \nu, \vec{n}) = \chi(\xi, \nu)\phi(\xi, \nu, \vec{n}). \tag{1.11}$$

Let us change the variable $r \to \nu$ in the inner integral, where

$$\nu = 2\sqrt{r^2 + |\vec{u}|^2 + \lambda - 2r(\vec{n}, \vec{u})}, \quad \frac{\partial \nu}{\partial r} = \frac{4[r - (\vec{n}, \vec{u})]}{\nu}. \tag{1.12}$$

We denote

$$r_{\pm}(\lambda, \vec{u}, \nu, \vec{n}) = (\vec{n}, \vec{u}) \pm \frac{\nu}{2}\sqrt{1 - \frac{4}{\nu^2}[\lambda + |\vec{u}|^2 - (\vec{n}, \vec{u})^2]}. \tag{1.13}$$

It is easy to see that if $(\vec{n}, \vec{u}) < 0$ in (1.10) then $\nu$ is in the set $(2\sqrt{|\vec{u}|^2 + \lambda}, \infty)$, and $r = r_+$, and if $(\vec{n}, \vec{u}) > 0$ and $0 \leqslant r \leqslant (\vec{n}, \vec{u})$ then $r = r_-$ and $2\sqrt{|\vec{u}|^2 + \lambda - (\vec{n}, \vec{u})^2} < \nu < 2\sqrt{|\vec{u}|^2 + \lambda}$. Finally, if $(\vec{n}, \vec{u}) > 0$ and $r > (\vec{n}, \vec{u})$, then we have $r = r_+$ and $\nu \in (2\sqrt{\lambda + |\vec{u}|^2 - (\vec{n}, \vec{u})^2}, \infty)$ (Fig. 23).

Taking into account all these facts, let us rewrite (1.10) as

$$\begin{aligned}(\mathcal{F}_\chi, \phi) = 2 \Bigg( \psi(\lambda, \vec{u}), \int_{S^2} \Bigg\{ \int_{2\sqrt{|\vec{u}|^2+\lambda}}^{\infty} \phi_1\left(-\frac{\nu}{4} + \frac{r_+^2}{\nu}, \nu, \vec{n}\right) \\ \frac{r_+ \, d\nu}{\nu^2 \sqrt{1 - \frac{4}{\nu^2}[|\vec{u}|^2 + \lambda - (\vec{n}, \vec{u})^2]}} + \\ + \theta[(\vec{n}, \vec{u})] \int_{2\sqrt{|\vec{u}|^2+\lambda-(\vec{n}, \vec{u})^2}}^{2\sqrt{|\vec{u}|^2+\lambda}} \left[ \phi_1\left(-\frac{\nu}{4} + \frac{r_+^2}{\nu}, \nu, \vec{n}\right) r_+ + \phi_1\left(-\frac{\nu}{4} + \frac{r_-^2}{\nu}, \nu, \vec{n}\right) r_- \right] \times \\ \times \frac{d\nu}{\nu^2 \sqrt{1 - \frac{4}{\nu^2}[|\vec{u}|^2 + \lambda - (\vec{n}, \vec{u})^2]}} \Bigg\} d\vec{n} \Bigg).\end{aligned} \tag{1.14}$$

Note that in the domain of integration of the second integral in (1.4) we have $r_{\pm}^2 \leqslant 4(\vec{n}, \vec{u})^2$, and therefore

$$\xi = -\frac{\nu}{4} + \frac{r_{\pm}^2}{\nu} \leqslant -\frac{\nu}{4} + \frac{4(\vec{n}, \vec{u})^2}{\nu} \leqslant -\frac{\nu}{4} + \frac{4(1 + \epsilon)}{\nu},$$

if $|\vec{u}| \leqslant 1 + \epsilon$, $\epsilon > 0$. Taking into account that, according to (1.23) of Section 13.1,

$$\operatorname{supp} \psi \subset T_0 = [(\lambda, \vec{u}) \colon \lambda \geqslant 0, |\vec{u}| \leqslant 1], \tag{1.15}$$

and that $a > 2$, we conclude that the second integral in (1.14) is equal to zero. The lower bound of integration over $\nu$ in the first integral is, in fact, the point $2\sqrt{|\vec{u}|^2 + \lambda + 2\epsilon_1}$, where $\epsilon_1 > 0$ is defined by the condition

$$-\frac{\nu}{4} + \frac{r_+^2}{\nu} \geqslant -\frac{\nu}{4} + \frac{a^2}{\nu},$$

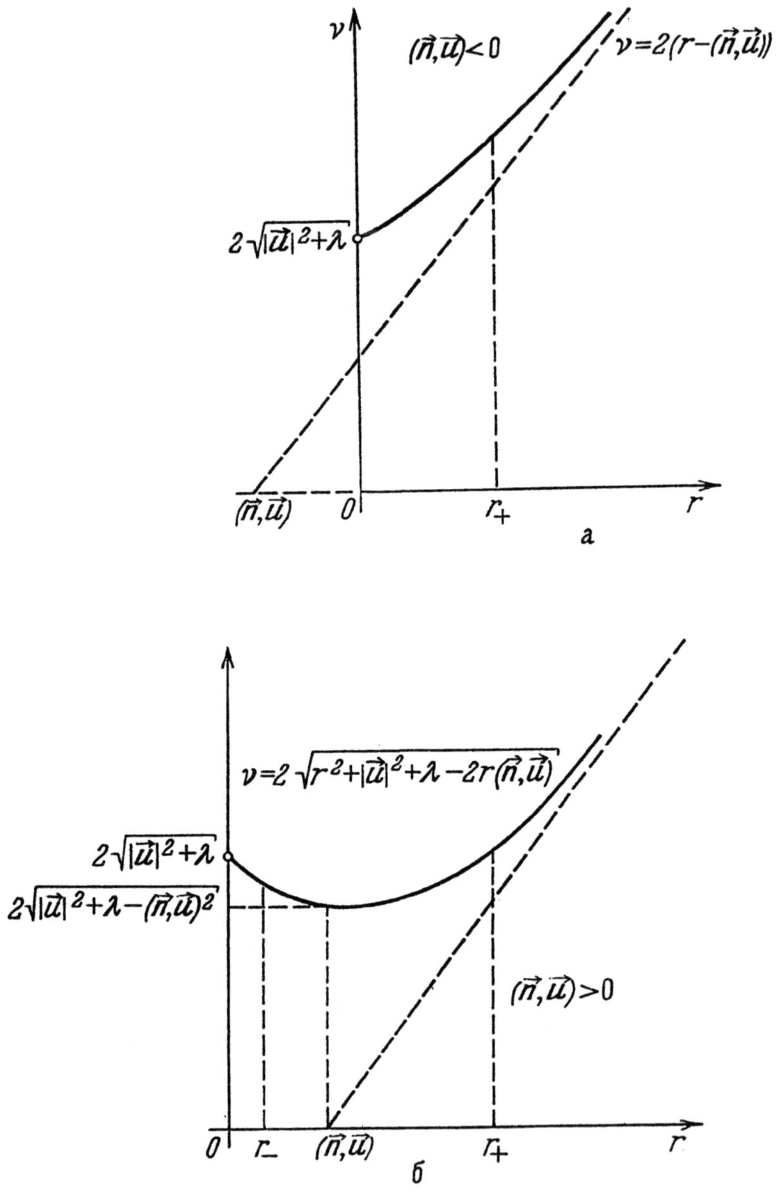


Figure 23.

and, due to (1.13) and (1.15), depends only on $a > 2$.

Let $\eta$ denote a $C^\infty$-function such that

$$\eta(t) = \begin{cases} 1, & t > \epsilon_1, \\ 0, & t < 0. \end{cases} \tag{1.16}$$

Introducing the 'cutting' function $\eta((\nu^2/4)-\lambda-|\vec{u}|^2)$, we rewrite formula (1.14) in the form

$$(\mathcal{F}_\chi,\phi)=\left(\psi(\lambda,\vec{u}),\int_{S^2}\int_{-\infty}^{+\infty}\chi(\xi_1(\lambda,\vec{u};\nu,\vec{n}))\phi(\xi_1(\lambda,\vec{u};\nu,\vec{u}),\nu,\vec{n})\times\right.$$
$$\left.\times\,\omega(\lambda,\vec{u};\nu,\vec{n})\,d\nu\,d\vec{n}\right). \tag{1.17}$$

where

$$\omega(\lambda,\vec{u};\nu,\vec{n})=\left\{\frac{2(\vec{n},\vec{u})}{\nu^2\sqrt{1-\frac{4}{\nu^2}[|\vec{u}|^2+\lambda-(\vec{n},\vec{u})^2]}}+\frac{1}{\nu}\right\}\times$$
$$\times\,\eta\left(\frac{\nu^2}{4}-\lambda-|\vec{u}|^2\right),\quad \xi_1(\lambda,\vec{u};\nu,\vec{n})=$$
$$=(\vec{n},\vec{u})\sqrt{1-\frac{4}{\nu^2}[|\vec{u}|^2+\lambda-(\vec{n},\vec{u})^2]}+\frac{2(\vec{n},\vec{u})-|\vec{u}|^2-\lambda}{\nu}. \tag{1.18}$$

The right-hand side of (1.17) does not depend on the function $\eta$ with properties (1.16).

Now let us put

$$\phi(\xi,\nu,\vec{n})=\phi_0(\xi)\phi_2(\nu,\vec{n}), \tag{1.19}$$

where

$$\phi_0\in\mathcal{S}(\mathbf{R}^1),\quad \phi_2\in\mathcal{S}(\mathbf{R}^1)\times\mathcal{E}(S^2)$$

and

$$\operatorname{supp}\phi_2\subset[(\nu,\vec{n})\in\mathbf{R}^1\times S^2:\nu>0].$$

Substituting (1.19) into (1.17) we get

$$(\mathcal{F}_\chi,\phi_0\phi_2)=\left(\psi(\lambda,\vec{u}),\int_{S^2}\int_{-\infty}^{+\infty}\chi(\xi_1(\lambda,\vec{u};\nu,\vec{n}),\nu)\times\right.$$
$$\left.\times\,\phi_0(\xi_1(\lambda,\vec{u};\nu,\vec{n}))\phi_2(\nu,\vec{n})\omega(\lambda,\vec{u};\nu,\vec{n})\,d\nu\,d\vec{n}\right). \tag{1.20}$$

Note that the function

$$\chi(\xi_1(\lambda,\vec{u};\nu,\vec{n}),\nu)\phi_0(\xi_1(\lambda,\vec{u};\nu,\vec{n}))\phi_2(\nu,\vec{n})\omega(\lambda,\vec{u};\nu,\vec{n})$$

belongs to $\mathcal{S}(T_0)$ as a function of $(\lambda,\vec{u})$ for any fixed $(\nu,\vec{n})$, $\nu>0$, $\vec{n}\in S^2$ and it is a $C^\infty$-function of $(\nu,\vec{n})$ in the topology of $\mathcal{S}(T_0)$. Moreover, it is easy to see that the integral over $(\nu,\vec{n})$ in the right-hand side of (1.20) can be calculated before the action of the functional so that

$$(\mathcal{F}_\chi, \phi_0\phi_2) = \int_{S^2}\int_{-\infty}^{+\infty} \phi_2(\nu, \vec{n})(\psi(\lambda, \vec{u}), \chi(\xi_1(\lambda, \vec{u}; \nu, \vec{n}), \nu) \times$$

$$\times \phi_0(\xi_1(\lambda, \vec{u}; \nu, \vec{n}))\omega(\lambda, \vec{u}; \nu, \vec{n}))\, d\nu\, d\vec{n}.$$

From this there follows the representation

$$(\mathcal{F}_\chi(\xi, \nu, \vec{n}), \phi_0(\xi)) = (\psi(\lambda, \vec{u}), \chi(\xi_1(\lambda, \vec{u}; \nu, \vec{n}), \nu) \times$$
$$\times \phi_0(\xi_1(\lambda, \vec{u}; \nu, \vec{n}))\omega(\lambda, \vec{u}; \nu, \vec{n})). \tag{1.21}$$

Now, it is clear that the function $(\mathcal{F}_\chi(\xi, \nu, \vec{n}), \phi_0(\xi))$ is infinitely differentiable with respect to $(\nu, \vec{n})$.

Let $p_1$ denote the order of $\psi$ in $\mathcal{S}'(T_0)$. Then

$$|(\mathcal{F}_\chi(\xi, \nu, \vec{n}), \phi_0(\xi))| \leqslant C_1 \|\chi(\xi_1(\lambda, \vec{u}; \nu, \vec{n}), \nu) \times$$

$$\times \phi_0(\xi_1(\lambda, \vec{u}; \nu, \vec{n}))\omega(\lambda, \vec{u}; \nu, \vec{n})\|_{p_1, T_0}, \tag{1.22}$$

where $\|\cdot\|_{p_1, T_0}$ is the norm in $\mathcal{S}_{p_1}(T_0)$ (see Section 1.1). On the other hand, there are natural numbers $p$ and $m$ such that

$$\|\chi(\xi_1(\lambda, \vec{u}; \nu, \vec{n}), \nu)\phi_0(\xi_1(\lambda, \vec{u}; \nu, \vec{n}))\omega(\lambda, \vec{u}; \nu, \vec{n})\|_{p_1, T_0} \leqslant$$
$$\leqslant C_2(1 + |\nu|)^m \|\phi_0\|_p, \tag{1.23}$$

where $\|\cdot\|_p$ is some norm in $\mathcal{S}(\mathbf{R}^1)$. From (1.22) and (1.23) the estimate (1.9) follows.

Let us give the following definition.

**Definition.** We say that a multiplier $\chi(\xi, \nu)$ in $\mathcal{S}(\mathbf{R}^2)$ (that is, $\chi \in \theta_M(\mathbf{R}^2)$) is admissible if it has property (1.3) with $a > 2$ and for some $a' > a$

$$\chi(\xi, \nu) \equiv 1, \quad \xi \geqslant -\frac{|\nu|}{4} + \frac{(a')^2}{|\nu|}. \tag{1.24}$$

It is clear that

$$\chi(\xi, \nu) \to 1(\xi), \quad \nu \to \pm\infty$$

in $\theta_M(\mathbf{R}^1)$.

Theorem 1 leads to the following definition.

**Definition 1.** We say that a causal distribution $F$, satisfying the condition $(S)$, has an automodel asymptotics (with respect to an automodel function $\rho$ in the direction $\vec{n}_0 \in S^2$ if for some admissible multiplier $\chi$ there exists a neighbourhood $U \subset S^2$ of the vector $\vec{n}_0$ such that

$$\frac{\mathcal{F}_\chi(\xi, \nu, \vec{n})}{\rho(|\nu|)} \xrightarrow{\vec{n} \in U} F_\pm(\xi, \vec{n}), \quad \nu \to \pm\infty \tag{1.25}$$

in $\mathcal{S}'(\mathbf{R}^1_\xi)$ and $(F_+, F_-) \not\equiv 0$. (The mapping $F \to \mathcal{F}_\chi$ is given by formulae (3.15)-(3.16).) If in (1.25) $U = S^2$, we say that $F$ has an automodel asymptotics with respect to $\rho$. In this case we formally write

$$F(q) \sim \rho(|\nu|)F_{\pm}(\xi,\vec{n}), \quad \nu \to \pm\infty.$$

**Remark.** If the limits in (1.25) exist for one admissible multiplier then they also exist for other admissible multipliers.

Indeed, let $\chi_1$ and $\chi_2$ be admissible multipliers. Then $\mathcal{F}_{\chi_1} - \mathcal{F}_{\chi_2} = \mathcal{F}_{\chi_1 - \chi_2}$ and, due to (1.4),

$$\operatorname{supp}(\mathcal{F}_{\chi_1} - \mathcal{F}_{\chi_2}) \subset \left[(\xi,\nu):\ |\nu| \geqslant 2,\ \xi \leqslant -\frac{|\nu|}{4} + b\right] \times S^2, \tag{1.26}$$

where $b = 1/2\max(a_1^2, a_2^2)$, $a_1$ and $a_2$ are the constants $a'$ in (1.24) for $\chi_1$, and $\chi_2$, respecitvely. Therefore, denoting by $\eta$ the multiplier in $\mathcal{S}(\mathbf{R}^2)$ with the properties

$$\eta(\xi,\nu) = \begin{cases} 1, & \xi \leqslant -\dfrac{|\nu|}{4} + b, \quad |\nu| \geqslant 1, \\ 0, & -\dfrac{|\nu|}{8} + b, \quad |\nu| \geqslant 1, \end{cases} \tag{1.27}$$

we get from (1.9) and (1.26)

$$|(\mathcal{F}_{\chi_1} - \mathcal{F}_{\chi_2}, \phi)| = |(\mathcal{F}_{\chi_1-\chi_2}(\xi,\nu,\vec{n}), \eta(\xi,\nu)\phi(\xi))_{\xi}| \leqslant$$
$$\leqslant c(1 + |\nu|)^m \|\eta(\xi,\nu)\phi(\xi)\|_p, \tag{1.28}$$

where the norm $\|\cdot\|_p$ is taken with respect to $\xi \in \mathbf{R}^1$. But

$$\|\eta(\xi,\nu)\phi(\xi)\|_p = \sup_{\xi \leqslant -(|\nu|/8)+b} (1 + |\xi|)^j |(\eta(\xi,\nu)\phi(\xi))^{(j)}| \leqslant$$
$$\leqslant \frac{C_k}{1 + |\nu|^k}$$

for any natural number $k$. From this, and from (1.28), it follows that the function $(\mathcal{F}_{\chi_1} - \mathcal{F}_{\chi_2}, \phi)$ decreases when $|\nu| \to \infty$ faster than any power of $1/|\nu|$. Hence, the statement has been proved.

Now, for the distributions from $\mathcal{S}'(\overline{V})$ we introduce the definition of a quasi-asymptotics on the light cone (more precisely, on its boundary). First we note that, according to (1.10) of Section 13.1, any distribution $\tilde{F}$ from $\mathcal{S}'(\overline{V})$ can be represented in the form

$$\tilde{F}(x) = \epsilon(x_0)\Phi_a(x^2,\vec{x}) + |x_0|\Phi_s(x^2,\vec{x}),$$

where $\Phi_a$ and $\Phi_s$ belong to $\mathcal{S}_+^1(3)$. The exact definition of the distributions in the right-hand side of this equality is given by formulae (1.6)-(1.8) of Section 13.1.

**Definition 2.** We say that a distribution $\tilde{F} \in \mathcal{S}'(\overline{V})$ has a quasi-asymptotics in the neighbourhood of a light cone with respect to an automodel function $\rho$ of order $\alpha$ if the distributions $\Phi_a(\kappa,\vec{x})$ and $\Phi_s(\ ,\vec{x})$ have a quasi- asymptotics at the origin over $\kappa$ with respect to $\rho$. That is, the following limit relations hold (see Section 3.6):

$$\frac{1}{\rho(k)}\Phi_j\left(\frac{\kappa}{k},\vec{x}\right) \to f_{-\alpha+1}(\kappa)\times g_j(\vec{x}),\ k\to\infty \tag{1.29}$$

in $\mathcal{S}'_+(3)$, where $j = a, s$ and $(g_a, g_s) \not\equiv 0$. In this case we shall formally write

$$\tilde{F}(x) \sim \frac{\rho[(x_+^2)^{-1}]}{\Gamma(1-\alpha)}[\epsilon(x_0)g_a(\vec{x}) + |x_0| g_s(\vec{x})],\ x^2 \sim 0.$$

Let us show that if $\tilde{F}$ is the Fourier transform of a (causal) distribution $F(q)$ satisfying conditon (5) then definition 2 is equivalent to another which seems, at first, to be more restrictive. Recall that according to (1.11) of Section 13.1

$$\Phi_a = 4\pi^2 i F_{\vec{u}}[B_\lambda[\psi_a(\lambda,\vec{u})]], \tag{1.30}$$

$$\Phi_s = 4\pi^2 F_{\vec{u}}[B_\lambda[\psi_s(\lambda,\vec{u})]].$$

Since in our case $\operatorname{supp}\psi_j \subset T_0$, $j = a, s$ (see (1.23), Section 13.1) - that is, supports of $\psi_j$, $j = a, s$ are compact with respect to $\vec{u}$ - then the distributions $\Phi_j(kappa, \vec{x})$, $j = a, s$, are in fact the entire functions over $\vec{x}$. Therefore, these distributions can be considered as distributions with respect to the variable $\kappa$, with $\vec{x}$ being a parameter. In this case we shall sometimes write $\Phi_{a,\vec{x}}(\kappa)$ and $\Phi_{s,\vec{x}}(\kappa)$, instead of $\Phi_a(\kappa,\vec{x})$ and $\Phi_s(\kappa,\vec{x})$, respectively. More precisely, for any $\phi \in \mathcal{S}_+$

$$(\Phi_{j,\vec{x}},\phi) = C_j(\psi_j(\lambda,\vec{u}), b[\phi](\lambda)e^{i(\vec{u},\vec{x})}),\ j = a, s, \tag{1.31}$$

where $C_a = 4\pi^2 i$, $C_s = 4\pi^2$.

The following theorem holds:

**Theorem 2.** *Let $\tilde{F}$ be the fourier transform of a causal distribution $F$ satisfying the condition $(S)$. Then $\tilde{F}$ has a quasi-asymptotics in the neighbourhood of a light cone with respect to an automodel function $\rho$, if and only if the distributions $\Phi_{j,\vec{x}}$, $j = a, s$, $\vec{x} \in \mathbf{R}^3$, have the following quasi-asymptotics at zero with respect to $\rho$:*

$$\frac{1}{\rho(k)}\Phi_{j,\vec{x}}\left(\frac{\kappa}{k}\right) \to f_{-\alpha+1}(\kappa)\times g_j(\vec{x}),\ k\to\infty$$

*in $\mathcal{S}'_+$, where $j = a, s$, $\alpha$ is the order of the function $\rho$, $(g_a, g_s) \not\equiv 0$, and there exists an integer $n \geqslant 0$ such that for any $\phi \in \mathcal{S}_+$*

$$\left|\left(\frac{1}{\rho(k)}\Phi_{j,\vec{x}}\left(\frac{\kappa}{k}\right),\phi(\kappa)\right)\right| \leqslant C_\phi(1+|\vec{x}|^n),\ j = a, s,\ k\to\infty. \tag{1.32}$$

**Proof.** If the distributions $\Phi_{j,\vec{x}}$, $j = a, s$, $\vec{x} \in \mathbf{R}^3$ have quasi-asymptotics at zero with respect to $\rho$ and estimate (1.32) hold, then by the Lebesgue theorem in the equalities

$$\left(\frac{1}{\rho(k)}\Phi_j\left(\frac{\kappa}{k},\vec{x}\right),\phi(\kappa,\vec{x})\right) = \frac{1}{\rho(k)}\int\left(\Phi_{j,\vec{x}}\left(\frac{\kappa}{k}\right),\phi(\kappa,\vec{x})\right)_\kappa d\vec{x},\ j = a, s,$$

it is possible to take the limit when $k \to \infty$ before integrating. Hence $\Phi_a(\kappa,\vec{x})$ and $\Phi_s(\kappa,\vec{x})$ have quasi-asymptotics in $\kappa$ at zero with respect to $\rho$.

Conversely, according to theorem 5 of Section 3.4, and to formula (1.30), $\Phi_j(\kappa,\vec{x})$, $j = a, s$, have a quasi-asymptotics in $\kappa$ at zero with respect to $\rho(k)$ if and only if $\psi_j(\lambda,\vec{u})$, $j = a, s$ have a quasi-asymptotics in $\lambda$ at infinity with respect to $k^{-2}\rho(k)$. Since $\operatorname{supp}\psi_j \subset T_0$, $j = a, s$, the sequence

$$\left\{\frac{k^2}{\rho(k)}\psi_j(k\lambda,\vec{u}),\ k \to \infty\right\}$$

converges in $\mathcal{S}'(T_0)$. On the other hand, according to (1.31) and (3.9) of Section 2.3

$$\left(\frac{1}{\rho(k)}\Phi_{j,\vec{x}}\left(\frac{\kappa}{k}\right), \phi\right) = C_j\left(\frac{k^2\psi_j(k\lambda,\vec{u})}{\rho(k)}, b[\phi](\lambda)e^{i(\vec{u},\vec{x})}\right),\quad j = a, s, \tag{1.33}$$

for any $k > 0$ and $\phi \in \mathcal{S}_+$. But since $b[\phi](\lambda)e^{i(\vec{u},\vec{x})} \in \mathcal{S}(T_0)$ it follows from (1.33) that $\Phi_{j,\vec{x}}(\kappa)$, $j = a, s$ have quasi- asymptotics at zero with respect to $\rho$ for any fixed $\vec{x}$. Moreover for some $n$ and each $\omega \in \mathcal{S}(T_0)$ the following inequality holds:

$$\left|\left(\frac{k^2\psi_j(k\lambda,\vec{u})}{\rho(k)}, \omega\right)\right| \leqslant C\,\|\omega\|_{n,T_0}.$$

Hence, due to (1.33)

$$\left|\left(\frac{1}{\rho(k)}\Phi_{j,\vec{x}}\left(\frac{\kappa}{k}\right).\phi(\kappa)\right)\right| \leqslant C\,\|b[\phi](\lambda)e^{i(\vec{u},\vec{x})}\|_{n,T_0},\quad k \to \infty,$$

and estimate (1.32) follows (see Section 2.3).

Let us now formulate our main theorems.

**2. The main theorems.** In this subsection we shall prove theorems which will link the automodel behaviour of causal distributions with the singularity of their Fourier transforms near the light cone and the quasi-asymptotics properties of weight functions in the Jost-Lehmann-Dyson representation.

**Theorem 1.** *Let a causal distribution $F$ satisfy the condition $(S)$ and let $(\psi_a, \psi_s)$ be its weight functions in the Jost-Lehmann-Dyson representation. If $F$ has an automodel asymptotics with respect to some automodel function $\rho$ into a direction $\vec{n}_0 \in S^2$ then there exists a neighbourhood $U \subset S^2$ of vector $\vec{n}_0$ such that for any $\phi \in \mathcal{S}(\mathbf{R}^1)$ there exist the limits*

$$\lim_{k\to\infty}\left(\frac{1}{\rho(k)}\cdot\psi_a(k\lambda,\vec{u}), \phi(-\lambda + (\vec{n},\vec{u}))\right) = \left(\frac{F_-(\xi,\vec{n}) - F_-(\xi,\vec{n})}{2}, \phi(\xi)\right).$$

$$\lim_{k\to\infty}\left(\frac{1}{\rho(k)}\psi_s(k\lambda,\vec{u}), \phi(-\lambda + (\vec{n},\vec{u}))\right) = -\left(\left[\frac{F_+(\xi,\vec{n}) + F_-(\xi,\vec{n})}{2}\right]^{(-1)}, \phi(\xi)\right) \tag{2.1}$$

*uniformly in $\vec{n} \in U$, where $F_\pm$ are defined in (1.25) and $1/2[F_+ + F_-]^{(-1)}$ is the first primitive of $1/2[F_+ + F_-]$ with respect to $\xi$, whose support belongs to the set $[\xi \in \mathbf{R}^1: \xi \leqslant 1] \times S^2$.*

**Proof.** We can assume without loss of generality that $F(q)$ is odd in $q_0$. Indeed, considering its even part $F_s(q)$ and its odd part $F_a(q)$ it is easy to see that $F_s$ and $F_a$ have an automodel asymptotics along $\vec{n}_0$ with respect to $\rho$. Hence, the odd distribution $F_1(q) = (1/2q_0)F_2$, which is defined by (1.17), has an automodel asymptotics with respect to $k\rho(k)$. Hence, if we prove that for its weight function $\psi_1(\lambda, \vec{u})$ the limit

$$\lim_{k\to\infty}\left(\frac{1}{k\rho(k)}\psi_1(k\lambda,\vec{\cdot}),\phi(-\lambda+(\vec{n},\vec{u}))\right) \not\equiv 0,$$

exists uniformly on $\vec{n} \in U$, then due to (1.19) of Section 13.1, and due to the fact that the mapping $\phi \to \phi'$ is an automorphism of the space $\mathcal{S}((-\infty, 1])$, it will follow that $\psi_1$ satisfies the limit relation (2.1).

Thus, we shall assume that $F = F_a$, $\psi_s = 0$, $\psi_a = \psi$. Let us use theorem 2 of Section 13.2, according to which

$$\left(\frac{1}{\rho(k)}\psi(k\lambda,\vec{u}),\phi(-\lambda+(\vec{n},\vec{u}))\right) =$$

$$= \frac{1}{\pi\sqrt{\pi}k^2\sqrt{k}\,\rho(k)}\left(F(q), q_0\phi_1^{(3/2)}\left(-\frac{k}{4}-\frac{q_0^2}{4k}+(\vec{n},\vec{u})\right)\right), \quad \phi \in \mathcal{S}(\mathbf{R}^1), \tag{2.2}$$

where $\phi_1(\lambda)$ is any function from $\mathcal{S}(\mathbf{R}^1)$ which is equal to $\phi(\lambda)$ for $\lambda \leqslant 1$ and satisfies the condition

$$\int_{-\infty}^{+\infty}\lambda^k\phi_1(\lambda)\,d\lambda = 0, \quad k = 0, 1, \ldots, N \tag{2.3}$$

for a sufficiently large $N$.

Let $\chi(\xi, \nu)$ be an admissible multiplier in $\mathcal{S}(\mathbf{R}^2)$. We rewrite equality (2.2) in the following way:

$$\left(\frac{1}{\rho(k)}\psi(k\lambda,\vec{u}),\phi(-\lambda+(\vec{n},\vec{u}))\right) = \tag{2.4}$$

$$= \Omega_k(\vec{n}) + \frac{1}{\pi\sqrt{\pi}k^2\sqrt{k}\,\rho(k)}\left(F(q)\chi\left(-\frac{q^2}{2|q_0|}, 2q_0\right) \cdot q_0\phi_1^{(3/2)}\left(-\frac{k}{4}-\frac{q_0^2}{4k}+(\vec{n},\vec{q})\right)\right),$$

where

$$\Omega_k(\vec{n}) = \frac{\pi^{-3/2}k^{-5/2}}{\rho(k)}\left(F(q)\left(1-\chi\left(-\frac{q^2}{2|q_0|}, 2q_0\right)\right) \cdot q_0\phi_1^{(3/2)}\left(-\frac{k}{4}-\frac{q_0^2}{4k}+(\vec{n},\vec{q})\right)\right).$$

Let us prove that

$$\Omega_k(\vec{n}) \xrightarrow[\longrightarrow]{\vec{n}\in S^2} 0, \quad k \to \infty, \tag{2.5}$$

provided $N$ in (2.3) is sufficiently large.

Denoting the order of $F$ in $\mathcal{S}'(T)$ by $p$, and using the fact that the function

$$\phi_1^{(3/2)}\left(-\frac{k}{4}-\frac{q_0^2}{4k}+(\vec{n}.\vec{q})\right)$$

belongs to $\mathcal{S}_p(T)$ if $N$ is sufficiently large (the lemma of Section 2.4), let us estimate $|\Omega_k(\vec{n})|$. We have

$$|\Omega_k(\vec{n})| \leqslant$$

$$\leqslant C_1\frac{k^{-5/2}}{\rho(k)}\left\|\left[1-\chi\left(-\frac{q^2}{2|q_0|},2q_0\right)\right]q_0\phi_1^{(3/2)}\left(-\frac{k}{4}-\frac{q_0^2}{4k}+(\vec{n}.\vec{q})\right)\right\|_{p,T} \leqslant$$

$$\leqslant C_2\frac{k^{-5/2}}{\rho(k)}\sup_{|\vec{q}|\leqslant 1+|q_0|,\, |\vec{q}|\leqslant a',\, 0\leqslant j\leqslant p}(1+|q|^{p'}\left|\phi_1^{(3/2+j)}\left(-\frac{k}{4}-\frac{q_0^2}{4k}+(\vec{n}.\vec{q})\right)\right|,$$

where $p'$ depends on $\chi$ and $a'$ is taken from condition (1.24).

According to the lemma of Section 2.4,

$$|\phi_1^{(3/2+j)}(\lambda)| \leqslant C_3(1+|\lambda|)^{-N'},\ j=0,1,\ldots,p \tag{2.6}$$

where $N'=N'(N)\to\infty$ when $N\to\infty$. Hence, taking into account the estimates of lemma 1 of Section 3.2 we get

$$|\Omega_k(\vec{n})| \leqslant C_4 k^{-5/2-\alpha+\epsilon}\sup_{|\vec{q}|\leqslant|q_0|+1,\, |\vec{q}|\leqslant a'}\frac{(1+|q_0|^{p'})}{\left[1+\left|\frac{k}{4}+\frac{q_0^2}{4k}-(\vec{n}.\vec{q})\right|\right]^{N'}},$$

where $\alpha$ is of the order of the automodel function $\rho$, and $\epsilon>0$ is an arbitrarily small numbers, $k>k_\epsilon$. Thus

$$|\Omega_k(\vec{n})| \leqslant C_5 k^{-5/2-\alpha+\epsilon+N'}\sup_{q_0\in\mathbb{R}^1}\frac{1+|q_0|^{p'}}{\left(1+k^2+q_0^2\right)^{N'}} \leqslant$$

$$\leqslant C_6 k^{-5/2-\alpha+\epsilon-N'+p'} \to 0,\ k\to\infty$$

if $N'+5/2+\alpha > p'+\epsilon$. Thus (2.5) has been proved.

Now let us turn to the second term in (2.4). According to (1.7), this can be represented in the form

$$\frac{\pi^{-3/2}k^{-5/2}}{8\rho(k)}\Bigg(\mathcal{F}_\chi(\xi,\nu,\vec{n}'),\nu^2\sqrt{\frac{\nu^2}{4}+\xi|\nu|}$$

$$\phi_1^{(3/2)}\left(-\frac{k}{4}-\frac{\nu^2}{4k}+(\vec{n},\vec{n}')\sqrt{\frac{\nu^2}{4}+\xi|\nu|}\right)\Bigg). \tag{2.7}$$

To go further, let us prove the following lemma.

**Lemma.** *Let the hypothesis of theorem 1 hold. Then there exists a polynomially bounded function* $f(\xi,\nu,\vec{n})$, *integers* $m$ *and* $n$, *and an admissible multiplier* $\chi_0(\xi,\nu)$, *such that*

$$\mathcal{F}_\chi(\xi,\nu,\vec{n}) = \chi_0(\xi,\nu)\partial_\xi^m f(\xi,\nu,\vec{n}), \tag{2.8}$$

$$\frac{1}{\rho(|\nu|)} f(\xi,\nu,\vec{n}) \xrightarrow{\vec{n}\in U} f_\pm(\xi,\vec{n}), \quad \nu \to \pm\infty, \tag{2.9}$$

$$\frac{1}{\rho(|\nu|)} |f(\xi,\nu,\vec{n})| \leqslant C(1+|\xi|)^n, \quad \vec{n} \in u, \quad |\nu| \geqslant 2, \tag{2.10}$$

*for some* $C > 0$, *and*

$$\operatorname{supp} f \subset \left[(\xi,\nu,\vec{n}) \in \mathbf{R}^2 \times S^2 \colon |\nu| \geqslant 2, \xi \leqslant 1 + \frac{1}{|\nu|}\right]. \tag{2.11}$$

**Proof.** Let us put

$$f(\xi,\nu,\vec{n}) = (-1)^m \mathcal{F}_\chi(\xi,\nu,\vec{n}) * f_m(-\xi), \tag{2.12}$$

where the convolution is taken over $\xi$, and $f_m$ is the kernel of fractional integration and differentiation (Section 2.4). Due to (1.8), $f$ satisfies condition (2.11). Moreover,

$$\mathcal{F}_\chi = \partial_\xi^m f. \tag{2.13}$$

If $\chi_0$ is an admissible multiplier with the parameters in (1.3) of Section 14.1 and (1.24) of Section 14.1 ($a_0 > 2$ and $a_0' < a$, respectively, $a$ being the corresponding parameter for $\chi$ in (1.3) of Section 14.1), then multiplying both sides of (1.13) by $\chi_0$ we get equality (2.8).

Let us prove that for a sufficiently large $m$ the function (2.12) is continuous, polynomially bounded and satisfies conditions (2.9) and (2.10). Indeed, it follows from theorem 1 of Section 14.1 that the family of distributions of one variable $\xi$ ($\vec{n}$ and $\nu$ are parameters)

$$\left\{ f_{\nu,\vec{n}}(\xi) = \frac{1}{(1+|\nu|)^q} \mathcal{F}_\chi(\xi,\nu,\vec{n}), (\nu,\vec{n}) \in \mathbf{R}^1 \times S^2 \right\} \tag{2.14}$$

is bounded in $\mathcal{S}'(\mathbf{R}^1)$, and that they are continuous in $\nu$ and $\vec{n}$. Since $\operatorname{supp} f_{\nu,\vec{n}} \subset [\xi \in \mathbf{R}^1 \colon \xi \leqslant 2]$, there is a natural number $p_1$ such that the family of distributions (2.14) is bounded in $\mathcal{S}'_{p_1}((-\infty,2])$, and the dependence on $(\nu,\vec{n})$ is continuous in the topology of $\mathcal{S}'_{p_1}((-\infty,2])$.

On the other hand, there is a natural number $m'$, such that the family of functions

$$\{\phi_a(\xi) = f_m(-(a-\xi)), m > m'\} \tag{2.15}$$

belongs to $\mathcal{S}_{p_1}((-\infty,2])$, where the dependence on $a$ is continuous, and the norm $\|\phi_a\|_{p_1,(-\infty,2]}$ is polynomially bounded in $a$. Thus, it follows that for $m > m'$ the function (2.12) is continuous with respect to all variables and is polynomially bounded.

By hypothesis, the sequence of distributions

$$\left\{g_{\nu,\vec{n}}(\xi) = \frac{1}{\rho(|\nu|)}\ \chi(\xi,\nu,\vec{n}),\ \nu \to \pm\infty\right\}$$

converge in $\mathcal{S}'((-\infty,2])$ uniformly in $\vec{n} \in V$ and therefore they converge uniformly in some $\mathcal{S}'_{p_2}((-\infty,2])$. Choosing $m'$ in such a way that the functions (2.15) belong to $\mathcal{S}_{p_2}((-\infty,2])$ we conclude that for $m > m'$ the function $f$ in (2.12) satisfies conditions (2.9) and (2.10).

Let us continue the proof of theorem 1. Substituting (2.8) into (2.7) we obtain

$$\frac{\pi^{-3/2}k^{-5/2}}{16\rho(k)}\left(\partial_\xi^m f(\xi,\nu,\vec{n}'),\ |\nu|^3\chi_0(\xi,\nu)\sqrt{1+\frac{4\xi}{|\nu|}}\right.$$

$$\left.\phi_1^{(3/2)}\left(-\frac{k}{4}-\frac{\nu^2}{4k}+(\vec{n},\vec{n}')\sqrt{\frac{\nu^2}{4}+\xi|\nu|}\right)\right) =$$

$$= \frac{(-1)^m\pi^{-3/2}k^{-5/2}}{16\rho(k)}\int_{\mathbf{R}^2}\int_{S^2} f(\xi,\nu,\vec{n}')|\nu|^3\partial_\xi^m\left[\chi_0(\xi,\nu)\sqrt{1+\frac{4\xi}{|\nu|}}\right.$$

$$\left.\phi_1^{(3/2)}\left(-\frac{k}{4}-\frac{\nu^2}{4k}+(\vec{n},\vec{n}')\sqrt{\frac{\nu^2}{4}+\xi|\nu|}\right)\right]d\vec{n}'\,d\xi\,d\nu =$$

$$= \frac{\sqrt{k}}{\rho(k)}\sum_{j=0}^{m}\int_{\mathbf{R}^2}\int_{S^2} f_j(\xi,\nu,\vec{n}')\frac{|\nu|^3}{k^3}$$

$$\phi_1^{(3/2+j)}\left(-\frac{k}{4}-\frac{\nu^2}{4k}+(\vec{n},\vec{n}')\sqrt{\frac{\nu^2}{4}+\xi|\nu|}\right)d\vec{n}'\,d\xi\,d\nu, \tag{2.16}$$

where all the functions $f_j$, $j = 0,1,\ldots,m$ are continuous, polynomially bounded and satisfy conditions (2.9)-(2.10). Moreover,

$$\operatorname{supp} f_j \subset \left\{(\xi,\nu,\vec{n}):\ |\nu| \geqslant 2,\ -\frac{|\nu|}{4}+\frac{a_0^2}{|\nu|} \leqslant \xi \leqslant 1+\frac{1}{|\nu|}\right\}. \tag{2.17}$$

Passing to the spherical coordinates with the $Z$-axis chosen along the $\vec{n}$-direction, and taking into account (2.17) we get for a typical term in (2.16)

$$\frac{\sqrt{k}}{\rho(k)}\int_{|\nu|\geqslant 2} d\nu \int_{-\frac{|\nu|}{4}+\frac{a_0^2}{|\nu|}}^{1+\frac{1}{|\nu|}} d\xi \int_{-1}^{1} d\mu \int_0^{2\pi} f_j(\xi,\nu,\mu,\phi)\frac{|\nu|^3}{k^3}$$

$$\phi_1^{(3/2+j)}\left(-\frac{k}{4}-\frac{\nu^2}{4k}+\mu\sqrt{\frac{\nu^2}{4}+\xi|\nu|}\right)d\phi. \tag{2.18}$$

Let us consider the integral (2.18) over the domain where $\mu < 0$. In this case we have

$$-\frac{k}{4}-\frac{\nu^2}{4k}+\mu\sqrt{\frac{\nu^2}{4}+\xi|\nu|} \leqslant -\frac{k}{4}-\frac{\nu^2}{4k}.$$

Therefore, if $k$ is sufficiently large the integral (2.18) over such a domain is estimated by the expression (see estimate (2.6))

$$\frac{1}{\rho(k)k^{5/2}}\int_{-\infty}^{+\infty}\frac{P(|\nu|)\,d\nu}{\left(1+ak+b\frac{\nu^2}{k}\right)^{N'}}. \tag{2.19}$$

where $a > 0$ and $b > 0$ and $P(\nu)$ is some polynomial independent of $k$. Moreover, $N'$ can be assumed to be arbitrarily large, provide $N$ in (2.3) is sufficiently large. After a change of variables, (2.19) can be written in the form

$$\frac{1}{k^2\rho(k)}\int_{-\infty}^{+\infty}\frac{P(\sqrt{k}\,|\nu|)\,d\nu}{(1+ak+b^2\nu^2)^{N'}}. \tag{2.20}$$

It is obvious that if $N'$ is sufficiently large, expression (2.20) tends to zero when $k \to \infty$.

Let $\delta \in (0, 1)$ be a given number. Consider the integral (2.18) over the domain where $0 \leqslant \mu \leqslant 1-\delta$ or $|\nu| \leqslant k(1-\delta)$, $\mu \geqslant 0$.

If $0 \leqslant \mu \leqslant 1-\delta$, then, by the inequality which holds in this domain,

$$0 \leqslant \mu\sqrt{\frac{\nu^2}{4}+\xi|\nu|} \leqslant \mu\frac{|\nu|}{2}+2 \tag{2.21}$$

we get

$$-\frac{k}{4}-\frac{\nu^2}{4k}+\mu\sqrt{\frac{\nu^2}{4}+\xi|\nu|} \leqslant -\delta\left(\frac{k}{4}+\frac{\nu^2}{4k}\right)-(1-\delta)\left(\frac{k}{4}+\frac{\nu^2}{4k}\right)+$$

$$+\mu\frac{|\nu|}{2}+2 \leqslant -\delta\left(\frac{k}{4}+\frac{\nu^2}{4k}\right)+2-(1-\delta)\left(\frac{k}{4}+\frac{\nu^2}{4k}-\frac{|\nu|}{2}\right).$$

But for any $k > 0$

$$\frac{k}{4}+\frac{\nu^2}{4k} \geqslant \frac{|\nu|}{2}. \tag{2.22}$$

Therefore, if $0 \leqslant \mu \leqslant 1-\delta$ the following inequality holds:

$$-\frac{k}{4}-\frac{\nu^2}{4k}+\mu\sqrt{\frac{\nu^2}{4}+\xi|\nu|} \leqslant -\delta\left(\frac{k}{4}+\frac{\nu^2}{4k}\right)+2. \tag{2.23}$$

Thus, it follows that if $k$ is sufficiently large the integral (2.18) over the domain under consideration can also be estimated by an expression of the form (2.19). Hence, it tends to zero when $k \to \infty$. Similarly, if $|\nu| \leqslant k(1-\delta)$, $\mu \geqslant 0$, then, noting that for $k \geqslant |\nu|$ the left-hand side of (2.22) is an increasing function of $k$, we get the inequality

$$\frac{k}{4}+\frac{\nu^2}{4k} \geqslant \frac{|\nu|}{4(1-\delta)}+\frac{\nu^2(1-\delta)}{4|\nu|} = \frac{|\nu|}{2}\cdot\frac{2-2\delta+\delta^2}{2(1-\delta)} =$$

$$= \frac{|\nu|}{2}(1+\epsilon),\quad \epsilon = \frac{\delta^2}{2(1-\delta)} > 0. \tag{2.24}$$

Hence, due to (2.21) and (2.24), in this domain we have

$$-\frac{k}{4}-\frac{\nu^2}{4k}+\mu\sqrt{\frac{\nu^2}{4}+\xi|\nu|}\leqslant -\epsilon_1\left(\frac{k}{4}+\frac{\nu^2}{4k}\right)-(1-\epsilon_1)\left(\frac{k}{4}+\frac{\nu^2}{4k}\right)+\mu\frac{|\nu|}{2}+2\leqslant$$
$$\leqslant -\epsilon_1\left(\frac{k}{4}+\frac{\nu^2}{4k}\right)+2.$$

where $\epsilon_1=(\epsilon/1+\epsilon)$. We have obtained an estimate that is similar to (2.23). As before we conclude that this part of integral (2.18) tends to zero when $k\to\infty$.

So it remains only to consider integral (2.18) over the domain where $\mu>1-\delta$ and $\nu>k(1-\delta)$ (the case $\nu<-k(1-\delta)$ can be considered in a similar way):

$$\frac{\sqrt{k}}{\rho(k)}\int_{k(1-\delta)}^{\infty}d\nu\int_{-\frac{\nu}{4}+\frac{a_0^2}{\nu}}^{1+\frac{1}{\nu}}d\xi\int_{1-\delta}^{1}d\mu\int_0^{2\pi}f_j(\xi,\nu,\mu,\phi)\frac{\nu^3}{k^3}\phi_1^{(3/2+j)}$$
$$\left(-\frac{k}{4}-\frac{\nu^2}{4k}+\mu\sqrt{\frac{\nu^2}{4}+\xi\nu}\right)d\phi. \tag{2.25}$$

Assuming that $k(1-\delta)>5$ we have, in the domain of integration,

$$\nu\geqslant 5,\ -1<\frac{4\xi}{\nu}<\frac{5}{\nu}.$$
$$\sqrt{\frac{\nu^2}{4}+\xi\nu}=\frac{\nu}{2}+\xi-\theta(\xi,\nu),\ \ 0\leqslant\theta(\xi,\nu)\leqslant\frac{4\xi^2}{\nu}. \tag{2.26}$$

Now let us make a change of variables in (2.25): $\nu=k+\sqrt{k}x$, $\mu=1-y/k$. Taking (2.26) into account we obtain

$$\int_{-\delta\sqrt{k}}^{\infty}dx\int_{-\frac{k+\sqrt{k}x}{4}}^{1+\frac{1}{k+\sqrt{k}x}}d\xi\int_0^{k\delta}dy\int_0^{2\pi}\left(1+\frac{x}{\sqrt{k}}\right)^3\frac{1}{\rho(k)}f_j\left(\xi,k+\sqrt{k}x,1-\frac{y}{k},\phi\right)\times$$
$$\times\phi_1^{(3/2+j)}\left(-\frac{x^2}{4}-\frac{y}{2}\left(1+\frac{x}{\sqrt{k}}\right)+\left(1-\frac{y}{k}\right)[\xi-\theta(\xi,k+\sqrt{k}x)]\right)d\phi. \tag{2.27}$$

In the domain of integration in (2.27) we have

$$1+\frac{x}{\sqrt{k}}\geqslant 1-\delta,\ \ 1\geqslant 1-\frac{y}{k}\geqslant 1-\delta,\ \ 0\leqslant\theta(\xi,k+x\sqrt{k})\leqslant\frac{4\xi^2}{k+x\sqrt{k}}. \tag{2.28}$$

Thus, from estimate (2.6) it follows that

$$\left|\phi_1^{(3/2+j)}\left(-\frac{x^2}{4}-\frac{y}{2}\left(1+\frac{x}{\sqrt{k}}\right)+\left(1-\frac{y}{k}\right)[\xi-\theta(\xi,k+x\sqrt{k})]\right)\right|\leqslant$$
$$\leqslant C_7[(1+x^2)(1+y)(1+|\xi|)]^{-(1/3)N'},\ k\geqslant\frac{5}{1-\delta}. \tag{2.29}$$

Let us now assume that

$$[\vec{n}'\in S^2:(\vec{n},\vec{n}')\geqslant 1-\delta]\subset V. \tag{2.30}$$

This is the case when $\delta$ is sufficiently small. Due to the lemma (estimate (2.10)) and lemma 1 of Section 3.2 we get an estimate

$$\left|\frac{1}{\rho(k)} f_j\left(\xi, k + x\sqrt{k}, 1 - \frac{y}{k}, \phi\right)\right| = \frac{\phi(k + x\sqrt{k})}{\rho(k)}\left|\frac{f_j(\xi, k + x\sqrt{k}, 1 - \frac{y}{k}, \phi)}{\rho(k + x\sqrt{k})}\right| \leqslant$$

$$\leqslant C_8(1 + |\xi|)^n\left(1 + \frac{|x|}{\sqrt{k}}\right)^{|\alpha|+\epsilon} \tag{2.31}$$

Besides

$$\frac{1}{\rho(k)} f_j\left(\xi, k + x\sqrt{k}, 1 - \frac{y}{k}, \phi\right) \to f_{j,+}(\xi, 1, \phi),\ k \to \infty. \tag{2.32}$$

Taking into account estimates (2.29) and (2.31) and limit relation (2.32) we conclude that integral (2.27) has a limit when $k \to \infty$, and that this limit is equal to

$$\int_{-\infty}^{+\infty} dx \int_{-\infty}^{1} d\xi \int_0^{\infty} dy \int_0^{2\pi} f_{j,+}(\xi, 1, \phi)\phi_1^{(3/2+j)}\left(-\frac{x^2}{4} - \frac{y}{2} + \xi\right) d\phi =$$

$$= 8\pi^{3/2}\int_{-\infty}^{1} \phi_1^{(j+3/2)}(\xi) f_{j,+}(\xi, \vec{n})\, d\xi. \tag{2.33}$$

From this and from (2.4), (2.5) and (2.16) it follows that there exist the limit (2.1) for any $\phi \in \mathcal{S}(\mathbf{R}^1)$ uniform in $\vec{n} \in U$. Taking into account (2.33), (2.16) and the fact that integration over the region $\nu < -k(1-\delta)$ gives (after taking the limit) the same result as (2.33), we derive the relation

$$\lim_{k\to\infty}\left(\frac{\psi_a(k\lambda, \vec{u})}{\rho(k)}, \phi(-\lambda + (\vec{n}, \vec{u}))\right) = \left(\frac{F_+(\xi, \vec{n}) - F_-(\xi, \vec{n})}{2}, \phi(\xi)\right), \tag{2.34}$$

where the $f_\pm$ are defined in (1.25). From(1.17), (1.19) of Section 13.1 and (2.34) we get a similar relation for $\psi_s$ :

$$\lim_{k\to\infty}\left(\frac{\psi_s(k\lambda, \vec{u})}{\rho(k)}, \phi(-\lambda + (\vec{n}, \vec{u}))\right) = \left(\left[\frac{F_+(\xi, \vec{n}) + F_-(\xi, \vec{n})}{2}\right]^{(-1)}, \phi(\xi)\right). \tag{2.35}$$

where $[F_+ + F_-]^{(-1)}$ is the first primitive of $F_+ + F_-$ which has support in $[\xi \in \mathbf{R}^1: \xi \leqslant 1] \times S^2$. The right-hand sides of (2.34) and (2.35) are not equal to zero simultaneously.

Theorem 1 together with theorem 3 of Section 3.4 gives the following corollary.

**Corollary.** *Suppose that a causal distribution $F$ satisfies the condition $(S)$ and has an automodel asymptotics with respect to $\rho$ in the directions $\vec{n}_0$ and $-\vec{n}_0$. Then for some neighbourhood $U \subset S^2$, of the vector $\vec{n}_0$ the distributions $\Phi_{j,t\vec{n}}(\kappa)$ $j = a, s$, $t \in \mathbf{R}^1$, $\vec{n} \in U$, defined by formula (1.31) have quasi-asymptotics at zero with respect to $k^2\rho(k)$.*

**Proof.** Let us introduce distributions $\psi_{j,\vec{n}}(\lambda,\xi)$, $j = a, s$, $\vec{n} \in S^2$ from $\mathcal{S}'_+(1)$ by

$$(\psi_{j,\vec{n}}, \phi) = (\psi_j(\lambda,\vec{u}), \phi(\lambda, (\vec{n},\vec{u}))), \quad \phi \in \mathcal{S}_+(1).$$

By the hypothesis and theorem 1 it follows that for any $\phi \in \mathcal{S}(\mathbf{R}^1)$ and $\vec{n} \in U$ there exists limits

$$\lim_{k\to\infty}\left(\frac{\psi_{j,\vec{n}}(k\lambda,\eta)}{\rho(k)}, \phi(-\lambda\pm\eta)\right) \equiv 0, \quad j = a, s.$$

Using theorem 3 of Section 3.6 we conclude that $\psi_{j,\vec{n}}(\lambda,\eta)$, $j = a, s$, $\vec{n} \in U$, have quasi-asymptotics at infinity over $\lambda$ with respect to $\rho$. This means that due to (2.26), for any $\phi \in \mathcal{S}_+(1)$ and $\vec{n} \in U$ there exist the limits

$$\lim_{k\to\infty}\left(\frac{\psi_j(k\lambda,\vec{u})}{\rho(k)}, \phi(\lambda, (\vec{n},\vec{u}))\right) \equiv 0, \quad j = a, s.$$

Since the support of $\psi_j(\lambda,\vec{u})$ is bounded with respect to $\vec{u}$ one can put $\phi(\lambda,\eta) = \phi_0(\lambda)e^{i\eta t}$, $\phi_0 \in \mathcal{S}_+$. Taking into account the formula (see (1.31))

$$\left(\frac{\Phi_{j,t\vec{n}}\left(\frac{\kappa}{k}\right)}{k^2\rho(k)}, \phi_1(\kappa)\right) = C_j\left(\frac{\psi_j(k\lambda,\vec{u})}{\rho(k)}, \phi_0(t)e^{it(\vec{n},\vec{u})}\right), \quad j = a, s,$$

where $\phi_1 = b[\phi_0]$ is an arbitrary function from $\mathcal{S}_+$, we conclude that $\Phi_{j,t\vec{n}}(\kappa)$, $j = a, s$, have quasi-asymptotics at zero with respect to $k^2\rho(k)$.

The next theorem brings about the one-to-one correspondence between the automodel asymptotics (along all directions) of a causal distributions, the quasi-asymptotics of its Fourier transform in the neighbourhood of a light cone, and the quasi-asymptotics of the weight function in the Jost-Lehmann-Dyson representation.

**Theorem 2.** *Let a causal distribution $F$ satisfy the condition $(S)$, with $\tilde{F}$ being its Fourier transform. Let $\rho$ be some automodel function of order $\alpha$. The following statements are equivalent:*

(A) *$F(q)$ has an automodel asymptotics with respect to $\rho$:*

$$F(q) \sim \rho(|\nu|)F_\pm(\xi,\vec{n}), \quad \nu \to \pm\infty, \quad (F_+, F_-) \not\equiv 0.$$

(B) *$\tilde{F}(x)$ has a quasi-asymptotics in the neighbourhood of a light cone with respect to $k^2\rho(k)$*

$$\tilde{F}(x) \sim \frac{(x_\mp^2)^{-2}\rho[(x_\mp^2)^{-1}]}{\Gamma(-1-\alpha)}[\epsilon(x_0)g_a(\vec{x}) + |x_0|g_s(\vec{x})], \quad x^2 \sim 0, \quad (g_a, g_s) \not\equiv 0.$$

(C) *There exists a strictly space-like hypersurface $\Sigma_a$ of class $C^\infty$ in $\mathbf{R}^4$ such that the weight function $\psi_{\Sigma_0}(\lambda,u)$ in the Jost-Lehmann-Dyson representation of $F$ has a quasi-asymptotics at infinity in $\lambda$ with respect to $\rho$.*

(D) *For any strictly space-like hypersurface $\Sigma$ of class $C^\infty$ in $\mathbf{R}^4$ the weight function $\psi_\Sigma(\lambda,u)$ has a quasi-asymptotics at infinity in $\lambda$ with respect to $\rho$.*

In particular, if $\Sigma = [u \in \mathbf{R}^4 : u_0 = 0]$ and $(\psi_a, \psi_s)$ are the weight functions for $F$ the following relation hold:

$$g_a(\lambda\vec{n}) = -4^{\alpha+2}\pi^2 e^{i(\pi/2)\alpha}(\lambda - i0)^{\alpha+1}\frac{\tilde{F}_+(\lambda,\vec{n}) - \tilde{F}_-(\lambda,\vec{n})}{2}, \tag{2.37}$$

$$g_s(\lambda\vec{n}) = 4^{\alpha+2}\pi^2 e^{i(\pi/2)\alpha}(\lambda - i0)^{\alpha}\frac{\tilde{F}_+(\lambda,\vec{n}) + \tilde{F}_-(\lambda,\vec{n})}{2}, \tag{2.38}$$

*where $\tilde{F}_\pm(\lambda,\vec{n})$ are the Fourier transforms of $F_\pm(\xi,\vec{n})$ over $\xi$.*

$$g_a(\vec{x}) = 4^{\alpha+2}\pi^2 i\,\tilde{\psi}_{a,0}(\vec{x}), \tag{2.39}$$

$$g_s(\vec{x}) = 4^{\alpha+2}\pi^2 \tilde{\psi}_{s,0}(\vec{x}), \tag{2.40}$$

$$\frac{F_+(\xi,\vec{n}) - F_-(\xi,\vec{n})}{2} = {}^{(-\alpha-1)}\check{\psi}_{a,0}(\xi,\vec{n}), \tag{2.41}$$

$$\frac{F_+(\xi,\vec{n}) + F_-(\xi,\vec{n})}{2} = {}^{(-\alpha)}\check{\psi}_{s,0}(\xi,\vec{n}). \tag{2.42}$$

*Here $\psi_{j,}0(\vec{u})$, $j = a, s$ are defined by the formulae*

$$\frac{1}{\rho(k)}\psi_j(k\lambda,\vec{u}) \to f_{\alpha+1}(\lambda) \times \psi_{j,0}(\vec{u}), \quad k \to \infty \tag{2.43}$$

*in $\mathcal{S}'(T_0)$, $\check{\psi}_{j,0}(\xi,\vec{n})$ is the Radon transform of the distribution $\psi_{j,0}(\vec{u})$ with compact support* (see Section 2.9), *$\tilde{\psi}_{j,0}(\vec{x})$ is Fourier transform of $\psi_{j,0}(\vec{u})$; and ${}^{(B)}\check{\psi}_{j,0}(\xi,\vec{n})$ is the left derivative (primitive) of $\check{\psi}_{j,0}(\xi,\vec{n})$ with respect to $\xi$* (see Section 2.4).

**Proof.** The implication (C) $\Longleftrightarrow$ (D) follows immediately from theorem 6 of Section 13.3. Let us prove the implication (D) $\Longleftrightarrow$ (A). To do this, let us suppose that the weight functions $\psi_a(\lambda,\vec{u})$ and $\psi_s(\lambda,\vec{u})$ have a quasi-asymptotics at infinity in $\lambda$ with respect to $\rho$. Let us assume, first, that $F$ is odd with respect to $q_0$; that is, $F = F_a$, $\psi = \psi_a$, $\psi_s = 0$. Let $\chi(\xi,\nu)$ be some admissible multiplier (see Section 14.1). Then according to (1.17) we have for any $\phi \in \mathcal{S}(\mathbf{R}^1)$

$$\left(\frac{1}{\rho(\nu)}\mathcal{F}_\chi(\xi,\nu,\vec{n}),\phi(\xi)\right) = \left(\frac{1}{\rho(\nu)}\psi(\nu\lambda,\vec{u}),\phi_\nu(\lambda,\vec{u},\vec{n})\right), \tag{2.44}$$

where, due to (1.17)–(1.18),

$$\begin{aligned}
\phi_\nu(\lambda,\vec{u},\vec{n}) &= \nu\chi(\xi(\nu\lambda,\vec{u};\nu,\vec{n}),\nu)\phi(\xi(\nu\lambda,\vec{u};\nu,\vec{n}))\omega(\lambda\nu,\vec{u};\nu,\vec{n}) = \\
&= \chi\left((\vec{n},\vec{u})\sqrt{1 - \frac{4}{\nu^2}[\nu\lambda + |\vec{u}|^2 - (\vec{n},\vec{u})^2]} + \frac{2(\vec{n},\vec{u})^2 - |\vec{u}|^2}{\nu} - \lambda, \nu\right) \times \\
&\times \phi\left((\vec{n},\vec{u})\sqrt{1 - \frac{4}{\nu^2}[\nu\lambda + |\vec{u}|^2 - (\vec{n},\vec{u})^2]} + \frac{2(\vec{n},\vec{u})^2 - |\vec{u}|^2}{\nu} - \lambda\right) \times \\
&\times \left(\frac{2(\vec{n},\vec{u})}{\nu\sqrt{1 - \frac{4}{\nu}[\nu\lambda + |\vec{u}|^2 - (\vec{n},\vec{u})^2]}} + 1\right)\eta\left(\frac{\nu^2}{4} - \nu\lambda - |\vec{u}|^2\right).
\end{aligned} \tag{2.45}$$

From this and from the properties of $\chi$ and $\eta$ it follows that

$$\phi_\nu(\lambda, \vec{u}, \vec{n}) \xrightarrow{\vec{n} \in S^2} \phi(-\lambda + (\vec{n}, \vec{u})), \quad \nu \to \infty$$

in $\mathcal{S}(T_0)$. Therefore, by (2.44) and by condition (D) we have

$$\left(\frac{1}{\rho(\nu)} \mathcal{F}_\chi(\xi, \nu, \vec{n}), \phi(\xi)\right) \xrightarrow{\vec{n} \in S^2} (f_{\alpha+2}(\lambda) \times \psi_0(\vec{u}), \phi(-\lambda + (\vec{n}, \vec{u}))), \quad \nu \to \infty,$$

that is just the relation which we wanted to prove. Moreover, denoting

$$\lim_{\nu \to \infty} \frac{1}{\rho(\nu)} \mathcal{F}_\chi(\xi, \nu, \vec{n}) = F(\xi, \vec{n})$$

in $'(\mathbf{R}^1)$ from (2.44) we derive

$$F(\xi, \vec{n}) = {}^{(-\alpha-1)}\check{\psi}_0(\xi, \vec{n}). \tag{2.46}$$

This equality is simply (2.41) for the odd distribution $F$.

In general case we have

$$F(q) = F_a(q) + 2q_0 F_1(q) \tag{2.47}$$

where $F_1$ is an odd distribution with respect to $q_0$ defined by equality (1.17) of Section 13.1 and the corresponding weight functions $\psi_1$ is defined by formula (1.19) of Section 13.1: $\psi_1 = \partial_\lambda \psi_s$. It is obvious that if $F$ has an automodel asymptotics with respect to $\rho(\nu)$, then $F_1$ has an automodel asymptotics with respect to $\nu^{-1}\rho(\nu)$. Using the already proved statement for odd distributions, and the fact that $\psi_1(\lambda, \vec{u})$ has a quasi-asymptotics at infinity in $\lambda$ with respect to $k^{-1}\rho(k)$ if and only if $\psi_s(\lambda, \vec{u})$ has a quasi-asymptotics at infinity in $\lambda$ with respect to $\rho$, we complete the proof of the implementatin (D) $\Longrightarrow$ (A). Moreover, formulae (2.41) and (2.42) follow from (2.46) and (2.47).

(B) $\Longleftrightarrow$ (C). Due to (1.11) of Section 13.1 and (3.9) of Section 2.3, we have

$$\frac{\psi_j(k\lambda, \vec{u})}{\rho(k)} = C_j F_{\vec{x}}^{-1}\left[B\left[\frac{\Phi_j\left(\frac{\kappa}{k}, \vec{x}\right)}{k^2\rho(k)}\right]\right], \quad j = a, s, \tag{2.49}$$

where $C_a = (1/4\pi^2 i)$, $C_s = (1/4\pi^2)$. From (2.48) we conclude that $\psi_a(\lambda, \vec{u})$ and $\psi_s(\lambda, \vec{u})$ have quasi-asymptotics at infinity in $\lambda$ with respect to $\rho(k)$ if and only if $\Phi_a(\kappa, \vec{x})$, and $\Phi_s(\kappa, \vec{x})$ have quasi-asymptotics at zero in $\kappa$ with respect to $k^2\rho(k)$. Letting $k \to \infty$ in (2.48), and using the limit relation (2.43), (1.29) and formula (3.14) of Section 2.3, we get formulae (2.39) and (2.40). Finally, formulae (2.37) and (2.38) follow from (2.39)-(2.42), from the inverse formula for the Radon transform (9.13) of Section 2.9 and from the equality (1.5) of Section 2.1.

Now it remains only to prove the implication (A) $\Longrightarrow$ (C) for $\Sigma_0$ equal to $[u \in \mathbf{R}^4 : u_0 = 0]$. By theorem 1, for any $\phi \in (\mathbf{R}^1)$ there exist the limits

$$\lim_{k \to \infty}\left(\frac{\psi_j(k\lambda, \vec{u})}{\rho(k)}, \phi(-\lambda + (\vec{n}, \vec{u}))\right), \quad j = a, s,$$

uniformly in $\vec{n} \in S^2$, which are not equal to zero simulatenously. So, by theorem 3 of Section 3.6 for $n = 3$ the distributions $\psi_j$ have quasi-asymptotics at infinity in $\lambda$

with respect to $\rho$.

Let us point out some corollaries to theorem 2. The first is that the asymptotics in the Björken domain of a causal distribution satisfying the condition $(S)$ is entirely determined by the behaviour of its Fourier transform in an arbitrarily small neighbourhood of the boundary of the light cone. In a more explicit form this can be expressed as:

**Corollary 1.** *Let a causal distribution $F$ satisfy the condition $(S)$, with $\tilde{F}$ its Fourier transform. Further, let $\eta$ be a $C^\infty$ function such that*

$$\eta(t) \equiv 0, \ t \leqslant \epsilon; \ \eta(t) \equiv 1, \ t \geqslant 2\epsilon,$$

*where $\epsilon$ is a positive number. Let us put*

$$\tilde{F}_2(x) = \eta(x^2)\tilde{F}(x),$$

*with $F_2$ being the inverse Fourier transform of $\tilde{F}_2$: $F_2 = F^{-1}[\tilde{F}_2]$. Then, for any admissible multiplier $\chi(\xi, \nu)$, and test function $\phi(\xi)$ from $\mathcal{S}(\mathbf{R}^1)$, the function*

$$(\mathcal{F}_{2,\chi}(\xi, \nu), \phi(\xi))_\xi$$

*decreases when $\nu \to \infty$ faster than any power of $1/|\nu|$.*

**Proof.** First note that $F_2(q)$ is a causal distribution. Moreover, it satisfies the condition $(S)$. Indeed, the causal distribution $F$ satisfies the condition $(S)$ if and only if

$$\operatorname{supp} \psi_j(\lambda, \vec{u}) \subset T_0 = [(\lambda, \vec{u}) \in \mathbf{R}_+^1 \times \mathbf{R}^3 : |\vec{u}| \leqslant 1], \ j = a, s, \tag{2.50}$$

where $\psi_a$ and $\psi_s$ are the weight functions in the Jost-Lehmann-Dyson representation (corollary of Section 13.1). But, due to (1.11) of Section 13.1, if conditions (2.49) hold for the weight functions $\psi_a$ and $\psi_s$ then this is also the case for the weight functions $\psi_{2,a}$ and $\psi_{2,s}$ in the Jost-Lehmann-Dyson representation for $F_2(q) = F^{-1}[\eta(x^2)\tilde{F}(x)]$. On the other hand, the corresponding distributions $\Phi_{a,s}(\kappa \vec{x})$ and $\Phi_{2,s}(\kappa, \vec{x})$ vanish if $\kappa < \epsilon$. Therefore, due to the lemma of Section 3.4, their oder at zero in $\kappa$ is equal to $-\infty$. This remark, together with theorem 2, proves the corollary.

The second corollary shows that no pair of distributions $(F_+, F_-)$ is an automodel asymptotics of a causal distribution.

**Corollary 2.** *A pair of distributions $(F_+, F_-)$ is an automodel asymptotics of a causal distribution $F$ satisfying the condition $(S)$ with respect to an automodel function $\rho$ of order $\alpha$ if and only if the distributions*

$$F_a(\xi, \vec{n}) = {}^{(\alpha+1)}\left[\frac{F_+(\xi, \vec{n}) - F_-(\xi, \vec{n})}{2}\right],$$

$$F_s(\xi, \vec{n}) = {}^{(\alpha)}\left[\frac{F_+(\xi, \vec{n}) + F_-(\xi, \vec{n})}{2}\right]$$

*satisfy the following conditions*

(a) $F_j \in \mathcal{E}'(\mathbf{R}^1) \otimes C^\infty(S^2)$, $\operatorname{supp} F_j \subset [-1, 1] \times S^2$, $j = a, s$;

(b) $F_j(\xi, \vec{n}) = F_j(-\xi, -\vec{n})$, $j = a, s$;

(c) $\sum_{|\beta|=k} a_\beta^{(j)} \vec{n}^\beta = (F_j(\xi,\vec{n}),\xi^k),\ k = 0,1,\ldots,\ j = a,s,$

*where* $\beta = (\beta_1,\beta_2,\beta_3) \in \mathbf{Z}_+^3$ *and* $a_\beta^{(j)}$ *are the moments of a distribution* $\psi_j(\vec{u})$ *with support in the unit ball* $\overline{B}_1$*:*

$$a_\beta^{(j)} = (\psi_j(\vec{u}),\vec{u}^\beta);\ \vec{u}^\beta = u_1^{\beta_1}u_2^{\beta_2}u_3^{\beta_3},\ j = a,s.$$

*In particular, the following conditions ('sum rules') are necessary:*

$$\int_{\vec{n}\in S^2} (F_j(\xi,\vec{n}),\xi^k)P_m(\vec{n})\,d\vec{n} = 0,\ m > k;\ k = 0,1,\ldots,\ j = a,s,$$

*where* $P_m(\vec{x})$ *is a harmonic polynomial of degree* $m$.

**Proof.** The necessity follows from equalities (2.41), (2.42) and from the analogue of the Palye-Wiener theorem for Radom transform (see Section 2.9). To prove sufficiency, let us define a distribution $F$ by means of the Jost-Lehmann-Dyson representation with the weight functions

$$\psi_j(\lambda,\vec{u}) = \partial_\lambda^N[\theta(\lambda)\lambda^N\rho(\lambda)] \times \psi_j(\vec{u}),\ j = a,s,$$

where the function $\rho(\lambda)$ is assumed to be continued in some way to a continuous function on $[0,\infty)$. The number $N$ satisfies the condition $N + \alpha > -1$. According to theorem 2 of Section 3.4, $\psi_a(\lambda,\vec{u})$ and $\psi_s(\lambda,\vec{u})$ have quasi- asymptotics at infinity in $\lambda$ with respect to $\rho$ with the limit functions $\psi_a(\vec{u})$ and $\psi_s(\vec{u})$, respectively. Due to theorem 2, $F$ has an automodel asymptotics with respect to $\rho$, for which the limit functions satisfy conditions (a) - (c) due to formulae (2.41) and (2.42).

Note that simple examples show that the quasi-asymptotics at the light cone of $\tilde{F}(x)$ with respect to $k^2\rho(k)$ does not guarantee any automodel asymptotics of $F(q)$ with respect to $\rho$ in the so-called 'physical' region $\xi > 0$.

Indeed, let us consider the function

$$F(q) = \epsilon(q_0)\delta(q^2).$$

Its Fourier transform has the form (see Section 2.1)

$$\tilde{F}(x) = 4\pi^2\epsilon(x_0)\delta(x^2).$$

So $F(q)$ is causal and satisfies the conditon $(S)$. Moreover, $\tilde{F}(x)$ has a (nontrivial) quasi-asymptotics in the neighbourhood of a light cone with respect to the automodel function $\rho(k) = k$, because $\Phi_a = 4\pi^2 i\,\delta(\kappa)$, $\Phi_s(\kappa,\vec{x}) = 0$. On the other hand,

$$\mathcal{F}_\chi(\xi,\nu,\vec{n}) \equiv 0,$$

when $\xi > 0$.

As we can see, the reason for this phenomenon is such that $F_\pm(\xi,\vec{n})$ may vanish for $\xi > 0$. The next statement gives the condition under which this is impossible.

**Corollary 3.** *Let a causal distribution* $F(q)$ *satisfy the condition* $(S)$ *and let* $\tilde{F}(x)$ *have a quasi-asymptotics in the neighbourhood of a light cone with respect to the automodel function* $k^2\rho(k)$, *and the limit functions* $g_a(\vec{x})$ *and* $g_s(\vec{x})$ *are not polynomials simultaneously. Then* $F(q)$ *has an automodel asymptotics with respect to* $\rho$ *in* *the phyiscal region* $\xi > 0$.

**Proof.** It is sufficient to prove that the distributions $F_{\pm}(\xi, \vec{n})$ are not equal to zero simultaneously for $\xi > 0$. Since $g_a$ and $g_s$ are not polynomials simultaneously, then according to (2.39) and (2.40) the supports of distributions $\psi_{a,0}(\vec{u})$ and $\psi_{s,0}(\vec{u})$ are not the point $\vec{u} = 0$ simultaneously. Therefore, the Radon transforms $\psi_{a,0}(\xi, \vec{n})$ and $\psi_{s,0}(\xi, \vec{n})$ do not vanish simultaneously for $\xi > 0$ (see Section 2.9). Now from (2.41) and (2.42) it follows that $F_+(\xi, \vec{n})$ and $F_-(\xi, \vec{n})$ do not vanish simultaneously for $\xi > 0$. (One must take into account that

$$\operatorname{supp} F_{\pm} \subset [(\xi, \vec{n}): \xi \leqslant 1, \vec{n} \in S^2].)$$

Finally, the last corollary gives the condition which guaranties an automodel asymptotics of $F$ only for $\xi > b$, $b > 0$.

**Corollary 4.** *Let the weight functions of a causal distributions* $F$ *satisfying the condition* (*S*) *have a quasi-asymptotics at infinity in* $\lambda$ *with respect to* $\rho$ *only when* $|\vec{u}| > b$. *In other words, for any* $\phi \in \mathcal{S}_+(3)$ *such that*

$$\operatorname{supp} \phi \subset [(\lambda, \vec{u}): \lambda \in \mathbf{R}^1_+, |\vec{u}| > b]$$

*let there exists the limits*

$$\left(\frac{\psi_j(k\lambda, \vec{u})}{\rho(k)}, \phi(\lambda, \vec{u})\right) \to (f_{\alpha+1}(\lambda) \times \psi_{j,0}(\vec{u}), \phi(\lambda, \vec{u})), \; k \to \infty; \; j = a, s.$$

*Then* $F$ *has an automodel asymptotics with respect to* $\rho$ *in the region* $\xi > b$. *This means that for any admissible multiplier* $\chi(\xi, \nu)$ *and any* $\phi(\xi)$ *from* $\mathcal{S}(\mathbf{R}^1)$ *such that*

$$\operatorname{supp} \phi \subset [\xi \in \mathbf{R}^1: \xi > b]$$

*there exists the limits*

$$\left(\frac{1}{\rho(\nu)} F_\chi(\xi, \nu, \vec{n}), \phi(\xi)\right) \xrightarrow{\vec{n} \in S^2} (F_{\pm}(\xi, \vec{n}), \phi(\xi)), \; \nu \to \pm\infty,$$

*and* $(F_+, F_-)$ *are not equal to zero simultaneously for* $\xi > b$ *if* $\psi_{j,0}(\vec{u})$, $j = a, s$, *are not equal to zero simultaneously for* $|\vec{u}| > b$.

**Proof.** The proof follows directly from formulae (2.44) and (2.45) if one notes that if $\operatorname{supp} \phi \subset [\xi \in \mathbf{R}^1: \xi > b]$ then there is some $\nu_0$ such that $\operatorname{supp} \phi_\nu \subset [(\lambda, \vec{u}, \vec{n}): \lambda \geqslant 0, |\vec{u}| > b]$ for $\nu > \nu_0$.

**3. On forbidden asymptotics in the Björken domain.** First, note that definition 1 of Section 14.1 is satisfied in the case of an arbitrarily continuous and positive function $\rho(t)$ for $t \to +\infty$ (not necessarily automodel). Hence, we say that a causal distribution $F$ satisfying the condition ($S$) has an asymptotics in the Björken domain with respect to a positive and continuous function $\rho$ for $t \to +\infty$ if for any admissible multiplier relation (1.25) of Section 14.1 hold uniformly in $\vec{n} \in U = S^2$.

From the previous results it follows that any automodel asymptotics in the Björken domain does not contradict the general principles of local quantum field theory. In other words, for any automodel function $\rho$ there exists a causal distribution $F$ satisfying the condition $(S)$ such that it has an asymptotics in the Björken domain with respect to $\rho$ (see Corollary 2 to theorem 2 of Section 14.2). Moreover, one can find a distribution which has an automodel asymptotics with respect to $\rho$ with the given limit functions $F_+$ and $F_-$ provided they satisfy conditions (a) - (c) of corollary 2 to theorem 2 of Section 14.2.

In this context there arises the following question: Is any given asymptotics in the Björken domain admissible, or are there some asymptotics which contradict the general principles of local quantum field theory?

The next theorem shows that there are, so to speak, some forbidden asymptotics which are not compatible with the causal and the special conditions.

**Theorem.** *Let a causal distribution $F$ satisfy the condition $(S)$ and have an asymptotics in the Björken domain with respect to a continuous and positive function $\rho$ such that the distribution $\theta(t)t^N\rho(t)$ has a quasi-asymptotics at infinity with respect to some automodel function $\rho_1$ of order $\alpha + N$ (here $N \geqslant 0$ is an integer number such that $N + \alpha > -1$). Then*

$$\frac{t^N\rho(t)}{\rho_1(t)} \to C \neq 0, \ t \to +\infty,$$

*so that $\rho$ is an automodel function of order $\alpha$.*

**Remark 1.** Due to theorem 3 of Section 3.5, the formulation of the previous theorem is correct. Indeed, if the hypothesis of the theorem holds then $F$ has an asymptotics in the Björken domain with respect to any continuous and positive function $\rho_2$, such that

$$\frac{\rho(t)}{\rho_2(t)} \to C_1 = 0, \ t \to +\infty. \tag{3.1}$$

But in this case, due to theorem 3 of Section 3.5, the distribution $\theta(t)t^N\rho_2(t)$ also has a quasi-asymptotics at infinity with respect to $\rho_1$.

**Remark 2.** The theorem excludes such asymptotics which, roughly speaking, have quasi- asymptotics but which are not automodel. The class of such functions is rather wide. It include certain oscillating functions; for instance,

$$\rho(t) = 1 + \alpha \sin t^\beta, \quad |\alpha| < 1, | \, |\beta > 0.$$

Indeed, this function has a quasi-asymptotics with respect to $\rho_1 \equiv 1$. But it is not automodel (see Section 3.1).

**Proof.** Now let us prove the theorem. One can assume without loss of generality that $N = 0$. Indeed, if this is not the case one can consider the function $F_1 = q_0^N F$ which is obviously causal, satisfies the condition $(S)$, and has an asymptotics in the Björken domain with respect to $t^N\rho(t)$. Moreover, one can consider only the case of an odd distribution $F$ with respect to $q_0$; that is, $F = F_a$, $\psi = \psi_a$, $\psi_s = 0$. Due to theorem 2 of Section 14.2, it is sufficient to prove that the weight function $\psi(\lambda, \vec{u})$ has a quasi-asymptotics at infinity in $\lambda$ with respect to $\rho_1$. To prove this let us use the inversion formula (2.8) of Section 13.2, which gives

$$\left(\frac{\psi(k\lambda, \vec{u})}{\rho_1(k)}, \phi(\lambda, \vec{u})\right) = \frac{\pi^{-3}k^{-4}}{\rho_1(k)}\left(F(q), q_0\int^{(3/2)}\phi_1^{(3/2)}\left(\frac{q_0^2 - |\vec{y}|^2}{k}, \vec{q} - \vec{y}\right) d\vec{y}\right). \tag{3.2}$$

where $\phi_1(\lambda, \vec{u}) \in \mathcal{S}(\mathbf{R}^4)$ is equal to $\phi(\lambda, \vec{u})$ for $\lambda \geqslant 0$ and satisfies the conditions

$$\int_{-\infty}^{+\infty} \lambda^k \phi_1(\lambda, \vec{u})\, d\lambda = 0, \quad k = 0, 1, \ldots, N \tag{3.3}$$

for sufficiently large $N$. We have

$$\int {}^{(3/2)}\phi_1^{(3/2)}\left(\frac{q_0^2 - |\vec{y}|^2}{k}, \vec{q} - \vec{y}\right) d\vec{y} =$$

$$= \int {}^{(3/2)}\phi_1^{(3/2)}\left(\frac{q_0^2}{k} - \frac{|\vec{z}|^2 - 2(\vec{q}, \vec{z}) + |\vec{q}|^2}{k}, \vec{z}\right) d\vec{z}. \tag{3.4}$$

Let us write the integrand inthe form

$${}^{(3/2)}\phi_1^{(3/2)}\left(\frac{q_0^2}{k} - \frac{|\vec{z}|^2 - 2(\vec{q}, \vec{z}) + |\vec{q}|^2}{k}, \vec{z}\right) =$$

$$= \left(\frac{k^2}{2|\vec{q}|^2}\right)^l \Delta_{\vec{\sigma}}^l\, {}^{(3/2)}\phi_1^{(3/2-2l)}\left(\frac{q_0^2}{k} - \frac{|\vec{z}|^2 - 2(\vec{q}, \vec{\sigma}) + |\vec{q}|^2}{k}, \vec{z}\right)\Bigg|_{\vec{\sigma} = \vec{z}}.$$

Now, integrating by parts we get

$$\int {}^{(3/2)}\phi_1^{(3/2)}\left(\frac{q_0^2 - |\vec{y}|^2}{k}, \vec{q} - \vec{y}\right) d\vec{y} =$$

$$= \left(\frac{k^2}{|\vec{q}|^2}\right)^l \sum_{j \geqslant 0} \frac{1}{k^j} \int \psi_j \left(\frac{q_0^2}{k} - \frac{|\vec{z}|^2 - 2(\vec{q}, \vec{z}) + |\vec{q}|^2}{k}, \vec{z}\right) d\vec{z} \tag{3.5}$$

where the sum consists of a finite number of summands; $l \geqslant 0$ is some integer number and $\psi_j(\lambda, \vec{z})$ are some functions from $\mathcal{S}_p(\mathbf{R}^4)$, where $l$ and $p$ are arbitrarily large (if $N$ in (3.3) is sufficiently large).

Similarly, putting in (3.4)

$${}^{(3/2)}\phi_1^{(3/2)}\left(\frac{q_0^2}{k} - \frac{|\vec{z}|^2 - 2(\vec{q}, \vec{z}) + |\vec{q}|^2}{k}, \vec{z}\right) =$$

$$= \partial_{\sigma_1}^l\, {}^{(3/2)}\phi_1^{(3/2)}\left(\frac{q_0^2}{k} - \frac{|\vec{z}|^2 - 2(\vec{q}, \vec{z}) + |\vec{q}|^2}{k}, \sigma_1, z_2, z_3\right)\Bigg|_{\sigma_1 = z_1},$$

and asusming that

$$\int \phi_1(\lambda, \vec{u}) u_1^k\, du_1 = 0, \quad k = 0, 1, \ldots, N \tag{3.6}$$

(which can be obtained by changing $\phi_1(\lambda, \vec{u})$ for $|\vec{u}| > 1$) we get, integrating by parts,

$$\int {}^{(3/2)}\phi_1^{(3/2)}\left(\frac{q_0^2 - |\vec{y}|^2}{k}, \vec{q}, \vec{y}\right) d\vec{y} =$$

$$= \sum_{0 \leqslant j \leqslant l} \left(\frac{q_1}{k}\right)^j \frac{1}{k^{l-1}} \int {}^1\psi_j \left(\frac{q_0^2}{k} - \frac{|\vec{z}|^2 - 2(\vec{q}, \vec{z}) + |\vec{q}|^2}{k}, \vec{z}\right) d\vec{z}, \tag{3.7}$$

where ${}^1\psi_j(\lambda, \vec{z})$ are some functions from $\mathcal{S}_p(\mathbf{R}^4)$, and $l$ and $p$ are arbitrarily large if

$N$

in (3.3) and (3.6) is sufficiently large.

Now let $\chi(\xi,\nu)$ be an admissible multiplier (see Section 14.1). From (3.2) we find that (cf. (2.4), Section 14.2)

$$\left(\frac{\psi(k\lambda,\vec{u})}{\rho_1(k)},\phi(\lambda,\vec{u})\right) =$$

$$= \Omega_k + \frac{\pi^{-3}k^{-4}}{\rho_1(k)}\left(F(q)\chi\left(-\frac{q^2}{2|q_0|},2q_0\right),q_0\int^{(3/2)}\phi_1^{(3/2)}\left(\frac{q_0^2-|\vec{y}|^2}{k},\vec{q}-\vec{y}\right)d\vec{y}\right), \quad (3.8)$$

where

$$\Omega_k = \frac{\pi^{-3}k^{-4}}{\rho_1(k)}\left(F(q)\left(1-\chi\left(-\frac{q^2}{2|q_0|},2q_0\right)\right),q_0\int^{(3/2)}\phi_1^{(3/2)}\left(\frac{q_0^2-|\vec{y}|^2}{k},\vec{q}-\vec{y}\right)d\vec{y}\right).$$

Taking into account that

$$\operatorname{supp}[1-\chi] \subset [q \in \mathbf{R}^4: |\vec{q}| \leqslant a]$$

and using the representation (3.7) for a sufficiently large $l$ we conclude that

$$\Omega_k \to 0, \quad k \to \infty \quad (3.9)$$

if $N$ in (3.3) and (3.6) is sufficiently large.

Now let $\eta$ be a $C^\infty$-function such that

$$\eta(t) \equiv 1, \quad |t| \leqslant 1; \quad \eta(t) \equiv 0, \quad |t| \geqslant 2. \quad (3.10)$$

Let us write the second term in (3.8) as the sum of two summands:

$$\frac{\pi^{-3}k^{-4}}{\rho_1(k)}\left(F(q)\chi\left(-\frac{q^2}{2|q_0|},2q_0\right),q_0\eta\left(\frac{|\vec{q}|^2}{k^2}\right)\int^{(3/2)}\phi_1^{(3/2)}\left(\frac{q_0^2-|\vec{y}|^2}{k},\vec{q}-\vec{y}\right)d\vec{y}\right) +$$

$$+ \frac{\pi^{-3}k^{-4}}{\rho_1(k)}\left(F(q)\chi\left(-\frac{q^2}{2|q_0|},2q_0\right),q_0\left[1-\eta\left(\frac{|\vec{q}|^2}{k^2}\right)\right]\right.$$

$$\left.\int^{(3/2)}\phi_1^{(3/2)}\left(\frac{q_0^2-|\vec{y}|^2}{k},\vec{q}-\vec{y}\right)d\vec{y}\right). \quad (3.11)$$

For the first summand, using (3.7) and (1.7) of Section 14.1, we get the expression

$$\frac{\pi^{-3}k^{-4}}{8\rho_1(k)}\sum_{0\leqslant j\leqslant l}\left(\mathcal{F}_\chi(\xi,\nu,\vec{n}),\nu^2\sqrt{\frac{\nu^2}{4}+\xi|\nu|}\,\eta\left(\frac{\frac{\nu^2}{4}+\xi|\nu|}{k^2}\right)\left(\frac{1}{k}\right)^{l-j}\times\right. \quad (3.12)$$

$$\left.\times\left(\frac{n_1\sqrt{\frac{\nu^2}{4}+\xi|\nu|}}{k}\right)^j\int^1\psi_j\left(-\frac{\xi|\nu|}{k}-\frac{|\vec{z}|^2-2(\vec{n},\vec{z})\sqrt{\frac{\nu^2}{4}+\xi|\nu|}}{k},\vec{z}\right)d\vec{z}\right.$$

Similarly for the second summand in (3.11), using (3.5), we get the expression

$$\frac{\pi^{-3}k^{-4}}{8\rho_1(k)}\sum_{j\geq 0}\frac{1}{k^j}\left(\mathcal{F}_\chi(\xi,\nu,\vec{n}),\nu^2\sqrt{\frac{\nu^2}{4}+\xi|\nu|}\left[1-\eta\left(\frac{\frac{\nu^2}{4}+\xi|\nu|}{k^2}\right)\right]\times\right.$$

$$\left.\times\left(\frac{k^2}{\frac{\nu^2}{4}+\xi|\nu|}\right)^l\int\psi_j\left(-\frac{\xi|\nu|}{k}-\frac{|\vec{z}|^2-2(\vec{n},\vec{z})\sqrt{\frac{\nu^2}{4}+\xi|\nu|}}{k},\vec{z}\right)d\vec{z}\right) \quad (3.13)$$

Note that

$$\mathcal{F}_\chi = \mathcal{F}_\chi^+ + \mathcal{F}_\chi^-$$

where, according to (1.8) of Section 14.1 (recall that $a > 2$),

$$\operatorname{supp}\mathcal{F}_\chi^\pm \subset \left[(\xi,\nu)\in\mathbf{R}^2: \pm\nu > 0,\ |\nu| \geqslant 2(a-1),\right.$$

$$\left.\frac{|\nu|}{4}+\frac{a^2}{|\nu|} \leqslant \xi \leqslant 1+\frac{1}{|\nu|}\right]\times S^2. \quad (3.14)$$

Making, in (3.12), the change of variables $\nu = kt$, and considering only the part with $\mathcal{F}_\chi^+$ (the part with $\mathcal{F}_\chi^-$ can be considered in a similar way) we get

$$\frac{1}{8\pi^3}\sum_{0\leqslant j\leqslant l}\left(\frac{1}{\rho_1(k)}\mathcal{F}_\chi^+(\xi,kt,\vec{n}),t^2\sqrt{\frac{t^2}{4}+\frac{\xi t}{k}}\,\eta\left(\frac{t^2}{4}+\frac{\xi t}{k}\right)\times\right.$$

$$\times\,\omega\left(\frac{k^2t^2}{4}+k\xi t\right)\left(\frac{1}{k}\right)^{l-j}\left(n_1\sqrt{\frac{t^2}{4}+\frac{\xi t}{k}}\right)^j\int^1\psi_j\left(-\xi t+\right.$$

$$\left.\left.+2(\vec{n},\vec{z})\sqrt{\frac{t^2}{4}+\frac{\xi t}{k}}-\frac{|\vec{z}|^2}{k},\vec{z}\right)d\vec{z}\right). \quad (3.15)$$

where $\omega$ is a $C^\infty$-function with the properties

$$\omega(t)\equiv 1,\ t>\epsilon;\quad \omega(t)\equiv 0,\ t<\frac{1}{2}\epsilon, \quad (3.16)$$

where $0<\epsilon<a^2$, $a^2$ are taken from (3.14).

Similarly we get for (3.13)

$$\frac{1}{8\pi^3}\sum_{j\geqslant 0}\frac{1}{k^j}\left(\frac{1}{\rho_1(k)}\mathcal{F}_\chi^+(\xi,kt,\vec{n}),t^2\sqrt{\frac{t^2}{4}+\frac{\xi t}{k}}\left[1-\eta\left(\frac{t^2}{4}+\frac{\xi t}{k}\right)\right]\times\right.$$

$$\times\,\omega\left(\frac{k^2t^2}{4}+k\xi t\right)\left(\frac{t^2}{4}+\frac{\xi t}{k}\right)^{-l}\int\psi_j\left(-\xi t+\right.$$

$$\left.\left.+2(\vec{n},\vec{z})\sqrt{\frac{t^2}{4}+\frac{\xi t}{k}}-\frac{|\vec{z}|^2}{k},\vec{z}\right)d\vec{z}\right). \quad (3.17)$$

Note again that

$$\psi_j,\ {}^1\psi_j \in \mathcal{S}_p(\mathbf{R}^4). \tag{3.18}$$

in (3.16) and (3.17), where $p$ is arbitrarily large if $N$ in (3.3) and (3.6) is sufficiently large.

Now use the representation (2.8) and the properties (2.9)-(2.10) (note that the automodelity of $\rho$ was not used in the proof of the lemma of Section 14.2). Repeating, verbatim, the proof of theorem 3 of Section 3.5, and taking account of the parameters $(\xi,\vec{n})$, one obtains that $\mathcal{F}_X^+(\xi,\nu,\vec{n}) \in \mathcal{S}'[[\xi \leqslant 2]\times \mathbf{R}_+^1 \times S^2]$ has a quasi-asymptotics at infinity in $\nu$ with respect to $\rho_1$. Therefore,

$$\frac{1}{\rho_1(k)}\mathcal{F}_X^+(\xi,kt,\vec{n}) \to \Gamma(\alpha+1)f_{\alpha+1}(t)\times F_+(\xi,\vec{n}),\ k\to\infty \tag{3.19}$$

in $\mathcal{S}_q'([\xi \leqslant 2]\times \mathbf{R}_+^1\times S^2)$ for some $q$.

We have to prove that there are $l$ and $p$ in (3.15), (3.17) and (3.18) such that expressions (3.15) and (3.17) have limits when $k\to\infty$. First, let us consider (3.15). We put

$$\delta = \frac{t^2}{4}+\frac{\xi t}{k},\quad \xi t = k\left(\delta - \frac{t^2}{4}\right). \tag{3.20}$$

It is clear that if $\xi < 0$ then

$$t > 2\sqrt{\delta}. \tag{3.21}$$

From this using (3.10), we get that there is $j_0$ such that the sequence

$$\left\{t^2\sqrt{\frac{t^2}{4}+\frac{\xi t}{k}}\,\eta\left(\frac{t^2}{4}+\frac{\xi t}{k}\right)\theta\left(\frac{t^2}{4}+\frac{\xi t}{k}\right)\left(n_1\sqrt{\frac{t^2}{4}+\frac{\xi t}{k}}\right)^j \int {}'\psi_j\left(-\xi t + \right.\right.$$

$$\left.\left. + 2(\vec{n},\vec{z})\sqrt{\frac{t^2}{4}+\frac{\xi t}{k}} - \frac{|\vec{z}|^2}{k},\vec{z}\right)d\vec{z},\, k \to\infty\right\} =$$

$$= \left\{\theta(\delta)\sqrt{\delta}\,\eta(\delta)(n_1\sqrt{\delta})^j\, t^2\int {}'\psi_j\left(-\xi t + 2(\vec{n},\vec{z})\sqrt{\delta} - \frac{|\vec{z}|^2}{k},\vec{z}\right).d\vec{z},\ k\to\infty\right\}$$

converges in $\mathcal{S}_q([\xi \leqslant 2]\times \mathbf{R}_+^1\times S^2)$ if $p$ is sufficiently large and $j > j_0$.

On the other hand, there is $l_0$ such that the sequence

$$\left\{\frac{1}{k^l}t^2\sqrt{\frac{t^2}{4}+\frac{\xi t}{k}}\,\omega\left(\frac{k^2t^2}{4}+k\xi t\right)\left(n_1\sqrt{\frac{t^2}{4}+\frac{\xi t}{k}}\right)^j \int {}'\psi_j\left(-\xi t + \right.\right.$$

$$\left.\left. + 2(\vec{n},\vec{z})\sqrt{\frac{t^2}{4}+\frac{\xi t}{k}} - \frac{|\vec{z}|^2}{k},\vec{z}\right)d\vec{z},\ k\to\infty\right\}$$

tends to zero in $\mathcal{S}_q([\xi \leqslant 2] \times \mathbf{R}^1_+ \times S^2)$, if $l > l_0$ and $p$ is sufficiently large.

So, if $p$ is sufficiently large and $l > j_0 + l_0$ then (3.15) tends to

$$\frac{\Gamma(\alpha+1)}{8\pi^3}\left\{f_{\alpha+1}(t) \times F_+(\xi,\vec{n}),\ \frac{t^3}{2}\eta\left(\frac{t^2}{4}\right)\left(\frac{n_1 t}{2}\right)^l \int \psi_l(-\xi t + t(\vec{n},\vec{z}),\vec{z})\, d\vec{z}\right\} \tag{3.22}$$

when $k \to \infty$.

Finally, using (3.20), (3.21), and (3.10), we conclude that for large $l$ and $p$ the sequence

$$\left\{t^2\sqrt{\frac{t^2}{4}+\frac{\xi t}{k}}\left[1-\eta\left(\frac{t^2}{4}+\frac{\xi t}{k}\right)\right]\left(\frac{t^2}{4}+\frac{\xi t}{k}\right)^{-l}\int \psi_j\left(-\xi t + 2(\vec{n},\vec{z})\sqrt{\frac{t^2}{4}+\frac{\xi t}{k}} - \right.\right.$$

$$\left.\left. - \frac{|\vec{z}|^2}{k},\vec{z}\right) d\vec{z},\ k \to \infty\right\} = \left\{t^2\sqrt{\delta}[1-\eta(\delta)]\delta^{-l}\right.$$

$$\left.\int \psi_j\left(-\xi t + 2\sqrt{\delta}(\vec{n},\vec{z}) - \frac{|\vec{z}|^2}{k},\vec{z}\right) d\vec{z},\ k \to \infty\right\}$$

in (3.17) converges in $\mathcal{S}_q([\xi \leqslant 2] \times \mathbf{R}^1_+ \times S^2)$. Therefore, for such $p$ and $l$ (3.17) tends to

$$\frac{\Gamma(\alpha+1)}{8\pi^3}\left\{f_{\alpha+1}(t) \times F_+(\xi,\vec{n}),\ \frac{t^2}{2}\left[1-\eta\left(\frac{t^2}{4}\right)\right]\left(\frac{4}{t^2}\right)^l \int \psi_0(-\xi t + t(\vec{n},\vec{z}),\vec{z}),\, d\vec{z}\right\}. \tag{3.23}$$

So, we have proved that the left-hand side of (3.8) has a limit when $k \to \infty$ for any $\phi \in \mathcal{S}_+(3)$. Hence, due to Theorem 2 of Section 13.2,

$$\frac{\rho(t)}{\rho_1(t)} \to C,\ t \to +\infty.$$

By Theorem 3 of Section 3.5, $C \neq 0$. □

**Remark.** Let us show that formulae (3.22) and (3.23) do not contradict formula (2.41). First note that (3.22) and (3.23) lead to (2.41) only if the distribution $F_a = {}^{(\alpha+1)}F_+$ (which is $1/2^{(\alpha+1)}F_+ - 1/2^{(\alpha+1)}F_-$ in our case since $F_+ = -F_-$) satisfies conditions (a) - (b) of Corollary 2 to Theorem 2 of Section 14.2. That is, it is the Radon transform of some distribution $\psi_{a,}\ 0(\vec{u})$ with support in the unit ball.

Let us return to the proof of the theorem. Remembering the origin of the functions $\psi_l$ and $\psi_0$ in (3.22) and (3.23) (see (3.5) and (3.7)) and integrating by parts in inverse order, we get the following limit of the left-hand side of (3.8)* (note that $F^-_\chi(\xi,\nu,\vec{n})$ gives the same contribution as $F^+_\chi(\xi,\nu,\vec{n})$)

*Remember that, in our case, $\alpha > -1$.

$$\lim_{k\to\infty}\left(\frac{\psi(k\lambda,\vec{u})}{\rho_1(k)},\phi(\lambda,\vec{u})\right) = \left(f_{\alpha+1}(\lambda)\times\psi_0(\vec{u}),\phi(\lambda,\vec{u})\right) =$$

$$= \frac{1}{8\pi^3}\left(F_+(\xi,\vec{n}),\int_0^\infty t^{\alpha+3}\,dt\right.$$

$$\left.\int^{(3/2)}\phi_1^{(3/2)}(-\xi t+(\vec{n},\vec{z})t,\vec{z})\,d\vec{z}\right),\quad \phi\in{}_+(3), \tag{3.24}$$

where $\phi_1$ is equal to $\phi$ for $\lambda \geqslant 0$ and $|\vec{u}| \leqslant 1$ and satisfies conditions (3.3) and (3.6) for a sufficiently large $N$. Under these conditions the function

$$\int_0^\infty t^{\alpha+3}\,dt\int^{(3/2)}\phi_1^{(3/2)}(-\xi t+(\vec{n},\vec{z})t,\vec{z})\,d\vec{z}$$

belongs to $\mathcal{S}_p(\mathbf{R}^1\times S^2)$. Moreover, it is the limit (in this space) of the sequence

$$\left\{\int_\epsilon^T t^{\alpha+3}dt\int^{(3/2)}\phi_1^{(3/2)}(-\xi t+(\vec{n},\vec{z})t,\vec{z})\,d\vec{z},\quad \epsilon\to+0,\ T\to+\infty\right\}.$$

Therefore, putting

$$\phi(\lambda,\vec{u}) = g_0(\lambda)g_1(\vec{u})$$

and assuming that the functions $g_0(\lambda)$ and $g_1(\vec{u})$ are changed in a proper way outside the sets $\lambda\geqslant 0$ and $|\vec{u}|\leqslant 1$, respectively so that conditions (3.3) and (3.6) hold, we get from (3.24)

$${}^{(-\alpha-1)}g_0(0)(\psi_0,g_1) =$$

$$= \frac{1}{8\pi^3}\lim_{\epsilon\to+0,T\to+\infty}\left(F_+(\xi,\vec{n}),\int_\epsilon^T t^{\alpha+3}\,dt\int^{(3/2)}g_0^{(3/2)}(-\xi t+(\vec{n},\vec{z})t)g_1(\vec{z})\,d\vec{z}\right). \tag{3.25}$$

Now, substituting into the right-hand side of (3.25) the expression

$$F_+(\xi,\vec{n}) = {}^{(-\alpha-1)}\psi_{a,0}(\xi,\vec{n}),$$

where $\psi_{a,}0(\xi,\vec{n})$ is the Radon transform of $\psi_{a,0}(\vec{u})$, and the left derivation of order $-\alpha-1$ is taken over $\xi$ (see Section 2.4), we have

$$\frac{1}{8\pi^3}\lim_{\epsilon\to+0,T\to+\infty}\left(\psi_{a,0}(\xi,\vec{n}),\int_\epsilon^T t^2\int^{(1/2-\alpha)}g_0^{(3/2)}(-\xi t+(\vec{n},\vec{z})t)g_1(\vec{z})\,d\vec{z}\,dt\right) =$$

$$= \frac{1}{8\pi^3}\lim_{\epsilon\to+0,T\to+\infty}\left(\psi_{a,0}(\vec{u}),\int_\epsilon^T t^2\,dt\int^{(1/2-\alpha)}g_0^{(3/2)}[(\vec{n},\vec{z}-\vec{u})t]g_1(\vec{z})\,d\vec{z}\right) =$$

$$= \frac{1}{4\pi^2}\lim_{\epsilon\to+0,T\to+\infty}\left(\psi_{a,0}(\vec{u}),\int_\epsilon^T t^2\,dt\int_{-1}^{+1}d\mu\int^{(1/2-\alpha)}g_0^{(3/2)}(t\mu|\vec{z}|)g_1(\vec{z}+\vec{u})\,d\vec{z}\right). \tag{3.26}$$

In the last expression one can take a limit when $\epsilon \to +0$ due to the fact that the support of $\psi_{a,0}(\vec{u})$ is compact. We have, further,

$$\frac{1}{4\pi^2}\int_0^T t^2\,dt\int_{-1}^{+1} d\mu \int {}^{(1/2-\alpha)}g\,\delta^{(3/2)}(t\mu\,|\vec{z}\,|)g_1(\vec{z}+\vec{u})\,d\vec{z} =$$

$$= \frac{1}{4\pi^2}\int_0^T\int_{-1}^{1}\int_{S^2}\int_0^\infty {}^{(1/2-\alpha)}g\,\delta^{(3/2)}(t\mu\rho)g_1(\rho\vec{n}+\vec{u})t^2\rho^2\,d\rho\,d\vec{n}\,d\mu\,dt =$$

$$= \frac{1}{4\pi^2}\int_0^1\int_{-1}^{1}\int_{S^2}\int_0^\infty {}^{(1/2-\alpha)}g\,\delta^{(3.2)}(t\mu\rho)g_1\left(\frac{\rho}{T}\vec{n}+\vec{u}\right)t^2\rho^2\,d\rho\,d\vec{n}\,d\mu\,dt =$$

$$= \frac{1}{4\pi^2}\int_{S^2}\int_0^\infty\left[{}^{(1/2-\alpha)}g\,\delta^{(-1/2)}(\rho) + {}^{(1/2-\alpha)}g\,\delta^{(-1/2)}(-\rho)\right]g_1\left(\frac{\rho}{T}\vec{n}+\vec{u}\right)d\rho\,d\vec{n} -$$

$$- \frac{1}{4\pi^2}\int_{S^2}\int_0^\infty\frac{1}{\rho}\left[{}^{(1/2-\alpha)}g\,\delta^{(-3/2)}(\rho) - {}^{(1/2-\alpha)}g\,\delta^{(-3/2)}(-\rho)\right]g_1\left(\frac{\rho}{T}\vec{n}+\vec{u}\right)d\rho\,d\vec{n}.$$

The first integral in the last expression tends to zero in the topology of $\mathcal{S}_q(\overline{B}_1)$ if a sufficient number of moments of $g_0$ is equal to zero. The second integral tends to the expression (in the same topology)

$$-\frac{1}{\pi}g_1(\vec{u})\int_0^\infty\frac{1}{\rho}\left[{}^{(1/2-\alpha)}g\,\delta^{(-3/2)}(\rho) - {}^{(1/2-\alpha)}g\,\delta^{(-3/2)}(-\rho)\right]d\rho =$$

$$= -\frac{1}{\pi}g_1(\vec{u})\left(\mathrm{P}\frac{1}{x},\ {}^{(1/2-\alpha)}g\,\delta^{(-3/2)}(x)\right) =$$

$$= g_1(\vec{u})^{(-\alpha-1)}g_0(0).$$

Here we have used the formula

$$f_{-(3/2)}(-x) * f_{3/2}(x) = -\frac{1}{\pi}\mathrm{P}\frac{1}{x},$$

which can be easily verified by taking the Fourier transforms (see Section 2.1 and Section 2.4).

Thus, the right-hand side of (3.25) is equal to

$${}^{(-\alpha-1)}g_0(0)(\psi_{a,0}, g_1).$$

Comparing this expression with the left-hand side of (3.25) we conclude that $\psi_0 = \psi_{a,0}$. That is what we wanted to prove.

4. **Asymptotic properties of the two-point Wightman function.** The two-point Wightman function has the form (see R.E. Streater and A.S. Wightman (1964)).

$$W_2(x_1 - x_2) = \langle\psi_0|A(x_1)B(x_2)|\psi_0\rangle = F[d\mu](x_1 - x_2),$$

where $d\mu(p) \neq 0$, $p = (p_0, \vec{p})$ is a nonnegative, $L_+^\uparrow$-invariant tempered measure with support in the future light cone $\overline{V}_3^+$; $L_+^\uparrow = L_+^\uparrow(3)$ is a proper, orthochronous Lorentz group in $\mathbf{R}^4$ (see Section 0.5); and $\mu(p)$ is the generating function of the measure $d\mu$, $\mu = [d\mu] * \theta_{\overline{V}_3^+}$ (see Section 3.1), where $\theta_{\overline{V}_3^+}$ is the characteristic function of the cone $\overline{V}_3^+$. The cone $\overline{V}_3^+$ is self-dual, regular and homogeneous. Moreover (see Section 2.6 and Section 2.8)

$$\mathcal{K}_{V_3^+}(z) = 8\pi(z^2)^{-2}, \quad \theta_{V_3^+}{}^* = \frac{1}{8\pi}\Box_3^2, \; z^2 = z_0^2 - z_1^2 - z_2^2 - z_3^2. \tag{4.1}$$

The Wightman function $W_2(x)$ can be holomorphically continued into the future tube domain $\tau^+ = \mathbf{R}^4 + iV_3^+$ and this continuation is given by the Laplace transform of the measure $d\mu$ (see Section 2.5):

$$W_2(z) = L[d\mu](z) = F[e^{-yp}\, d\mu(p)](x), \; z = x + iy.$$

(Note that the operations of Laplace and Fourier transforms are defined here on the basis of the Lorentz scalar product $zp = z_0p_0 - (\vec{z}, \vec{p})$ see Section 13.1.)

Let us apply the many-dimensional generalization of the Hardy-Littlewood theorem for measures (see Section 4.3) to the distribution $W_2$. As a result, we obtain the following theorem on the asymptotic properties of the Wightman function $W_2$.

**Theorem.** *Let $W_2 = F[d\mu]$ be a two-point Wightman function, and let $\rho$ be an automodel function of order $\alpha$. The generating function $\mu$ of the measure $d\mu$ has an asymptotic in the cone $\overline{V}_3^+$ with respect to the function $t^4\rho(t)$ if and only if the following condition holds: there is an open subcone $C^1 \subset V_3^+$ such that*

$$\frac{r^4}{\rho\left(\frac{1}{r}\right)} W_2(iry) \to h(y)\; 0, \; r \to +0, \; y \in C'. \tag{4.2}$$

*In this case the measure $d\mu$ has a quasi-asymptotics in $\overline{V}_3^+$ with respect to $\rho$:*

$$\frac{1}{\rho(k)}[d\mu](k\rho) \to [d\nu](p), \; k \to \infty \; \text{ in } \mathcal{S}'(\overline{V}_3^+), \tag{4.3}$$

*where the generating function $\nu(p)$ of the measure $d\nu(p)$ is continuous in $V_3^+$, and*

$$\frac{\mu(kp)}{k^4\rho(k)} \overset{p \in K}{\Longrightarrow} \nu(p), \; k \to \infty, \; \forall K \subset\subset V_3^+ \tag{4.4}$$

*There are only a few possibilities:*

$$\alpha = -4, \; h(y) = h > 0, \; [d\nu](p) = h\delta(p), \; \nu(p) = h\theta_{V_3^+}(p); \tag{4.5'}$$

$$\alpha = -2, \; h(y) = h\cdot(y_+^2)^{-1}, \; [d\nu](p) = \frac{h}{2\pi}\delta_+(p^2), \; \nu(p) = \frac{h}{9}p_+^2; \tag{4.5''}$$

$$\alpha > -2, \; h(y) = h\cdot(y_+^2)^{-(\alpha/2)-2},$$

$$[d\nu](p) = h\frac{(\alpha+4)^2(\alpha+2)(\alpha+6)\Gamma\left(\frac{\alpha+9}{2}\right)}{\pi\sqrt{\pi}\,\Gamma(\alpha+8)\Gamma\left(\frac{\alpha+6}{2}\right)}(p_+^2)^{\alpha/2},$$

$$\nu(p) = h\frac{8\Gamma\left(\frac{\alpha+9}{2}\right)}{\sqrt{\pi}\Gamma(\alpha+8)\Gamma\left(\frac{\alpha+6}{2}\right)}(p_+^2)^{(\alpha/2)+2}. \tag{4.5'''}$$

*The cases $-\infty < \alpha < -4$ and $-4 < \alpha < -2$ are impossible.*

**Proof.** First note that the measure $d\mu$ is invariant under the space reflection $\vec{p} \to -\vec{p}$ (see Section 1.5). Therefore, in the definition of the Wightman function $W_2$ the Lorentz scalar product can be replaced by the Euclidean one. After that it is possible to apply theorem 1 of Section 4.3. It remains only to verify formulae (4.5) for $\alpha = -4$ and $\alpha \geqslant -2$, and to show that the hypothesis of the theorem is impossible for any other $\alpha$.

First, let us show that the case $\alpha < -4$ is impossible. Indeed, the function $t^4\rho(t)$ is automodel of order $\alpha + 4$. Therefore, $(1/t^4)\rho(t) \to \infty$ when $t \to \infty$ (see Section 3.2). Thus, from the condition $d\mu \geqslant 0$ ($d\mu \not\equiv 0$) it follows that

$$\frac{r^4}{\rho\left(\frac{1}{r}\right)}W_2(iry) = \frac{r^4}{\rho\left(\frac{1}{r}\right)}\int_{V_3^+} e^{-ryp}\, d\mu(p) \to \infty, \; r \to +0, \; y \in C'.$$

This contradicts condition (4.2).

Now let $\alpha \geqslant -4$. From (4.2) and from the invariance of $d\mu$ under $L_+^\uparrow$ and the space reflection it follows that the function

$$h(y) = \lim_{r\to+0}\frac{r^4}{\rho\left(\frac{1}{r}\right)}\int_{V_3^+} e^{-ryp}\, d\mu(p) = \lim_{r\to+0}\frac{r^4}{\rho\left(\frac{1}{r}\right)}\int_{V_3^+} e^{-r(y,p)}d\mu(p)$$

is defined for all $y \in V_3^+$ and that is $L_+^\uparrow$-invariant and homogeneous of degree $-\alpha - 4$. Moreover, since $h(y) = l[d\nu](iy)$ (theorem 1, Section 4.3) then $h \in C^\infty(V_3^+)$. Hence, $h(y)$ can be represented in the form of (4.5):

$$h(y) = h \cdot (y_+^2)^{-(\alpha/2)-2}, \; y \in V_3^+,$$

or by (4.1) in the form

$$h(y) = \frac{h}{(8\pi)^{(\alpha/4)+1}}\,\mathcal{K}_{V_3^+}^{(\alpha/4)+1}(iy), \; y \in V_3^+. \tag{4.6}$$

On the other hand, due to theorem 1 of Section 4.5 the generating function $\nu(p)$ of the measure $d\mu$ satisfies the Fredholm integral equation of the first type

$$\mathcal{K}_{V_3^+}(iy)h(y) = \Gamma(8+\alpha)\int_{\mathrm{pr}\, V_3^+}\frac{\nu(\sigma)d\sigma}{(y,\sigma)^{8+\alpha}}, \; y \in V_3^+,$$

or by (4.6) the equation

$$\frac{h}{(8\pi)^{(\alpha/4)+1}}\mathcal{K}_{V_3^+}^{(\alpha/4)+2}(iy) = \Gamma(8+\alpha)\int_{\mathrm{pr}\, V_3^+}\frac{\nu(\sigma)\,d\sigma}{(y,\sigma)^{8+\alpha}}. \tag{4.7}$$

The solution of the integral equation (4.7) is unique if it exists. Let us find it. Note that (see Section 2.8)

$$K_{V_3^+}^{(\alpha/4)+2}(iy) = \int_{V_3^+} \theta_{V_3^+}^{(\alpha/4)+2}(p) e^{-(y,p)} dp = \int_{\text{pr} V_3^+} d\sigma \int_0^\infty r^3 \theta_{V_3^+}^{(\alpha/4)+2}(r\sigma) e^{-r(y,\sigma)} dr =$$

$$= \int_{\text{pr} V_3^+} \theta_{V_3^+}^{(\alpha/4)+2}(\sigma) \int_0^\infty r^{\alpha+7} e^{-r(y,\sigma)} dr\, d\sigma =$$

$$= \int_{\text{pr} V_3^+} \theta_{V_3^+}^{(\alpha/4)+2}(\sigma) \frac{d\sigma}{(y,\sigma)^{\alpha+8}} \int_0^\infty t^{\alpha+7} e^{-t}\, dt =$$

$$= \Gamma(\alpha+8) \int_{\text{pr} V_3^+} \frac{\theta_{V_3^+}^{(\alpha/4)+2}(\sigma)\, d\sigma}{(y,\sigma)^{\alpha+8}}, \quad y \in V_3^+. \tag{4.8}$$

Here we have used the homogeneity of the function $\theta_{V_3^+}^{(\alpha/4)+2}$ where the degree is equal to $4((\alpha/4)+2)-4 = \alpha+4$. Comparing (4.8) with (4.7) we conclude that

$$\nu(p) = \frac{h}{(8\pi)^{(\alpha/4)+1}} \theta_{V_3^+}^{(\alpha/4)+2}(p). \tag{4.9}$$

The function $\theta_{V_3^+}^{(\alpha/4)+2}(p)$ is $L_+^\uparrow$-invariant and homogeneous of degree $\alpha+4$ with support in the cone $\bar{V}_3^+$. Therefore,

$$\theta_{V_3^+}^{(\alpha/4)+2}(p) = C_\alpha (p_+^2)^{(\alpha/2)+2}, \quad p \in \bar{V}_3^+. \tag{4.10}$$

Let us calculate the coefficient $C_\alpha$. Substituting (4.10) into (4.8), and putting $y = (1, \vec{0})$ we get

$$K_{V_3^+}^{(\alpha/4)+2}(i, \vec{0}) = (8\pi)^{(\alpha/4)+2} = \Gamma(8+\alpha) C_\alpha \int_{\text{pr} V_3^+} (\sigma^2)^{(\alpha/2)+2} \sigma_0^{-\alpha-8}\, d\sigma =$$

$$= C_\alpha \Gamma(8+\alpha) \int_{\sqrt{2}|\vec{\sigma}|<1} (1 - 2|\vec{\sigma}|^2)^{(\alpha/2)+2} (1 - |\vec{\sigma}|^2)^{-(\alpha+9/2)}\, d\vec{\sigma} =$$

$$= 4\pi C_\alpha \Gamma(8+\alpha) \int_0^{1/\sqrt{2}} (1 - 2r^2)^{(\alpha/2)+2} (1-r^2)^{-(\alpha+9/2)} r^2\, dr.$$

Making a change in the variable of integration

$$\frac{r^2}{1-r^2} = u, \quad r^2 = \frac{u}{1+u}, \quad rdr = \frac{du}{2(1+u)^2},$$

we get

$$(8\pi)^{(\alpha/4)+2} = 2\pi C_\alpha \Gamma(8+\alpha) \int_0^1 (1-u)^{(\alpha/2)+2} \sqrt{u}\, du =$$

$$= 2\pi C_\alpha \Gamma(8+\alpha) B\left(\frac{3}{2}, \frac{\alpha}{2}+3\right) =$$

$$= \pi^{3/2} C_\alpha \Gamma(8+\alpha) \frac{\Gamma\left(\frac{\alpha}{2}+3\right)}{\Gamma\left(\frac{\alpha+9}{2}\right)}.$$

Thus, we derive

$$C_\alpha = \frac{(8\pi)^{(\alpha/4)+2}\Gamma\left(\frac{\alpha+9}{2}\right)}{\pi^{3/2}\Gamma(\alpha+8)\Gamma\left(\frac{\alpha+6}{2}\right)}.$$

From this and from (4.10) and (4.9) formulae (4.5) follow for $\nu(p)$ if $\alpha \geqslant -4$.

To obtain formulae (4.5) and $d\,\nu(p)$ let us use the equalities

$$[d\,\nu](p) = \nu * \theta_{V_3^+}^{-1} = \frac{h}{(8\pi)^{(\alpha/4)+1}}\theta_{V_3^+}^{(\alpha/4)+1} = \frac{1}{8\pi}\Box_3^2\nu(p) \qquad (4.11)$$

which follow from (4.9) and (4.1). Now, taking into account the equalities

$$\Box_3(p_+^2)^\lambda = 4\lambda(\lambda+1)(p_+^2)^{\lambda-1}, \; p^2 > 0, \; \lambda > 0, \qquad (4.12)$$

$$\Box_3^2(p_+^2)^\lambda = 16\lambda^2(\lambda^2-1)(p_+^2)^{\lambda-2}, \; p^2 > 0, \; \lambda > 0, \qquad (4.13)$$

we get formulae (4.5) for $[d\,\nu](p)$ from formulae (4.5) for $\nu(p)$ if $\lambda = (\alpha/2)+2$ and $\alpha > -2$.

Let us consider the case $\alpha = -2$. Due to (4.11) and (4.12) we have

$$[d\,\nu](p) = \frac{h}{64\pi}\Box_3^2 p_+^2 = \frac{h}{8\pi}\Box_3\theta_{V_3^+}(p) = \frac{h}{2\pi}\delta_+(p^2).$$

Here we have used the first of the formulae (see, for instance, V.S. Vladimirov, 1966)

$$\Box_3\theta_{V_3^+}(p) = 4\delta_+(p^2), \; \frac{1}{2\pi}\Box_3\delta_+(p^2) = \delta(p). \qquad (4.14)$$

The case $-4 < \alpha < -2$ is impossible because in this instance the measure $d\,\nu$ is negative due to (4.11), (4.5) and (4.13) for $\lambda = (\alpha/2)+2$.

It remains only to consider the case $\alpha = -4$. Formulae (4.5) for $\nu(p)$ holds, so that

$$\nu(p) = h\frac{8\Gamma\left(\frac{5}{2}\right)}{\sqrt{\pi}\Gamma(4)\Gamma(2)}\theta_{V_3^+}(p) = h\,\theta_{V_3^+}(p).$$

Therefore, due to (4.11) and (4.14)

$$[d\,\nu](p) = \frac{1}{8\pi}\Box_3^2\theta_{V_3^+}(p) = h\,\delta(p).$$

**Remark.** In the work of K.A. Bukin (1979) a connection was established between the quasi-asymptotics of a measure $d\,\mu(p)$ at infinity and the quasi-asymptotic behaviour of its Fourier transform in the neighbourhood of a light cone $x^2 = 0$. The following result was also formulated in this work:

$$\lim_{p^2\to+\infty}\frac{\mu(p)}{(p^2)^{\alpha/2}} = H > 0, \; \alpha \geqslant 2,$$

if and only if the condition

$$\lim_{y^2\to+0} (y^2)^{d/2} W_2(iy) = h\,,$$

holds, when the constants $h$ and $H$ are connected by the relation

$$H = \frac{8\Gamma\left(\frac{\alpha+5}{2}\right)}{\sqrt{\pi}\,\Gamma(\alpha+4)\Gamma\left(\frac{\alpha+2}{2}\right)} h.$$

This result directly follows from the theorem that we have just proved.

**Comments on Chapter 5.**

Many physical works have been devoted to the theoremtical justification of the asymptotic behaviour of quantities (form-factors) describing the processes of deep inelastic scattering of elementary particles. The first of these were the works of J.D. Bjŏrken, A.A. Logunov, M.A. Mestvirishvili, Nguen Van Heu, V.A. Matveev, R.M. Muradjan, A.N. Tavkhelidze, H. Leutwyler and R.A. Brandt, among others. An extensive list of related literature can be found in the survey by B. Geyer, D. Robaschik and E. Wieczorek (1979).

In the work of N.N. Bogolyubov, V.S. Vladimirov and A.N. Tavkhelidze on the basis of general principles of local quantum field theory, the casuality of form-factors was proved and the conditions of the weight function which guarantee a proper asymptotic behaviour of form-factors in the Bjŏrken domain and a proper asymptotic at the light cone were pointed out.

In the works of B.I. Zavialov (1973a, b, 1974) the equivalence of power quasi-asymptotics of a weight function at infinity and canonical quasi- asymptotics at a light cone was established. It was also shown that the existence of quasi-asymptotics was a sufficient condition for power asymptotics of form-factors in the Bjŏrken domain. Under some additional assumptions this condition is also necessary.

In the works of V.S. Vladimirov and B.I. Zavialov (1979, 1980) (the case of rotationally-invariant causal functions) and (1981, 1982) (the general case) with no additional assumptions, the equivalence of the quasi- asymptotic behaviour of a causal distribution near to the light cone and its asymptotic behaviour in the Bjŏrken domain with respect to an arbitrary automodel functions (Abelian and Tauberian types of theorems) was stated.

The question of forbidden asymptotics of form-factors (which are incompatible with the axioms of local quantum field theory) was first considered in the work of B.I.Zvavialov (1977) with some additional assumptions. In the works of V.S. Vladimirov and B.I. Zavialov (1979, 1980) these assumptions were omitted. Thus, any positive and continuous function for $t \to +\infty$ is an admissible asymptotics (automodel function) or a forbidden one (nonautomodel function with its first primitive being automodel) or is insignificant. The restrictions on the automodel asymptotics of causal functions were obtained by V.S. Vladimirov and B.I. Zavialov (1981).

## BIBLIOGRAPHY


Bajsanki, B. and Karamata, J. (1968, 1969). 'Regularly Varying Functions and the Principle of Equicontinuity', *Publ. Ramanujan Inst.* **1,** 235–246.

Beltrami, E.J. and Wohlers, M.R. (1966). *Distributions and the Boundary Values of Analytic Functions*, Academic Press, New York London.

Bierstone, E. (1978). 'Extension of Whitney Fields from Subanalytic Sets', *Invent. Math.* **46,** No. 3, 277–300.

Binghma, N.H., Goldie, C.M. and Tengels, J.L. (1983). 'Regular Variation' (mimeographed).

Björken, I.D. (1967). 'Lectures in Varenna School', Course 41, Varenna, Italy.

Bremermann, H. (1965). *Distributions, Complex Variables, and Fourier Transforms*, Addison-Wesley, Reading, Mass.

Bogolyubov, N.N., Logunov, A.A. and Todorov, I.T. (1974). *Axiomatic Quantum Field Theory*, Benjamin, New York.

Bogolyubov, N.N. and Vladimirov, V.S. (1973). 'Representation of $n$-Point Functions', *Trudy Mat. Inst. Steklov*, **122,** 5–21; English transl. *Proc. Steklov Inst. Math.* **122** 1–18.

Bogolyubov, N.N., Vladimirov, V.S. and Tavkhelidze, A.N. (1972). 'On Scaling Asymptotics in Quantum Field Theory, II', *Theoret. Mat. Fiz.* **12,** 305–330; English transl., *Theoret. Math. Phys.* **12,**

Brüning, E. and Stiehel, P. (1974). 'On the Equivalence of Scaling, Light Cone Singularities and Asymptotic Behavour of the Jost-Lehmann Spectral Function', *Comm. Math. Phys.* 36, No. 2, 137–156.

Bukin, K.A. (1979). 'Asymptotic Behaviour of the Two-Point Wightmann Function', *Teoret. Mat. Fiz.* **40,** No. 1, 28–37; English transl., *Theoret. Math. Phys.* **40,** 581–587.

Bukin, K.A. (1981). *The Estimate of the Remainder Term in the Multi-dimensional Tauberian Theorem*, Dep. VINITI, 2.07.81, No. 3278–81, pp. 213–217.

Danilov, L.I. (1985). 'On the Regularity of an Open, Acute Cone in $\mathbb{R}^n$ ', *Sibir. Mat. Z.*, **26**, No. 2, 198-201.

Delange, H. (1963a). 'Théorèmes tauberiens relatifs à l'integrale de Laplace', *J. Math.* XLII Fasc.

Delange, H. (1963b). 'Théorèmes tauberians pour les series multiples de Dirichlet et les integrales multiples de Laplace', *Ann. Scient. Ecole* norm. super ser. 3, XLII Fasc.

Drozzinov, Yu.N. (1969). 'Stabilization of the Solution of the Generalized Cauchy problem for an Ultraparabolic Equation, *Izv. Akad. Nauk SSSR, ser. Mat.* **33**, 368-379; English transl., *Math. USSR Izv.* **33**.

Drozzinov, Yu.N. (1981). 'A Multi-dimensional Tauberian Theorem for Holomorphic Functions with nonnegative Imaginary Part, *Diklady Akad. Nauk SSR* **258**, No. 3; English transl., *Soviet Math. Dokl.* **23**, No. 3, 545-548.

Drozzinov, Yu.N. (1983a). 'A Multi-dimensional Tauberian Theorem for Holomorphic Functions of Bounded Argument and the Quasi-asymptotics of Passive Systems, *Mat. Sb.* **117**, No. 1; English transl., *Math. USSR Sb.* **45**, No. 1, 45-61.

Drozzinov, Yu.N. (1983b). 'Passive Linear Systems of Partial Differential Equations;, *Mat. Sb.* **116**, No. 3; English transl., *Math. USSR Sb.* **44**, No. 3, 269-278.

Drozzinov, Yu.N. and Zavialov, B.I. (1977). 'Quasi-Asymptotics of Generalized Functions and Tauberian Theorems in the Complex Domain, *Mat. Sb.* **102**, 372-390; English transl., *Math. USSR Sb.* **31**,

Drozzinov, Yu.N. and Zavialov, B.I. (1980). 'Tauberian Theorems for Generalized Functions with Supports in Cones', *Mat. Sb.* **108**, 78-90; English transl., *Math. USSR Sb.* **36**,

Drozzinov, Yu.N. and Zavialov, B.I. (1982). 'On a Multi-dimensional Analogue of a Theorem of Lindelöf, *Doklady Akad. Nauk SSSR*, **262**, No. 2, 296-270; English transl., *Soviet Math. Dokl.* **25**, No. **1**, 51-52.

Drozzinov, Yu.N. and Zavialov, B.I. (1984). 'Comparison Tauberian Theorems', *Doklady Akad. Nauk SSSR*, **279**, No. 3, 532-535, English transl., *Soviet Math. Dokl.* **30**. No. 3, 680-682.

Drozzinov, Yu.N. and Zavialov, B.I. (1986a). 'Multi-dimensional Comparison of Tauberian Theorems for Distributions', *Mat. Sb.* **126**, No. 4, 515-542; English transl., *Math. USSR. Sb.* **54**, No. 2, 499-524.

Drozzinov, Yu.N. and Zavialov, B.I. (1986b). 'Asymptotic Properties of Some Classes of Distributions', *Izv. Akad. Nauk SSSR, Ser. Mat.* **49**, No. 1, 81-140. English transl., *Math. USSR Izvestiya* **26**, No. 1, 77-131.

Dyson, F.J. (1958). 'Integral Representation of Causal Commutators', *Phys. Rev.* **110**, 1460-1464.

Frennemo, L. (1965). 'On General Tauberian Remainder Theorems', *Math. Scand.* **17**, 77-80.

Frennemo, L. (1966). 'Tauberian Problems for the $n$-dimensional Laplace Transform, I', *Math. Scand.* **19**, 41-53; (1967) II, **20**, 225-239.

Freud, G. (1951). 'Restglied eines Tauberschen Satzes, I', *Acta Math, Acad. Scient. Rung.* **2**, 299-308.

Galeev, R.H. (1981). 'Multi-dimensional Linear Passive Systems in a Hilbert Space', *Differentsial'nye Uravneniya* **17**, No. 2, 278-285; English transl., *Differential Equations* **17**, 191-196.

Galeev, R.H. (1982). 'The Cauchy problem for Passive Systems in a Hilbert Space', *Differentsial'nye Uravneniya* **18**, No. 10, 1718-1724; English transl., *Differential Equations* **18**, 000-000.

Ganelius, T.H. (1971). 'Tauberian Remainder Theorems', *Lect. Notes Math.* **232**

Garding, L. and Lions, J.L. (1959). *Functional Analysis*, No. 1 del Supplemento al Vol. 14, Serie X, del *Nuovo Cimento*, 9-66.

Gel'fand, I.M. and Shilov, G.E. (1964). *Generalized Functions*, Vol. I: *Operations on Them*, Fizmatgiz, Moscow, 1958; English transl., Academic Press, New York.

Gel'fand, I.M. and Shilov, G.E. (1968). *Generalized Functions*, Vol. II: *Spaces of Fundamental and Generalized Functions*, Fizmatgiz, Moscow, 1958; English transl., Academic Press and Gordon, New York.

Gel'fand, I.M., Graev, M.I. and Vilenkin, N.Ya. (1966). *Generalized Functions, Integral Geometry and Representation Theory*, Vol. 5, Fizmatgiz, Moscow, 1962; English transl., Academic Press, New York.

Geyer, B., Robaschik, D. and Wieczorek, E. (1979). 'Theory of Deep Inelastic Lepton-Hadron Scattering', *Fortschriftte der Physik* **27**, No. 2-3, 75-168.

Gradshtein, I.S. and Ryzhik, I.W. (1965). *Table of Integrals, Series and Products*,

Fizmatgiz, Moscow, 1963; English transl., Academic Press, New York.

Gushchin, A.K. and Mikhailov, V.P. (1971a). 'On the Stabilization of the Solution of the Cauchy Problem for a Parabolic Equation', *Differentsial'nye Uravneniya* **7**, 297-311; English transl., *Differential Equations* **7**.

Gushchin, A.K. and Mikhailov, V.P. (1971b). 'On the Stabilization of the Solution of the Cauchy Problem for a Parabolic Equation with One Spacial Variable', *Trudy Mat. Inst. Steklov* **112**, 181-202; English transl., *Proc. Steklov Inst. Math.* **112**.

Gushchin, A.K. and Mikhailov, V.P. (1984). 'On the Uniform Stabilization of Solutions of the Cauchy Problem for the Second-Order Hyperbolic Equation', *Trudy Mat. Inst. Steklov* **166**, 76-90.

Haan, L. de and Omey, E. (1983). 'Integrals and Derivatives of Regularly Varying Functions in $\mathbf{R}^d$ and Domains of Attraction of Stable Distributions, II', *Stoch. Proc. Appl.* **13**, 157-170.

Haan, L. de and Resnick, S.I. (1979). 'Derivatives of Regularly Varying Functions in $\mathbf{R}^d$ and Domains of Attraction of Stable Distributions', *Stoch. Proc. Appl.* **8**, 349-355.

Hardy, G.H. (1949). *Divergent Series*, Oxford Univ. Press, London and New York.

Hardy, G.H. and Littlewood, J.E. (1913). 'Contributions to the Arithmetic Theory of Series', *Proc. London Math. Soc.* **2**, No. 11, 411-478.

Hardy, G.H. and Littlewood, J.E. (1914). 'Tauberian Theorems Concerning Power Series and Dirichlet's Series whose Coefficients are Positive', *Proc. London Math. Soc.* **13**, 174-191.

Helgason, S. (1980). *The Radon Transform*, Birkhäuser, Boston, Basel, Stuttgart.

Hörmander, L. (1963). *Linear Partial Differential Operators*, Springer-Verlag.

Hörmander, L. (1971). 'Fourier Integral Operators, I', *Acta Math.* **127**, 79-183.

Hörmander, L. (1983). 'Analysis of Linear Partial Differential Operators I. Distribution Theory and Fourier Analysis', Springer-Verlag.

Jost, R. and Lehmann, H. (1957). 'Integral Darstellung Kausaler Kommutatoren', *Nuovo Cimento* **5**, 1598-1610.

Karamata, J. (1930a). 'Sur un mode de croissance reguliere des fonctions', *Mathematica (Cluj)*, **4**, 38-53.

Karamata, J. (1930b). 'Sur certains "Tauberian Theorems" de M.M. Hardy et Littlewood', *Mathematica (Cluj)*, **3**, 33-48.

Knopp, K. (1939). 'Limitirungs Umkehrsätze fur Doppelfolgen', *Math. Z.* **45**, No. 4, 15-28.

König, H. and Meixner, T. (1958). 'Lineare Systeme und Lineare Transformationen', *Math. Nachr.* **19**, 265-322.

Korevaar, J. (1951). 'An Estimate of the Error in Tauberian Theorems for Power Series', *Duke Math. J.* **18**, 723-734.

Korevaar, J. (1954). 'A Very General Form of Littlewood's Theorem', *Indig. Math.* **16**, 36-45.

Kostjuchenko, A.G. and Sargs'jan, I.S. (1979). 'The Distribution of Eigenvalues', *Self adjoint Ordinary Differential Operators.*

Kozlov, S.M. (1983). 'Multi-dimensional Spectral Asymptotics for Elliptic Operators', *Dokl. Akad. Nauk SSSR* **268**, No. 4, 789-793; English transl., *Soviet math. Dokl.*

Kuhn, J. (1974). 'On the Connections between Singularity Structure of Causal Distributions and Behaviour in the Björken Limit', *Nuovo Cimento*, **23A**, 420-430.

Leutwyler, H. and Otterson, P. (1973). 'Theoretical Problems in Deep Inelastic Scattering', in Gatto, R. (ed.), *Scale and Conformal Symmetry in Hadron Physics*, Wiley, New York, pp. 1-30.

Lojasiewicz, S. (1957). 'Sur la valuer et limite d'une distribution dans une point', *Studia Math.* **16**, No. 1, 1-36.

Loomis, L. (1943). 'The Converse of the Fatou Theorem for Positive Harmonic Functions', *Trans. Amer. Math. Soc.* **53**, 239-250.

Ludwig, D. (1963). 'Examples of the Behaviour of Solutions of Hyperbolic Equations for Large Times', *J. Math. Mech.* **12**, No. 4.

Malgrange, B. (1966). *Ideals of Differential Functions*, Oxford Univ. Press.

Methée, P.D. (1954). 'Sur les distributions invariantes dans les group des rotations de Lorentz', *Comm. Math. Helv.* **28,** 225-269.

Natanson, I.P. (1950). *Constructive Function Theory, Fizmatigiz,* Moscow.

Omey, E. (1981). 'Regular Variation and its Applications to Second Order Linear Differential Equations', *Bull. Soc. Math. Belgique* **23,** 207-223.

Omey, E. (1982). 'Multivariate reguliere variatie en toepassingen in Kanstheorie', Ph.D. Thesis, Catholic University of Louvain.

Ovcharenko, I.E. (1983). 'Moment Sequences with Measures with Gaps in Some Domains', *Doklady Akad. Nauk SSSR,* **273,** No. 1, 41-45; English transl., *Soviet Math. Dokl.,*

Postnikov, A.G. (1979). 'Tauberian Theory and its Applications', *Trudy Mat. Inst. Steklov* **144,** 1-147; English transl., *Proc. Steklov Inst. Math.* **144,** 1-138; AMS (1980), Issue 2.

Riesz, M. (1949). 'L'integral de Riemann-Liouville et le probleme de Cauchy', *Acta Math.* **81,** 1-222.

Rothaus, O.S. (1960). 'Domains of Positivity', *Abh. Math. Sem. Univ. Hamburg,* **24,** 189-235.

Schwartz, L. (1950-1951). *Theorie des Distributions,* I-II, Paris.

Seneta, E. (1976). 'Regularly Varying Functions', *Lecture Notes in Math.* **508,** Springer Verlag, Berlin.

Shabat, B.V. (1976). *Introduction to Complex Analysis* (2nd edn.), Part II, Moscow, 'Nauka', Moscow.

Smirnov, V.A. (1976). 'Asymptotic Expansions of Generalized Functions with Singularities on the Light Cone', *Teoret. Mat. Fiz.* **29,** No. 3, 336-346; English transl. *Theoret. Math. Phys.* **29,** 1108-1114.

Smirnov, V.A. (1977). 'Connection Between the Behaviour of Electromagnetic Form Factors in the Björken Limit and Singularities of their Fourier Transforms on the Light Cone', *Teoret. Mat. Fiz.* **33,** No. 3, 319-326; English transl., *Theoret. Math. Phys.* **33,** 1046-1051.

Stadtmüller, U. (1981). 'A Defined Tauberian Theorem for Laplace Transforms in Dimension $\alpha > 1$, *J. Reine Angew. Math.* **328,** 72-83.

Stadtmüller, U. (1982). 'Tauberian Theorems for Laplace and Stiltjes Transforms', *J. Math. Anal. Appl.* **86,** No. 1, 146–156.

Stadtmüller, U. and Trautner, R. (1979). 'Tauberian Theorems for Laplace Transforms', *J. Reine Angew. Math.* **311-312,** 283–290.

Stadtmüller, U. and Trautner, R. (1981). 'Tauberian Theorems for Laplace Transforms in Dimension $\alpha > 1$', *J. Reine Angew. Math.* **323,** 127–138.

Stam, A.J. (1977). 'Regular Variations in $I\mathbb{R}^d_+$ and the Abel-Tauber Theorem', Preprint Math. Inst., Rijksuniv. Groningen, Report 189.

Stankovic, B. (1982). 'Theorems of the Generalized Tauberian Type for Measures', *Rev. Research Faculty of Science, Univ. of Novi Sad*, Vol. 13.

Stankovic, B. (1985a). 'Theorems of Tauberian Type for Measures', *Glasnik Mat.* 20 (40).

Stankovic, B. (1985b). 'Abelian and Tauberian Theorems for Stiltjes Transform of Distributions', *Uspekhi Mat. Nauk* **40,** No. 4, 91–103.

Streater, R.E. and Wigtman, A.S. (1964). *PCT, Spin and Statistics, and All That*, Benjamin, New York.

Subhankulov, M.A. (1976). *Tauberian Theorems with Remainder*, Nauka, Moscow.

Tauber, A. (1897). 'Ein Satz aus der Theorie der unendlichen Reihen', *Monatsh Mat. Phys.* **8,** 273–277.

Tidten, M. (1979). 'Fortsetzungen von $C^\infty$-Funktionen, welche auf einer abgeschlossenen Mengen in $\mathbb{R}^n$ definiert sind', *Manuscripta Math.* **27,** 291-312.

Vladimirov, V.S. (1966). *Methods of the Theory of Functions of Many Complex Variables*, MIT, Press, Cambridge, Mass.

Vladimirov, V.S. (1976). 'Many-Dimensional Generalization of the Hardy- Littlewood Tauberian Theorem', *Izv. Akad. Nauk SSSR, Ser. Mat.* **40,** 1084-1101; English transl., *Math. USSR Izv.* **10,** No. 5; AMS (1978).

Vladimirov, V.S. (1979). *Generalized Functions in Mathematical Physics*, Mir Publishers, Moscow.

Vladimirov, V.S. (1981). *Equations of Mathematical Physics*, Mir Publishers, Moscow (1st edn., Marcel Dekker, New York, 1971).

Vladimirov, V.S. and Drozzinov, Yu.N. (1967). 'The Generalized Cauchy Problem for the Ultraparabolic Equation', *Izv. Akad. Nauk SSSR Ser. Mat.* **31,** 1341-1360; English transl., *Math. USSR Izv.* **1,** 1285.

Vladimirov, V.S., Drozzinov, Yu.N. and Zavialov, B.I. (1984a). 'Theorems of Tauberian Type for Distributions', *Trudy Mat. Inst. Steklov* 163, 42-48; English transl., *Proc. Steklov Inst. Math.* **163,** (1985), No. 4, 53-60.

Vladimirov, V.S., Drozzinov, Yu.N. and Zavialov, B.I. (1984b). 'Tauberian Theorems for Distributions', in *Proc. International Conf. on Complex Analysis and Applications, Varna 1981*, Sofia, Bulg. Ac. Sc., pp. 524-532.

Vladimirov, V.S. and Volovich, I.V. (1982). 'The Ising Model with Magnetic Field and the Diophantine Moment Problem', *Teoret. Mat. Fiz.* **53,** No. 1, 3-15; English transl. *Theoret. Math. Phys.* **53.**

Vladimirov, V.S. and Zavialov, B.I. (1979). 'On the Tauberian Theorems in Quantum Field Theory', *Teoret. Mat. Fiz.* **40,** No. 2, 155-178; English transl., *Theoret. Math. Phys.* **40.**

Vladimirov, V.S. and Zavialov, B.I. (1980). 'Tauberian Theorems in Quantum Field Theory', Itogi Nauki: *Modern Problems of Mathematics*, Vol. 15, VINITI, Moscow, pp. 95-130.

Vladimirov, V.S. and Zavialov, B.I. (1981). 'Automodel Asymptotes and Light Cone Behaviour in Quantum Field Theory', *Doklady Akad. Nauk SSSR*, **258,** No. 4, 842-846; English transl., *Soviet Math. Dokl.*

Vladimirov, V.S. and Zavialov, B.I. (1982). 'Automodel Asymptotes of Causal Functions and their Light Cone Behaviour', *Teoret. Mat. Fiz.* **50,** No. 2, 163-194; English transl., *Theoret. Math. Phys.* **50.**

Winer, N. (1932). 'Tauberian Theorems', *Ann. Math.*, (2) **33,** 1-100.

Wiener, N. (1959). *The Fourier Integral and Certain of its Applications*, Cambridge Univ. Press, 1933; reprint, Dover Publ., New York.

Whitney, H. (1934a). 'Analytic Extensions of Differentiable Functions Defined in Closed Sets', *Trans. Amer. Math. Soc.* **36,** No. 1, 63-89.

Whitney, H. (1934b). 'Functions Differentiable on the Boundaries of Regions', *Ann. Math.* **35,** No. 3, 482-485.

Yakymiv, A.L. (1981). 'Multi-dimensional Tauberian Theorems and their

Application to the Bellmann-Harris Branching Processes', *Mat. Sb.* **115**, No. 3, 463-477; English transl., *Math. USSR Sb.* **43**, No. 3, 413-425.

Yakymiv, A.L. (1983). 'Multi-dimensional Tauberian Theorems of Karamata, Keldysh and Littlewood Type', *Doklady Akad. Nauk SSSR*, **270**, No. 3, 558-561; English transl., *Soviet Math. Dokl.* **27**, No. 3, 657-660.

Zavialov, B.I. (1973a). 'On the Behavour of the Electromagnetic Form Factors in the Neighbourhood of the Light Cone', *Teoret. Mat. Fiz.* **16**, No. 1, 61-65; English transl., *Theoret. Math. Phys.* **16**, 672-675.

Zavialov, B.I. (1973b). 'Automodel Asymptotics of Electromagnetic Form Factors and the Behaviour of their Fourier Transforms near the Light Cone', *Teoret. Mat. Fiz.* **17**, No. 2, 178-188; English transl., *Theoret. Math. Phys.* **17**, 1074-1081.

Zavialov, B.I. (1974). 'Quasi-Asymptotics of Generalized Functions and Scaling of Electromagnetic Form Factors', *Teoret. Mat. Fiz.* **19**, No. 2, 163-171; English transl., *Theoret. Math. Phys.* **19**, 426-432.

Zavialov, B.I. (1977). 'Björken Asymptotics of Form Factors of Deep Inelastic Scattering and General Principles of Field Theory', *Teoret. Mat. Fiz.* **33**, No. 3, 310-318; English transl., *Theoret. Math. Phys.* **33**.

Zavialov, B.I. (1981). 'Jost-Lehmann-Dyson Representation in $fs(S)'_\alpha$ Spaces', *Teoret. Mat. Fiz.* **49**, No. 2, 147-155; English transl., *Theoret. Math. Phys.* **49**.

Zemanian, A. (1965). *Distribution Theory and Transform Analysis*, McGraw-Hill.

Zemanian, A. (1972). *Realizability Theory for Continuous Linear Systems*, Academic Press, New York.

Zharinov, V.V. (1980). 'Quasi-Asymptotic Behaviour of Fourier Hyperfunctions', *Teoret. Mat. Fiz.* **43**, No. 1, 32-38; English transl., *Theoret. Math. Phys.* **43**, 302-306.

GPSR Compliance

*The European Union's (EU) General Product Safety Regulation (GPSR) is a set of rules that requires consumer products to be safe and our obligations to ensure this.*

*If you have any concerns about our products, you can contact us on ProductSafety@springernature.com*

In case Publisher is established outside the EU, the EU authorized representative is:

Springer Nature Customer Service Center GmbH
Europaplatz 3
69115 Heidelberg, Germany

**Batch number: 10607222**

Printed by Printforce, the Netherlands